T0186463

Revitalizing Manufacturing
Text and Cases

Janice A. Klein
Harvard Business School
Harvard University

CRC Press
Taylor & Francis Group
Boca Raton London New York

CRC Press is an imprint of the
Taylor & Francis Group, an **informa** business

CRC Press
Taylor & Francis Group
6000 Broken Sound Parkway NW, Suite 300
Boca Raton, FL 33487-2742

© 1990 by Taylor & Francis Group, LLC
CRC Press is an imprint of Taylor & Francis Group, an Informa business

No claim to original U.S. Government works

Visit the Taylor & Francis Web site at
http://www.taylorandfrancis.com

and the CRC Press Web site at
http://www.crcpress.com

Acknowledgments

This text is a reflection of the toil and creativity of the men and women who spend their lives striving to improve the productivity and worklife within operations. I thank the case subjects (many of whose names have been disguised) and their companies for their time and candor in sharing their thoughts and experiences. I also owe an enormous debt to my many co-workers within industry who first taught me about the "real world" of manufacturing; their patience and support in helping "a young kid out of school" pushed me to want to further explore their joys and frustrations in coping with their world.

Many people have contributed to the cases in this text. My thanks go to the Harvard Business School administration and the Division of Research for their support in developing the "Management of Operations (MOPS)" course and this subsequent casebook and to my students in the first three classes of MOPS who helped to shape the course material. I am also grateful to those case authors who have allowed me to use their work in this book. Special thanks goes to the research assistants who helped in developing many of the cases—Kevin Davis, Darryl Lavitt, Steve Rogers, and especially Sabra Sherry—and to my many colleagues who have reviewed the material.

And finally, many thanks to my editor, Barbara Feinberg, for invaluable assistance and advice, to Fran Charon for her administrative help and patience through countless revisions, and, last but not least, to my husband Dan for the support and encouragement he provided me.

Jan Klein

Contents

Introduction

Any real change implies the breakup of the world as one has always known it, the loss of all that gave one an identity, the end of safety. And at such a moment, unable to see and not daring to imagine what the future will now bring forth, one clings to what one knew, or thought one knew; to what one possessed or dreamed that one possessed. Yet, it is only when a man is able, without bitterness or self-pity, to surrender a dream he has long cherished or a privilege he has long possessed that he is set free—he has set himself free—for higher dreams, for greater privileges.

—James Baldwin
Nobody Knows My Name

In today's environment, flexibility and continual improvement are critical to competitiveness in operations; rigid operations atrophy and wither. Flexibility and improvement, however, imply an openness to new ideas, technology, ways of doing things—in a word, change. Ideally, particular companies "design" flexibility—change—in accordance with their particular strategies and goals. But as companies rush in with improvements, they often fail to take into account that introducing change into the operating level can introduce dislocation. The more radical the changes proposed (or mandated), the more severe the "real change." Executives cannot just snap their fingers and expect people below to step in line; operations have identities, and asking people to go through "real change" threatens their sense of security ("what one possessed or thought one possessed") and their sense of the future ("the world as one has always known it").

Systemic change within operations generally occurs in one of three ways: shutting down an operation and moving to a greenfield site, retrofitting an existing facility (i.e., temporarily shutting down the operation, gutting out the inside, installing new equipment and systems, and retraining the work force), or transforming an ongoing operation. The greenfield option is probably the easiest for introducing change, but if American manufacturing is to regain its competitiveness, retrofitting and transforming operations must become the norm. Before instituting such changes, however, an operations manager must first understand the capabilities and constraints of the current environment: One needs to appreciate what needs to be changed and what is involved in making that change before one initiates it. Operating managers must, therefore, understand the history of their existing operations and know how and why things are done as they are. This requires knowing where knowledge and information currently reside within the operation, knowing how to tap into that knowledge, and then knowing how to effectively utilize and/or redirect it. Only then can an operations manager know where to begin.

The cases in this book were designed to help future operations managers along that journey. The focus throughout is on the hands-on, day-to-day management of an operation. The setting is the operating unit, typically the manufacturing plant, where at least 50 percent of the people in many organizations work. This is also where a corporation dedicates a significant portion of its resources—facilities, equipment, and inventories. Hence, the operating unit is the point where decisions must be made about how effectively to deploy those people, equipment, and materials. The majority of the cases in this book examine "traditional" manufacturing operations, that is, fabrication/machining, assembly, or continuous processes in "low" technology industries, what is commonly referred to as the "rust belt" industries; these are the operations most in need of "new dreams." *Revitalizing Manufacturing*, therefore, examines how the people, materials, and technology/equipment are integrated to improve the performance of such operations.

Many believe that the only way one can learn to manage an operation is through the "school of hard knocks," which to a great extent is true. That process is the only way to feel, see, smell, and develop a sixth sense about what is going on. Nonetheless, although the ultimate understanding of a particular production environment comes through field exposure, the cases in this text provide a structure for identifying the capabilities and constraints of an operation; they signal what to look for when entering a production facility.

Overall, the cases focus on understanding the requirements for becoming an effective operations manager and are designed to help future managers understand the ramifications of change at the workplace. The cases and readings will help students ask such questions as:

1. What are the critical problems inherent in traditional operations?
2. What is the range of choices and implementation tactics available to operations managers?
3. How can an operations manager overcome resistance to change?

The cases are action-oriented: Case protagonists must make decisions about what they would do today, not six months or a year out. They must factor in the day-to-day activities going on all around them, yet remember that they must produce or ship a product as well as introduce change. They must keep in mind the company's overall strategy and goals; at the same time they must consider the long-term impact of their short-term actions: what signals are sent, what precedents are being set, and how their actions will lead to their longer-term objectives.

MAJOR THEMES

Understanding the management of operations requires an appreciation of the operation's history and its current technology (e.g., product and processes), systems (e.g., information and materials), people, and organization—both formal and informal. Thus, the text approaches change within manufacturing from both a production and operations management and a human resource management perspective. Several themes flow throughout the cases.

1. *Managing Knowledge*—Most organizations have become so functionalized that a key factor in a manager's success is finding and coordinating timely and accurate data. Who are the knowledge holders? How well do they understand the process? How accurate is their information? How willing are they to share it? Most often, day-to-day operational knowledge resides quite low within the operation, with the operators and supervisors who manage the daily production process. But with changing technology, the locus can shift, and often become diffused. An operations manager must be cognizant of these shifts in order to locate, utilize, and expand the knowledge.

2. *Managing from the Middle*—The manager of an operating unit is typically stuck in the precarious position between corporate management and the "bottom" of the organization, that is, the employees within the operating unit. Managing at levels within a manufacturing plant—plant manager down to first-line supervisor—therefore requires understanding the pressures and concerns of managers, subordinates, and peers. Proper expectations must be set and clear lines of communication assured—upward, downward, and sideways. Functional turf boundaries are typically quite problematic, and although there is a formal organization, the informal network is usually much more powerful. As with acquiring knowledge, becoming a part of that network is often difficult for outsiders, especially those whom others view as trainees on their way to the top.

3. *Operating Level Pressures*—Traditional manufacturing operations frequently go from one crisis to another, often because of an overemphasis on short-term measures and pressures to meet delivery promises at any cost. In addition, as a result of turf disputes, the various functions within and supporting operations often have separate agendas and different—at times conflicting—measures for success. This short-term, narrow mentality has a negative effect on productivity and has led to many of the problems associated with decaying factories.

4. *Improving Manufacturing Performance*—Improving manufacturing performance and responsiveness to market demands requires continual incremental improvement. Continual learning and improvement can come only from more flexible operations; flexibility is also needed to provide speedy design-to-market for new products and an ability to produce in small quantities at a low cost to meet diverse market needs. Increased flexibility can come from more flexible technology (CIM, etc.) or from changes in systems or organizational structures; maximum competitiveness occurs, however, when the three are consistently aligned. Each will be examined separately and then in concert.

At times, however, major change, or radical surgery, is necessary to regain competitiveness; *Revitalizing Manufacturing* considers both, but focuses on the latter. This means balancing the management of the day-to-day operation and meeting short-term objectives (i.e., shipping the product and dealing with daily employee, material, and equipment problems) with introducing new technology, systems, or structures, which affects the attitudes and behaviors of the entire work force. The typical result is confusion and a temporary (assuming the change is appropriate and implemented properly) decline in productivity. The worth of good operations managers often rests on their ability to minimize this temporary dip and maximize overall performance.

5. *Manufacturing Management Development*—Traditionally, manufacturing has been viewed as the stepchild within industry. Managers are often home-grown, that is, coming up through the ranks, but seldom arrive at the executive suite. In order to make manufacturing more competitive and integrate new technology, systems, and structures, manufacturing operations must develop managers with broader skills and visions. While doing this, however, the wealth of knowledge and ability within the operation that comes from experience, in contrast to schooling, cannot be overlooked or underestimated.

OUTLINE OF MODULES

The cases begin with understanding typical manufacturing operations, then explore the introduction of new concepts, and finally examine the revitalization of decaying operations.

Module 1: The Manufacturing Operation

This module surveys traditional manufacturing plants and the functions/roles within them. Particular emphasis is placed on plant managers and first-line supervisors as key focal points within operations. The sequence of cases, therefore, starts at the bottom of the organization with the first-line supervisors, the traditional "man in the middle." Subsequent cases introduce the major functional roles: quality engineering, manufacturing engineering, materials management, and facilities and maintenance. These cases explore the various pressures and forces affecting each function and how each interacts with the others.

After the lower level of the organization has been surveyed, the role of the facility or plant manager is considered, particularly in the face of crises. Operations managers are ultimately responsible for the safety of their employees and the public/environment in general. Recent disasters, such as Bhophal and Chernobyl, have underscored the importance of disaster prevention and contingency planning.

Labor disputes create another type of crisis within operations. Plant managers must not only help negotiate an equitable settlement which addresses employee concerns and avoids unwanted precedents that would create problems for the long-term viability of the operation, but also find ways to continue critical production operations.

The module concludes by exploring what is necessary to develop manufacturing managers for the future. The case material reviews the training currently done by two leading manufacturing companies and the military, which has increasingly become a recruiting ground for many manufacturing firms. The question posed is whether or not this training is sufficient and/or appropriate for what is needed to manage operations in the twenty-first century.

Module 2: Implementing Workplace Change

Three major categories of change—new technology, new systems, and new structures—are addressed, separately and together.

New technology: The cases examine the impact of new technology on a current operation or work force. What steps need to be taken to successfully introduce a new piece of equipment or a new process? By whom? What skills are needed? What training is required? The cases also explore how the new technologies change jobs and interrelationships among the various functions studied during the first module.

New Systems: This series of cases look at new quality (statistical process control), inventory (just-in-time), and product/process grouping (group technology and cellular manufacturing) philosophies. What is needed to implement new systems? Who needs to do what? What are the typical sources of resistance and why? What types of skills are needed? How do the new systems affect the jobs of people within an operation?

New Structures: This third group explores changes within the organizational infrastructure, such as measurement systems, organizational structure, job design, and human resource policies. From an operating manager's perspective, what technological and/or systems issues need to be addressed? What is their impact on different levels and functions of the manufacturing organization? And how do such changes affect the role of the first-line supervisor and how can managers overcome resistance to change at that level?

The module concludes with a discussion on the interaction of the various topics raised throughout the module. (See Exhibit 1.) The cases focus on the introduction of multiple changes. Although an operations manager strives for synergy, conflicts often arise. Hence, the cases emphasize that while implementing individual change programs, operations managers must be aware of the impact of those changes on the entire operating system.

Module 3: Turnaround Management

The earlier modules are tied together in this final module on the revitalization of American industry at the factory level. Here, the cases trace the history of several operations, highlighting the fact that many of the existing problems in decaying facilities stem from past actions. The series of cases shows the pervasiveness of similar conditions, for example, the lack of product/process focus, adversarial labor-management relations and noncompetitive labor costs, deteriorating facilities, lack of shop floor controls, and inept management. The module investigates different strategies for turning these operations around and the characteristics and skills of turnaround managers at the operating unit level.

EXHIBIT 1   Three Ways to Improve Operations Performance

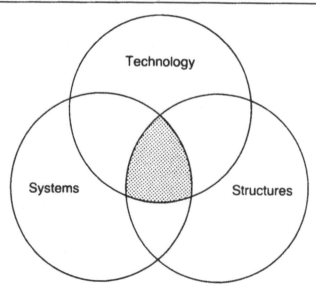

The Manufacturing Organization

The heart of a manufacturing organization is the plant—the physical location where the raw materials and purchased parts are transformed into a finished product. To improve competitiveness of manufacturing operations, operations managers must thoroughly understand the operation they are trying to improve: what is good about the current operation and needs to be retained, plus what is ailing and needs to be treated. In addition, to formulate manufacturing and corporate strategy that can be implemented, managers need to understand current operations both to determine the fit of that strategy with the existing environment and to set realistic objectives, goals, and timetables. This module, therefore, provides a look at life within the four walls of a manufacturing facility.

The first step is to learn where knowledge resides within an operation; an operations manager must have an appreciation of what kinds of knowledge should or do exist, how organizations have typically structured tasks and functions (thereby dividing up areas of expertise or knowledge), and how people within operations tend to share that knowledge and information—both formally and informally. The manager must then acquire the skills to tap into that knowledge and orchestrate it to improve organizational performance.

TYPES OF KNOWLEDGE[1]

There are many types of knowledge, depending upon the job. However those individuals who perform or who have daily contact with those who perform a specific task usually have the most knowledge relevant to that

---

[1] This section is adapted from Klein and Bescher, 1985.

1

job. That is, they possess specialized knowledge of that specific job; each individual (if competent) is an expert in his or her job. This expertise can be labeled either *functional* or *operational* knowledge.

Functional experts live in all the staff positions within a manufacturing organization. They have been schooled in process-specific or *functional knowledge*, such as manufacturing engineering or quality control: An engineer knows exactly what temperature is optimal for a particular material in an annealing furnace and just what a 25° temperature change will do to the crystalline structure of that material. Other functional experts reside in production planning and scheduling areas or maintenance. (See below under description of typical manufacturing organizations.)

Operational knowledge holders are directly involved in making the product. They are the machinists who know the proper settings for each piece of equipment, the assembly workers who know just the right way to position a wire on a circuit board, or the welders who know just how to play the weld puddle around a critical corner to avoid a crack in the weld when the job is finished. This is knowledge gained from doing the task day after day. Although one individual can possess both functional and operational knowledge about a particular job, the two types of knowledge are typically shared by the functional expert and the person physically doing the task. An engineer can provide the technical expertise or theory for doing an operation, but it is typically the individual assigned the task who possesses the most operational knowledge relevant to performing the job on a day-to-day basis.

A third type of knowledge resides in systems or procedures, that is, *procedural knowledge*. This is the knowledge which at one time resided in functional or operational experts but became routinized to the extent that it became proceduralized. Procedural knowledge can be found in various forms—it can be programmed into a robot's logic or a scheduling system, or it can be step-by-step planning instructions found in job sheets. Although it no longer resides in individuals, it is still a critical form of knowledge that must be managed.

Last, there is the knowledge of how all the pieces of the system fit together, referred to here as overall *manufacturing know-how*. Manufacturing know-how is the ability to integrate and use the other three types of knowledge—functional, operational, and procedural; it is the knowledge of what it takes to make the desired output of the firm. It requires a working knowledge of all the relevant aspects of the business coupled with an in-depth understanding of the hands-on requirements of producing a product or service on a daily basis. Most importantly, it is an awareness of the interrelationships among all the experts in each of the operational and functional areas. It is the knowledge that operations managers *must* possess.

To possess manufacturing know-how, a manager must have a thorough understanding of the design and process technologies of the product or service being produced. It includes an understanding of equipment capabilities, material requirements and flows, and human resource needs—for today and the future. It is the knowledge of how each element of the system—product design, individual tasks, organizational functions, and long-term strategy—interrelates to make the whole.

Embodied in manufacturing know-how is the ability to match short-term daily needs with the longer-term goals of the organization. This does not necessitate being an active participant in setting strategic corporate goals, but it does mean being sufficiently aware of those goals to make competent decisions on equipment investments or work force levels. Similarly, producing a quality product or service requires a familiarity with the competitor's product design and relative position within the marketplace, even though a detailed analysis of the competition is typically best left to those individuals possessing in-depth marketing know-how. Likewise, the product's structural design is best done from an engineering know-how perspective; but engineering must appreciate manufacturing's capabilities and constraints, and manufacturing must anticipate engineering design changes. The need for overlap in knowledge and increased coordination between manufacturing and engineering (or research and development) intensifies with the need to get products quickly to market with shortening product life cycles. Another dimension of manufacturing know-how is an appreciation for the financial end of the business. For example, manufacturing know-how does not require an in-depth expertise in accounting principles or the latest tax laws but does demand an understanding and appreciation of what financial restraints exist and the boundaries within which the operations manager is able to make decisions.

THE EVOLUTION OF MANUFACTURING ORGANIZATIONS

The early days of American industry were characterized by small factories, patterned basically along European lines: Authority flowed directly from the owner/boss to the workers, except where a foreman might supervise factory workers. In those situations, the foreman became the surrogate owner/boss, with total authority over the workplace. As organizations grew, additional levels of the hierarchy were added; the foreman, however, still remained the "boss" of the factory.

In the early 1900s, Frederick Winslow Taylor noted that industry was experiencing great "difficulty—almost the impossibility—of getting suitable foreman and gang bosses" (Taylor, 1947) due to the job's broad responsibilities. As a solution, he suggested the introduction of "functional foremanship," whereby the foreman's job would be divided into eight separate roles:

1. Route clerks
2. Instruction card clerks
3. Cost and time clerks

4. Gang bosses
5. Speed bosses
6. Inspectors
7. Repair bosses
8. Shop disciplinarian

Under functional foremanship, each worker would receive direction from eight supervisors rather than one. Because of that feature, functional foremanship was not generally adopted, but the development of staff or functional departments to deal with different aspects of the business and to relieve the foremen of many of the peripheral duties was a direct outgrowth (Lansburgh, 1928). The parallels are fairly easy to see: Route clerks have become manufacturing/process engineers; cost and time accounting is typically done by financial analysts; inspectors fall under the quality control function, etc.

Although organizations are always in a state of flux and no two operations have identical organizational structures, Exhibit 1 is typical of many traditional manufacturing operations. Generally, the duties and responsibilities of each of the functional areas are as follows:

Shop Operations.   The manager of shop operations, often called a superintendent or general supervisor, directs the activities of several first-line supervisors. These first-line supervisors are responsible for planning, organizing, directing, and controlling the work flow and the activities of the direct labor work force. They, as well as their manager, are measured against cost, quality, and delivery objectives; the latter has traditionally received the highest priority.

EXHIBIT 1   Typical Organization Chart for Traditional Manufacturing Operation

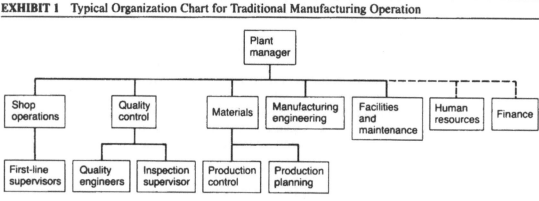

Quality Control. This function assures that the manufactured product or service meets the customers' quality requirements. A staff of quality engineers typically develop the system of controls necessary to achieve the intended quality, plan and implement specific process controls, design needed measurement equipment, and audit the rest of the organization to assure that the controls are being used. Nonexempt inspectors, under the direction of a quality control supervisor, are charged with checking the product or service against the standards developed by the quality control engineers.

Materials. The materials function is responsible for assuring that materials (raw materials, work-in-process inventories, and finished goods) are in the right place at the right time. This includes the creation of a master schedule; the purchase of raw materials and component parts; management of stockrooms and warehouses; planning and scheduling the material flow through the operation on a daily or weekly basis; and the distribution of finished goods to the customer. These activities are typically divided into a number of subgroups, as is shown in Exhibit 1.

Manufacturing Engineering. Manufacturing engineering personnel plan, develop, install, and maintain the tools and equipment necessary to produce the product. They determine how the product or service will be produced, that is, they translate product specification into processes and methods understandable to the shop floor. They are also responsible for organizing or designing the jobs of the direct labor work force and developing the work measurement systems (job standards). Depending on the organization and/or production process, the function falls under various names, for example, industrial engineering or process engineering.

Facilities and Maintenance. Often referred to as plant engineering, the facilities and maintenance function is responsible for the design, construction, and upkeep of the equipment, buildings, and grounds. This includes providing needed utilities (water, power, heating and cooling, telephones, waste removal, etc.) and maintaining roadways and parking areas.

FUTURE TRENDS

As organizations look at their worldwide competition and ways to become more competitive, many of the functional lines described above are blurring. Changes in product design and technological advances are also forcing operations managers to reexamine the way work is organized and where knowledge resides. Some of the current and future trends include:

1. Merging of Functions. With the rate of change escalating and the need for flexibility to deal with that change, functional boundaries often get in the way of progress. As a result, many companies are moving toward more multidisciplinary job design for both exempt and nonexempt employees. For example, line operators or machinists are assuming responsibility for minor maintenance and upkeep of their equipment and for picking up their own material from the stockroom; rather than operations making the product and quality inspecting it, operators are now inspecting their own work. Engineers are also having to broaden their area of expertise to include information systems, process engineering, quality engineering, etc. In some cases, engineers are moving into the materials or purchasing to work closer with vendors in product/process designs.

2. Shifting of Knowledge/Information Bases. When functions merge, so do knowledge and information data bases. In addition, as organizations streamline their operations to become more competitive, organizational structures are becoming flatter; this, in turn, leads to better communications and decision making at lower levels of the operation. Many operations are being decentralized to focus on specific products and structured so that individual departments or groups have all the needed resources close at hand to manage their day-to-day activities, as would a small entrepreneurial company. As a result, there is a broader skill and knowledge base at the bottom of organizations. The distinctions between the knowledge purview and responsibility between labor and management are also being blurred. For example, line operators in some plants are assuming the role of traditional industrial engineers, establishing their own work standards and laying out the work flow.

3. Change in Focus and Measurements. The traditional focus of many operations has been to "make the product that engineering designed and deliver it when marketing and sales said to." Just as functions are merging within operations, similar changes are occurring between operations and other functions within a corporation. Rather than leaving customer satisfaction up to the marketing and sales groups, a customer focus is permeating throughout the entire operation. Rather than narrow goals and measures which optimize the performance of one function, often at the expense of others and the entire corporation, there is greater integration and interaction between functions (internal and external to operations) to optimize overall firm performance. Rather than short-term static goals, the focus is shifting to continuous improvement which derives from increased understanding of the operation through process controls.

DEVELOPING OPERATIONS MANAGERS FOR THE FUTURE

With the multitude of changes occurring at the workplace, it follows that operations managers must possess a number of new skills and abilities. Since manufacturing know-how, to a great extent, is experience-based, the selection of entry-level operations professionals and their training and development are critical to organizations that strive for worldwide competitiveness. The training, especially in technical areas, that operations managers receive in Japan and Germany has contributed largely to those countries' ability to be viewed as outstanding manufacturers (Hayes and Wheelwright, 1984).

The challenge facing manufacturing organizations is knowing from where future managers should be recruited and how best they should be equipped for today's and tomorrow's changing demands. Traditionally, the source of operations professionals was up-from-the-ranks; but internally promoted managers often lacked the broad technical skills to manage sophisticated technologies. Increasingly, companies are looking externally to find these skills, partly because it is generally assumed that new recruits are more flexible and receptive to change than internal people, who have years of experience to undo. This trend intensifies at the first-line supervisory level, for there is a perceived lack of qualified internally promoted candidates who want jobs in operations management. Manufacturing companies have found it increasingly difficult to encourage top-notch nonexempt employees to move into management because of the pressures, pay, job security, and hours associated with being a first-line supervisor (Schlesinger and Klein, 1987). Bringing in large numbers of "outsiders," however, can also result in a number of negative consequences, including high attrition of recruits, morale problems, and conflicts between traditional internally promoted managers and supervisors and the newcomers (Klein, 1986).

OVERVIEW OF MODULE

The cases within this module provide a tour through the functions of a "traditional" operating unit; particular emphasis is placed on the role of plant managers and first-line supervisors because they are, in essence, microcosms of the entire operation. Plant managers are the coordinators of all functions, both internal (shop operations, quality control, materials, manufacturing engineering, and facilities and maintenance) and external (marketing and sales, research and development, planning and administration, etc.) to the plant; first-line supervisors are the ultimate "funnel" of all activities within the plant.

The appropriate place to begin the tour of the operation is, therefore, at the role of the first-line supervisor. The "Barbara Newell" case, along with the reading "Let First-Level Supervisors Do Their Job," provide an overview of the pressures and problems facing incumbents of the traditional man-in-the-middle position. The next stop is the quality function. "Daniel Industries (A)" looks at the role of the quality control engineer, while "Hank Kolb" investigates the top of the function, the quality assurance manager. The reading on "Quality Problems, Policies, and Attitudes in the United States and Japan" provides background on the general state of quality management as viewed by first-line supervisors in the United States and Japan.

The manufacturing engineering and materials roles are then explored, respectively, in "Millford Corporation" and "Kool King (B)." The functional tour winds up with the facilities and maintenance group in "Robinson Laboratories." The accompanying reading on "The Role of Facilities and Maintenance" presents a more global view of facilities and maintenance as a career and the increasing importance of the function in light of new technology, just-in-time inventory practices, and other changes that will be discussed in the following module on workplace change.

The module then moves on to a look at the role of the plant manager, particularly in the face of crisis. Operations managers are responsible for much more than the deployment of the physical and human resources within the operations; they are also responsible for the health and well-being of their employees, the community, and the environment. The case of the "Explosion at Dover Plant" and the reading on "Managing Disasters," therefore, examine actions needed in the prevention and handling of crisis situations. The case on "Black Diamond Coal Company (B)" looks at another problem confronting many operations managers: strike management. The strike is a bit unusual; the section foremen (first-line supervisors) have gone on strike allegedly due to the termination of a fellow foreman charged with safety violations. The mine superintendent must uncover the real reason, determine the economic impact of each lost day of production, and find a way to get the foremen back to work. Thus, the module returns full-circle to the critical role of the first-line supervisor.

The module concludes with an examination of the selection and training of operations managers in light of today's changing demands. "The Development of Manufacturing Managers for the Future" describes the management development programs at General Electric and Procter & Gamble, two companies long admired for their manufacturing management training efforts. The final case on "The Development of Air Force Maintenance Officers" investigates the training of military officers. Recently, many firms are finding that former military officers provide the best of both worlds: advanced technical training plus management and leadership experience. The question is whether the training described in these two cases is sufficient to move operations into the 21st century.

REFERENCES

Hayes, Robert H., and Steven C. Wheelwright. *Restoring Our Competitive Edge: Competing through Manufacturing.* New York: John Wiley & Sons, 1984.

Klein, Janice A. "The Changing Role of First-Line Supervision and Middle Management." Washington, D.C.: U.S. Department of Labor, 1986.

Klein, Janice A., and Bescher, Robert F. "Managing Knowledge within Operations." Harvard Business School Working Paper #1-786-002, May 1985.

Lansburgh, Richard. *Industrial Management.* New York: John Wiley & Sons, 1928.

Schlesinger, Leonard A., and Janice A. Klein. "The First-Line Supervisor: Past, Present, and Future." In *Handbook of Organizational Behavior,* ed. J. Lorsch, Englewood Cliffs, N.J.: Prentice-Hall, 1987.

Taylor, Frederick W. *Scientific Management.* New York: Harper & Row, 1947.

Barbara Newell—First-Line Supervisor

Monday, June 7, 1976, 7 A.M.: "Welcome to the plant. I'm really looking forward to working with you. I'm here if you need anything. I think the best way for you to get your feet wet is to follow Tom Brown, the supervisor you are replacing, for a week before he leaves the plant. We'll have plenty of time to chat later. He'll show you around and introduce you to people."—Bob Lawrence, first-shift superintendent

With less than fifteen minutes of discussion, Bob Lawrence took Barbara Newell, newly appointed first-line supervisor for assembly line #1, out to the assembly line to meet Tom Brown. Newell had joined DigiWatch Company's management training program nine months earlier after receiving an MBA from a major business school. (She also had a BS in electrical engineering from a Midwestern state university, but no business experience other than sales work at her home-town department store during summer breaks.) While at DigiWatch she had been introduced to the technology and manufacturing process as a process engineer in Plant #1. She knew this new assignment would last a minimum of twelve months.

DIGIWATCH COMPANY

DigiWatch Company, a wholly owned subsidiary of a large multinational corporation, manufactured digital watches.[1] (See Exhibit 1 for an organization chart.)

In the early 1970s, Peter Brody, operations manager, determined that he needed to develop future managers within the organization. As a result, DigiWatch began recruiting bright, young MBAs with technical undergraduate degrees and placing them in a two- to three-year training program. The length of the program (and specific assignment rotation) varied with each recruit, depending upon his or her work experience. At a minimum, each trainee was expected to complete assignments as a manufacturing or process engineer and as a first-line supervisor. Additional assignments could be in the areas of quality control, production planning and control, purchasing, or maintenance.

FIRST WEEK AS A FIRST-LINE SUPERVISOR

Newell spent much of the first week on her new assignment following Tom Brown around as he performed the job she was about to take over. Exhibit 2 shows an excerpt from her personal notes on Brown's daily activities. She knew that after Friday he would be unavailable to help her if any questions arose. (He had just been promoted to a job in a plant in Puerto Rico.) Both operators and upper management regarded Brown as one of the best supervisors in the plant, and Newell found him very helpful in giving her a rundown on the people with whom she would have to work. (See Exhibit 3 for Brown's comments.) She wished, though, that she had had more time to spend with him.

Newell spent over half the week in meetings with various managers within the plant to learn what they did. Late Friday afternoon, after she had bid Brown farewell, she began reflecting on what she had been told and had observed during the week.

Personnel.   Most of Tuesday had been spent in the personnel department. The day started with a two-hour meeting with Roger Simpson, personnel manager, who began by handing her a book entitled *Union-Free Supervisor*[2] and saying, "The most important part of your job is to keep

---

[1]The technology and manufacturing process were quite similar to those at Texas Instruments–Time Products Division, although this in-basket exercise is not based on Texas Instruments. Rather, it has been designed to accompany "Texas Instruments–Time Products Division" (9-677-043).

[2]James L. Dougherty, *Union-Free Supervisor* (Houston: Gulf Publishing Company, 1974).

EXHIBIT 1 DigiWatch, Inc. Organization Chart

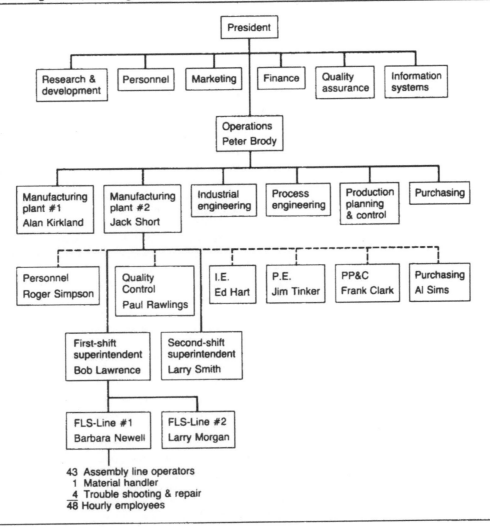

the employees feeling like they are being treated properly so they don't want a union. My organization is here to help you in any way we can." Throughout the meeting he stressed the need to develop a participative management style. Newell spent the rest of the day with other members of the personnel department, who explained personnel practices and procedures in the plant. (See Exhibit 4 for excerpts from the policies manual.)

EXHIBIT 2
Excerpts from Newell's Notes on Tom Brown's Daily Activities

Thursday, June 10

6:30 A.M.	Read mail; reviewed 2nd shift supervisor's notebook; checked raw materials against daily requirements. (Noted shortages in order to call Purchasing at 8 A.M.)
7:00 A.M.	Checked attendance and asked Larry Morgan for fill-ins; assigned operators to work stations. (Note: Tom wished Barb Sampson happy birthday and asked Lisa Swift how her sick boy was. He also noted that Stella Carpenter asked for a light job because she injured her elbow playing tennis last week.)
7:15 A.M.	George Kramer reported that the label applicator at operation #30 was clogged; assigned extra operator to hand sponge labels; called Maintenance.
7:25 A.M.	Patti Carter reported in late; found work for her.
7:45 A.M.	Wanda Casey complained that the lighting was insufficient at her work station; called Personnel to report complaint and Maintenance to have the lights replaced.
8:00 A.M.	Called Purchasing concerning material shortages; Purchasing said that material would be unavailable for one week; notified Production Planning and Control of shortage and got substitute schedule; told George Kramer of change (George handles all model changeovers as assigned by Tom).
8:10 A.M.	Alice Knight complained that incoming switches were defective; called Quality Control to have someone inspect them.
8:20 A.M.	Lisa Smart requested reassignment to another work station because of disagreement with coworkers; found another job for her temporarily.
8:30 A.M.	Reviewed quality (inspection) reports on heat-staking operations for quality circle discussion next week.
8:50 A.M.	Met with Bob Lawrence prior to 9 A.M. production meeting to review yesterday's production.
9:00 A.M.	Weekly production meeting with Production Planning and Control; reviewed previous week's output and received schedule for forthcoming week (listing of number of watches by model, color, etc.); meeting usually lasts at least one hour.
10:30 A.M.	Calculated overtime requirements for weekend for report due to Production Planning and Control by noon; reviewed overtime needs with Bob Lawrence before submission.
11:00 A.M.	Lunch.

EXHIBIT 3
Tom Brown's Comments on People in the Organization

Bob Lawrence (first-shift superintendent)—35 years old; BS in electrical engineering (part-time from local college); two years on current job; nice guy; does whatever Short says.

Jack Short (manufacturing manager)—40 years old; BS in electrical engineering; two and a half years on current job; likes to make you think you are part of the team, asks for inputs but ignores them.

Larry Morgan (first-line supervisor on line #2)—55 years old; high school graduate; employed by DigiWatch since the plant opened in 1972; thinks all first-line supervisors should be promoted from hourly ranks; "father confessor" to operators.

Roger Simpson (personnel manager)—38 years old; BS in psychology; four years on current job; always takes the employee's side; paranoid about a union organizing the plant.

Ed Hart (industrial engineer)—31 years old; BS in industrial engineering; one year on current job; straight numbers guy; "Mr. Clipboard and Stop Watch."

Jim Tinker (process engineering manager)—40 years old; BS in electrical engineering; four years on current job; nice guy; open to ideas and suggestions; overworked.

Frank Clark (production planning and control manager)—32 years old; BS in industrial engineering; one and one half years on current job; always changing the production schedule; thinks he runs the place.

Al Sims (purchasing)—58 years old; high school graduate; has been in purchasing function for as long as anyone can remember; nice guy but too easy on suppliers of parts and materials.

Paul Rawlings (quality control manager)—36 years old; BS in electrical engineering; six months on current job; "by the book" manager; trying to make a name for himself.

Quality Control. Newell's meeting with the quality manager, Paul Rawlings, had been relatively short. He stated that there was a severe quality problem on the line, handed her a memorandum (Exhibit 5), and told her that he expected her to fix the problems. He also mentioned that he thought Tom Brown had been too easy on his employees.

Production Planning and Control. Frank Clark, manager of production planning and control, spent most of his meeting with Newell explaining how hard it was to schedule the plant with constantly changing demands from marketing. He stated that the number one job of the supervisor was to meet the schedule that his organization issued and that the best way to achieve this objective was to make sure that all parts and materials were available when needed and to keep the line running at all times.

EXHIBIT 4
Excerpts from Company Personnel Policies and Procedures*

Policy on Rules of Employee Performance and Conduct:

These rules pertaining to employee conduct, performance, and responsibilities have been established so that all personnel can conduct themselves according to certain rules, good behavior, and conduct. The purpose of these rules is not to restrict the rights of anyone, but rather to help people work together harmoniously according to the standards our company has established for efficient and effective operations.

| | Penalties | | |
	First Offense	Second Offense	Third Offense
Rules			
1. Being tardy habitually without reasonable cause	Twice in 30-day period. Written correction	Three times in 30-day period. Three-day disciplinary layoff	Four times in 30-day period. Dismissal
2. Being absent without notification or excuse	Three times in 12-month period. Written correction	Five times in 12-month period. Three-day disciplinary layoff	Ten times in 12-month period. Dismissal
3. Leaving your job or your regular working place during working hours for any reason without authorization from your supervisor, except for lunch, rest periods, and going to the restrooms	Written correction	One-day disciplinary layoff	Dismissal
4. Leaving work before end of shift or not ready to go to work at the start of shift	Written correction	One-day disciplinary layoff	Dismissal
5. Inefficiency or lack of application of effort on the job	Written correction	Written correction	Dismissal
6. Violating a safety rule or safety practice	Written correction	Written correction	Dismissal

---

EXHIBIT 4   (continued)

Policy on Probationary Employees

All hourly rated employees hired will be classified as probationary employees during the first 90 days of employment. If employment continues beyond such time, they will then be considered regular employees and the record of service will revert back to the most recent hiring date.

A "Probationary Period Rating Report" will be completed on each new employee by the first-line supervisor after he/she has been employed four weeks (30 days).

Should any new employee fail to meet our standards of performance, attitude, attendance, and cooperation before the 90-day period, then that employee should be separated from our payroll as soon as possible.

*Modified from "How to Develop a Company Personnel Policy Manual," A Dartwell Management Guide.

Purchasing.   Newell's meeting with the purchasing manager, Al Sims, had been a pleasant break from the others. He had been very friendly, inquired about her family, and offered help whenever she needed it.

Industrial Engineering.   Ed Hart, plant industrial engineer, had started the meeting by saying, "I'm here to help you get the most out of your people. The way we plan the jobs, the operators don't need much training and must go out of their way to screw up the process." He then went on to explain the learning curve and its importance in meeting customer objectives and the target schedule.

Maintenance.   Newell spent over two hours in the office of the manager of process engineering, Jim Tinker, but their actual conversation lasted only about twenty minutes in total; it seemed as though every time he began a sentence someone interrupted them with a new emergency. Besides process engineering (physical design and tooling of the production process), Tinker was also in charge of the maintenance operation, which appeared to Newell to involve one crisis after another. He warned her to "keep an eye out for potential machine problems. I need as much lead time as possible to schedule my maintenance people. We're understaffed and overworked!"

ASSEMBLY LINE #1

Assembly line #1 consisted of 48 employees: 43 assembly operators (AO), four troubleshooting and repair people (TS), and one material handler (MH). (See Exhibits 6 and 7 for an operations list and employee roster.) During their walks around, Brown provided Newell with information about line employees. Most, he noted, were hard workers who rarely gave him much trouble. He did, however, caution her about a few problems:

> Jane Cook was in a car accident in February and leaves early every Tuesday afternoon for a doctor's appointment. Someone has to fill in for her.

> Maria Whitney has some personal problems at home. She appears to be quite moody, sometimes giddy, other times very quiet. She may have some type of drug problem.

> Wanda Casey, a possible union sympathizer, went to the medical office on Wednesday complaining about a rash on her hands which she said was caused by the tape she handles on her job.

> George Kramer, the most knowledgeable person on the line, is being considered for a first-line supervisor's job on the second shift.

In addition, Brown mentioned that he was getting pressure from Roger Simpson to start a second quality circle on the line.[3] The quality circle program had been initiated in March as a vehicle to increase operator participation. Brown had stalled in responding to Simpson because he was having trouble with one quality circle. As he explained to Newell:

> Since you can't shut the line down and Jack Short has decided that we can't let the operators receive overtime for quality circle activities, it is extremely difficult to free up operators to attend a meeting one hour per week. There are 48 people on the line, and it takes about 40 of them to keep it running. Therefore, because of limited staffing and absenteeism, you can't let eight or ten operators go off into a quality circle meeting unless you can borrow some people to fill in on the line from Larry Morgan, first-line supervisor on line #2.

> In addition, quality circle projects in this area usually involve reducing downtime or improving productivity, which translates into making the line run faster. The eight or ten line members end up being resented by the other operators. It's okay for a member of management to improve productivity, but not a line operator. As a result, the only people interested in participating in the quality circles are the troubleshooting and repair people.

As Newell was shuffling through the papers on her desk (see Exhibits 5–11), Bob Lawrence stopped by on his way out.

> Well, how did your first week go? I'm really interested in your impressions while they are still fresh. Let's meet on Monday morning at 8 A.M. I'd especially like your thoughts on how we can reduce the high absenteeism on the line. Have a nice weekend!

---

[3]At DigiWatch, a quality circle was a group of eight to ten operators who met one hour a week with their supervisor to discuss quality or other process problems associated with their line. This was quite similar to Texas Instruments' People and Asset Effectiveness (P&AE) program.

EXHIBIT 5
Memo

Date: June 9, 1976

To: Barbara Newell

From: Paul Rawlings

Yesterday's sampling (N = 100) of finished goods inventory from line #1* has turned up the following problems:

No. of Defects

4	mislabeled boxes: incorrect color of band
2	loose batteries
2	broken lens
1	band not attached
3	nonoperative watches

*Production from both first and second shift

I want a report outlining your corrective actions by Monday morning, June 14.

EXHIBIT 6   Operation List

Operation Number	Description	Standard Time* (seconds)	Training Time† (minutes)	Desirability‡
1	Prepare module	20	60	+
2	Functional test	21(avg.)	90	+
3	Frequency test	21(avg.)	90	+
4	Scrape contacts	20	45	−
5	Clip post (502 only)	18	30	0
6	Apply tape	14	30	−
7	Remove cover	16	15	0
8	Heat-stake lens	21	60	+
9	Inspect bezel	41	120	+
10	Clean holes	11	30	−
11	Install set switch	18	45	0
12	Install command switch	18	45	0
13	Check switch	12	15	0
14	Clean bezel	12	30	−
15	Install module	21	45	+
16	Install battery clip	20	60	+
17	Heat-stake clip	15	60	0
18	Install batteries	21	15	+
19	Switch check	10	10	+
20	Date-code	10	15	0
21	Place o-ring	14	15	0
22	Install back	15	30	0
23	Functional test	21(avg.)	150	+
24	Install band	42	180	+
25	Cosmetic check	17(avg.)	45	0
26	Final test	35	90	+
27	Quality control§	—	—	
28	Place on cuff	36	90	+
29	Place in box	19	60	0
30	Apply label ⎫	14	30	0
31	Place cover ⎭			
32	Place manual	30	15	+

*Standard time per operation (direct labor seconds).
†Time to train average worker with experience elsewhere on the line, so that operator could perform the operation within the specified standard, 99 percent of the time (estimate by Ed Hart).
‡Tom Brown's perception of operators' attitude toward each operation or station.
§Performed by members of the quality control organization.

EXHIBIT 7 Operator List

Name	Seniority	Job*	Attitude†	Workmanship†	Normally Assigned Operation
Kathy Morris	2- 2-76	AO	Excellent	Good	1
Sylvia Brown	3-15-76	AO	Good	Acceptable	1
Ginny King	1-12-76	AO	Excellent	Excellent	2
Barb Sampson	1-19-76	AO	Excellent	Excellent	2
Bill Richardson	6-16-75	AO	Good	Good	3
Ruth Taylor	4-12-76	AO	Fair	Acceptable	4
Don James	4-12-76	AO	Fair	Acceptable	5
Wanda Casey	4- 5-76	AO	Poor	Good	6
Helen Day	4- 5-76	AO	Fair	Acceptable	7
Louise Jackson	1-19-76	AO	Excellent	Excellent	8
Betty Wiliams	3- 1-76	AO	Excellent	Good	8
Lisa Smart	9-20-75	AO	Excellent	Excellent	9
Susan McHugh‡	8-13-73	AO	Excellent	Excellent	9
Lisa Swift‡	6-20-74	AO	Excellent	Excellent	9
Gail Swanson	4-19-76	AO	Fair	Acceptable	10
Mary Jones	4-12-76	AO	Fair	Good	11
Phyllis McGuire	4-19-76	AO	Poor	Acceptable	12
Alice Knight	4-12-76	AO	Fair	Good	13
Margaret Higgins	4-26-76	AO	Fair	Acceptable	14
Sally Mitchell	3- 8-76	AO	Good	Good	15
Jackie Leslie	3-29-76	AO	Good	Good	15
Lynn Black	2- 2-76	AO	Excellent	Good	16
Judy Lee	3- 1-76	AO	Good	Good	16
Patti Carter	4- 5-76	AO	Fair	Good	17
Richard Parker	2-23-76	AO	Excellent	Acceptable	18
Liz Slate	4-12-76	AO	Excellent	Good	18
Janet Clark	3- 1-76	AO	Good	Good	19/20
Maria Whitney	4-26-76	AO	Fair	Poor	21
Elaine Wayne	3-15-76	AO	Good	Acceptable	22
Frank Post	4- 5-76	AO	Excellent	Excellent	23
Chuck Page	4-12-76	AO	Good	Good	23
Lillian Doe	10-13-75	AO	Good	Excellent	24
Diane Rockwell	11-17-75	AO	Excellent	Acceptable	24
Jane Cook	1-26-76	AO	Excellent	Excellent	24
Stella Carpentar	4- 5-76	AO	Fair	Good	25
Peggy Green‡	1- 5-76	AO	Excellent	Excellent	26
Nancy Weathers‡	11-17-75	AO	Excellent	Good	26
Linda Peters	1-12-76	AO	Excellent	Excellent	28
Edith Coleman	10- 6-75	AO	Excellent	Excellent	28
John Mello	3- 1-76	AO	Good	Excellent	29
Paula Doyle	4- 5-76	AO	Excellent	Poor	30/31

EXHIBIT 7   (continued)

Name	Seniority	Job*	Attitude†	Workmanship†	Normally Assigned Operation
Anne Ross	3-29-76	AO	Good	Good	32
Steve Cooper	5-31-76	AO	Good	Acceptable	32
George Kramer‡	3-27-72	TS	Excellent	Excellent	
Michael Gray‡	5- 1-72	TS	Good	Excellent	
Joe Little‡	4- 3-72	TS	Excellent	Excellent	
Tom Smith‡	5-15-72	TS	Excellent	Excellent	
Allen Kirby	5-10-76	MH	Fair	Acceptable	

*AO = assembly operators; TS = troubleshooting; MH = material handler.
†Tom Brown's general impressions of employees' attitude and workmanship on the job.
‡Quality circle members.

EXHIBIT 8
Memo

Date: June 1, 1976

To: Tom Brown

From: Roger Simpson

Attached is the absenteeism report for your line. The 7.5 percent overall rate is too high, but 10.7 percent for probationary employees is abominable. Something must be done!

Make sure you review any actions you plan to take with Lois Chaney, employee relations specialist. We need to fix the problem, but don't want to upset the work force.

Absenteeism Report

Name	Seniority	Days Absent*	Days Available†	Percent
George Kramer	3-27-72	0	250	0.0
Joe Little	4- 3-72	2	250	0.8
Michael Gray	5- 1-72	4	250	1.6
Tom Smith	5-15-72	5	250	2.0
Susan McHugh	8-13-73	0	250	0.0
Lisa Swift	6-20-74	9	250	3.6
Bill Richardson	6-16-75	12	250	4.8
Lisa Smart	9-20-75	0	182	0.0

EXHIBIT 8 (continued)

Name	Seniority	Days Absent*	Days Available†	Percent
Edith Coleman	10- 6-75	0	171	0.0
Lillian Doe	10-13-75	0	166	0.0
Diane Rockwell	11-17-75	0	141	0.0
Nancy Weathers	11-17-75	0	141	0.0
Peggy Green	1- 5-76	7	106	6.6
Linda Peters	1-12-76	4	101	4.0
Ginny King	1-12-76	6	101	5.9
Barb Sampson	1-19-76	2	96	2.1
Louise Jackson	1-19-76	1	96	1.0
Jane Cook	1-26-76	25	91	27.5
Kathy Morris	2- 2-76	5	86	5.8
Lynn Black	2- 2-76	3	86	3.5
Richard Parker	2-23-76	3	71	4.2
Betty Wiliams	3- 1-76	10	66	15.2
Janet Clark	3- 1-76	3	66	4.5
John Mello	3- 1-76	2	66	3.0
Judy Lee	3- 1-76	3	66	4.5
Sally Mitchell	3- 8-76	5	61	8.2
Sylvia Brown	3-15-76	3	56	5.4
Elaine Wayne	3-15-76	4	56	7.1
Jackie Leslie	3-29-76	6	46	13.0
Anne Ross	3-29-76	5	46	10.9
Helen Day	4- 5-76	6	41	14.6
Patti Carter	4- 5-76	5	41	12.2
Frank Post	4- 5-76	1	41	2.4
Wanda Casey	4- 5-76	6	41	14.6
Paula Doyle	4- 5-76	5	41	12.2
Stella Carpentar	4- 5-76	4	41	9.8
Ruth Taylor	4-12-76	5	36	13.9
Don James	4-12-76	6	36	16.7
Alice Knight	4-12-76	2	36	5.6
Liz Slate	4-12-76	5	36	13.9
Mary Jones	4-12-76	7	36	19.4
Chuck Page	4-12-76	4	36	11.1
Gail Swanson	4-19-76	4	31	12.9
Phyllis Mcguire	4-19-76	6	31	19.4
Maria Whitney	4-26-76	5	26	19.2
Margaret Higgins	4-26-76	3	26	11.5
Allen Kirby	5-10-76	1	16	6.3
Steve Cooper	5-31-76	0	1	0.0
Line average				7.5

*Number of days absent over prior 12 months.
†Number of work days on payroll over prior 12 months.

EXHIBIT 9
Memo

Date: May 28, 1976

To: Tom Brown

From: Jim Tinker

Our preventive maintenance schedule for line #1 calls for the following equipment to be overhauled in June as noted below:

Date	*Operation Number*
Tuesday 6/8	26—Final test
Tuesday 6/15	3—Frequency test
Tuesday 6/22	8 and 17—Heat-stake

In addition, we hope to test out some new tooling for installing bands (operation 24) during the week of June 21. This may require us to shut the line down for an hour or so.

EXHIBIT 10
Memo

Date: May 31, 1976

To: Tom Brown

From: Ed Hart

Since the assembly process has changed considerably since the line started in February, I think it is time to time-study all the operations again. I will begin at operation #1 on Monday, June 14, and work my way down the line.

EXHIBIT 11
Memo

Date: June 5, 1976

To: Tom Brown

From: Jim Tinker

    We have just about finished a redesign of the tooling for the heat-staking of the lens at operation #8 and plan to try it out by the end of the month. We will need an additional inspector at that operation to monitor the process for at least two weeks. Please assign someone who knows the process, rather than just a fill in.

Barbara—

    The quality circle has been working on the heat-stake problem for the past two months. I haven't had time since I got this memo (since we didn't hold the circle meeting this week) to tell the circle members that engineering has been working on this problem too. They (the quality circle) have really been working hard at finding a solution and will be upset.

Tom
6/12

Let First-Level Supervisors Do Their Job

W. Earl Sasser, Jr.

Frank S. Leonard

"Our supervisors can probably have more influence on our productivity, worker absenteeism, product quality, morale of our work force, labor relations, and cost reduction than any other group in the company," the vice president of personnel at a manufacturing company recently told us. We were there to do research on the function of first-level supervisors.

"If we don't do something soon, we're going to lose our best foremen—it's no wonder that they're turned off, given the pressures they have to live with," the plant manager at the same company said.

Being a first-level supervisor is one of the most difficult, demanding, and challenging jobs in any organization. Buried in an organizational web, this person must be adroit at administering a unit and at perceiving which, among all the daily tasks delegated downward, are the most important to accomplish. Through such administrative competence, he or she must be able to link the unit's accomplishments to the functioning of other organizational subunits.

Even at the first level, a supervisor must be able to think and act in terms of the total system of operation.[1] This includes defining and assigning priorities, planning and organizing, and programming and coordinating the operating tasks of a department so that the objectives of both the department and the company as a whole are achieved.

Furthermore, the first-level supervisor must excel in interpersonal skills. More and more, the trend is for employees to be a heterogeneous group of individuals, many of whom are not especially dedicated to their jobs, their departments, or their companies. Handling the variety of attitudes and values in this multiple-generation worker base has become extremely difficult. Also, the work force is aging as the post-World War II babies reach middle age, and challenges to mandatory retirement are widening the age spread.

Along with age, the increase in working women and minorities has become a factor in the work force. Supervisors must learn to deal with these new workers and yet guard against discriminatory practices. Also, the fact that the educational level of the work force has continued to rise means that the supervisor does not often maintain an educational advantage over the worker. (In 1977, more than 90% of the U.S. population between 20 and 29 were high school graduates, and 8 million Americans were enrolled in colleges and universities.)

Source: W. Earl Sasser, Jr., and Frank S. Leonard, "Let First-Level Supervisors Do Their Job," *Harvard Business Review* 58, no. 2 (March–April 1980), pp. 113–21.

[1] See Robert Dubin et al., *Leadership and Productivity: Some Facts of Industrial Life* (Novato, Calif.: Chandler & Sharp, 1965), p. 75.

Challenge of the First Level

In addition to the increasing pressure for administrative and clerical efficiency at the first level, two areas of supervisory competence that are continually problematic are human relations and technical knowledge. Workers are no longer conformists who without question accept the rules and procedures that management lays down. No longer do they take authority at face value.

Human Relations

Many workers view their jobs as necessary evils to provide the resources for fulfilling their lives in leisure time, which they are pressing harder and harder to increase. It is the first-level supervisor who must cope with such workers face to face and day to day. Being able to communicate effectively is vital. In a recent study of 25 middle managers, the materials manager of an electrical company expressed a theme common to the group: "Being able to work with people is the most important characteristic a first-level supervisor can have. I can buy technological expertise, but it's hard to find someone with good, basic communication skills."[2]

Technical Competence

First-level supervisors must of course have technical competence in the areas they supervise. The supervisors must be able to perform the specific task they ask their workers to do and must, to some degree, understand the equipment and the process technology they manage.

Technological changes continue to occur rapidly, though, and supervisors can no longer hope to understand completely all the complex equipment and processes they are in charge of. New products and new processes abound—computers, plastic molding, electronic test equipment, temperature- and pressure-sensitive distillation, component machining, complex metal alloy foundries,

acoustic devices, and synthetic rubber, to name but a few.

Having good technical skills gives supervisors both enough understanding to deal with the many specialists brought in to accomplish the units' objectives and the ability to train subordinates in their tasks.

Mix of Skills

Despite the difficulty and challenge of the first-level supervisor's job, many upper-level managers fail to appreciate its merits or its requirements. Although most of them agree that the human relations aspect of the job is important, they often promote a supervisor for such skills as record keeping. Although the mix of skills needed for each position varies from situation to situation, managers often fail to perceive the particularity of the task required, the type of people being supervised, or the stage the organization is going through.

One general supervisor at an auto manufacturing plant said: "For some reason, our supervisors just aren't able to switch between our departments; they may be great in materials handling, but they have a hell of a time in welding. It's almost like it's a different job!" Supervising highly skilled welders requires a different blend of skills than supervising semiskilled laborers.

To overcome the growing pains and technical difficulties of starting up production or of making a major product changeover, the supervisor must emphasize technical skills rather than interpersonal relations, which must be downplayed in the rush to finish technical tasks. During stable periods, however, administrative and interpersonal skills rise in importance in the first-level supervisor's order of priorities.

Decline of the Position

Although a person serving as a first-level supervisor is performing a major function, the

---

[2]See Thomas De Long, "What Do Middle Managers Really Want from First-Line Supervisors?" *Supervisory Management*, September 1977, p. 8.

position has often been labeled "the man in the middle," "the forgotten man," "the master and victim of double talk," and "the marginal man." Such descriptions not only indicate the male domination of the position but also its degeneration to one of "being on the edge," "being victimized," and "fading in importance."[3]

Confusion of Roles

The causes for the decline of the first-level position are manifold. Over the years confusion has developed about what to expect of supervisors and what role to give them. The position has two very separate roots.

One root is the master craftsman of the past. He was a real entrepreneur—bidding on jobs, hiring employees to perform required tasks, and managing their progress. Like the subcontractor of today, the master craftsman took on the difference between the revenues for jobs completed and the costs associated with those jobs as his own profits or losses. The master craftsman's skill and knowledge of the job were the key ingredients on which these profits or losses depended.

The other root is the "lead man," the foreman of a gang of workers performing manual labor. Like the lead dog or lead horse of a work team, the lead man served as an example for other work-crew members. He often set the pace by calling out a cadence to synchronize the crew's physical movements. The lead man was part of the actual work, and yet he was responsible for the behavior of the whole group.

The amalgamation of these two roles has resulted in today's confusing hybrid. Peter Drucker notes: "From the master craftsman the supervisor of today has largely inherited what is expected of him. From the lead man he has, however, largely inherited his actual position."[4]

The word *supervisor* has conflicting connotations. A supervisor not only commands, directs, controls, and inspects but also takes responsibility for, leads, shepherds, administers, guides, consults, and cares for. Just how the connotation varies from situation to situation and from person to person is in itself a reason for the ambiguity—and the decline—of the first-level supervisor's role.

Specialization of Skills

Another cause of the decline is the rise of staff service departments in such fields as quality control, production planning and control, industrial engineering, personnel, maintenance, and cost accounting. Most of these staff service departments were created to handle the new demands of scientific management in the 1920s and 1930s. The more recent growth of specialization and professionalization within companies has been noted as an important trend of the twentieth century.[5]

Each staff group wants to have a say in the job, to establish a power base, and to protect its area of expertise. Its success in meeting these needs has eroded the authority of first-level supervisors. As Thomas Patten points out: "The foreman found himself in effect sur-

---

[3]See F. J. Roethlisberger, "The Foreman: Master and Victim of Double Talk," *Harvard Business Review,* September–October 1965, p. 22; Thomas A. Patten, Jr., *The Foreman: Forgotten Man of Management* (New York: American Management Associations, 1968); and Donald E. Wray, "Marginal Man of Industry: The Foreman," *American Journal of Sociology,* January 1949, p. 298.

[4]Peter F. Drucker, *The Practice of Management* (New York: Harper & Row, 1954), p. 321.

[5]See James G. March and Herbert A. Simon, *Organization* (New York: John Wiley & Sons, 1958); H. L. Wilensky, "The Professionalization of Everyone?" *American Journal of Sociology* 70 (1964), p. 137; and Charles A. Myers and John G. Turnbull, "Line and Staff in Industrial Relations," *Harvard Business Review,* July–August 1956, p. 113.

rounded by specialists who were taking over parts of what had formerly been his job. He was left with little to do except administer the plans and programs devised by the service departments."[6]

Rise of Unions

A further influence has been the rise of the unions, which have stripped supervisors in unionized plants of much of their remaining authority. Rather than always dealing directly with workers, the supervisors have become more dependent on, and are quite often the target of, the union. It has become increasingly difficult to hire or fire without union involvement. Hiring often has to come from the union list; firing has to follow a strict interpretation of the contract, often requiring a number of warnings. Layoffs are normally by seniority, not according to productivity. Disciplinary action was formally taken away from the prerogative of the first-level supervisor's judgment and set down in black and white.

And, even when the strict letter of the contract is followed, grievances are often filed by the union steward. The company, in some instances, has failed to support the supervisor in a legitimate claim against the union. When such actions have eroded the power base of first-level supervisors, they have been bypassed by workers and union officials, and workers have taken their problems to the union steward instead. The first stage of any grievance procedure—talking with the supervisor in charge—has become lip service. "Don't talk to him, he doesn't know anything" has become a self-fulfilling prophecy on the shop floor.

The union has been a co-conspirator in usurping the first-level supervisor's prerogative to set work standards. Setting work standards—the one domain supervisors had

prided themselves on and that had been considered *their* territory—has become the domain of the industrial engineering and industrial relations departments working with the union.

The union has also served to lower the prestige of the first-level supervisor by winning large wage increases, improved working conditions, and job security for its members. First-level supervisors have seen workers' wages rise more rapidly than their own; they have not had the same job security that the workers have fought for; they can be fired or demoted at a moment's notice; and the Taft-Hartley Act effectively precludes them from organizing.

Crossfire of Demands

The first-level supervisor is a "person caught between"—primarily between middle management and the work force. Both groups have very different values and priorities. Middle managers tend to be interested in cost, efficiency, and performance; workers tend to be more interested in wage rates, security, and comfort. Managers usually believe that hard work leads to advancement; workers often see little point in exerting themselves. To management, the labor contract and work rules seem restrictive; to labor, they seem protective from unreasonable management demands. Managers are concerned about the status of their positions; workers want recognition for work well done. Managers usually identify strongly with the company; workers often have little company loyalty.

The first-level supervisor is caught directly in a crossfire of values and priorities:

- The supervisor often does not know the objectives and policies of top management but heavily influences what management can accomplish.
- The supervisor is not part of the work force but depends heavily on its acceptance.

---

[6]Patten, *The Foreman*, p. 18.

- The supervisor is in the first line of management but has little authority.
- The supervisor is a member of management but is far removed from the locus of decision making.
- The supervisor is limited by precedents and company culture but serves as the agent of change, without whose action little occurs in the company.
- The supervisor establishes standards and precedents but has little information or knowledge on which to base decisions.
- The supervisor is supposed to spend much time on interpersonal relationships but finds that much of that time is needed for record keeping.
- The supervisor is supposed to have a position of leadership but feels that leadership traits are suppressed because of the low self-image associated with the position.
- The supervisor is asked to identify with the values and aspirations of management but is at a dead end in career progress and development.
- The supervisor is usually young and deals with a young, diverse, new type of working person but is evaluated, trained, and rewarded by older, more conservative, more authoritarian supervisors.

This combination of role confusion, increase in staff services, overlap of power with the unions, and conflicting demands has reduced the position of first-level supervisor to just a shadow of its earlier form.

Success at the First Level . . .

As we can see from the often ambiguous and contradictory findings, success is difficult to identify. Sometimes it means productivity, sometimes satisfaction, and sometimes quality of work life. What is successful to employees is not always the same as what is successful to management, and that is not necessarily the same as what is successful to the first-level supervisor.

. . . According to Outside Observers

Our knowledge of what makes a successful supervisor is still quite incomplete. However, several studies have been carried out since the end of World War II.

A pioneering effort was a three-year study (1947–1949) conducted by Aaron Q. Sartain and Alton W. Baker in the offices of Prudential Insurance Company in Newark, New Jersey.[7] Two samples of matched pairs of work groups, 12 in each sample, were carefully selected. The samples were statistically alike with regard to number of men and women, marital status, average age, education, years of experience, salary grade, average distance from job to home, and average score on a battery of psychological tests for each pair of work groups.

However, the productivity differences between the two samples of work groups were statistically significant. Prior to the study, characteristics of the group leaders (supervisors) such as age, education, experience, and salary were thought to explain the differences in group productivity, but the study did not show that any of these factors makes a difference.

What it did show is that the high-productivity groups had more pride in their work than the low-productivity groups. Supervisors of high-productivity groups usually supervised in a more general manner than did supervisors of low-productivity groups. These latter supervisors closely watched their workers and gave greater amounts of instruction to them. Overinstruction was an easy way to oversupervise.

The supervisors of the low-productivity groups made a larger number of requests for promotions and salary increases, but a lower percentage of their requests was approved.

---

[7]See Aaron Q. Sartain and Alton W. Baker, *The Supervisor and His Job* (New York: McGraw-Hill, 1972) for a description and analysis of this study.

The supervisors of the high-productivity groups were more critical than their counterparts.

Finally, and perhaps most important, the high-productivity supervisors talked about their people; the low-productivity supervisors talked about their jobs. The researchers classified the former group as "employee oriented," the latter as "work centered."

Sartain and Baker concluded their analysis of this study by noting that there are no iron-clad rules for supervising.

The Institute for Social Research, under the direction of Rensis Likert, followed the Prudential study with a number of similar studies in a variety of settings. The findings from these studies can be summarized as follows:

- Supervisors viewed themselves more favorably than did their subordinates.
- Employees in the high-productivity groups liked their work less than their counterparts in the low-productivity groups. (A happy worker is not always the most productive worker.)
- As a general rule, the better supervisors spent more time in meetings with their employees.
- The supervisors of the more productive groups were judged by their employees to have greater influence with top management.
- Keeping subordinates informed, thinking of them as individuals, taking an interest in them, soliciting their opinions, and developing an atmosphere of trust were traits of the better supervisors.

In a later study, Saul W. Gellerman analyzed the jobs of 12 first-line supervisors in the packaging plant of a major food-processing company.[8] Gellerman followed each of the supervisors through the plant for an entire shift, noted every move, and questioned each course of action.

Gellerman found three supervisors (A, B, and C) particularly interesting. For each of these supervisors Gellerman detected a number of important elements of substance (what is done) and style (how it is done). How their superiors described their ways of supervising is shown in the *Exhibit*.

What is certain is that the job of first-line supervisor is an extremely difficult and demanding one that requires shifting sets of information, skills, and abilities. A successful supervisor seems to have the ability to balance the demands of task, employees, union, and management with his or her own needs for esteem and respect. But this balancing act takes place in a ring where not all of these demands can be met at once.

. . . According to Subordinates

Employees' attitudes often reveal the quality of supervision. To see what good first-level supervisors are like, it is useful to hear what subordinates want from their leaders. And to see what kind of supervision encourages the development of first-level supervision, it is helpful to hear what they think of their managers.

A 1969–1970 survey of working conditions shows workers' satisfaction to be significantly correlated with the adequacy of resources and the competence of their supervisors.[9] Workers said that "people orientation" is important to them but is not the only thing that contributes to their satisfaction and productivity. Supervisors whom the workers viewed as effective combined "people management" with competence at the job, maintenance of high performance standards, and ability to supply

---

[8]Saul W. Gellerman, "Supervision: Substance and Style," *Harvard Business Review*, March–April 1976, p. 89.

[9]*The 1969–1970 Survey of Working Conditions: Chronicles of an Unfinished Enterprise*, ed. Robert P. Quinn and Thomas W. Mangione (Ann Arbor: University of Michigan, 1973).

EXHIBIT Gellerman's Study on Quality of Supervision

Supervisor	A	B	C
Rating by superiors	Mediocre	Good	Excellent
Substance:	Checks location but not activity of subordinates; avoids insistence on prescribed procedures.	Frequently checks that employees are following correct procedures.	Concentrates attention where it will be most valuable.
Style:	Regularly (albeit mechanically) uses first names in all employee contacts, tends to stay on foot rather than sit, and does much energetic rushing about.	Mixes gentle humor and reassurance with genuine pride in subordinates and has an easy, positive relationship with people.	Reassures the discouraged, jollies along the angry, and leaves those who need no help pretty much alone; gives support as it is needed and only where it is needed.

workers with adequate help, equipment, and information related to their jobs.

In 1977, a national restaurant chain we interviewed undertook a confidential survey of its employees as part of an attempt to find the cause of a corporate sales plateau. The results of the survey were quite revealing:

- The employees felt they had little job security since most of their rights depended on the esteem in which particular managers held them. The rate of turnover was high because the company had a policy of moving managers throughout the organization.
- This rapid turnover of management personnel created another concern—each new manager seemed to expect a different standard or a different type of performance from employees.
- There was a general feeling that managers who had been hourly employees in the past could deal better with the hourly employees than managers without hourly experience who had come straight through the corporate training program.
- Managers were very quick to correct but slow to reward. Employees felt a definite lack of encouragement and praise.
- Most employees felt that their performance had never been evaluated and that they did not know where they stood with their managers.
- Several employees felt that it was common for them to get "bad shifts" as punishment and that they never found out whether they had done something wrong until the schedule was posted. Only when they inquired of management did they find out what had gone wrong.
- Some expressed the feeling that they would like to go into management except that they saw the constant squeeze on managers—the conflict between the desire to be a "good guy" and the ability to produce the results that upper management demanded.

Overall, there was a strong correlation between how the employees ranked their unit managers (supervisors) and the performance of their units. The better supervisors produced better operating results.

. . . According to First-Level Supervisors

James W. Driscoll, Daniel J. Carroll, Jr., and Timothy A. Sprecher, in recent research, asked first-level supervisors about the amount of control they had over factors that motivated their subordinates. Their findings reaffirm the generalizations we presented earlier:

"Unfortunately, these first-level supervisors are still 'the man in the middle.' (The only change is a semantic update in gender.) They report no more control over the things they consider important than over the things they consider unimportant. It is quite likely this lack of control generates very high levels of frustration in first-level supervisors. They are held responsible for producing organizational results through their subordinates, but they lack control over the means to motivate these workers."[10]

Research we recently completed at several plants confirms that this attitude is widespread. A quote from a general foreman summarizes a common complaint: "They [upper management] have completely taken away our ability to get things done. We are still responsible for things that we have little control over—absenteeism, purchasing parts, quality, labor relations, maintenance. When we go to them with some problem, to get some help, all we get is, 'Fix it, make it go away.'"

By giving control of these factors to first-level supervisors, middle and top management could help the supervisors motivate their sub-

---

[10]James W. Driscoll, Daniel J. Carroll, Jr., Timothy A. Sprecher, "The First-Level Supervisor: Still the Man in the Middle." *Sloan Management Review,* Winter 1978, p. 34.

ordinates. Driscoll, Carroll, and Sprecher discovered that the higher-level managers very accurately perceived this control discrepancy between what is important and what the first-level supervisors control: "Basically, these first-level supervisors seem to be in an unwinnable situation. They need help, and their bosses seem to know it."

Such a finding suggests that an important starting point in designing a program to make supervision effective is not changing the behavior of first-level supervisors but convincing those who manage them to yield some control.

Several researchers have recently studied what first-level supervisors want from their jobs. In their sample of 300 first- and second-level supervisors at Allied Chemical Company, Michael J. Abhoud and Homer Richardson found that, out of 10 factors evaluated, first-level supervisors ranked interesting work first and salary second.[11] The second-level managers ranked salary first and interesting work second. Other factors ranked evenly by both levels of managers include, in descending order, chance for promotion, appreciation for work done, good working conditions, job security, loyalty of supervisors, "feeling in on things," tactful discipline, and help on personal problems.

In a survey of 65 first-level supervisors, Paul W. Cummings asked the respondents to list their motivations for accepting a first-level position. He noted that 90 percent of the respondents listed more money as a reason for accepting, 38 percent listed advancement, 48 percent listed the challenge of a new position, and 40 percent said they enjoy leadership positions.[12]

In a study of the attitudes of plant supervisors and salesmen, Sartain and Baker found that 78 percent of the salesmen rated their work favorably, whereas only 56 percent of the supervisors did so. The survey also indicated that the supervisors felt that they had fewer opportunities in their jobs for personal growth, development, and advancement than the salesmen.[13]

Obviously, the background of first-level supervisors has a lot to do with what they expect of this position, and their aspirations change as their service lengthens. A young process engineer, placed in a first-level supervisory position to be groomed for management, wants different things from the job than the 40-year-old lathe operator who finally cracks this lower rung of management.

Getting onto the lower rung is a time for remolding, but managers must be careful not to foster it in the form of stagnation. To expect that good wages is all that first-level supervisors want is a gross, misleading simplification. They may learn, however, that this is all they can expect from management. The opportunity is there for management to encourage these individuals to see the first-level job as a transition and to expect some career development from it.

Improving the Situation

If first-level supervisors are to succeed, they must first establish the informal authority and interpersonal influence to back up the responsibility that comes with their position. Then first-level supervisors must continue to deal with their immediate supervisors and their work force in a manner that minimizes the conflicts between the two groups and permits them to retain the authority to perform effectively.

---

[11]Michael J. Abhoud and Homer Richardson, "What Do Supervisors Want from Their Jobs?" *Personnel Journal*, June 1978, p. 308.

[12]Paul W. Cummings, "Occupation Supervisor," *Personnel Journal*, August 1975, p. 448.

---

[13]See Sartain and Baker, *The Supervisor and His Job.*

A major reason that a first-level supervisory job seems so difficult to master is the decrease in its traditional authority, an increase in dependency on other people to get the job done, and an apparent lack of other operating levers. Many see this erosion of formal authority and increasing dependence as a condition to be straightened out by increasing the first-level supervisor's authority. This is an unrealistic remedy. The decreasing power base of this lower-level manager is due to two pervasive organizational phenomena—division of labor into specializations and scarce resources of all types. Influencing people has to take forms other than exercising formal authority.

Supervisors to Use New Levers

Levers can be thought of as tools for influencing people in specific situations; none are applicable, however, to every situation. Levers such as job assignment, overtime, work conditions, equipment repair, and even hiring and firing are now often out of the supervisors' control. Very few discretionary items exist in the operating budget. What are some of the available levers today? How can first-level supervisors exercise influence and get the job done without using the more traditional levers? They can:

- Use positive reinforcement in the form of incentive schemes, job redesign, and awareness of psychological needs, including peer group acceptance and pride.
- Try negative reinforcement—both the traditional type (write up, fire, suspend) and more indirect means (job reassignment, job redesign, forced overtime).
- Delegate the resolution of a sticky problem to a shop steward or another union official.
- Appeal to workers for support on the basis of having gone out on a limb for them or having given over some prerogative to them in the past.
- Appeal to workers on the basis of under-

standing their position, since first-level supervisors once stood in their shoes.
- Appeal to workers on the basis of previously agreed-on goals and plans for achieving them.

As can be seen from this list, the available levers have shifted from administrative and technical competence to competence in interpersonal and group relations. This employee-oriented area requires the development of nontraditional authority and power bases and an understanding of the subtle processes of influence and persuasion. The first-level supervisor must have the ability to analyze and resolve the various dependencies that management and workers have.

As one director of production for a large defense subcontractor we visited said: "I know in my gut that the real key to productivity that the general manager is pushing on and [the key] to better labor relations that the union is yelling about is my supervisors. Any investment in them in training, communications, time, energy, attention, or plain listening gets one of the best returns in this company."

Managers to Shore Up the Position

Rather than contribute to the continued erosion of the first-level supervisory position, upper management should shore up the position by encouraging and training first-level supervisors to use available power sources to energize their situation. The end result would be an environment in which satisfaction and productivity abounds.

Upper management should recognize the difficulties associated with the position and help these supervisors develop a power base. Power can come from many sources: a mandate from management, personal confidence, a reputation for being able to tackle tough situations, loyalty of the work force, and dependence of the work force and management on the first-level supervisor's knowledge and skills.

Both middle and top managers should strive to create an organizational environment in which first-level supervisors can perform their function most effectively. There are a number of steps that managers above first-level supervisors can take to help:

- Become aware of the actual working conditions of first-level supervisors. Don't assume that the key to present-day first-level supervisory effectiveness is the same as it was 10 or 20 years ago. "When *I* ran that assembly line, I did things this way" is a meaningless and misleading appeal. Things aren't the same!
- Keep first-level supervisors informed about the corporate perspective as it relates to their operation. To relate to upper-level management and to present the management viewpoint to the work force, first-level supervisors must know some of the long-term goals of the corporation.
- Keep first-level supervisors aware of upper-level managers' priorities. Without a clear idea of these priorities, first-level supervisors risk disapproval of their actions.
- Educate first-level supervisors about new technological developments that might affect their job. Knowledge of the equipment and process technology they are supervising is essential for gaining credibility with both management and the work force and for exposing areas of potential improvement.
- Provide feedback on how well first-level supervisors are meeting management's expectations.

- Provide first-level supervisors an opportunity on company time to work together on specific problems affecting their job. Such teamwork not only generates solutions where the problems are but also allows peer interaction and learning as part of the job—something that most managers take for granted and that first-level supervisors' day-to-day routines lack.
- Assist first-level supervisors in keeping the work force up to date on any information that may affect their job. A good in-plant communications program administered through first-level supervisors can pay handsome dividends.[14]
- Provide training for first-level supervisors to improve their skills in dealing with people. Such a training program should include sessions on topics like being an effective listener, performance appraisal, motivation, disciplinary procedures, and labor relations.
- Encourage first-level supervisors to stand up for and express their beliefs to upper management.

In essence then, the first-level supervisor must become more political in both skill and outlook. The real key is the ability to understand, influence, and merge the two worlds of management and workers. First-level supervisors are forced to walk the high wire and, like the circus, their act is now in the center ring.

---

[14]Louis I. Gelfand, "Communicate through Your Supervisors," *Harvard Business Review,* November–December 1970, p. 101.

Daniel Industries (A)

In August of 1980, Bill Close was transferred to the electrical components department of Daniel Industries, as chief quality control engineer. Daniel had lost several major orders in June, as a result of quality problems which had been traced back to faulty glass cases used in the assembly of electronic displays. The cases were made in the components department's sealing section. Close's primary task was to take immediate action to improve the outgoing quality of cases and to install a long-range system for controlling the operation. He had been given authority to reject any of the sealing section's output; however, the actual operation of the department was the responsibility of the general supervisor, and Close was not authorized to shut down processes or equipment.

Eight million cases were manufactured by the sealing section each week. They were made from two component parts: small lengths of glass tubing and small pieces of wire to which a small glass bead had been fused in another section of the department (see Exhibit 1). The components were loaded into hoppers and were gravity fed into sealing machines. The sealing machine basically was a rotary table which contained 16 rotating heads, or holding fixtures. As the table turned, the heads passed under the two hoppers. At the first hopper, a glass tube dropped onto a locating pin in the head. At the second, a beaded lead was inserted into the tube so that the top of the short end of the lead rested on top of the pin and the back edge of the bead was flush with the top of the tube. The head grasped the tube and lead and carried the device through five successive gas sealing flames, each of which was directed toward the case at a dif-

This case, based on Kopar Industries (A), 9-613-009, was prepared as the basis for class discussion rather than to illustrate either effective or ineffective handling of an administrative situation.

EXHIBIT 1   Case Assembly and Sealing Machine

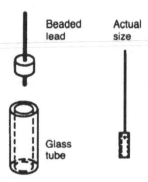

Beaded lead   Actual size

Glass tube

The beaded lead is inserted into the tube and flame sealed, glass to glass.

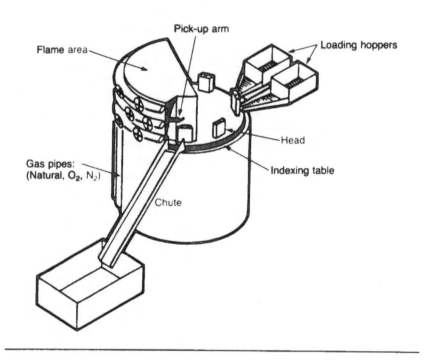

Pick-up arm

Flame area

Loading hoppers

Gas pipes: (Natural, O_2, N_2)

Head

Indexing table

Chute

---

ferent height (zone). The case was then carried past a tempering jet of nitrogen, and then to a transfer arm which removed the case and dropped it down a chute into a box. Output was collected by the operators when the boxes were filled, weighed to determine the physical count, and placed in an accumulation area. The cases were then put through a chemical cleaning tank in batches of 20,000 (tank capacity) to clean the leads. The output from the cleaning tank was then sample-inspected according to procedures outlined in Exhibit 2.

The section ran its 30 machines for three full shifts, five days per week. Normally, two operators and one flame-setter looked after a group of six machines; however, the size of the group and the number of machines assigned varied frequently due to high absenteeism and turnover rates. The operators were responsible for loading the machines, for periodically checking the quality of output (no formal procedures had been estab-

---

EXHIBIT 2
Acceptance Inspection Plan

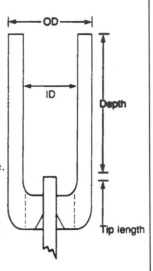

I. List of defects
    A. Mechanical
        1. Outside diameter: greater than 0.100″ or less than 0.092″.
        2. Inside diameter: greater than 0.064″ or less than 0.058″.
        3. Depth: outside of 0.181″–0.189″.
        4. Tip length: less than 0.005″.
        5. Re-entry: angle, formed at termination of seal between metal and glass, other than acute.
        6. Underseal: seal length less than 20 mils on glass-to-glass portion of seal.
        7. Cracked Case: a visible break or smooth line in the glass.
        8. Undercut-seal: etched such that less than 0.020″ of seal remains.
        9. Concentricity: wire not centered in case.
      10. Bubbles.
          a. Sliver bubble: greater than 10 mils in "thickness" and 30 mils in length (radially, around glass-to-glass seal).
          b. Bubble: greater than 0.010″ diameter.
      11. Cracked Seal: any visible crack in sealed region.
    B. Cleanliness
        1. Burnt wire: blistered appearance of wire.
        2. Burnt tip: black appearance of copper portion on the tip.
        3. Dirty glass: foreign matter on glass exceeding standard samples.
        4. Discoloration of wire: bright in appearance per engineering samples.

EXHIBIT 2   (continued)

II.   Equipment
    1.   Microscope, binocular, 30X.
    2.   Gauge: #325M.
    3.   Micrometer: Calipers.
    4.   Stamps: Accept/Reject.
III.   Definitions
    A.   Unit of Product: one case.
    B.   Sampling Plan:

Table 1

Sample Size	Defects	Total Allowable Defects	Average Outgoing Quality Limit*	Combined Allowable Defects	Average Outgoing Quality Limit*
	Sect. 1 A.1–A.4	2	2.5%		
	A.5–A.10	4	5.0%	4	5.0%
	A.11	1	1.0%		
50					
	Sect. 1 B.1–B.3	3	3.5%	3	3.5%
	B.4	1	1.0%		

*Average Outgoing Quality Limit: the maximum percent of defective items that will pass on to the next stage of the production process assuming that all rejected lots are screened and the defective parts in them replaced by good parts. The defective parts in the accepted lots, however, will not be screened out.

    C.   Defective unit of product: any unit containing one or more defects.
    D.   Acceptable lot: lot whose sample meets the requirements of the sampling plan.
    E.   Inspection lot: production lot.
    F.   Frequency of inspection: each lot.
IV.   Inspections
    A.   Select the sample randomly from the lot.
    B.   Inspect the sample for defects in the order presented in Section I.
V.   Action
    A.   Determine disposition of lots in accordance with Section III–B.
        1.   Accepted lots—stamp the lot traveler "Accepted" and return to Production.
        2.   Rejected lots—stamp the lot traveler "Rejected" and return to the Production Supervisor for rework, screening or scrap.
    B.   Rejected material must be resubmitted for reinspection prior to release to the stock room.

lished), and for weighing and carrying output to the accumulation area. Most of their time was spent clearing jams in the hopper feeding mechanisms. The flame-setters were responsible for adjusting the flame height and gas mixtures (both very critical operations) and for any maintenance which could be performed without disassembling the mechanism. The team was paid a group incentive as described in Exhibit 3. Base rates for operators and flame-setters were $10.50 and $12.25, respectively. Although groups were supposed to be paid only for good work, this policy had largely been ignored.

EXHIBIT 3
Sealing Section Incentive Rate

No. People on Team	No. Machines Assigned	Pieces/ Standard Hour	Standard Hours per Million Pieces	Factor
3	6	15,165	65.941	.18
3	7	17,407	57.448	.15
3	8	19,570	51.099	.13
3	9	21,651	46.187	.12
3	10	23,666	42.260	.11
4	8	20,220	49.456	.13
4	9	22,469	44.506	.12
4	10	24,662	40.548	.11
4	11	26,794	37.332	.10

Total group production (theoretically only good parts) was divided by the appropriate pieces/standard hour to obtain a standard hour content of output. In addition, every machine downtime (for maintenance purposes) of more than .1 hour was recorded, the total amount of such downtime was computed, and this total was multiplied by the factor. This figure was then added to the standard hour content and the sum was multiplied by each employee's base rate to yield direct labor earnings for each individual. Base rate was guaranteed. Whenever all of a group's machines went down simultaneously, the time was recorded and an amount equivalent to each employee's average earnings times the downtime period, was added to direct labor earnings.

Standards were derived by deducting from theoretical machine output:

1. Random misses—work sampling data indicated that heads were not filled 7½% of the time.

EXHIBIT 3   (continued)

2. Time to clear jams—work sampling and time study data indicated 3.06 minutes between jams and 0.73 minutes average time to clear jams. The largest portion of the 0.73 minutes represented the time it took for an operator to notice the jam and walk to it. Usually, only a few seconds were required to clear the parts which stuck in the hopper feed mechanism.

3. An allowance for interference (i.e., operator was not available when machine was down) due to:
    a. Busy cleaning jams
    b. Loading
    c. Recording data
    d. Examining quality
    e. Cleaning machines
    f. Adjusting fire
    g. Minor maintenance
    h. Removing and weighing output
    i. Communication on quality and machine problems
    j. Personal time

Figures had been computed for different combinations of operators and machines because the department required this flexibility to compensate for absenteeism.

One supervisor managed the section on each shift. Two maintenance workers reported to each supervisor. They were paid $13.70 per hour and were responsible for major repair work. Machine downtime for major repairs averaged 40 hours per day, with approximately 50 percent of this occurring on the third shift. Three inspectors who worked part-time for the section reported to the quality control office.

Exhibit 4 presents a representative sample of the most recent inspection data available to Close. According to Close's best estimates, only 25 percent of the rejected lots were actually scrapped. The final assembly section reported that a minimum of 10 percent of the cases they received were defective.

Standard manufacturing cost per thousand good cases was broken down into $7.50 for direct labor, $15.00 for material and $15.00 for overhead (200 percent of direct labor). These figures included all operations necessary to manufacture the finished cases. Labor and material costs had been computed by the accounting department from historical output and cost figures. Close estimated the major variable components of overhead to be $12,500 per month for gases and power, and $75,000 per month for maintenance parts and labor. He treated the sealing section's inspectors' wages of $11.00 per hour as fixed costs since these people were

EXHIBIT 4   Acceptance Sampling Data

Defect Description	Percent Defective					Resubmitted Work Percent Defective			
	8/26	9/12	9/9	9/16	Avg.*	8/26	9/12	9/9	9/16
O.D.	.5	.3	.6	1.0	.6	.7	.2	—	.2
I.D.	—	—	.2	.2	.1	—	—	—	—
Overall depth of case	.4	.3	.4	.5	.4	.6	.5	.2	—
Reentry	.9	.8	1.1	.7	.9	.6	1.2	.5	.8
Concentricity	1.0	1.3	1.7	1.5	1.4	1.2	.7	.3	.6
Cracked seal	.4	.3	.4	.5	.4	.3	.2	—	.2
No tip	.1	.2	.1	.1	.1	—	—	.2	.1
Cracked case	1.3	.9	1.1	1.0	1.1	1.3	2.6	3.7	1.5
Underseal	.6	.5	.7	.8	.7	.6	.4	.9	.9
Bubble	—	.1	—	.1	.1	—	.1	—	.1
Malformed	.4	.4	.4	.9	.5	.3	.3	.3	.5
Broken case	1.4	1.0	.9	.8	1.0	1.1	1.9	1.5	.8
Reversed lead	.4	.3	.2	—	.2	.2	.2	—	1.6
Burnt wire	—	—	—	.1	—	—	—	—	—
Machine bent wire	1.2	1.2	1.4	1.5	1.3	1.2	1.6	1.4	.8
Cleanliness	.6	.6	.8	.5	.6	1.0	1.5	.5	1.6
Undercut	—	—	—	—	—	.2	.2	—	.1
Total process average	9.2	8.2	10.0	10.2	9.4	9.3	11.6	9.5	9.8
No. lots produced	605	583	471	622	570	51	52	13	33
No. lots rejected for maverick†	40	20	11	28	25	4	2	—	4
No. lots rejected (total)	104	63	67	117	88	16	11	1	8
Percent of lots rejected	17.2	10.8	14.2	18.8	15.3	31.4	21.2	7.7	24.2
Shift average I	8.4	8.8	10.6	10.2					
Shift average II	9.8	8.6	9.6	10.4					
Shift average III	7.4	6.8	8.0	8.0					
Total‡	8.6	8.2	9.6	9.6					

*Unweighted average of figures in four preceding columns.
†Not included in overall process average.
‡Shift average totals differ from overall process averages due to significant variations in output from shift to shift. Typically, third shift was understaffed and maximum production was obtained on the first.

shared with two other sections (including the one manufacturing the beaded leads) and would be required whether or not the cases were inspected.

Close first attempted to isolate the variables in the sealing process and to identify the reasons for each type of reject listed in Exhibit 4. He began with the basic assumption that a good process must produce good parts, and that parts would either be very good or very bad. His inventory of reject causes is presented in Exhibit 5.

EXHIBIT 5   Description and Reasons for Reject Causes

O.D. — outside diameter of case too large.

1. "Glass we use is 0.096" ± 0.003" in diameter. We used to use 0.096" ± 0.001" but saved $135,000/year by making the switch. Our process is controllable to ±0.003". Therefore, we are bound to get many seals with O.D.'s greater than of engineering's maximum tolerance. However, I have experimented and found that the assembly process will take up to 0.105" O.D. If we changed to a maximum of 0.102", we would have less than 1 percent rejects, but engineering doesn't want to change."

2. Overheating due to poor flame setting. The case is sealed in an upside down fashion If it is overheated, we get a bulge which starts to run.

3. Improper location of flames gives same results as above.

I.D. — if the inside diameter is less than 0.058", the solder preform won't lie flat. It pinches in.

1. Flame temperature

Over-all depth of case — necessary to insure mating with tungsten whisker.

1. Damage or wear to locating pin in machine head.

2. Scale formation of pin material — "We should change to less corrosive pin metals."

Re-entry — 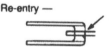 this is a weak seal and it shears when the wire lead is moved. There should always be an acute angle at the juncture of the bead and the lead.

1. Low flame setting

2. Case length

Concentricity — ⊙ leads to cracks and assembly problems.
Beaded lead ⟍ Tube

1. Undersize bead or poor bore in glass tubing

2. Bulge on O.D. due to overheating

Cracked seal — small cracks or faults in seal — may lead to failure later. Cases age and may disintegrate several days later.

1. Flame setting

2. Jarring of part while cooling

No tip — lead must be at least 0.005" exposed inside the case to provide good contact with solder preform.

1. Poor beaded leads

Cracked case — crack in case itself (not at seal).

1. Poor adjustment of flame zones (heights)

2. Damage from pick-up arm

EXHIBIT 5 (continued)

Underseal — weak seal — low strength, loss of hermeticity.

1. Flame control
2. Flame position

Bubble — air bubbles in seal region.

1. Poor flame zones
2. Poor adjustment of nitrogen jet (improper cooling)

Malformed.

1. Excessive heat
2. Machine fails to index
3. Flame goes out

Broken case.

1. Damage from pickoff arm
2. Broken tubing

Reversed leads — beaded lead improperly inserted (upside down).

1. Feed mechanism malfunction

Burnt wire.

1. Too hot
2. Nitrogen cooling jets improperly placed

Machine bent wire — leads are bent.

1. Transfer arm needs adjustment

Cleanliness — of lead and plug.

1. Overheating
2. Impurities or improper chemical strength in cleaning tanks

Undercut — weak seal.

1. Low heat
2. Flame zones

Maverick — odd parts.

1. Machine malfunctions

Notes:

1. "A good supervisor or operator can usually tell what caused the defect by its location and conformation. The inspectors only identify the defects. They are not familiar with causes."

2. Most of the defects relating to flame zones, cooling jets, and temperatures come from changes that take place during the process, not from incorrect initial settings. The same is true for problems involving pin wear or transfer arm adjustments except that these changes usually are gradual.

In addition to these Close noted a number of other variables which affected quality. Whenever the weather changed significantly and temperatures dropped (32°F seemed to be critical) the gas company changed the gas mixture in the lines, and the number of rejects soared until flames were readjusted. The same was true at various times of day when more users drew from the city mains. In addition, the location of specific machines was often important, as drafts from open doors, etc., could affect quality.

Machine speed also seemed to be an important variable, especially in regard to problems caused by the operation of the transfer arm. Most machines were set to operate at their maximum speed of 3.5 revolutions per minute. However, it was believed that the quality level of output at this speed was significantly lower than it might have been at a slower pace.

Close's initial visit to the section was greeted with considerable hostility, by both the supervisors and the workers. The supervisors did not want him "infringing on their rights." They believed that the operation was an art which could not be pinned down to numbers and standard procedures. The flame-setters had learned their skill from older workers, and each had his or her special way of adjusting the flames and special tools. Shortly thereafter, a group of workers met with the general supervisor and made it clear that they did not want a change in the status quo. The company had no union.

Close knew that any new inspection personnel he required would have to be drawn from a number of employees who had been laid off recently in another section, and he felt that their skills and abilities were questionable. He did not think there would be any possibility of using the inspectors or operators who were already familiar with the sealing section's operations.

Hank Kolb, Director Quality Assurance

Hank Kolb was whistling as he walked toward his office, still feeling a bit like a stranger since he had been hired four weeks before as director, quality assurance. All that week he had been away from the plant at an interesting seminar, entitled "Quality in the '80s," given for quality managers of manufacturing plants by the corporate training department. He was now looking forward to digging into the quality problems at this industrial products plant employing 1,200 people.

Kolb poked his head into Mark Hamler's office, his immediate subordinate as the quality control manager, and asked him how things had gone during the past week. Hamler's muted smile and an "Oh, fine," stopped Kolb in his tracks. He didn't know Hamler very well and was unsure about pursuing this reply any further. Kolb was still uncertain of how to start building a relationship with him since Hamler had been passed over for the promotion to Kolb's job—Hamler's evaluation form had stated "superb technical knowledge; managerial skills lacking." Kolb decided to inquire a little further and asked Hamler what had happened; he replied: "Oh, just another typical quality snafu. We had a little problem on the Greasex line last week [a specialized degreasing solvent packed in a spray can for the high technology sector]. A little high pressure was found in some cans on the second shift, but a supervisor vented them so that we could ship them out. We met our delivery schedule!" Since Kolb was still

This case was prepared by Frank S. Leonard as a basis for class discussion rather than to illustrate either effective or ineffective handling of an administrative situation.

relatively unfamiliar with the plant and its products, he asked Hamler to elaborate; painfully, Hamler continued:

> We've been having some trouble with the new filling equipment and some of the cans were pressurized beyond our AQL [acceptable quality level] on a psi rating scale. The production rate is still 50% of standard, about 14 cases per shift, and we caught it halfway into the shift. Mac Evans [the inspector for that line] picked it up, tagged the cases "hold," and went on about his duties. When he returned at the end of the shift to write up the rejects, Wayne Simmons, first-line supervisor, was by a pallet of finished goods finishing sealing up a carton of the rejected Greasex; the reject "hold" tags had been removed. He told Mac that he had heard about the high pressure from another inspector at coffee break, had come back, taken off the tags, individually turned the cans upside down and vented every one of them in the eight rejected cartons. He told Mac that production planning was really pushing for the stuff and they couldn't delay by having it sent through the rework area. He told Mac that he would get on the operator to run the equipment right next time. Mac didn't write it up but came in about three days ago to tell me about it. Oh, it happens every once in a while and I told him to make sure to check with maintenance to make sure the filling machine was adjusted; and I saw Wayne in the hall and told him that he ought to send the stuff through rework next time.

Kolb was a bit dumbfounded at this and didn't say much—he didn't know if this was a big deal or not. When he got to his office he thought again what Morganthal, general manager, had said when he had hired him. He warned Kolb about the "lack of a quality attitude" in the plant, and said that Kolb "should try and do something about this." Morganthal further emphasized the quality problems in the plant: "We have to improve our quality, it's costing us a lot of money, I'm sure of it, but I can't prove it! Hank, you have my full support in this matter; you're in charge of these quality problems. This downward quality-productivity-turnover spiral has to end!"

The incident had happened a week before; the goods were probably out in the customer's hands by now, and everyone had forgotten about it (or wanted to). There seemed to be more pressing problems than this for Kolb to spend his time on, but this continued to nag him. He felt that the quality department was being treated as a joke, and he also felt that this was a personal slap from manufacturing. He didn't want to start a war with the production people, but what could he do? Kolb was troubled enough to cancel his appointments and spend the morning talking to a few people. After a long and very tactful morning, he learned the following information.

1. *From personnel.* The operator for the filling equipment had just been transferred from shipping two weeks ago. He had had no formal training in this job but was being trained by Simmons, on-the-job, to run the equipment. When Evans had tested the high pressure cans the op-

erator was nowhere to be found and had only learned of the rejected material from Simmons after the shift was over.

2. *From plant maintenance.* This particular piece of automated filling equipment had been purchased two years ago for use on another product. It had been switched to the Greasex line six months ago and maintenance had had 12 work orders during the last month for repairs or adjustments on it. The equipment had been adapted by plant maintenance for handling the lower viscosity of Greasex, which it had not originally been designed for. This included designing a special filling head. There was no scheduled preventive maintenance for this equipment and the parts for the sensitive filling head, replaced three times in the last six months, had to be made at a nearby machine shop. Nonstandard downtime was running at 15% of actual running time.

3. *From purchasing.* The plastic nozzle heads for the Greasex can, designed by a vendor for this new product on a rush order, were often found with slight burrs on the inside rim, and this caused some trouble in fitting the top to the can. An increase in application pressure at the filling head by maintenance had solved the burr application problem or had at least forced the nozzle heads on despite burrs. Purchasing agents said that they were going to talk to the sales representative of the nozzle head supplier about this the next time he came in.

4. *From product design and packaging.* The can, designed especially for Greasex, had been contoured to allow better gripping by the user. This change, instigated by marketing research, set Greasex apart from the appearance of its competitors and was seen as significant by the designers. There had been no test of the effects of the contoured can on filing speed or filling hydrodynamics from a high-pressured filling head. Kolb had a hunch that the new design was acting as a venturi (carrier creating suction) when being filled, but the packaging designer thought that was unlikely.

5. *From manufacturing manager.* He had heard about the problem; in fact, Simmons had made a joke about it, bragging about how he beat his production quota to the other supervisors and shift managers. The manufacturing manager thought Simmons was one of the "best supervisors we have . . . he always gets his production out." His promotion papers were actually on the manufacturing manager's desk when Kolb dropped by. Simmons was being strongly considered for promotion to shift manager. The manufacturing manager, under pressure from Morganthal for cost improvements and reduced delivery times, sympathized with Kolb but said that the rework area would have vented with their pressure gauges what Wayne had done by hand. "But, I'll speak with Wayne about the incident," he said.

6. *From marketing.* The introduction of Greasex had been rushed to market to beat competitors and a major promotional-advertising campaign

was underway to increase consumer awareness. A deluge of orders was swamping the order-taking department and putting Greasex high on the back-order list. Production had to turn the stuff out; even being a little off spec was tolerable because "it would be better to have it on the shelf than not there at all. Who cares if the label is a little crooked or the stuff comes out with a little too much pressure? We need market share now in that high-tech segment."

What bothered Kolb most was the safety issue of the high pressure in the cans. He had no way of knowing how much of a hazard the high pressure was or if Simmons had vented them enough to effectively reduce the hazard. The data from the can manufacturer, which Hamler had showed him, indicated that the high pressure found by the inspector was not in the danger area. But, again, the inspector had only used a sample testing procedure to reject the eight cases. Even if he could morally accept that there was no product safety hazard, could Kolb make sure that this would never happen again?

Skipping lunch, Kolb sat in his office and thought about the morning's events. The past week's seminar had talked about the role of quality, productivity and quality, creating a new attitude, and the quality challenge, but where had they told him what to do when this happened? He had left a very good job to come here because he thought the company was serious about the importance of quality, and he wanted a challenge. Kolb had demanded and received a salary equal to the manufacturing, marketing, and R&D directors, and he was one of the direct reports to the general manager. Yet he still didn't know exactly what he should or shouldn't do, or even what he could or couldn't do under these circumstances.

Quality Problems, Policies, and Attitudes in the United States and Japan: An Exploratory Study

David A. Garvin

Product quality is both a problem and an opportunity for U.S. manufacturers: a problem because foreign competitors are often far ahead in offering products of superior quality, and an opportunity because American consumers are increasingly concerned about the quality of the goods and services that they buy. The result is a heightened interest in quality management at many U.S. companies.

Quality problems might arise from a number of sources, including poor designs, defective materials, shoddy workmanship, and poorly maintained equipment. There is little evidence on whether the mix of these problems is the same at companies with varying levels of performance. Do companies with poor quality face the same problems as those with superior quality? Do Japanese companies face the same set of problems as U.S. firms? These issues are important because they have a direct bearing on the policies companies should adopt in trying to improve their quality, and on the relevance of Japanese approaches to quality management for U.S. companies.

For example, much of the debate about quality management has focused on the relative responsibilities of management and labor.

Conflicting views are common. One leading analyst argues that "there is no possibility for the workforce to make a major contribution to solving [a] company's quality problems";[1] another asserts that "the major driving force behind [Japan's] productivity [and quality] movement came from millions of rank-and-file workers who have taken the initiative to suggest changes to their superiors."[2] Such widely varying interpretations reflect the limited evidence now available on the mix of quality problems in the United States and Japan.

A related issue concerns the association between quality performance and organizational commitment to quality. Most analysts agree that a necessary condition for successful quality performance is a management dedicated to that goal. Support for this claim, however, is primarily anecdotal. Systematic studies relating management attitudes to quality performance are rare. There is a similar lack of evidence on the association between workforce attitudes and quality performance. While a small number of studies have dealt with the subject, they have produced conflicting results.

[1] J. M. Juran, "Product Quality—A Prescription for the West; Part II Upper Management Leadership and Employee Relations," *Management Review*, July 1981, p. 59.

[2] Hirotaka Takeuchi, "Productivity: Learning from the Japanese," *California Management Review*, Summer 1981, p. 8.

This note will use data from a single, broadly representative industry—room air conditioning—to explore the causes of quality problems and the contributors to quality performance in the United States and Japan. It will focus on two issues: the changing mix of quality problems as quality performance improves, and the relationship between management policies, worker attitudes, and quality performance.

One reason for the paucity of evidence on these topics has been the difficulty of securing reliable and representative data. Few companies keep comprehensive records of the causes of their quality problems; still fewer periodically assess their organization's commitment to quality. While such data might be collected through surveys, the possibility of bias remains. The responses of workers and managers to questions about the causes of their companies' quality problems are likely to reflect some degree of self-interest. For example, blame is likely to be assigned by one group to the other. In response to a question asking which groups are primarily responsible for America's lagging productivity, industrial executives cited labor unions 74 percent of the time and major corporations 41 percent of the time. By contrast, union leaders cited labor unions 36 percent of the time and major corporations 62 percent of the time.[3]

HYPOTHESES

The great emphasis that Japanese companies place on manufacturing and labor relations is now widely acknowledged. In both areas, they appear to be well ahead of comparable U.S. firms. Most Japanese manufacturers pay great attention to process control, equipment design and maintenance, housekeeping, product handling, and other aspects of production that might adversely affect quality. Manufacturing excellence is viewed as an important goal. Most Japanese companies are also strongly committed to continuing improvement in their operations. Defect rates are closely monitored, often with the aid of statistical techniques; unanticipated problems result in formal analysis and remedial action. The goal of this activity is the accumulation of knowledge that will lead to a reduction in process-related errors.

In the United States, less attention is directed to these issues. Statistical controls have been adopted only recently by many U.S. firms. Defect rates are evaluated against "acceptable quality levels" rather than against standards of continual improvement. Fine-tuning of the production process is a secondary concern. Among U.S. companies, manufacturing excellence is seldom regarded as a top management priority.

Japanese companies also place greater emphasis on workforce involvement and harmonious labor relations than do most U.S. companies. Their use of quality control circles for problem-solving is one example of this approach; another is their interest in bottom-up communication and consensual decision making. Because Japanese unions are aligned with individual companies, they have an important stake in the future of their firms. The result is a supportive and cooperative work force that identifies closely with corporate goals.

These considerations suggest that Japanese companies will face fewer quality problems that arise from deficiencies in production management or workmanship than will comparable U.S. companies. They are also likely to experience a smaller proportion of problems in these categories. While Japanese companies are now pursuing quality improvements in areas removed from production, such as vendor management and product design, their earliest efforts were process and workforce-

---

[3]"IEs Offer Views on Reasons behind and Possible Cures for Declining U.S. Productivity," *Industrial Engineering*, October 1981, pp. 38–44.

related. The history of the Japanese quality movement suggests that the techniques of statistical process control, manufacturing management, and worker involvement in problem solving were the primary basis for Japan's post–war improvement in quality. Attention to vendors and designs came more recently, with the advent of the Total Quality Control (TQC) and Company-Wide Quality Control (CWQC) movements. Because Japanese companies have devoted significantly more time to the process and workforce-related aspects of quality, they are likely to experience a smaller proportion of quality problems in these areas, especially when compared with U.S. firms. These arguments lead to the following hypothesis:

Hypothesis 1: The mix of quality problems will be different in the United States and Japan. Japanese companies will experience a smaller proportion of quality problems that are attributable to poor process design, insufficient maintenance, and weak production control than will comparable American firms. They will also experience fewer problems due to poor workmanship.

A second hypothesis follows from the first. When the techniques of statistical quality control were first introduced in Japan, they were accompanied by a massive training program. Most early efforts focused on upper management. These training programs were well attended, and the principles of quality control were quickly disseminated. These principles emphasized the close connection between quality improvement, productivity gains, and cost reduction, as well as the desirability of focusing on the former for motivational purposes. A number of success stories demonstrated the usefulness of this approach. It soon became the standard for much of Japanese industry, and the driving force behind managers' efforts to upgrade manufacturing. Training programs were later established to teach the same principles to supervisors and production

workers. Once these principles gained wide acceptance, a strong commitment to quality emerged.

At many U.S. companies, a different ethic developed. In the United States today, quality is often considered secondary to other goals. Few managers or workers are trained in the principles of quality control. The connection between quality, productivity, and cost is often poorly understood. In these circumstances, the commitment of managers and workers to quality improvement is likely to be much weaker than it is at comparable Japanese companies. This argument suggests the following hypothesis:

Hypothesis 2: Japanese managers and workers will display a stronger commitment to producing goods of high quality than will managers and workers at comparable U.S. companies.

Such high levels of commitment might be sustained by a number of measures, including programs to inculcate values, extensive training, and tight evaluation and control systems. Several scholars have emphasized the success of Japanese companies in influencing employee behavior through statements of company philosophy and values. Because these statements frequently stress the importance of quality, formal controls may play a lesser role. As Ouchi points out, in the presence of widely shared values and beliefs, clan controls often dominate, rather than market or bureaucratic mechanisms.[4] This would suggest that Japanese supervisors will be less frequently evaluated against formal quality goals than will American supervisors, and might be experiencing less day-to-day pressure to meet specified performance standards. Pressure to perform however, is a subjective phenome-

---

[4] W.G. Ouchi, "A Conceptual Framework for the Design of Organizational Control Mechanisms," *Management Science* 25, No. 9, 1979, pp. 836–838, 841.

non: It reflects employees' *perceptions* of the stresses they endure. The strong informal controls employed by Japanese companies might thus be perceived by workers and supervisors as creating great pressures for improvement, even if formal targets were absent. An emphasis on continual improvement, rather than the "acceptable quality levels" typical of American firms, could well reinforce such feelings. For these reasons, it is difficult to predict *a priori* whether Japanese workers and supervisors will feel themselves to be under greater or less pressure to improve quality than their U.S. counterparts. These arguments lead to the following tentative hypothesis:

Hypothesis 3: Japanese supervisors will be under less day-to-day pressure to produce goods of high quality and will be evaluated less frequently on their performance against quality goals than will supervisors at comparable U.S. companies.

The final two hypotheses expand the preceding discussion to include differences in company behavior that are independent of culture. Several scholars have argued that Japanese approaches to quality management are transferable to the United States because they are as much a reflection of differing management priorities as they are a reflection of national or cultural traits. One implication of this argument is that companies with superior quality—whether they are Japanese or American—will share a number of common features. In both countries, high levels of quality will be associated with managers and workers who are strongly committed to that goal. Superior performers will be more likely to evaluate their supervisors against tight quality standards, and more likely to apply pressure for quality improvement. Companies with poorer quality, on the other hand, are likely to display less management and work force commitment to quality, less pressure for improvement, and a different mix of quality problems,

with management failings likely to be a special source of difficulty.

These considerations suggest that United States firms with the best and the poorest quality performance will differ in many of the same ways that comparable Japanese and American companies are expected to differ. In both cases, successful performers will be distinguished by the intensity with which they have embraced particular policies and attitudes. The observed differences will be primarily matters of degree—for example, differing amounts of pressure that supervisors are feeling to improve quality, or differing levels of workforce commitment to quality—rather than fundamental differences in kind. These arguments suggest the following hypothesis:

Hypothesis 4: U.S. companies with the best quality performance will differ from comparable U.S. companies with poor quality performance in many of the same ways that Japanese companies with superior quality differ from American firms with poorer results. In particular, the best U.S. performers will experience a different mix of quality problems, with a smaller proportion of problems that are attributable to management errors; will have managers and workers who display a stronger commitment to producing goods of high quality; and will have supervisors who are under greater pressure to produce goods of high quality and who are evaluated more frequently on their performance against quality goals.

Similar reasoning leads to the idea of a "quality spectrum." A firm's position on the spectrum would reflect its quality performance; rankings would be in either ascending or descending order. To the extent that quality performance reflects differences in quality policies and attitudes, and to the extent that policies and attitudes can be adjusted incrementally, changes in the three variables should move together. For example, a company with slightly better quality performance than average should display slightly more management and work force commitment to quality; one with vastly superior quality performance should have significantly higher levels of quality commitment. A similar argument can be made about changes in the mix of quality problems as one moves along the spectrum. Moreover, these conclusions should be independent of culture, since the explanatory variable is quality performance and not the country of manufacture. The fifth hypothesis thus focuses on the common denominators that distinguish high quality performers from those with poorer results, whether they are Japanese or American firms.

Hypothesis 5: Quality problems, policies, and attitudes will vary systematically along a spectrum of quality performance. The higher a company's quality performance, the smaller the proportion of quality problems that will be attributable to management errors, the greater the management and work force commitment to producing goods of high quality, the greater the pressure on supervisors to produce goods of high quality, and the more frequent the evaluation of supervisors on quality goals.

METHODOLOGY

To test these hypotheses, survey responses were collected from first-line supervisors at U.S. and Japanese manufacturers of room air conditioners. Respondents were confined to a single industry in order to minimize the impact that product and process differences might have on the results. The room air conditioning industry was selected for study because it contains companies of varying size and character, implying a wide range of quality policies and performance; offers relatively standardized products, which facilitates intercompany comparisons; and employs a simple assembly-line process, which is representative of other mass production industries.

Nine U.S. companies and seven Japanese companies participated in the study. Plants were selected as the unit of analysis, rather than companies, because practices differed within firms. Two of the American companies operated two plants apiece; otherwise, each company employed a single plant. In total, 18 plants were involved in the study, 11 of them American and 7 of them Japanese. Together, they account for approximately 90 percent of the shipments of room air conditioners in each country.

All participants were provided with several questionnaires. These were designed to collect background information on each plant's product line, production practices, vendor management practices, quality policies, and quality performance, while also surveying the attitudes of first-line supervisors. At U.S. plants, all first-line production supervisors were surveyed; at Japanese plants, surveys were confined to a small sample of production supervisors. Only six of the Japanese plants agreed to participate in this part of the study. At all plants, the first-line supervisors' questionnaire was administered separately from the other questionnaires, and was left unsigned. Background data were collected in 1981 and 1982, and all supervisors were surveyed in the last half of 1982.

To aid in interpreting the results, plants were first classified by quality performance. Several measures were employed to insure consistency, for companies did not always employ identical practices in recording information about quality. The classification scheme includes such measures as the assembly-line defect rate, the percentage of all defects requiring off-line repair, the ratio of repairmen to assembly-line direct laborers, the service call rate under the first year of warranty coverage, and the service call rate less "customer instruction calls: on which no product problems were found. These measures reflect both internal and external quality. Internal quality includes all defects observed before the product leaves the factory; external quality includes all failures incurred in the field after the unit has been installed. The measures resulted in five separate categories of quality performance, which are summarized in Table 1.[5] Failure rates generally rise monotonically as one moves from the Japanese manufacturers to the poor U.S. plants. The only exception is external failures for the better and average U.S. plants; that discrepancy reflects the impact of the other quality measures described above in determining the overall quality ranking assigned to a plant.

RESULTS

Hypothesis 1 predicts differences in the mix of quality problems at comparable U.S. and Japanese companies. Table 2 presents the basic

[5]David A. Garvin, "Quality on the Line," *Harvard Business Review*, September–October 1983, pp. 64–75.

data bearing on that hypothesis. Eight causes of quality problems were identified on the supervisors' questionnaire. In two categories, problems due to poor product design and problems due to materials/purchased parts, Japanese supervisors assigned significantly higher percentages than did U.S. supervisors. In two other categories, problems due to workmanship/workforce and problems due to maintenance/adjustment of process or equipment, Japanese supervisors assigned significantly lower percentages. In the remaining categories, Japanese supervisors assigned smaller percentages than U.S. supervisors, although the differences were not statistically significant. (The category "other" is excluded from further analysis because it contains a miscellaneous collection of unrelated problems.)

These findings provide strong support for the first hypothesis. Japanese and American firms appear to experience a different mix of quality problems. Japanese supervisors attributed a much smaller proportion of their companies' quality problems to process and work

TABLE 1   Quality Performance by Company Categories, Room Air Conditioning Industry, 1981–1982

Groupings of Companies	Internal Failures: Assembly-Line Defects per 100 Units (Median)	External Failures: Service Calls per 100 Units Under First-Year Warranty Coverage (Median)*	No. of Plants	No. of Supervisors
Japanese manufacturers	.95	.6	7†	29
Best U.S. plants	9.0	7.2	2	24
Better U.S. plants	26.0	10.5	3	88
Average U.S. plants	63.0	9.8	3	40
Poor U.S. plants	135.0	22.9	3	47

*Service call rates in the United States normally include calls where no product problems were found ("customer instruction calls"); those in Japan do not. All U.S. medians have been adjusted to exclude these calls, except for the poor U.S. plants, where the data was unavailable. For the other U.S. plants, customer instruction calls averaged 2.3 per 100 units under the first year of warranty coverage.
†Only six of these plants completed supervisors' questionnaires.

force-related factors, as predicted by their emphasis on training, harmonious labor relations, and the skills of manufacturing management.

These findings, however, are subject to an important caveat. Because the results are based on survey responses, they reflect supervisors' *perceptions* of the causes of quality problems rather than direct measures of problem incidence. Other groups within the firm might have different perspectives. Some indirect evidence on this point is found in Table 2, where the responses of U.S. manufacturing executives to a question about the causes of their companies' quality problems are reported from a separate study.[6] U.S. supervisors and U.S. manufacturing executives share

similar views about the causes of their companies' quality problems. Although supervisors placed greater emphasis on management errors, process design, and equipment maintenance while executives focused more heavily on purchased parts and materials, systems, and controls, the differences were small, typically only a few percentage points. These findings provide indirect support for the validity of U.S. supervisors' responses. Similar data were not available to check the responses of Japanese supervisors. It is therefore important to remember that the close identification of Japanese supervisors with their workers might result in an understatement of the proportion of quality problems at Japanese firms that were due to work force or workmanship.

Table 3 presents the responses of U.S. and Japanese supervisors to questions about their companies' quality performance, quality attitudes, and quality policies. Despite the superior quality performance of their companies,

---

[6]Frank S. Leonard and W. Earl Sasser, "The Incline of Quality," *Harvard Business Review*, September–October 1982, p. 164.

TABLE 2 Supervisors' Perceptions of the Causes of Their Companies' Quality Problems, U.S. versus Japan, and Perceptions of U.S. Manufacturing Executives

Causes of Quality Problems	*Mean Score (Percent)*		*Mean Score (Percent) U.S. Manufacturing Executives†*
	Japan	*U.S.*	
Poor product design	31%	11%	12%
Materials/purchased parts	25	15	21
Poor design of process or equipment	10	12	7
Workmanship/workforce	10	23	22
Maintenance/adjustment of process or equipment	8	16	11
Inadequate systems or controls	8	11	14
Management errors (including providing insufficient instructions to the work force)	8	10	6
Other	2	1	7
	102%*	99%*	100%*

*Columns may not add to 100 percent due to rounding.
†Leonard and Sasser, 1982, p. 164.

Japanese supervisors did not perceive their firms to have a better understanding of the causes of their quality problems, nor did they rate their firms' quality higher than did U.S. supervisors. These findings probably reflect the limited competition between Japanese and American products in this industry, and differing views of acceptable quality performance in the two countries. In assessing product quality, supervisors might simply be comparing their firms to domestic, rather than foreign, competitors. And because of their companies' emphasis on continuing improvement, Japanese supervisors might admit more readily to gaps in their understanding of quality even at relatively high levels of quality performance. Without further evidence, however, these explanations remain speculative.

Other data in Table 3 provide strong support for the second hypothesis, which predicts that

TABLE 3   Evaluations of Quality Performance, Quality Attitudes, and Quality Policies, U.S. versus Japanese Supervisors*

Survey Questions	Mean Score	
Quality Performance	*Japan*	*U.S.*
Rating of the quality of their company's products relative to competitors	5.5	5.7
Stage that their company has reached in understanding its quality problems	5.9	5.8
Quality attitudes		
Weight that management attaches to the following manufacturing objectives:		
Producing high quality (defect free) products	6.7	5.9
Low cost production	6.2	5.9
Improving worker productivity	6.2	5.8
Meeting the production schedule	6.1	6.5
Degree to which the company's production workers care about product quality	5.8	4.9
Quality policies		
Degree to which supervisors are feeling pressure to improve quality	5.7	6.1
Percentage of supervisors who are held responsible for:		
The defect rates in their area	76	78
The amount of work performed by workers in their area	31	77
The scrap costs in their area	17	80
Other quality measures	48	28
Percentage of supervisors who have seen the formal quality statement of their company.	86	64

*Items were scored from 1 to 7, with 1 indicating lower performance, less weight attached to that objective, etc., and 7 indicating higher performance or greater weight attached to that objective.

Japanese managers and workers will display a stronger commitment to producing goods of high quality than will managers and workers at comparable U.S. companies. Among management's four primary manufacturing objectives, producing high quality (defect free) products was perceived by Japanese supervisors to be the objective receiving the greatest weight. In the United States, supervisors believed that managers placed more emphasis on meeting the production schedule; it was the only manufacturing objective that received a higher score in the United States than in Japan. A similar disparity is evident in supervisors' ratings of the degree to which their company's production workers care about product quality. Here, again, the Japanese scored significantly higher.

These findings suggest a basic difference between United States and Japanese approaches to manufacturing. Companies in both countries appear to be driven by a single dominant goal. But the goals differ: In the United States, it is meeting the production schedule, and in Japan, it is producing high quality (defect free) products. These goals were ranked well above other manufacturing objectives in each country.

Table 3 also has a bearing on the third research question, concerning the pressures that U.S. and Japanese supervisors feel for quality improvement and the control systems that are employed. As expected, Japanese companies in the study relied more heavily on statements of company philosophy, while U.S. companies relied more often on formal evaluations against rework, scrap, and defect goals (the latter difference, however, was statistically insignificant). Only where other measures were involved were Japanese supervisors held accountable more frequently than their American peers. To some extent, these findings are likely to be a by-product of the small number of quality problems experienced by Japanese firms in this industry. Defect rates below one

percent imply extremely low rework and scrap costs; for many companies, such low rates would make other measures more useful as quality goals.

Japanese supervisors also reported feeling less pressure to improve quality than their U.S. counterparts. While this finding is consistent with the Japanese reliance on clan controls, rather than bureaucratic mechanisms, other factors may also play a role. For example, the kinds of pressures required to improve a manufacturing process already in statistical control, which is the case at many Japanese companies, may differ from the kinds of pressure required to bring a process under control in the first place, which is the situation at many U.S. companies. The type and amount of pressure exerted might therefore be a direct outgrowth of the type of quality improvement required.

Table 4 presents supervisor's responses grouped by their companies' quality performance. All performance categories are drawn from Table 1. These groupings permit an analysis of the quality spectrum, and a look at whether quality problems vary systematically as quality levels change.

According to the fourth hypothesis, U.S. firms with high and low levels of quality will experience a different mix of quality problems. In particular, management failings are predicted to be a more important cause of problems at the poorer performing plants. The fifth hypothesis expands this idea, predicting that the mix of quality problems will change systematically as one moves along the quality spectrum. Table 4 provides general support for the fourth hypothesis, but none for the fifth. For example, supervisors at the poorest plants ascribed 12.6 percent of their quality problems to management errors; supervisors at the best U.S. plants attributed only 6 percent of their quality problems to this cause. As one moves along the quality spectrum, however, from the best quality performers to the better, average, and poorest plants, the percentages do not increase uniformly. The same is true of the proportion of problems attributed to workmanship/workforce and inadequate systems or controls. The figures for the poorest U.S. plants are significantly different from those for the best U.S. plants, but the percentages do not rise or fall monotonically as one moves along the quality spectrum. Nor do the Japanese plants, which are ranked ahead of the best U.S. plants, fit into the pattern.

These findings suggest that the mix of quality problems experienced by manufacturing plants is only partially explained by their level of quality performance. While certain types of problems appear to be more common at poorer performers than at plants with higher levels of quality—management errors and inadequate systems and controls being the most prominent examples—these relationships are not found at all points along the quality spectrum. Only when the best and worst U.S. plants are compared separately do the predicted differences emerge.

The same is generally true of quality policies and attitudes. According to Table 5, in most cases supervisors at the best and worst U.S. performers report significant differences in the rating of their companies' quality performance, the level of their companies' quality understanding, and the quality practices and attitudes that their companies employ. The responses of supervisors at the better and average U.S. plants, however, frequently fail to conform to the predicted pattern. For example, the proportion of supervisors evaluated on rework, scrap, and defect measures is sig-

TABLE 4   Supervisors' Perceptions of the Causes of Their Companies' Quality Problems, by Company Categories

	Mean Score (Percent)				
	Japanese Manufacturers	*Best U.S. Plants*	*Better U.S. Plants*	*Average U.S. Plants*	*Poor U.S. Plants*
Poor product design	31%	12%	9%	14%	12%
Materials/purchased parts	25	15	12	25	13
Poor design of process or equipment	10	14	13	11	11
Workmanship/work force	10	30	25	19	21
Maintenance/adjustment of process or equipment	8	15	18	13	16
Inadequate systems or controls	8	9	12	8	14
Management errors (including providing insufficient instructions to the work force)	8	6	11	7	13
Other	2	0	0	1	2
Total	102%*	101%*	100%	98%*	102%*

*Columns may not add to 100 percent due to rounding.

TABLE 5 Supervisors' Evaluation of Quality Performance, Quality Attitudes, and Quality Policies, by Company Categories

Survey Question	*Japanese Manufacturers*	*Best U.S. Plants*	*Better U.S. Plants*	*Average U.S. Plants*	*Poor U.S. Plants*
Quality performance					
Rating of the quality of their company's products relative to competitors	5.5	6.2	6.1	5.0	5.2
Stage that their company has reached in understanding its quality problems	5.9	5.9	6.0	5.7	5.5
Quality attitudes					
Weight that management attaches to the following manufacturing objectives:					
Producing high quality (defect free) products	6.7	6.2	6.1	5.7	5.4
Low cost production	6.2	6.4	5.8	5.5	6.0
Improving worker productivity	6.2	6.3	5.9	5.4	5.5
Meeting the production schedule	6.1	6.5	6.5	6.5	6.4
Degree to which the company's production workers care about product quality	5.8	5.2	5.1	4.6	4.5
Quality policies					
Degree to which supervisors are feeling pressure to improve quality	5.7	5.9	6.2	6.4	5.7
Percentage of supervisors who are held responsible for:					
The defect rates in their area	76	88	74	80	79
The amount of rework performed by workers in their area	31	88	75	74	79
The scrap costs in their area	17	92	82	74	75
Other quality measures	48	42	32	18	23
Percentage of supervisors who have seen the formal quality statement of their company	86	57	64	54	75

nificantly higher at the best U.S. plants than at the poorest, but otherwise little systematic variation was shown. Supervisors at the best U.S. plants also rated the quality of their company's products relative to competitors and the stage that their company had reached in understanding its quality problems significantly higher than did supervisors at the poorest plants; the better and average U.S. plants, however, again failed to provide consistent results.

In two areas, scores do decline monotonically across the entire quality spectrum. Both areas involve quality attitudes. The weight that management attaches to producing high-quality (defect-free) products averages 6.7 at Japanese manufacturers, 6.2 at the best U.S. plants, 6.1 at the better U.S. plants, 5.7 at the average U.S. plants, and 5.4 at the poorest U.S. plants. The degree to which supervisors thought their company's production workers cared about product quality declines in a similar fashion, from a score of 5.8 at the Japanese manufacturers to a score of 4.5 at the poorest U.S. plants.

These findings prompt two observations. First, they suggest that where quality attitudes are concerned, differences between the best and poorest U.S. performers parallel those noted in Table 3 between Japanese and American manufacturers. In the realm of management priorities and worker attention, a quality spectrum does, in fact, appear to exist.

That spectrum, however, is subject to an important qualification. While the best and worst U.S. performers offer a distinguishable mix of quality problems, policies, and attitudes, the better and average U.S. plants are less clearly differentiated. Their scores are often quite close to those of plants in the adjoining quality categories. For example, on ratings of the quality of their company's products relative to competitors, the weight that management attaches to producing high quality (defect free) products, and the degree to which the company's production workers care about product quality, the best and better U.S. plants report scores that are statistically indistinguishable. The same is true of the average and poor U.S. plants. Moreover, experiments with other simple classification schemes, which divided U.S. plants into three categories based on their performance on measures of either internal or external quality, rather than a combination of the two, produced little change in the results. These findings suggest that for some purposes the quality groupings devised for U.S. firms in this industry may require rethinking; as one moves from the middle of the quality spectrum toward either of the poles, the boundaries between categories become less and less distinct.

DISCUSSION

The data presented here support the view that differences in quality performance are accompanied by differences in worker and management attitudes. High levels of quality are associated with strong commitments to that goal, whether the comparison is between U.S. and Japanese manufacturers or between U.S. manufacturers with varying levels of quality performance. In fact, management commitment often translates into pressure to perform, for there is a positive correlation between the pressure that supervisors felt to improve quality and the weight that they believed their management attached to producing high quality (defect free) products.

In most other areas, however, systematic relationships were not found across the entire quality spectrum, especially when the Japanese manufacturers were included. In fact, there is some evidence to suggest that Japanese manufacturers approach quality problems in a completely different way than their American counterparts. Table 6 presents intercorrelations of supervisors' perceptions of

TABLE 6 Intercorrelations of Supervisors' Perceptions of the Causes of Their Companies' Quality Problems, U.S. and Japan*

	Poor Product Design	Materials/ Purchased Parts	Poor Design of Process or Equipment	Workmanship/ Workforce	Maintenance/ Adjustment of Process or Equipment	Inadequate Systems or Controls	Management Errors (including providing insufficient instructions to the workforce)	Other
Poor Product Design	—	−.37‡	−.50†	−.36‡	−.47†	−.27	−.02	−.36‡
Materials/purchased parts	−.04	—	−.31	−.06	−.40‡	−.03	−.11	.22
Poor design of process or equipment	.13‡	−.18†	—	.06	.81†	−.08	−.27	.06
Workmanship/work force	−.22†	−.24†	−.39†	—	.12	−.12	.11	−.11
Maintenance/adjustment of process or equipment	−.36†	−.18†	−.04	−.27†	—	−.08	−.19	−.01
Inadequate systems or controls	−.17†	−.15‡	−.02	−.24‡	−.15‡	—	−.09	.23
Management errors (including providing insufficient instructions to the workforce)	−.11	−.21†	−.21†	−.18†	−.10	.14‡	—	−.12
Other	.05	−.07	.06	−.13‡	−.10	−.04	.01	—

*Upper right triangle presents coefficients for Japan and lower left triangle coefficients for the United States
†p < .01
‡p < .05

the causes of their companies' quality problems for separate samples of U.S. and Japanese supervisors. As one would expect, most of the correlations are negative; a large percentage of problems attributed to one category implies lower percentages in others because the total must add to 100 percent. An important pattern, however, emerges upon examination of the few positive correlations. The percentage of problems that Japanese supervisors attributed to poor process design has a significant positive correlation with the percentage they attributed to poor process maintenance; the correlation between these two categories is negative and insignificant among U.S. supervisors. By contrast, the percentage of problems that U.S. supervisors attributed to poor process design is positively correlated with the percentage they attributed to poor product design, but negatively correlated among Japanese supervisors.

These findings suggest, quite tentatively, that Japanese and American supervisors may be using different frameworks for organizing their views about quality. Among Japanese supervisors, the production process appears to be viewed in its entirety; both equipment design and equipment maintenance fall within this category, even though they involve different activities. Among U.S. supervisors, the focus appears instead to be on tasks and activities. Design issues, involving both product and process, form one identifiable category; another, captured by the positive correlation between the proportion of problems attributed to management errors and the proportion attributed to inadequate system and controls, involves those tasks that can be clearly ascribed to management failings.

Coupled with the findings of Tables 4 and 5, these correlations suggest that U.S. and Japanese manufacturers not only face a different profile of quality problems, but may also be approaching the task of quality management quite differently. The Japanese firms in this sample displayed a strong management commitment to quality, organized their thinking around process control and production management, and had workers who demonstrated a clear concern for quality improvement even without explicit goals to reduce prevailing levels of rework and scrap. While they still faced quality problems, most were found outside the shop floor, primarily in the areas of product design and purchased parts and materials.

The U.S. firms in the sample presented a more mixed picture. While plants with the best and worst quality performance were clearly distinguishable, those falling in the middle of the quality spectrum were not. Companies with the poorest quality generally displayed a weak management commitment to quality, little understanding of the causes of their quality problems, lack of worker attention to quality, and a larger proportion of quality problems reflecting management failings. The best U.S. companies showed a greater management commitment to quality (although their devotion to meeting the production schedule was even higher), greater pressure on supervisors to meet explicit quality goals, greater worker attention to quality, and a smaller proportion of quality problems in the areas of poor systems and controls and management errors.

CONCLUSION

Of the five hypotheses advanced at the beginning of this paper, the first three have been strongly supported by the empirical results. Japanese and American firms in this industry experienced different mixes of quality problems and displayed different degrees of worker and management commitment to producing goods of high quality. Japanese companies typically experienced a smaller proportion of quality problems due to poor process design, insufficient maintenance, and other weaknesses in production and workforce management but larger proportions due to poor product design and purchased parts and materials. Japanese supervisors also reported being under less pressure to produce goods of high quality than their American counterparts, and were evaluated less frequently on their performance against most quality goals.

The idea of a quality spectrum, however, was only partially confirmed. As one moved from the Japanese companies to U.S. companies with the best quality performance to U.S. companies with the poorest quality performance, worker and management commitment to quality declined as predicted. Where other quality policies or attitudes were concerned, shifts along the spectrum were less uniform. In most cases, only the best and worst U.S. companies differed as expected. The poorest firms generally displayed a greater proportion of quality problems due to management failings, less emphasis on quality goals, and slightly less pressure for quality improvement than their U.S. counterparts with the best quality.

These findings have several important im-plications for managers. First, they confirm the widely held view that high levels of quality performance are accompanied by organizational commitment to that goal. Attitudes appear to be quite important: Without a management and work force dedicated to quality, little is likely to be accomplished. Second, the evidence reported here suggests that attempts by U.S. firms to mimic Japanese quality practices without first adapting them to local conditions are unlikely to be completely successful. Not only does the mix of quality problems differ in the two countries, the framework for thinking about quality appears to differ as well. Finally, the results of this study underline the wide gap between the best U.S. performers and the Japanese, and also indicates considerable differences in their quality policies and attitudes. If the best U.S. companies hope to achieve quality levels comparable to those of their Japanese competitors, they may have to rethink their present approaches; the absence of a quality spectrum suggests that a strategy of "more of the same" is unlikely to close the gap.

Millford Corporation

Early in July 1987, Mark Gordon, a methods engineer at Millford Corporation, was reviewing the Cost and Variance Statement of July 1, 1987. As project engineer for the Sollomon SN29 computer numerical controlled (CNC) transfer machine, he was particularly concerned about the machine's performance. The transfer machine, installed in February 1987 at a cost of $522,000, had been acquired to cut labor costs and improve quality in parts production. However, the report had once again shown an unfavorable direct labor variance. The Sollomon work center had an actual direct labor expense of $5,190 against a budgeted expense of $3,171 for June, causing an unfavorable variance of $2,019 for the month and a cumulative unfavorable variance of $10,737 since operations had begun. Although an unfavorable variance had been expected during startup while unsolved problems caused downtime, and large amounts of maintenance and idle operator labor, the company's top management would view the continuing unfavorable variance as symptomatic of ongoing problems of integrating the machine into the factory.

COMPANY BACKGROUND

The Millford Corporation manufactured and distributed several lines of upholstered furniture; its principal product lines were sofas and chairs. These product lines were augmented by correlated furniture pieces to provide complete room "settings." There were several style groups and

This case was prepared as a basis for class discussion rather than to illustrate either effective or ineffective handling of an administrative situation. Names and data have been disguised.

within each could be as many as 15 different items including recliner and rocker chairs, sofas, and tables. In all, more than 300 chair and sofa designs were offered. The company's sales force sold nationwide to 10,000 dealers and had a volume of $78 million in 1986. This represented only a small percentage of the $15 billion industry sales from over 5,000 plants, but it was large enough to place Millford in the top 20 manufacturers. Although a high percentage of Millford accounts were small customers, the trend in the industry had been toward mass merchandisers. To meet this trend Millford's management were directing their efforts to developing and adding large accounts.

THE PURCHASE OF THE SOLLOMON SN29

One of the critical problems Millford top management had traditionally faced was matching capacity to sales volume. As sales volume trends increased, the Capacity Planning Group (the vice president of manufacturing, the two plant superintendents, and the production control manager) was responsible for determining possible capacity problems. One problem the group highlighted in the early 1980s was the machine capacity of the mill shop.

In addition, Millford faced the more general problem of labor shortage. It had become almost impossible to find new workers, and a study of the local area indicated the problem would continue. According to one manager, the problem was one of quality and not quantity:

> The population of our labor pool is expected to be 30 percent higher in 1990. However, the problem is finding enough people with the skills or even the capability we need. In the past year we've had 3,300 people apply for jobs; 2,800 were deemed acceptable. We hired 900, half of these for the upholstery department. Of those hired for the upholstery department, 80 percent terminated before the end of 12 months and it takes 12 months to train an upholsterer. The mill department has not really been a difficult problem. But even there the 12-month termination rate has been about 55 percent.

Faced with expanding capacity requirements and a diminishing labor market, Millford had made a concerted effort to increase productivity through automation. In 1984, Michael Swain, the manufacturing vice president, had established a methods engineering department to study and implement methods for improving their plant operations. A number of process changes had resulted from these efforts.

1. Automatic loading systems for some of the manual woodworking machines, enabling batch rather than piece-by-piece loading.
2. A material handling system that separated, facilitated sorting and grading, and stacked purchased lumber.

3. A waste recovery system for wood, scrap, chips, dust, and flour, for use as kiln fuel and resale.

The purchase of the new transfer machine was considered a step towards automating the mill shop and handling the labor shortage problem, and a more immediate response to the machine capacity problem.

The chain of events that led to the purchase of the transfer machine began in the spring of 1986. Brian Ross of Woodwork Products, U.S. representative for the Sollomon Company, wrote to Daniel Mintz, one of Millford's owners, describing the SN29 and its advantages. Mintz forwarded the letter to Swain, who turned it over to Gordon with instructions to evaluate the machine. According to Gordon:

> Michael Swain is always interested in types of equipment that could improve factory operations. We briefly discussed the feasibility of the SN29 for our operation. Because we were faced with a labor and machine shortage in the mill shop, the only question was economics. At Michael's instruction, I proceeded to make an economic evaluation of the Sollomon.

Gordon's economic evaluation consisted of finding nine high-volume, high-labor parts that could be produced on the transfer machine and calculating labor savings to be realized by producing these parts on the Sollomon. The results of this analysis were summarized in a letter to Swain in July 1986 (Exhibit 1) and showed a payback of 3.5 years and an internal rate of return of 21.5 percent, based on a cost of $627,000 and assuming a full two-shift operation. The estimate of 145 hours of machine time was obtained from the Sollomon Company on the basis of specifications and monthly requirements (for the nine parts) Gordon had provided. During the summer of 1986, Swain and others visited two U.S. plants using Sollomon Model 19 machines (an earlier version) and two European installations of the Model 29. Gordon recalled:

> Everyone was enthusiastic about the machine, but still an expenditure of $525,000 for a woodworking machine was more than we were used to investing, since the average machine in the mill shop cost about $28,000. We might not have purchased it then except for the fact that we were able to pick one up right away at a lower cost rather than going through all the import problems.
>
> Sollomon had shipped a 29 into the United States for display at an industry fair in Louisville. Daniel Mintz, Michael Swain, myself, and others went to the fair to see it during September. We watched the machine produce parts similar to those produced by Millford. We negotiated with Brian to buy the display model for $420,000, and the decision to purchase was made then and there. We were anxious to try it because it could mean a new direction for Plant One, and Brian was anxious to get one operating in this country. I believe this is still the only one he has sold.

The SN29 eventually became Gordon's responsibility. Gordon had graduated from Mississippi Technical Institute with a B.S. degree in industrial technology. Prior to graduation he had spent four years in the Air Force as an electronics technician, and before coming to Millford in 1984,

---

EXHIBIT 1

July 13, 1986
To: M. Swain   From: M. Gordon

Subject: Summary of Economic Evaluation, Sollomon SN29

1. Annual savings:

Based on production of nine parts, two months requirements of each part produced per set-up.

Process	Machine time	× Direct labor*	= Labor cost to produce
Present methods:	1,073 hrs.	× $8.40/hr.	=	$ 9,013.20
Sollomon SN29	145 hrs.	× $8.40/hr.	=	1,218.00
Direct labor savings/month, first shift		=	$ 7,800.00
Assume 2nd shift efficiency to be 90 percent of first shift,			
Direct labor savings/month, second shift		=	7,020.00
Direct labor savings/month, two shifts		=	$14,820.00
Indirect labor savings/month (one material handler)		=	2,000.00
Total estimated monthly savings, two-shift basis		=	$16,820.00

2. Costs:
    (a) One-time cost
      Machine cost: $525,000
      Freight & installation: 40,000
      Tooling & fixtures: 30,000
      Computer support equipment: 25,000
      Part programs (per part number):   7,000
                                           $627,000

*Includes: $7.37 (average operator rate) & 14 percent fringe benefits.
For standard cost purposes, Millford used a rate of $9.04 per SALH for variable overhead (not including labor fringes) in the Mill Department.

---

he worked for three years with a steel company as a production supervisor. Since coming to Millford, he had been involved with the operation of the lumber yard, rough end, and mill room. The management of the installation and development of the SN29 was his fifth project. Earlier projects included layout and construction of a new receiving yard; design,

EXHIBIT 1 (continued)
   (b) Training cost: Reflected in first eight-month operation as follows:

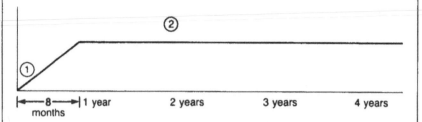

(1) First-year labor savings estimated at $134,560
(2) Thereafter, annual labor savings estimated at $201,840
    (16,820 × 12)
Training cost = 2 − 1 = $67,280

(c) Maintenance cost:
   No available facts or information. General opinion is that it will not
   exceed the costs for maintaining the machines it replaces.

3.   Payback & long-term (10-year) profitability

(a) Payback = 3.5 years, operating the Sollomon on a two-shift basis.

(b) True rate of recovery or equivocal annual earning rate = 21.5%
(Interest rate at which the $627,000 would have to be invested in an
annuity fund in order for that fund to be able to make payments equal
to and at the same time as the receipts from the project.)

layout, and construction of the lumber grader-stacker; layout and development of the plywood dimension plant; and development of the wood
waste reuse system.

MANUFACTURING PROCESS

Millford's manufacturing facilities were housed in two plants in Meridian,
Mississippi: one for the manufacture of frame parts and the other for assembly, upholstery, and shipping. The manufacturing process, diagrammed in Exhibit 2, involved the conversion of ungraded lumber, purchased metal parts, cotton and synthetic filling, cotton and synthetic
fabrics, and other materials into upholstered furniture. The frame manufacturing process consisted of four basic steps: lumber grading and
drying, rough mill, finish mill, and frame assembly. Once the frame was
assembled, sewing, upholstery, and covering were required to produce a

EXHIBIT 2 Manufacturing Process

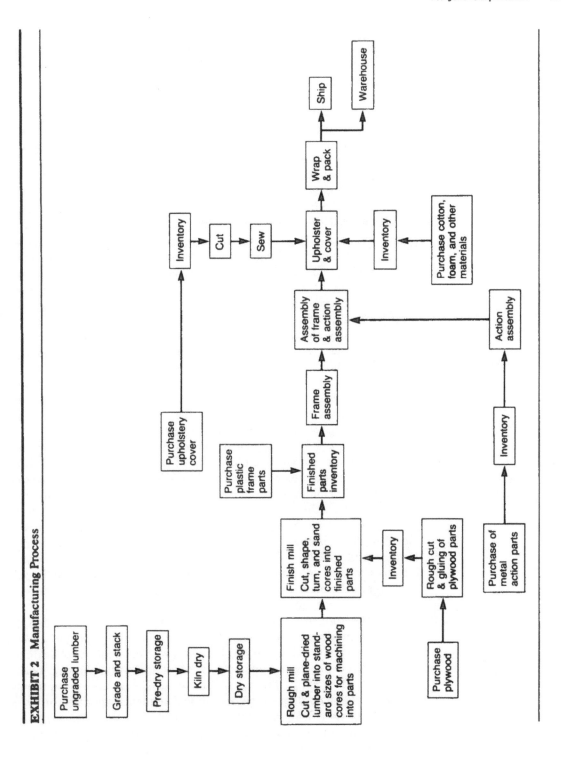

finished piece of furniture. Buffer inventories were maintained between the lumber yard and the rough mill process, and between finish mill and assembly. The labor content of direct manufacturing costs varied from process to process, with direct labor cost comprising one-third in the lumber yard and kiln area, one-half in the mill department, and two-thirds in assembly and upholstering. In 1987, Millford employed approximately 900 people in direct labor, including seven in the lumber yard, 188 in the mill shop, 240 in the assembly (frame and action) area, and the balance in the upholstery, cover, and shipping areas.

Production of Wood Parts: The SN29 was being integrated into the wood parts manufacturing process, which began with the receiving of ungraded lumber in five thicknesses from one inch to three inches. The lumber was graded and stacked on an automatic handling machine (that required one operator) and stored in the open air. The stacked lumber was processed through a drying kiln to reduce the moisture content of the lumber prior to entering the mill shop. Once dried, the lumber had to be protected from the weather. Because the wood's moisture content varied, extreme differences in times between drying and frame finishing could greatly affect its characteristics (size and shape).

From the dry storage area the wood entered the rough mill to be cut into "core stock" that the finish mill used to produce finished wood parts. Exhibit 3(a) shows the layout of the mill department and the location of the machinery. Machinery descriptions are listed in Exhibit 3(b). A piece of core stock was a finished rectangle of wood in one of the 250 possible stock dimensions, slightly larger than the maximum size of the largest finished part that would be produced from the core.

The lumber was first cut, planed, jointed, and sawed to a uniform rough thickness on a conveyor line consisting of equipment 1 through 6 in Exhibit 3(b). The resulting pieces varied in width and length as the operator cut the board to minimize scrappage. The pieces were then sorted on carts according to lengths; widths too thin for making cores were sent to the panel-flow machine to be glued into larger pieces. The sorted lengths were then processed through the line comprising equipment numbers 7 through 13 in Exhibit 3(b) to produce the finished core stock, which was the starting point for the finish mill operations. Ordinarily, the rough mill operations were scheduled by the supervisor to produce core stock needed for current finish mill requirements.

Using the core stock and rough cut parts from the plywood shop, the finish mill produced approximately 1,600 different frame parts. Work moved through the mill in varying lot sizes, with each lot traveling on one or more carts. The operations performed were sawing, shaping, boring, planing, rough sanding, turning, and finishing. The orders moved through the finish mill according to the operational sequence specified on an accompanying specification sheet (see Exhibit 4 for an example), which listed all operations required to produce the finished part, including the operations necessary to produce the core stock.

EXHIBIT 3(a) Layout of Mill Department*

Rough Mill Area

Incline belt conveyor for scrap

Finish Mill Area

Pattern Storage

Cafeteria

*See Exhibit 3(b) for definition of number codes.

EXHIBIT 3(b)   Equipment Description

Code	Name	Use	Cost
1	Lumber elevator	Raise lumber to working height	$ 24,800
2	Cut-off saws	Cut lumber to length	10,800
3	Bursting saw (rip)	Rip or "burst" cupped lumber pieces	23,200
4	Double surface planer	Surface both sides of lumber pieces	99,000
5	Edging (rip) saw	"Joint" both edges of lumber pieces	23,200
6	Canted roll conveyors	Properly direct lumber through edging saws	6,200
7	Rip saw	Cut lumber to widths	23,200
8	Panel flo	Glue pieces into panels	108,000
9	Single surface planer	Surface one side of lumber pieces	37,000
10	Lathe	Trim, bore, and dowell both ends of piece	99,000
11	Miter saw (variety saw)	Make individual miters, trims, or rips	7,700
12	Lathe	Trim and bore both ends of piece	74,000
13	Band saw	Saw irregular shapes	10,000
14	Double end trim	Trim both ends of pieces	25,000
15	Profiler	"Shape" or "profile" one side automatically	62,000
16	Single spindle shaper	Perform light shaping by hand	7,700
17	Wide belt sander	Rough and finish sand one side	99,000
18	Soft drum sander	Polish sand pieces	775
19	Disc sander	Polish sand turnings	775
20	Edge belt sander	Sand long lengths	19,000
21	Spindle carve	Perform light hand shaping	1,400
22	Sander	Automatically sand round turnings	46,500+
23	Automatic lathe	Automatically "turn" pieces	50,000
24	Router	Perform deep cuts	37,000+
25	Finger jointer	"Finger joint" one end of piece	19,000
26	Master carver	"Carve" irregular pieces	56,000
27	Shaper	Automatically shape or profile	62,000
28	Sollomon SN29	Perform various operations and shape both sides	522,000
29	Vertical boring machines	Bore holes in vertical plane	30,000
30	Horizontal boring machines	Bore holes in horizontal plane	12,500
31	Double end dowel driver	Dowel both ends of piece automatically	30,000
32	Single and dowel driver	Dowel one end of piece manually	14,000
33	Wood hog	"Grind" wood waste into boiler fuel	46,500
34	Wood chipper	Chip wood into form acceptable for resale to paper industry	77,500+

EXHIBIT 4 Specification Sheet

LUMBER SPECIES		CORE DIMENSION			PIECES PER CORE	FINISH DIMENSION			B.F. PER PIECE	PART NO.		
		THICK	LENGTH	WIDTH		THICK	LENGTH	WIDTH				
ELM										51-501 R		
FIN. HDWD.	6/4	27	6 5/8		1	1 5/32	26	p	1.8633	51-501 L		

SET UP AT 200		PIECES		PIECES ORDERED		PIECES RECEIVED		

MACH NO	NO OPER	QUAN	OPERATION DESCRIPTION	S.M.H.	S.U	S.A.L.H.
						1.804
P&CO			Plane and cut off			
RP			Rip to 3 5/16" width	1.890	15	1.945
RP			Joint to 3 3/16" width	.280	15	.355
PF			Glue 2 pcs. to 6 3/8" width	1.293	02	
BP			Plane 2 faces to 1 3/16" thickness (1 pass/face)	.436	20	.536
WBS			Rough sand 2 faces (1 pass/face)	.396	20	.496
B-424			Trim to length @ miter and bore 2-7/16" holes in each end and drive 2-7/16" x 2" dowels in each end (select face)	.282	95	.657
OP			Shape edge/patt.	.375	27	.510
SSS	(1)		Shape curved edge R&L	.386	15	.461
SSS	(2)		Joint and buckshape straight edge R&L	.463	15	.538
ROOT			Bore 4-7/16" & 3-3/16" holes in face R&L	.819	85	.644
R-170	D-1(1)		Counterbore 1-7/16" x 3/16" hole in edge R&L	.382	63	.697
R-170	D-1(2)		Counterbore 1-7/16" x 3/16" hole in edge R&L	.382	63	.697
ROUTER	(1)		Route 2 grooves in face R&L	1.499	15	1.574

MACH NO	NO OPER	QUAN	OPERATION DESCRIPTION	S.M.H.	S.U	S.A.L.H.
ROUTER	(2)		Round 2 grooves R&L	.715	15	.790
WBS			Polish sand 2 faces to 1 5/32" thickness	.632	20	.732
BuFF			Sand around router cuts R&L	.749	10	.799
PNS			Rough sand R&L	1.319	05	1.344
PNS			Polish sand R&L	1.319	05	1.344
				11.704		14.119

ISSUE #4 12-8-70 E.V.

S.C. = .334
E.B.S. = .644

STANDARD TOLERANCES
UNLESS OTHERWISE SPECIFIED

1. Tolerance for Dowel Hole Spread = ± 1/64".
2. Tolerance for Dowel Hole Diameter = ± .005".
3. Tolerance for Dowel Hole Depth = ± 1/16".
4. Tolerance/Maximum Thickness, Width, Length, = ± 1/32". Except Stretcher Length = ± 1/64" and ± 1/3°.
5. Tolerance/Maximum Angles or Lead-Off = ± 1°.

DWN BY	CK. BY

51-501
PART NO.

There was no standard movement pattern through the machine shop. The supervisor assigned a job to the machinist responsible for the first operation listed on the specification sheet. The operator would take the specification sheet, draw the core stock, draw any necessary tooling and/ or patterns, set up the machine, and run the lot. When the run was completed the operator informed the supervisor and was assigned another job. The completed job was then assigned to another operator, who sought the material and proceeded with the next operation. When all operations were completed, stock handlers moved the parts to the finished parts inventory areas to be drawn on as needed by frame assembly.

Recently, Millford had begun replacing some of the interior and exterior frame parts with injection molded plastic parts. These parts were produced by a vendor who used molds Millford supplied; these molds were purchased from another vendor for as much as $75,000 apiece. Under normal conditions, only one mold was necessary for any part number, with each mold capable of producing all of Millford's requirements for a given part at a contract cost of a few cents per part. Molded plastic parts did not make up a significant percentage of the parts inventory, but there were plans to replace other wood parts with plastic where volumes and unit costs made replacement feasible.

THE SOLLOMON SN29

Developed and produced by Sollomon, Ltd., an English machine tool company, the Sollomon SN29 was described by the manufacturer as a "23-station computer numerical controlled longitudinal and transverse automat." The machine featured a two-pallet transfer system allowing internal part loading during processing. It positioned each work-holding pallet into fixed stations along a horizontal rail to perform sawing, boring, mortising, square tenoning, shaping, and sanding. A computer program directed each operation; typically, parts did not require all operations.

The machine was designated to hold in position up to three cores per pallet, move them past 23 work stations, and return a finished part to the operator station. Operators loaded cores into simple fixtures and jigs attached to each pallet; clamps were used to hold each core in position. Programs were stored in machine memory and operated each process function independently. Servo[1] D.C. drive screws positioned both cutters (from overhead) and finishing sanding heads for each station. Once at the end of the rail, the work piece returned along the same process path to the load-unload station.

Operating the transfer machine required only loading and unloading the machine between each cycle, monitoring machine functions, checking product quality, and performing cleaning and minor maintenance. For a part previously produced on the Sollomon, set-up on the machine required loading the correct computer program, installing the fixture and jig on the pallet, and mounting correct cutters. These tasks needed four hours for set-up including producing and checking the first part. Minor

---

[1]A servo was a motor that could be finely controlled to rotate at a precise rate of speed.

adjustments (offsets) were made to the computer memory to correct process variations for cutter size and fixture alignment. This task required operator knowledge of the tooling and computer functions.

When a part was to be run for the first time on the Sollomon, it was necessary to design fixtures and jigs with special cutters. Computer programs could be produced off-line away from the machine in the engineering area and loaded by punched tape or magnetic disk into the computer controller. Approximately 20 hours of engineering were needed to design the tooling and produce the part program. During set-up and testing program, changes (included in the engineering time) were always required. Very little idle time was necessary since program changes could be made, on-line, at the machine.

THE STARTUP

Millford began constructing the foundation for the transfer machine in early October 1986; all parts of the transfer machine were received in Millford's plant by the end of the month except for the machining units (shapers, saws, etc.), which were not received until early February. By late December, Millford's tool department had completed the boring knives and cutter heads for the transfer machine's work stations.

During January 1987, Mark Gordon and Al Grillo, the tool maker who was to become the chief maintenance person, familiarized themselves with the machine with the aid of an operational handbook and supervised the setting of the SN29 on its foundation. Looking back, Gordon felt that they had been able to get the machine up quite rapidly:

> When the first Sollomon engineer arrived for startup on February 8, he was amazed to find that we had the machine almost completely set up; they usually had to unpack and set up the machine themselves.
>
> A principal reason for our success with the machine has been Grillo, one of our best tool makers, who handled the actual set-up and maintenance. He had been able to learn the machine very rapidly and by catching minor problems has eliminated the development of major problems.

A Sollomon engineer remained at Millford from February 8 until March 22 to provide technical aid during installation and startup. The first part was run on the machine on February 24 and two-shift production was started on March 20.

Although Millford experienced no major problems with the startup of the Sollomon, a series of minor problems prevented any sizable production until late March: Sensitivity to public electric power fluctuations interfered with operation of control relays; improper ventilation resulted in overheating of saw motors; improper gear settings caused rapid wearing of drive belts. All of these problems resulted primarily from the Millford personnel's lack of familiarity with the transfer machine. The following excerpts from Gordon's log book indicated that other problems resulted from efforts to tool for the first few parts.

3/3/87   Began set-up for part #43-1221.

3/4/87   Still setting up for 43-1221.

3/5/87   Tried to run a few 43-1221s. Software needed revision; feed rate too fast.

3/6/87   Grillo inventoried tools and copied set-up.

3/8/87   Discontinued, knives for 1221 must be reground; spent all day trying to line adjust vernier settings to no good end. Checked out template cutting units.

3/9/87   Discontinued efforts on 1221. Began set-up for part #SA-296. Had to make drastic changes in the hold-down bars due to the small diameter knives. Took knives to Hymann (outside tool and die maker) to be reworked.

3/10/87   Commenced setting back up on part #51-515. (Part #51-515 was being run on 3/2/87 prior to attempt to set up part #43-1221.) Damaged 3½" spread boring head, took it to Royce (outside contractor); will not have it until Tuesday of next week.

3/11/87   Royce has no gear stock on hand. Will be rest of week before boring head is returned.

*   *   *   *   *

3/17/87   Found it advantageous for mill room to (1) try new template for part #SA-296, and (2) run a few for production. Set up machine for SA-296. Required extensive template change. Began making new template.

3/18/87   Completed new template and put it on machine.

3/19/87   Ran a few SA-296s.

3/20/87   Ran SA-296 for eight hours. Machine ran OK. Trained Rick Schneider for 2nd shift operation.

3/22/87   Ran SA-296 two shifts. Machine performed OK. Herb Alexander, Sollomon engineer, left.

Initial operations disclosed one significant error in the assumptions used to estimate labor savings expected to result from the SN29. When the first part was set up at Millford, the Sollomon engineer pointed out that a single profiler station could not remove all the wood necessary to shape the finished part from available core stock. The depth of cut called for, typically, was as much as one inch, whereas the customary European practice was to shape the part roughly on a band saw, leaving one-quarter inch or less to be removed by the profiler knives. The machine hour estimates Sollomon provided had assumed the smaller depth of cut, whereas Millford, consistent with U.S. woodworking practice, would need either to add a preparatory operation or set up additional profiler stations on the SN29. In this situation, they chose to use three stations on the SN, which resulted in lengthening the operating cycle. By March 27, 1987, the new transfer machine was operating on a regular two-shift basis. Exhibit 5 summarizes the production of the SN29 from March 27 through June 30, 1987. Although production was somewhat sporadic at first, by June most of the bugs had been eliminated and Gordon felt that production grew as operators became more efficient.

EXHIBIT 5   SN29 Production 3/27/87–6/30/87

Date	Part Number	First Shift	Second Shift
Mar. 27	SA–296	200	
29	SA–296	437	331
31	51–515	224	414
Apr. 1	51–515	406	387
2	51–515	218	44
5	51–515	512	484
6	51–515	330	
13	SA–296	215	198
16	SA–296	250	233
19	SA–296	240	
28	43–1213 + 15	*	520
29	43–1213 + 15	800	830
30	43–1213 + 15	775	780
May 3	43–1213 + 15	820	830
4	43–1213 + 15	900	812
5	43–1213 + 15	818	810
6	43–1213 + 15	210*	
6	43–1209	387	138*
6	43–1224		416*
7	43–1212	525	530
8	43–1213	397	
9–26	Broken carriage casting		
26	43–1213		730
27	43–1213	600	775
June 1	SA–296		609
2	SA–296		480
3	SA–296	435	
4	SA–296	673	513
7	43–1216	628	690
8	43–1216	640	320*

*Set-up change begun during this shift.

EXHIBIT 5   (continued)

Date	Part Number	First Shift	Second Shift
8	43–1215		5
9	43–1215	701	775
10	43–1215	351	732
11	43–1209	730	605
12	43–1212	380	351
14	43–1212	645	579
15	43–1215	700	720
16	43–1215	675	607
18	51–501	304	320
19	51–501	125	
21	51–501	300	250
22	51–501	250	325
23	51–501	342	330
24	51–501	295	280
25	51–501	331	325
26	51–501	121	125
28	51–501	330	302
29	51–501	231	380
30	51–501	296	325
		18847	19378

SCHEDULING THE SOLLOMON

One of the major tasks Gordon faced was choosing the parts to be made on the Sollomon. It could cut, shape, bore, and sand, but it could not finish sand or perform lathe (turning) operations and had to work from core stock. This necessitated other machines' performing operations before and after the Sollomon operations. Since all the operations listed on the specification sheet prior to the SN29 operations involved the production of standard core stock, the Sollomon had a buffer stock of raw material from the rough mill.

In choosing parts to be made, Gordon worked closely with Gary Chamberlain in production scheduling. Gordon explained this process:

The computer explodes the Capacity Planning Group's forecast into a six-month parts forecast. I get a copy of this forecast from Gary and look for large volume parts. I check these parts for their labor content and calculate how many of the operations can be sequenced on the Sollomon and how much labor we can save on these parts. Once I have a group of parts to choose from, I get together with Gary and the mill shop supervisor. We decide which parts would be best to schedule through the Sollomon, both to meet requirements and to level the usage of other machines.

By the end of June, seven parts were being produced on the transfer machine. The specification sheets for one of these seven parts, showing the piece, the woodworking operations, the Standard Machine Hours (SMH) in hours per 100 pieces, the set-up (SU) time, and the Standard Allocated Labor Hours (SALH) in hours per 100 pieces, is shown in Exhibit 4. Normally, the first operation shown on the sheet, plane and cutoff (P&CO), was performed in the rough mill to produce the cores. The sheet shown in Exhibit 4 specifies operations and times for normal routing; the operations checked in the left-hand margin are the operations that can be sequenced on the SN29 transfer machine.

Gordon used a three-step process to select which parts should be run on the SN29. First, he located part numbers with high requirements. Next, he calculated which operations of these high volume parts could be performed on the SN29. Finally, he calculated the SALH required for the SN29 to perform these operations. He compared this with the SALH for the conventional routing shown on the specification sheet to judge if labor savings were large enough to justify using the transfer machine to produce the part. Gordon originally calculated the SALH for the operation to be performed on the new transfer machine as follows:

Part Number	SALH for Normal Routing	SN29 SMH
SA 296	5.469	1.666
43 1209	3.255	.833
43 1212	3.255	.833
43 1213	2.773	.833
43 1216	2.773	.833
51 501	5.612	1.666

Although Gordon made the original choice of parts to be run on the Sollomon, he shared the responsibility of keeping the machine loaded with Chamberlain and Lou Andrews (assistant supervisor in the mill shop).

In discussing the problems of scheduling the Sollomon, Chamberlain stated that parts could be divided into three groups—A, B, and C. A parts were high use and high (labor) cost parts, B parts were medium use and medium cost, while C parts were low cost and/or low use. A parts represented about 18 percent of the part numbers but 80 percent of the load of the mill shop load. Chamberlain explained how this was affecting the Sollomon:

> We are concentrating on putting A parts such as the SA-296 through the Sollomon, but in order to get larger lot sizes for the Sollomon, I am entering orders for eight week's usage based upon forecasts. Thus when we have forecasts for 3,000 185s, I can order 3,000 pairs of arms or 6,000 of the SA-296 parts.

He emphasized that there was no clear procedure for scheduling the Sollomon:

I coordinate with Mark to see what his loading requirements are as far as meeting operating costs, and Mark and I talk with Lou to see how loading the Sollomon can level the total mill load and help us still meet order requirements. Both Lou and I periodically walk down to check the machine's backlog. We don't want that machine standing idle because we didn't provide any orders for it.

One of the problems which Chamberlain had with the forecasts was that they were based on aggregates:

Aggregates are close, but the actual mix of styles and types are something else. When the forecast is broken down by style, we're doing well if we aren't off by more than 25 percent of actual. With the large volumes we are talking about on the Sollomon, 25 percent is pretty far off.

OPERATING COST AND EVALUATION

Because the machine had been evaluated for purchase on the basis of labor savings, the evaluation of operations was based upon labor costs. Three direct labor people were charged to the Sollomon. On the first shift, the Sollomon had a full-time operator at $7.14 per hour with no special skill requirements, and a full-time maintenance person at $12.54 per hour for die making and set-up. The maintenance person's time was loaned and charged out to other departments when that individual was not working on the SN29. The second shift had one full-time operator/set-up person at $11.63 per hour. Costs of fringe benefits were calculated at 14 percent of base.

The accounting department established a separate budget allowance for the Sollomon, based on its production cost, using the SALH the part would have if it had gone through manual machines. The details of the budget are shown in Exhibit 6. Each month a budget was calculated from the Sollomon's monthly production and compared to the actual labor cost to give a cumulative yearly control variance. This resulted in the Sollomon's having a deficit variance of $4,275 as of May 1, 1987, resulting from a budget allowance of $1,718 and actual expenses of $5,993. Gordon felt that the budgeting procedure set up for the Sollomon was in error, and the machine's value should be judged on the basis of the difference between the normal SALH and the SN29 SALH for the pieces produced, rather than on a comparison of actual labor costs and SALH for the parts produced.

Gordon outlined his accounting procedure for the manufacturing vice

EXHIBIT 6

February 26, 1987

TO: M. Swain   FROM: Accounting Dept.
     M. Gordon
     L. Andrews

SUBJECT: Accounting procedure for the Sollomon, Effective March 1, 1987

I.   The Sollomon account number will be 9909 in Dept. 16.

II. All direct labor (actual milling of parts) will be charged to this account. The employees assigned to this job should be transferred to this account on a permanent basis.

III. All set-up labor on the Sollomon will be charged to this account. This may be in the form of one person on a full-time basis. If this person works on any other equipment, that time should be transferred out of this account. If any outside help is required, that time should be transferred into this account.

IV. The budget allowance for this account will be the total SALH for all regular operations eliminated by the Sollomon, extended by all parts produced on the Sollomon during the accounting period. The 9901 Account budgeted hourly rate will be used to determine the dollar budget.

Example:

Part #	Reg. Opn. Eliminated	Reg. SALH/L	Pcs.	Budget Hrs.
51–501	B424	.282		
	Profiler OP. & root	.594		
	Router #1	1.499		
	SSS #1	.386		
	SSS #2	.463		
	WBS	1.028		
		4.252	5,000	212.6

V. The budget allowance for this account will, in effect, be a reduction of the budget allowance for both 9901 and 9908 accounts.

VI. No parts will actually be routed for the Sollomon during fiscal 1986–87 or 1987–88. This is a necessity, if the accounting procedure described here is to function properly.

president in the letter shown in Exhibit 7. Reviewing some of the other problems in the evaluation of the Sollomon, he observed:

Although we should be running high labor parts through the Sollomon, 20 percent of the time we are running "squares" through the machine. That is, we are simply squaring off core stock for the lathe. These are planing and sanding rework operations on parts 43-1209, 43-1212, 43-1213, 43-1215, and 43-1216. Unless these parts are perfectly squared before mounting in the lathe, they turn off center resulting in a nonsymmetrical pattern, and the piece has to be scrapped. The operations we do on the squares are of very low labor input but very high tolerance. This squaring operation should be done on a molder, which costs about $66,000. But we do not have a molder so they go through the SN29.

The location of the transfer machine is not optimal, resulting in a material flow problem.

No parts are routed through the Sollomon on the spec sheets. The set-up person must be able to convert the individual standard specs to "29" specs. Eventually we hope to have certain spec sheets reworked to show both standard and "29" specifications.

The Sollomon is a 23-station machine. On the parts we run through the Sollomon, it now replaces about five conventional machining steps. Thus the single operation "29" station replaces five operators.

EXHIBIT 7

May 18, 1987

TO: M. Swain   FROM: M. Gordon

SUBJECT: Sollomon SN29 Performance Evaluation

1. Machine installation was started February 8 and completed March 20. It began operating, production-wise, on a two-shift basis.

2. In order to have a monthly evaluation of the machine's performance, for simplicity's sake, it was decided to create a separate account number (9909) on the C&V Statement.

    The labor hours saved by the machine are multiplied by the budgeted direct labor rate (9901) and transferred into the new account to show "Savings." The "Actual Expense" column of the C&V Statement is the actual direct labor expense involved in operating the machine.

3. In order to arrive at a more accurate "Labor Rate," I will include the following Direct Labor and Expense Accounts, all of which are affected by the operation of the SN29.

EXHIBIT 7 (continued)

9901	Direct labor—Measured	$ 6.928
9905	Trainees	.163
9906	Incentive bonus	.540
99165	Other services	.467
9924	Overtime premium	1.740
9925	Shift premium	.125
9928	Fringe benefits	.677
Total		$10.640

4.(a)   March 10 through 31

SALH of parts, normal routing	250.339	
Labor cost of same (SALH × 10.64)		$ 2,663.61
SALH of parts, Sollomon	171.999	
Labor cost of same (SALH × 10.64)		$ 1,830.07
Savings involved, partial month of March:		$ 833.54

  (b)   April

SALH of parts, normal routing	1,321.108	
Labor cost of same (SALH × 10.64)		$14,056.59
SALH of parts, Sollomon	957.368	
Labor cost of same (SALH × 10.64)		$10,186.40
Savings involved, April		$ 3,870.19

5. Operation of the SN29 made it evident that too much material was being used to produce the top arm, 185 group. A job change, resulting in material and labor savings amounted to .3118 cents/piece, has since been instituted on the following parts. Forecasted annual savings are included and should be credited to the Sollomon.

SA 296	$11,039
309	7,672
310	2,450
337	448
339	5,026
340	2,429
Total annual savings due to job change	$29,064

THE FUTURE OF THE TRANSFER MACHINE

Because of the SN29's special problems in scheduling, set-up, costing, and maintenance, Gordon felt he would be connected with the existing SN29 operation well into the foreseeable future:

> Right now everything about the Sollomon is special. Although operation of the machine requires little training, its set-up, maintenance, and scheduling need special attention because it does not fit into the normal system. Grillo and I will continue to be responsible for these areas, while actual operation will come under the control of the supervisor of Department 16. Even when we have the set-up, maintenance, and scheduling on this machine standardized, we'll probably be adding more of these and similar machines in Department 16. Michael Swain and I have talked about the possibility of a series of CNC machines, with each one as a separate line. While I don't envision any particular start-up problems on additional machines, I'm sure there will be added complexities of scheduling.

Moreover, Gordon felt that the Sollomon SN29 should be the first of many such machines in Millford's production process, and he hoped to use his experience with this installation to anticipate the effects of future installations of new processing techniques.

Kool King Division (B)

Bob Kunz looked hastily at the desk calendar as he sat down in his new office chair at 7:15 A.M. on Tuesday, September 25th, 1979. He noticed that his predecessor in the job of Kool King Materials Manager, Mr. Russel Frank, had scheduled a 90-minute interview with a candidate for the open purchasing manager position, starting at 8:30 A.M. Kunz also noticed that representatives from FRC, Inc., a major vendor, were scheduled for a meeting lasting from 11:00 A.M. through the rest of the day.[1]

Kunz turned from the calendar to Frank's in-basket, and began to shuffle through the contents (see Exhibits 2 through 8).

BACKGROUND

At 4:00 P.M. on September 24th, Bob Kunz had entered the office of Mr. Jim Lewis, Vice President and General Manager of the Kool King Division. When he walked in, he had been a TIA Corporate Staff Operations Consultant. When he walked out several hours later, he did so with the title of Acting Materials Manager, Kool King Division, and with the knowledge that his predecessor might never return to work. In explaining the situation to him, Lewis indicated that Frank, 58, had suffered a severe heart attack on the last day of a scheduled week-long vacation.

This case was prepared by Jeffrey Miller and Ramchandran Jaikumar as the basis for class discussion rather than to illustrate either effective or ineffective handling of an administrative situation.

[1]FRC was an $80 million firm specializing in the design and manufacture of compressors of all types.

EXHIBIT 1   Kool King Division Organization

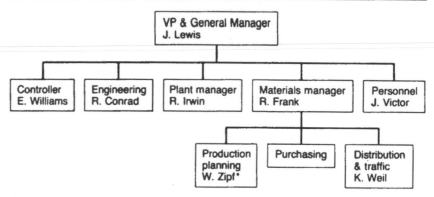

*Zipf was new in the production planning position, having recently transferred from the Budgets Department (reporting to the controller).

EXHIBIT 2

DATE 9/21/79   HOUR 11:40

TO R. F.

WHILE YOU WERE OUT

M Hank

OF Receiving / Inspection

PHONE 634-4327
   AREA CODE   PHONE NUMBER

TELEPHONED	X	RETURNED CALL		LEFT PACKAGE	
PLEASE CALL		WAS IN		PLEASE SEE ME	
WILL CALL AGAIN		WILL RETURN		IMPORTANT	

MESSAGE

They had to reject the entire shipment of 6249's that came in today (Friday). This is the third shipment in a row with quality problems.

SIGNED

The AICO-UTILITY Line Form No 55-058

EXHIBIT 3

DATE __9/21/79__   HOUR __4:10__

TO __R. F.__

WHILE YOU WERE OUT

M __Willard Homans__

OF __Order Entry__

PHONE __634-3851__
AREA CODE   PHONE NUMBER

TELEPHONED		RETURNED CALL		LEFT PACKAGE	
PLEASE CALL		WAS IN	X	PLEASE SEE ME	
WILL CALL AGAIN		WILL RETURN		IMPORTANT	

MESSAGE __National Department Stores placed an order for 3000 Mighty Midgets and 5000 Midgets on our early order discount plan. Willard doesn't think marketing included this order in their forecast (They didn't think they'd win that account). He thought you should know this.__

SIGNED _____

The AICO-UTILITY Line Form No 55-056

Kunz, 28, had worked for TIA Corporation since his graduation from the Harvard Business School. He had spent a year in various staff manufacturing positions at corporate headquarters before requesting a transfer to a line job in a division. Having observed a number of his colleagues "deadend" in corporate staff positions, he had set his sights on working up through an operating job. He had anticipated a move soon, but had been surprised by the emergency call to the Kool King Division. Lewis had commented to him in their meeting: "I'm not happy about the circumstances, but I'm glad to have you on my staff. I've got a feeling that a lot of our problems are rooted in the materials area."

EXHIBIT 4

September 12, 1979

Dear Mr. Frank,

I have discussed your request to increase the contract amounts for compressors from 108,000 to 122,000 this year with our production department. Since this is the second upward contract revision in two months, you can appreciate that this places them in a difficult position. However, since you are a good new customer with growing requirements, I have persuaded them to agree to the change, under the following conditions:

1. That you will take ¼ of your total requirements in each of the next four quarters in equal parts in order to allow us to smooth the load your requirements place on us;

2. That you provide at least 16 weeks advance notice of weekly shipment requirements;

3. That you provide at least 8 weeks advance notice for any changes in weekly shipment requirements, and that such changes be limited to ±15 percent.

I am sure that you can understand the reasons for these conditions. Industry capacity is tight and we must all work together to make the best use of it.

I look forward to the opportunity to discuss these matters with you in person. May I suggest that we meet before the engineering design review for the 4782 scheduled at your office on September 25th? I might note that it has been great to watch our engineers work together over the last six months on the development of this new Slim Line compressor design. My production people tell me that tooling will be available to make it in just three more months.

Sincerely,

M. L. Hamlish
Customer Representative
FRC Corporation

MLH/jmb

EXHIBIT 5

DATE 9/24/79   HOUR 4:40

TO Mr. Kunz

WHILE YOU WERE OUT

M Irwin

OF 634-3284

PHONE _____
AREA CODE   PHONE NUMBER

TELEPHONED	RETURNED CALL	LEFT PACKAGE
PLEASE CALL	WAS IN	PLEASE SEE ME ☒
WILL CALL AGAIN	WILL RETURN	IMPORTANT ☒

MESSAGE Would like you to come to his office as soon as you get in to talk about the 6249 compressor problem.

SIGNED

The AICO-UTILITY Line Form No. 85-056

EXHIBIT 6

Memorandum

TO: Distribution

FROM: R. Conrad

SUBJECT: Quality Problems on 6247 Compressors

DATE: 9/20/79

We have reviewed the problems on the last two shipments of 6249 Compressors from FRC with their engineers and our people from the manufacturing, inspection, and purchasing departments. The basic problem appears to be the ability of these compressors to meet our pressure specifications.

On the first lot rejected, the average incoming quality level for the shipment was approximately 50 percent below our standard.* After negotiations with FRC, they accepted the lot in return and promised to rework it as fast as possible. They committed to ship this repaired lot back by October 5th (along with the regularly scheduled lot).

The average incoming quality of the second lot was about 10 percent below standard. Although it seems apparent to us that poor manufacturing methods and quality control procedures probably led to the problems with this lot, FRC claims that the problems were due to improper handling by the railroad, and refuses to accept the lot in return. The railroad also disclaims any culpability in the matter.

At this point, several options have been suggested. One is to do a 100-percent inspection of all the units in the second lot. We may be able to find enough good ones to keep production supplied for awhile, although we don't know and it will cost us about $4.00 per compressor to set up and perform the tests. A second option is to change our definition of acceptable quality, at least in this instance. We feel that our acceptable quality levels are probably quite a big higher than those of competition. The slightly lower quality level shouldn't affect the basic function of the midget unit, although the compressor life may be affected. A third option is to continue our investigations with FRC, the railroad, and our people.

Distribution:

R. Irwin	Manufacturing
R. Frank	Materials
H. Williams	Receiving/Inspection
R. Williams	Purchasing

*Average incoming quality was measured by testing a random sample of 25 compressors from each shipment. Kool King engineers set minimum "standards" for these average incoming lot specifications on the basis of the performance characteristics and expected product life agreed on by marketing, engineering, manufacturing, and general management when the products had been designed.

EXHIBIT 7

Memorandum

TO: Russel Frank

FROM: Bill Zipf

SUBJECT: Revised Production Schedules

DATE: September 21, 1979

Enclosed are preliminary weekly production schedules reflecting our recent decision to add a full second shift in April of this fiscal year. The first two weeks (Sept. 7, 14) have already gone according to schedule, and it looks now as if production through this week will finish on schedule too.

The first week in which this schedule is different from the rough cut plan we had previously laid out is the week ending October 19th. Previously, we had planned to switch over from Midget to a run of Mighty Midget (2,750 per week for three weeks) during that week.* I have copied this schedule to purchasing, in advance since they will probably have to make some changes in materials orders if it is approved.

I am looking forward to receiving your comments soon so that we can develop the final schedule by September 28, as planned.

cc: Purchasing

*The schedule assumes a constant level workforce (one shift, 8 hours, 5 days/week) through April.

EXHIBIT 7   (continued)

Planned Production by Week—Fiscal 1980 (September 1, 1979–August 31, 1980)

	Midget	Mighty Midget	Breeze Queen	Breeze King	Islander	Super*	Slim Line†
				Chassis Series			
Estimated FY 1980 sales	46,500	11,000	41,000	5,000	5,000	1,500	10,000
Inventory, 9/1/79	0	1,420	2,165	0	2,604	312	1,512
Standard unit cost, this model	$100.70	$106.67	$143.43	$153.02	$192.86	$239.17	$104.81
Estimated FY 1980 production requirements	46,500	9,500	39,000	8,000	2,500	1,200	8,500
Production rate (units/day)	550	550	325	325	275	65	550
Week ending							
September 7	550			750			
14	1000						
21	2750						
28	2750						
October 5	2750						
12	2750						
19	2750						
26	2750						
November 2	2750						
9	2750						
16	2750						
23	2750						
30	2750						
December 7	2750						
14	2750						
21	2750						
28		2250					
January 4		2250					
11		2750					
18		2750					
25			1025				
February 1			1025				
8			1625				
15			1625				
22			1625				
29			1625				

*The Super units were assembled on a small assembly line which was not part of the main assembly line.
†The Slim Line was undergoing complete redesign—no production of Slim Line units until 1980.
‡Begin second shift operations

EXHIBIT 7   (continued)

Planned Production by Week—Fiscal 1980 (September 1, 1979–August 31, 1980)

		Midget	Mighty Midget	Breeze Queen	Breeze King	Islander	Super*	Slim Line†
					Chassis Series			
March	7			1625				
	14			1625				
	21			1625				
	28			1625				
April	4‡			2450				
	11			3250			195	
	13			3250			325	
	25			3250			325	
May	2			3250			325	
	9			3250				
	16			3250				
	23					1650		
	30				2150			
June	6				3250			
	13							3200
	20							5300
	27	3200						
July	4	2250						
	11	2250	500		Leave flexible for			
	18				changes; vacation			
	25				shutdown.			
August	1							
	8							
	15							
	22							
	29							

EXHIBIT 8

September 21, 1979

Dear Russ,

Enclosed is the monthly materials status report that our department has developed in the past. Although it is a little late this month due to our recent organizational changes, I thought it important to issue because it reflects the lead time stretch-outs many of our vendors are currently experiencing, despite the general fall in leads most industries are reporting (see *Purchasing World* report enclosed). It seems that our vendors are either "recession-proof" this time, or that they are lagging any downturns in the mild recession people are now predicting for this fall and winter.

You might make a special point to marketing that lead time stretch-outs pose special supply problems for standard materials that we attempt to dual source and obtain the lowest prices on, and that we do not have under long-term (annual) contract (especially steel and motors). They also indicate a decline in our ability to expedite the delivery of outstanding orders, and an increased tendency for our suppliers to deliver behind schedule.

Sincerely,

Ralph Williams

Ralph Williams
Senior Buyer
Acting Purchasing Manager

RW/jmb
enc.

EXHIBIT 8 (continued)   Major Materials Status Report As of 9/21/79

No.	Material‖	Vendor	Where Used	Units on Hand	Standard Cost	Current Lead Time*	On Order‡
1428	Motor	Acme	Midget, Mighty Midget, Slim Line	6,252	$ 19.55	20 weeks	3,000 per wk. thru 11/2/79
2915	Motor	Reserve	Breeze Queen, Breeze King	46	$ 27.39	20 weeks	3,000 due 11/9/79; 1800 per week thru 1/18 each week thereafter
8443	Motor	Acme	Islander, Super	427	$ 58.42	20 weeks	4,000 due 1/18
6249	Compressor	FRC	Midget#	3,512	$ 46.20	12 weeks†	3,000 per week thru 10/19
9347	Compressor	FRC	Mighty Midget	3,474	$ 52.80	12 weeks†	3,000 per week from 10/19–11/2
4782	Compressor	FRC	Slim Line	14	$ 51.00	12 weeks†	—
3476	Compressor	FRC	Breeze Queen	2,702	$ 78.56	12 weeks†	1,700 per week 11/9 thru 1/4
5428	Compressor	FRC	Breeze King	47	$ 88.15	12 weeks†	—
6090	Compressor	FRC	Islander	182	$ 94.87	12 weeks†	—
4328	Compressor	FRC	Super	56	$107.18	12 weeks†	—
7213	Sheet Steel	Various	All chassis series	2,802	$ 9.87§	14 weeks†	11,000 sq. ft. per wk. thru 12/7

*Lead times quoted by vendors as the expected minimum elapsed time between order placement and delivery of the quantity ordered as of 9/21/79 (for standard items).

†The contract with FRC identified the expected annual requirements for each compressor type, and stated that Kool King would give 12 weeks' advance notice of actual delivery requirements.

‡Purchasing placed orders for long lead items for use in Fiscal 1980 in the Spring of 1979 based on FY 1979 schedules plus about 20%. The purchasing manager explained that this type of extrapolation was necessary because of the timing of the divisional financial, marketing and production planning cycles. Planning wasn't started until late May, and marketing forecasts and production schedules weren't finally approved and released until late summer, when corporate approved the divisional budgets.

§Average standard cost per unit based on 3.6 sq. ft. of 3/16 sheet per average chassis unit.

‖The cost of miscellaneous materials averaged $19.20/unit across all models. Miscellaneous materials are not included in this list.

#Units on hand do not include any rejected materials.

EXHIBIT 8 (continued)   PW Monthly Procurement Status Report

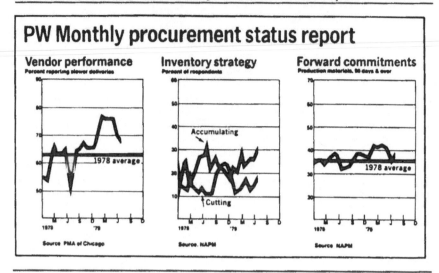

PW Monthly procurement status report

Reproduced with permission from *Purchasing World*, September 1979, p. 134

Robinson Laboratories

Ray Johnson, facilities manager for Robinson Laboratories, had been recruited nearly a year earlier in late May 1984, when his predecessor had suddenly resigned. When Johnson accepted the job, his supervisor, Akilah Grant, had told him he would be expected to contribute immediately. Though she had been in her position for nearly three years, Grant was trained as a chemical engineer and spent her time on technical process issues. Johnson would be left on his own.

As facilities manager for Robinson Laboratories, he supervised the department handling the installation and maintenance of production equipment; the department was also responsible for every aspect of the buildings and grounds. Should production employees be having difficulties with their machine lines, for example, Johnson's group would be called in to make repairs. If a water pipe burst or the temperature of a building were too low, Johnson's group would respond to the complaints. The facilities group at Robinson would also be responsible for design, layout, and building any research and development laboratories that might be needed.

Johnson described his department as "the group that serviced and supported all of the departments at Robinson Laboratories." Because most of these services were needed in emergencies, Johnson also described his group as "firefighters."

ROBINSON LABORATORIES

Robinson Laboratories, a growing firm in the health care industry located in Houston, manufactured diagnostic test kits to test blood samples. Exhibit 1 pictures a kit. Robinson reported 1985 sales of $50 million to customers such as clinical laboratories, blood banks, and hospitals.

To house its 450 employees, Robinson had four buildings totaling 192,000 square feet. See Exhibit 2 for a brief description of the facilities. Two were old, built in 1912, and the remaining two were constructed in the 1950s. The largest building, consisting of 72,000 square feet and five floors, housed most of the support offices (marketing, accounting, human resources, and research and development). Two other buildings of the main complex contained the production departments. The North Street facility was used for warehousing and shipping.

RAY JOHNSON

Johnson was an energetic 40-year-old when he joined Robinson Laboratories as witnessed by the fact that he was enrolled in an MBA program at night and expected to graduate in three years. He had received his undergraduate degree in the same way, attending evening classes while working full time.

His work experiences were varied. Before joining Robinson Laboratories he had worked for six other companies, five of which were in manufacturing, though none in the health care field. His job titles had ranged from maintenance mechanic to facilities engineer to maintenance manager. These positions were mostly related to facilities or production equipment and included several with management responsibilities.

At Robinson Laboratories the facilities manager was traditionally a "working" manager. He had to roll up his sleeves, pick up tools, and make repairs. Grant expected Johnson to do the same—not regularly, but when it was necessary. Johnson's production equipment knowledge and experience were also vital to the job. For example, he discovered shortly after his arrival that his predecessor had assigned one of the equipment mechanics to the bagging production area for an entire shift each day so when the bagging machine broke down—which it did several times a day—he would be able to repair it immediately. The mechanic had been doing this for a month, which Johnson considered "a gross waste of time." Instead Johnson and an equipment engineer analyzed the machine and quickly determined that because of design defects it should be shut down and replaced immediately.

EXHIBIT 1
Sample of Robinson Laboratories' Product Line

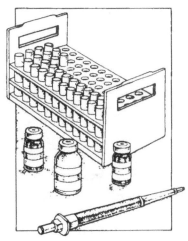

Top Quality Reagents

Our antibodies and reagents are produced to meet our strict performance standards, to make sure that all components work perfectly in unison. Even our assay tubes are molded to our rigid specifications. Once produced, the reagents undergo rigorous quality testing to assure you of consistency and dependability.

Convenience

Our kits feature ready-to-use reagents when possible. Each assay provides a simplified, highly accurate and specific procedure to save you time and money through greater efficiency.

Diagnostic Applicability

Normal range studies conducted by leading specialists help put your laboratory's results in perspective. Clinical monographs and data are available upon request.

A Decade of Lab Experience

We revolutionized the BC* field and set the standard for reliability and ease of use in coated-tube assays when we introduced Robinson's BC* kits. These were the first assays to offer the convenience of coated tubes while providing consistency and reproducibility. We lead the diagnostics field in the development of free hormone assays and other tests that are demanded by the medical community to improve the quality of diagnosis and therapy.

Service

Each of our kits is backed by Robinson expertise with technical service coordinators and representatives ready to help any time. We offer professional, personalized service, easy ordering and prompt delivery.

* Blood Chemistry

EXHIBIT 2
Description of Facilities

250 Main Street

72,000 square feet—5 floors

Houses: Executive, marketing, accounting, human resources, facilities, maintenance shop, shed for equipment, and various other laboratories

260 Main Street

55,000 square feet—3 floors

Houses: Manufacturing, technical production, hot lab inventory storage, shipping, and equipment maintenance shop

425 Broadway

25,000 square feet—2 floors

Houses: Research & development

50 North Street (5 miles from main complex)

40,000 square feet—2 floors

Houses: Warehouse and receiving

THE FACILITIES DEPARTMENT

Equipment

Johnson's department was responsible for 300 pieces of equipment that varied in age and condition. He described them as "ranging from brand new to old and decrepit, with some being upgraded, repaired, or replaced." Half was production equipment and half was facilities. The former consisted of items such as conveyor systems, deionizing water systems, bagging machines, and bottle capping machines. Facilities equipment included compressors, vacuum pumps, boilers, telephone systems, and heating and air conditioning systems.

Personnel

The department Johnson inherited experienced several immediate changes, the most significant of which was the increase in employees under his management. His predecessor's staff consisted of 17 employees; Johnson's staff quickly grew to 25.

The principal reason for increased staff resulted from moving the seven equipment maintenance employees from the process development department to the facilities department. Grant had initiated this move, with Johnson's concurrence, believing that the equipment and facilities maintenance areas should be in the same department. This change put them in the same department, but the equipment group's work area was not in the same building as the facilities'.

Johnson also added an engineer's position to his staff, believing that the department's heavy workload justified that move. The department was supported by a $3 million budget. (Exhibit 3 is Johnson's organization chart; Exhibit 4 is Robinson Laboratories' chart.)

EXHIBIT 3   Facilities and Maintenance Organization

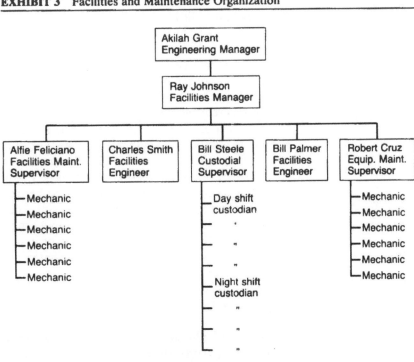

EXHIBIT 4   Robinson Laboratories' Organization Chart

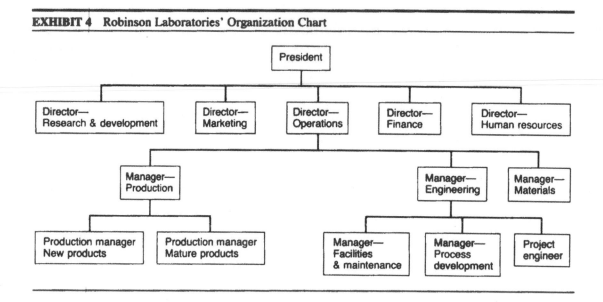

Seniority and Skill Levels.   The custodians and mechanics had various years of seniority and skill levels (see Exhibit 5). But the criteria used to determine hourly wages were related only to skill. For example, a skill level 5 mechanic earned more than a skill level 4 even if the latter mechanic had more seniority than the former. Both a facilities mechanic and a maintenance mechanic on the same level would earn the same, however, and for this reason, Johnson contemplated eliminating the separate titles. Mechanics could thereby be cross-trained and able to work on both assignments—equipment and facilities—instead of one or the other.

Johnson hesitated in making this change, for several reasons. He was not sure that one supervisor could manage a larger work force. He also realized a training program would have to be designed, which would be time-consuming; and even if he had the time, he was not certain all the mechanics had the ability or willingness to learn new skills. Furthermore, he was aware that the union had recently been decertified, and he did not wish to stir up any pro-union sentiments that might arise from the absence of skill demarcation.

Beyond their differences in skill levels and seniority, custodians and mechanics varied in their ethnic and language backgrounds.

Perceptions of Abilities.   Opinions differed about Johnson's employees' abilities and performance. For his part, Johnson believed the custodians and mechanics performed solidly, though he recognized both groups could improve. For example, half of the custodians worked the day shift,

EXHIBIT 5 Facilities and Maintenance Employee List

	Seniority Date
Facilities Manager	May 21, 1984
Facilities Supervisor	January 16, 1978
Facility Engineer	June 4, 1979
Equip. Maint. Supv.	July 6, 1976
Facility Engineer	January 29, 1984
Custodial Supervisor	November 30, 1981
Sr. Facilities Mech.	March 18, 1978
Sr. Facilities Mech.	March 21, 1983
Sr. Facilities Mech.	April 2, 1979
Sr. Facilities Mech.	April 29, 1980
Facilities Mech.-4	February 19, 1980
Custodian	December 4, 1984
Custodian	December 17, 1984
Custodian	September 24, 1981
Custodian	July 29, 1984
Custodian	July 30, 1984
Lead Custodian	January 30, 1978
Lead Custodian	March 7, 1977
Equipment Mech.-5	December 6, 1979
Equipment Mech.-4	July 5, 1983
Equipment Mech.-5	January 10, 1983
Equipment Mech.-4	May 11, 1981
Sr. Equip. Mech.	December 8, 1975
Sr. Equip. Mech.	September 15, 1980

but the supervisor was assigned to the night shift along with the other half of the custodian work force. Johnson also felt that the custodians' and mechanics' general performance was negatively affected by their heavy workloads, an opinion shared by the R&D laboratory manager Ariel Rogers. She considered the facilities department's nonsalaried work force too small. When they had time to perform work she requested, they did an excellent job, she believed. But finding time was often a problem.

Denise Moore, manager of new products production, was somewhat more critical. She, too, blamed understaffing for scheduled repair or maintenance work in her area that was not done on time. But she felt that understaffing was no excuse for mechanics not informing her that work would not be done; they never called, but simply did not show up. This would cause her to miss an entire day of production.

Moore also believed there was a language problem with some of the mechanics. Her production line would be affected because when breakdowns occurred, mechanics assigned to repair the machine often could not be given a verbal description of the problem if they had a weak understanding of English. The result was increased downtime. She also felt that many mechanics were unqualified and thought some of her production employees could do as well, if not better, as troubleshooters.

Salaried Staff.   In contrast, Johnson's salaried staff was highly regarded by everyone, including Rogers and Moore. Supervisors were considered always accessible, very responsive, and able to perform high quality work. Johnson agreed but was also able to identify weakness in the group. Specifically, he felt that all the salaried people were weak in administrative work (i.e., planning and documenting on paper). He described them as a "nuts and bolts group who were not accustomed to and did not enjoy paperwork." Johnson's evaluations differed by individual (see Exhibit 6).

All the supervisors spent 75 percent of their time managing employees, assigning them to specific jobs, and following up to ensure the work was done properly. Twenty percent of their time was spent ordering materials or tools and monitoring work contracted to outsiders. The remaining 5 percent was devoted to paperwork. The supervisors also attended basic and advanced supervision seminars to improve their management skills.

EXHIBIT 6   Johnson's Assessments of His Staff

	Strengths	*Weaknesses*
1. Cruz, R.	Engineering knowledge Planning Production equipment knowledge	People skills
2. Feliciano, A.	People skills Delegator Technical expertise	Scheduling Writing Planning
3. Palmer, G.	Electrical equipment knowledge Layout skills Scheduling Interpersonal skills	Administrative skills
4. Smith, C.	Equipment and facilities knowledge Space planning Interpersonal skills	
5. Steele, B.	Planning Scheduling	Will not take responsibility Poor disciplinarian

Both engineers spent 50 percent of their time assisting in the design, organization, and implementation of facilities and equipment construction and renovations. Twenty-five percent of the time was involved in the maintenance of buildings and equipment, including responding to emergency calls. Fifteen percent was devoted to such areas as specifying equipment voltages and utilities requirements. The remaining 10 percent was spent providing training and guidance to the mechanics on equipment installation, maintenance, and other tasks.

COMPUTERIZING THE MAINTENANCE DEPARTMENT

Energy Management System

Shortly after his arrival on the job, Johnson secured permission to buy and install a computerized energy management system that would monitor the temperature of every floor, boiler, freezer, and refrigerator in each building of Houston's operation. The system would also control the overhead lighting in each building. Johnson could control and adjust the temperatures and light on a centralized computer in his office, and the system would be monitored 24 hours a day by Johnson, his staff, or the security department.

Johnson was able to justify the $49,000 purchase price by showing that the system would pay for itself after two years because of energy savings. All lights would be timed to turn off at a certain time instead of relying on individuals as they left buildings. Further, the system could detect problems quickly. If an air conditioning system in a building were malfunctioning, for example, Johnson would know immediately and could respond before anyone realized the problem existed.

Preventive Equipment Maintenance System

In addition to the energy management system, Johnson brought on a computerized equipment maintenance system that had over 300 pieces of facilities and equipment programmed in it. Ready for introduction into the facilities department, it had been purchased to minimize "firefighting" by implementing for the first time a preventive maintenance program. Tasks performed included scheduled work such as equipment lubrication, electronic circuitry adjustment, filter changes, and equipment calibration. Johnson viewed the program as "a tool to ensure the continuing proper and efficient operation of equipment." In addition, a formal record would also provide Johnson with a means of determining how productive the mechanics were. As one of the outside consultants working on the project with Johnson said:

> The average maintenance crew member throughout the country provides 4.5–5 hours of efficient productivity in an eight-hour workday. But Johnson will never know how productive his work force is without a system that documents every task and the time it takes to complete that task.

Finally, the system could be used by supervisors to organize their mechanics' workday. Johnson realized that the potential benefits were endless.

System Design.   Representatives from an outside consulting firm had spent six days with mechanics analyzing each piece of equipment to determine the regularity and specific maintenance work that should be performed. (This was the mechanics' first introduction to the new system.) Information collected was then put into the consultants' computer system, and each month a separate card would be printed for each piece of equipment, describing the maintenance work required (see Exhibit 7). These cards were sent to Johnson, who then gave them to his supervisors. In turn, the supervisors divided the cards up for their staffs, with orders that the work on the cards was to be completed, documented on the cards, and returned to the supervisors by a certain date. Upon receiving the cards, supervisors returned them to the consultants. Johnson would be immediately notified of any cards that had not been returned or of those returned empty, and he would follow up by finding out who was responsible for work not completed. The system cost $1,800 for the set-up plus a monthly fee of $400.

The 300 pieces in the system accounted for 80 percent of Houston's equipment. Portable and prototype equipment, considered disposable, was not included, nor was equipment serviced by an outside vendor under a preventive maintenance contract.

System Implementation and Maintenance.   Johnson was aware that implementing and maintaining this new system would not be easy. The sales consultant had told him that several factors would determine the success rate. First and foremost, *everyone* must buy into the system, including top management from other departments and Johnson's own supervisors and mechanics.

Regarding other departments, Johnson needed assurance from engineering, production, and research and development that their management and work force would support the system. For example, would these departments permit the facilities department to work on equipment when it was not actually broken? Would this interfere with production? Johnson was not sure how to approach such issues. Should he meet with

EXHIBIT 7
Sample of Computerized Maintenance Card

Front

01378 ROBINSON LABS   01/20
00507 FACILITY MAINTENANCE   CHECK
01250 BLDG.600 3RD.FL.   WORK
                                                              PERFORMED

C. CHECK LIQUID LINE INDICATOR FOR LEVEL AND
   MOISTURE   ( )
C. CHECK OIL LEVEL IN COMPRESSOR   ( )
C. CHECK EVAPORATOR DRAINS   ( )
A. CHECK PRESSURES AND CLEAN CONDENSORS AND
   EVAPORATORS   ( )

CHECK WITH OPERATOR OR SUPERVISOR ON POSSIBLE
PROBLEMS.

TOTAL TIME. Hrs. _____ Min. _____

DATE _____ CLOCK NO. _____

Back

MATERIALS USED: _____

                    _____

                    _____

REG. TIME: _____ MAT'L. COST _____

OVERTIME: _____ LABOR RATE _____

RECOMMENDED REPAIRS: _____

_____

COMMENTS:

each department separately or together? Should he demand that they allow him to schedule work when it was needed or should he allow other departments to determine his schedules, which could lead to his department working overtime, evenings, and weekends?

Johnson needed his own department's support as well. He had to convince both mechanics and supervisors that the new system would lead to making their jobs easier, despite increased paperwork. And paperwork was a major hurdle to cross; supervisors especially were not good at it and loathed this aspect of their jobs. Mechanics might also respond negatively to paperwork for they never had had to document any of their whereabouts, times, or tasks; they were accustomed to working in an unstructured fashion, completing assignments at their own pace without anyone following up on them. And because some mechanics had difficulty with English, the required documentation might present other problems.

Johnson considered several alternatives. He could design a detailed training program to teach supervisors and mechanics how to complete the paperwork. Such a program would emphasize the virtues of the new system while conveying the impression that the paperwork could be "fun and easy." Or, Johnson could take the paperwork responsibilities out of the hands of supervisors and mechanics and assign them to a secretary or one of his staff employees. Johnson even considered requiring the supervisors to do all the paperwork—including that of the mechanics.

He recognized, however, that if the responsibility remained with the supervisors and mechanics (or even with the supervisors alone), he had to develop a process to handle workers who did not complete paperwork assignments. Should they be disciplined, retrained, or fired? How many chances would the employee receive?

THE NEW FACILITY

About six months after Johnson had become facilities manager, the National Institute of Health (NIH) gave approval to Robinson Laboratories to develop a facility that would house a P3 laboratory to research and produce a particularly virulent and contagious virus. A P3 laboratory was required by the government for facilities where viruses were cultivated for research. It was to protect the environment. If any trace of a virus escaped the controlled atmosphere, the virus would be quickly destroyed and stopped from growing. Robinson Laboratories planned that its new facility would house both the laboratory and production operations, for Robinson wished to develop diagnostic kits to isolate the virus.

The company believed it could beat competitors to the market with its diagnostic kits if the facility were built quickly. Four months were usually needed for such an undertaking, but Johnson was told to complete the task in half the time.

During his review of the information published by NIH, which detailed specifications for building the facility, Johnson realized this project would be a multiple challenge. First he had to find a location and then design and build the facility. Johnson and no one else at Robinson Laboratories had ever built or even seen such a facility. In addition to designing and constructing the facility, Johnson had to review his plans with local health and environmental agencies, including the fire department.

Design

The equipment in the facility would include incubators, magnetic stirrers, ultraviolet pass-through boxes, freezers, and refrigerators. For the production aspect, test tube filling and sterilization equipment would be needed. Given the technical requirements of the equipment and strict government specification, which included things like seamless floors, special air filtration systems, air handling systems, coved corners, epoxy painted walls, and all lights sealed into the ceiling, Johnson had to decide whether his department should design the facility or contract it out. Johnson had contacted one of the few consulting firms who specialized in such facilities, and they had responded that the minimum time required to design and build a facility would be six months.

Design costs would range from $10,000 if the work were done internally, to $15,000 if it were contracted. If done by Johnson's group, the work would be extremely time-consuming, especially for Johnson, who would be a member of the design team. Hence, he gave strong consideration to using an experienced design firm. On the other hand, he considered this task a challenge and also saw it as an opportunity to demonstrate to his manager his design skills.

Location

But before any design work began, Johnson had to decide where the facility would be located. The minimum space needed was 20,000 square feet, to be used for offices, laboratories, and production areas. There would be 30 employees, of which 12 would be salaried and the others hourly, working in the facility.

The only building owned by Robinson Laboratories in the Houston complex that would be appropriate was the research and development center which housed 45 employees. The center had 25,000 square feet, was already outfitted for laboratories, and its module walls could easily be adapted to government specifications. Converting the R and D center would cost an estimated $4 million. This contrasted with an estimated $5 million to rent and convert a new building. Such a tremendous cost savings easily led to the decision that the R&D center would be the facility. Everyone seemed pleased except the R&D employees, who had no say in the matter.

Construction

Johnson faced the same dilemma in construction that he had in design: should the construction of the new facility be performed by Johnson's staff or by an outside contractor? The only way that Johnson's group could do the work was if several new mechanics were hired. The present group was too understaffed to complete the project quickly while performing their regular assignments. New recruits would ease some of the pressure, but Johnson was concerned about the time it would take to train new people, even if they possessed the requisite skills. The cost of adding new staff and doing the work internally would be approximately $100,000 more than contracting it out. In addition, Johnson knew that focusing on the new facility meant that he would have to sacrifice support to other areas. Despite this, Johnson was giving serious consideration to allowing his group to do the work. It would be a great challenge.

The Role of Facilities and Maintenance

Janice A. Klein
Steven Rogers

Historically, facilities and maintenance[1] has been a nonglamorous field, often not receiving the attention and recognition it deserves. As defined by one company, the maintenance department's responsibilities are to "ensure a level of maintenance that will provide acceptable availability of operating systems and protection of corporate assets."[2] Specifically, the group must provide maintenance services for all of a company's buildings and equipment. In addition, as the group responsible for interfacing with the EPA and OSHA departments, it must ensure that the company is adhering to pollution and safety regulations. Also responsible for energy management, the maintenance group monitors electricity, gas, etc. use to make sure as little as possible is consumed without interfering with productivity.

Aside from these common responsibilities, the tasks of maintenance groups may vary from company to company. In some organizations the maintenance group is responsible for plant security systems and personnel, designing and building new equipment as well as facilities, space planning, and utilities layout.

While the maintenance group stays busy throughout the year with these responsibilities, a plant's one- or two-week annual shutdown is particularly important; in fact, this is the most important time. During those days, maintenance cleans the entire plant; performs preventive maintenance on major equipment (furnaces, etc.); repairs and paints existing equipment; installs new equipment; and repairs and paints the buildings and grounds. To use the time properly and efficiently, the maintenance manager must do a great deal of pre-planning—identifying the work to be done, ordering parts, and scheduling an adequate work force. Because the workload during the shutdown period requires additional resources, the maintenance group often employs machine operators or other employees not normally associated with the maintenance group, who do not have or want to take vacation time during shutdown.

The Maintenance Organization

The typical maintenance department consists primarily of hourly employees (custodians and high-skilled mechanics), supervisors, and facilities engineers. The department's manager usually reports to the plant manager. The group may be centralized, with workers dispatched to different areas as needed, or decen-

---

[1]Throughout the remainder of the note, the facilities and maintenance group will be referred to as simply the maintenance group.

[2]John Teresko, "Rusted Gears—Will Neglected Maintenance Undermine Recovery?" *Industry Week*, February 6, 1984, p. 47.

tralized, with workers assigned to certain areas of responsibility. Deciding to centralize or decentralize depends on several factors: whether there is sufficient demand for decentralized maintenance services, the time required to travel from a central shop, the degree of specialized skills required, and the seriousness of downtime to the operation.

Determining the optimum size of the department can be an even bigger problem. Because the skills required for maintenance work are often quite specialized, it is more a matter of how much of a particular skill is required than of simply figuring the number of workers needed. In some areas, state licensing is required of skilled maintenance workers such as electricians and plumbers; therefore, the availability of employees with the proper training and accreditation may have an effect on the size of the department. Another factor that may determine a department's size is any agreement between the company and its labor union regarding the demarcation of skilled labor positions. Typically, this requires multiple skilled trades job classifications such as electricians, plumbers, carpenters, millwrights, and riggers. Companies' attempts to reduce the number of classifications are often resisted by the skilled trades group. Nonetheless, steps in that direction are being made. For example, at the New United Motor Manufacturing Co. (the joint venture between Toyota and General Motors in Freemont, California) there are only four job classifications in the entire plant, but three are for skilled trades.

Ideally, a large staff would reduce downtime of equipment because repairs would be made quickly and preventive maintenance could be performed. But a large staff would result in a large overhead expense with potential worker idle time. On the other hand, a small staff would be fully occupied, with a probable backlog of repair jobs. Crew costs would be lower (assuming excessive overtime could be avoided), but the consequences of equipment failure would be heightened. Therefore, the optimum size of a maintenance crew should balance the cost of the crew with the cost of waiting for repairs with the seriousness (safety considerations) of the needed repair.

Maintenance Policies

The basic objective of any maintenance policy is simply to minimize the total cost and time of maintaining an acceptable level of equipment and facility reliability. Reliability, as defined by one company, is the "certainty or probability that a production system or a piece of equipment will function properly for a reasonable time after it is put to use."[3]

Companies can choose from three different maintenance policies: remedial, preventive, or conditional.[4] Under the remedial policy a piece will be completely overhauled, replaced, or repaired when it breaks down. The preventive policy entails performing maintenance, according to a certain schedule, before a breakdown. Lastly, the conditional policy refers to work performed on the basis of inspecting and measuring the condition of the equipment. Equipment that passes the test would be allowed to operate until either breaking down or reaching a certain number of operating hours, after which it would then be replaced, repaired, or overhauled.

All costs associated with any maintenance policy are considered indirect, including the costs of labor, parts and other items needed to make the repairs; idle time for production workers if they have to wait for the equipment to be repaired; and overtime required to make up for lost production.

---

[3] Richard B. Chase and Nicholas J. Aquilano, *Production and Operations Management* (Homewood, Illinois: Richard D. Irwin, Inc., 1973), p. 550.

[4] Leonard J. Garrett and Milton Silver, *Production Management Analysis* (New York: Harcourt Brace Jovanovich, 1966), p. 486.

In most instances, the work associated with maintenance policies will be performed by the employees of that department. But sometimes, due to understaffing, inexperienced mechanics, or a one-time major project, an outside contractor may be hired to perform special projects. There is also a growing trend to contract out custodial work to firms specializing in that area. Minor maintenance work may also be handled by the machine operators, if the task is not too complicated and they have been properly trained. Having machine operators perform their own maintenance work and troubleshooting is another growing trend, not being viewed favorably by skilled maintenance workers for various reasons. Some skilled workers simply want to protect their jobs, while others sincerely believe that operators ultimately do more harm to the equipment because they are not properly trained.

Attitudes toward Maintenance

An efficient maintenance operation has been the exception rather than the rule. One of the reasons is that maintenance has traditionally been ignored by corporate officials. From one consultant's perspective "top management has always seen maintenance as a stand-alone issue, usually as an uncontrollable, overhead cost with no added value."[5] Another consultant commented, "Corporate policymakers tend to view maintenance as no more than a drain on resources, primarily because costs as a percentage of sales have continued to increase."[6] In some companies these costs have been as much as 15 percent of sales. The av-

erage maintenance budget, however, ranges from 5 percent to 7 percent of sales.

A group of chief executive officers was recently surveyed regarding their opinions about maintenance. When asked, "What were the shortcomings of their maintenance group?" the responses were:

- Lack of long-range planning
- Inability to use current tools effectively
- Too much crisis management
- Failure to look at all options
- Poor staffing
- Inability to communicate schedule and cost information to users[7]

Costs of Poor Maintenance

Indeed, poor maintenance practices will have a negative effect on profits. For instance, the breakdown of a machine may result in, at best, idle time for production employees waiting for the machine to be repaired, and at worst, a complete shutdown of the plant because the equipment cannot be repaired. Thus, poor maintenance has led to an increase in production costs.

Poor maintenance can also lead to a company's inability to implement and maintain an effective just-in-time production scheduling system, which demands that machines operate and produce products according to a closely defined schedule and volume. If the equipment is broken, the system cannot work and raw inventories increase, waiting to be processed. The result is an increase in inventory costs as well as poor service to customers because of missed delivery deadlines.

Poor maintenance practices can also distort the picture of a company's, or even industry's, true production capacity. The statistics for some companies and industries show that ex-

---

[5]Teresko, "Rusted Gears—Will Neglected Maintenance Undermine Recovery?" p. 47.

[6]John Teresko, "Today's Maintenance Won't Fix Tomorrow's Factories" *Industry Week*, June 25, 1984, p. 62.

[7]*AIPE Journal*, January/February, p. 17.

cess capacity is available; rarely is there a statistic that identifies how much of the excess capacity is actually useable. "Between June and December, 1983, government statistics indicated, capacity utilization in the U.S. manufacturing sector rose from 74.9 percent to 79.4 percent. But it is likely that much of the remaining 20.6 percent is either so outdated or in such deplorable condition that it could never efficiently be brought back onstream."[8]

Benefits of Effective Maintenance

On the other hand, effective maintenance programs have been responsible for helping some companies establish themselves as the least-cost supplier of products in their industry. One expert's opinion was that "the maintenance discipline holds the key to reduced labor, material, energy, and production costs. In turn, these reductions often lead to improved service, productivity, and product quality."[9] The major result of this domino effect is increased profits.

An effective program can also make "disaster avoidance" easier. This is very important for companies that use or manufacture hazardous products such as chemicals, gases, or viruses. The implementation and maintenance of a good preventive maintenance program can lessen the chances of chemical leaks, via scheduled routine checks and adjustments to equipment.

The Ideal Maintenance Program

Properly run, a maintenance operation can greatly enhance the production system. An ideal maintenance program should "guarantee optimum use of available personnel, equipment, facilities, and funds."[10] This can be accomplished only when the maintenance function is managed in the same way that a direct manufacturing system is managed: As much work as possible must be scheduled; inventories of spare parts maintained; guidelines on minimum maintenance requirements established; prescribed quality standards met; employee training programs created; attention to preventive maintenance given; and a system to track data along with spare parts inventory put in place.

The development of an effective information system is one of the first steps in designing a maintenance program. An information system provides documented maintenance procedures and records, permitting the maintenance group to track maintenance effectiveness, estimate the expected lives of the equipment, and determine which critical parts to schedule for repair. Essentially, this takes a lot of the guesswork out of the maintenance function. For instance, an analysis of maintenance repair records might show that mechanics are doing poor work, leading to repeated breakdowns; that an operator is working incorrectly and causing breakdowns; or that machines contain design weaknesses which should be corrected. Managers must become intimately involved with the information system because "close managerial follow-up is necessary to ensure the timely receipt of accurate and complete data."[11]

Unfortunately, however, without qualified maintenance managers, a promising information system is almost useless. Maintenance managers must be selected for their ability to manage, and not because they were good mechanics. Historically, this has not been the case. People have been promoted to management positions primarily because they were good mechanics. The result: the loss of good mechanics and the addition of poor managers.

---

[8]Teresko, "Rusted Gears—Will Neglected Maintenance Undermine Recovery?" p. 47.

[9]Ibid., p. 45.

[10]Garrett and Silver, *Production Management Analysis*, p. 513.

[11]Teresko, "Rusted Gears—Will Neglected Maintenance Undermine Recovery?" p. 48.

But good maintenance managers are difficult to retain. Technical knowledge of the field is important, and typically that knowledge is gained through progression with the maintenance function. Once an individual progresses to the top of the maintenance department, further progression is often limited to moving to a larger company with a larger maintenance group. Qualified maintenance managers are important because they are responsible for efficiently managing large budgets. "In many companies the annual tab for maintenance exceeds net profit."[12] As one consultant stated, "The annual maintenance bill for U.S. business has been in excess of $200 billion, but about 33 percent of that has been wasted because of poor management."[13]

Along with these increasing maintenance costs are increasing costs of equipment and other assets. To ensure that these investments realize strong returns, the maintenance group must not only be involved in performing service on equipment after it is purchased but be included in the original evaluation of the equipment. This involvement can result in significant cost savings. Plant managers have been known to say that getting a new piece of equipment into the plant required making a hole in the roof—the equipment was too large to fit through the doors. An expensive surprise like this can be avoided if maintenance is involved on the front end. In addition, maintenance personnel can provide input as to the fit of a proposed investment with existing equipment relative to comparability of replacement parts, thereby reducing spare parts inventory and future repair costs. Lastly, if new investments require the acquisition of new skills for installation and/or maintenance of the new equipment, the maintenance group needs sufficient lead time to acquire needed knowledge and skills.

Companies often do not attempt to improve their maintenance programs for two major reasons: It can be expensive and time consuming. One consultant estimated that "it takes two to five years of intensive, consistent, and well-thought-out effort to introduce meaningful change. And it is not uncommon to incur costs from $500,000 to $1.5 million."[14] Nonetheless, improving maintenance is becoming more popular. Knowing the symptoms of maintenance neglect is a first step. These symptoms include excessive machine breakdowns, frequent emergency repairs, lack of an equipment replacement program, domination of maintenance by production, and poor maintenance facilities.

Future of Maintenance

Despite the need for improving maintenance programs today, the leading-edge technologies of tomorrow's factories will make current maintenance management and practices obsolete. According to one consultant, "In the most advanced technological scenario the maintenance people will be all that is left."[15] The maintenance workers in these automated facilities will no longer be merely mechanics; they will need to diagnose both hardware and software problems, demanding that the mechanics learn new skills in electronics and information systems. A survey of CEOs showed most of them believed that "in the next five to ten years, computer monitoring, control and simulation systems will be used increasingly in the facilities management and equipment maintenance functions."[16] Such computer integration may mean that much maintenance work will be performed remotely, at a site away from the equipment, with skeleton maintenance crews located at equipment sites or

[12]Ibid., p. 45.

[13]Ibid., p. 48.

[14]Teresko, "Today's Maintenance," 1984, p. 60.

[15]Ibid., p. 62.

[16]*AIPE Journal*, January/February, 1983, p. 16.

dispatched from a centralized office, possibly thousands of miles away.

But with these changes will be problems. A systems department probably will not sit idly by as a maintenance group, with its new skills and training in information systems, invades systems territory. Similar issues are likely to surface with other groups as the boundaries between functional departments become blurred. As problems arise, there will be a need for increased functional integration to reduce duplication of efforts or delays in coordination between functions. There will, no doubt, be a lot of jostling and debates regarding the role of maintenance in a totally computer-integrated environment.

Explosion in the Dover Plant

At 11:52 p.m. on Monday night, June 12, 1978, Unit A of the Xenon Corporation's Dover plant exploded. It was a rather big blast: heavy pieces of some of the equipment fell 900 feet away. The explosion was heard over a wide area, causing panic in Atlanta, Georgia, some 25 miles away. One eyewitness described the explosion as sounding like "a real crisp stroke of lightning, or breaking of a large piece of glass, followed immediately by another one, less intense."

The explosion started in the lower part of a 34-foot high steel recovery column. According to an employee who saw the event, "The column was blown out of the structure riding on what appeared to be a huge crimson ball of fire, which spread out and converged again."

In the next minute, one of the adjacent tanks caught fire. By this time all the utilities and power supplies to the plant had been cut off, and the entire plant was shut down. It was then that one of the supervisors called the plant manager, Ken Davis, at home.

THE DOVER PLANT

The Dover plant was one of the largest plants owned and operated by the Xenon Corporation. Xenon's sales in 1977 were over $5 billion, and it was involved in several industries besides chemicals. It had manufacturing operations in over 50 locations in the United States, and in 40 locations

This case was prepared by Professor Kasra Ferdows as the basis for class discussion rather than to illustrate either effective or ineffective handling of an administrative situation.

in foreign countries. Its products were sold internationally in almost all countries of the world.

The Dover plant was one of the plants in the Household Chemicals (HC) Division. With headquarters in Nashville, Tennessee, the HC Division had 15 manufacturing operations in the United States and six abroad. Exhibit 1 shows a partial organization chart of the division.

Davis, manager of the Dover plant, had been with the Xenon Corporation for 27 years. He joined Xenon in 1951 after graduating from the University of Tennessee (BSc. in chemistry). Before coming to Dover as

EXHIBIT 1 Partial Organization Chart of Household Chemicals Division

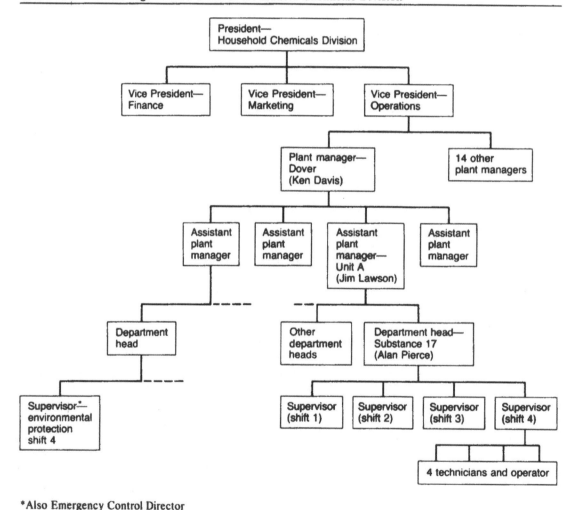

*Also Emergency Control Director

assistant manager in 1972, he had worked in three other major plants. In 1976 he was promoted to the position of plant manager. The Dover plant was located on a 600-acre lot in an industrial park surrounded by miles of uninhabited land.

In 1977 the plant produced approximately two billion pounds of chemicals, which was 14 percent more than 1976 and 50 percent more than 1974. The plant accounted for 20 percent of the sales of the HC Division and had been running very well during the last three years. All four units of the plant had been making positive contributions to profit. Actual performance had exceeded budget in each of the last three years. In 1978 there were 1,200 direct workers and about 250 indirect workers employed at Dover in four shifts. Like many other chemical plants, Dover was run round-the-clock, seven days per week. The total cost of a complete shutdown and restart of the plant was estimated to be $150,000. Management viewed the labor relations in the plant to have been excellent in the last few years, and foresaw no major problems in that area.

UNIT A

The unit involved in the explosion produced a highly reactive chemical known as "Substance 17." Like most other unstable chemicals, Substance 17 displayed explosive properties. In its pure state the substance was dangerously unstable. The process developed for its manufacture, therefore, did not permit its isolation in pure form. Instead, the process was intended to operate under conditions which sought to avoid an explosive concentration or temperature. Close monitoring and control were still required as this substance maintained its explosive tendency even in diluted forms, particularly as the temperature increased. Control of temperature (as well as concentration) was therefore crucial, particularly since the chemical process used to make Substance 17 generated its own heat.

The explosion in Unit A was the only major accident involving Substance 17 in the Dover plant in its ten years of operation. Minor incidents in this unit, such as utility interruptions, mechanical failures, and loss of refrigeration had been handled without difficulty by an emergency system which dumped all the reactive materials into an emergency catch basin. In each dump about $30,000 worth of materials (in 1978 prices) would be lost, but the danger was safely removed. In the last ten years more than a dozen process dumps had been recorded without mishap.

Unit A was one of the four production units of the Dover plant. It was located at the south end of the plant (see Exhibit 2, site layout). Some of the Substance 17 produced in this unit was shipped to customers, and the rest was processed into other products (in the same unit). Almost all the input materials used in the unit were purchased from outside.

EXHIBIT 2   Site Layout

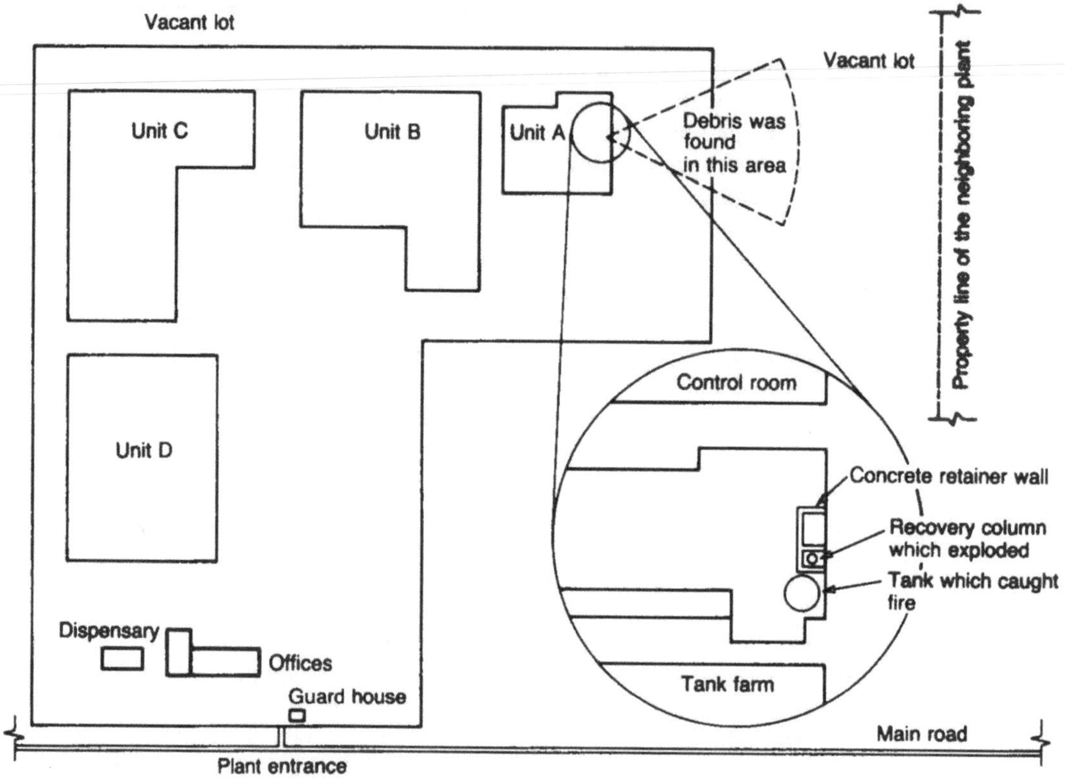

Unit A was under the supervision of Jim Lawson, one of Dover's four assistant managers. Lawson, about 50, had received his Bachelor's degree in engineering approximately 26 years previously, and had been working for Xenon Corporation ever since. He had been transferred to Dover in early 1976. Reporting to him was Alan Pierce, head of the Substance 17 Department. Pierce was 38 years old, held a BS degree in engineering, and had been with the company for 17 years, the last four of which were spent at Dover.

EMERGENCY CONTROL PLANS

Dover management prided itself for its efforts in preventing and dealing with emergencies. Despite stringent preventive measures, however, as in other large chemical complexes emergency alarms occurred at the plant fairly regularly. In a typical month it was not unusual to go through up to ten alarms. There were, therefore, elaborate measures for dealing with these emergency conditions.

According to the Emergency Control Manual, an emergency/disaster was defined as "an occurrence which would bring with it significant public and community interest and most probably significant concerns to our employees and their families." The manual, about 30 pages long, included detailed instructions regarding: (a) how an emergency plan was to be activated, and (b) which people were responsible for dealing with various aspects of the emergency. It contained checklists for helping these people attend to various important functions such as communications, safety, pollution, security, plant operations, accident investigation, damage assessment, rebuilding, and insurance. It also contained a call list with names and telephone numbers of the key people to be called if an emergency occurred during evenings, night shifts, or weekends. Each person notified was supposed to call two or three others (specifically assigned to him) so that a large number would be notified in a short time.

The pivotal responsibility during an emergency was carried by the Emergency Director on duty. This was not a full-time job. Rather, it was an additional assignment for the supervisor of the Environmental Protection section. These supervisors (one in each shift) had been trained specifically and extensively for handling emergency situations. They were regarded as the most senior supervisors in the plant.

On each shift, about 20 workers throughout the plant had been selected and trained to act as the Emergency Squad. There were 20 others, some in the same shift and some on-call, who served as the backup Emergency Squad.

The Emergency Director on duty was responsible for activating the so-called "disaster plan," which involved sounding the alarms and activating an emergency radio frequency, thereby sending the people responsible for implementing the disaster plan directly to their pre-designated places within the plant. All others, including managers, were to stay out of the accident area. No one (except authorized personnel) was allowed to move about the plant once the emergency alarm was sounded. During an emergency, the Emergency Director on duty was in charge. *Only* the plant manager had the authority to override his decisions.

Under normal conditions, the Emergency Directors frequently simulated emergency conditions at various locations in the plant and familiarized the Emergency Squad with the characteristics of various sections of the plant. Even senior managers, including Ken Davis and his four assistants, had been through numerous disaster control drills, and some, including Davis, had personally participated in fire-fighting exercises.

THE ACCIDENT

On the night of the explosion, Davis was asleep in his house (about 15 miles from the plant). He was awakened by the sound of the explosion. As he described the event:

It was a very loud noise—obviously a sonic boom or an explosion. I had heard this type of noise before, but this was the loudest that I had ever experienced. I lay there for a while, saying to myself, "God, I hope that that's not our plant."

When the phone rang after a minute or two, I knew we were in trouble. I thanked Bill, the supervisor who called to tell me that Unit A had blown up, and told him I'd be right there. I only had time to make one phone call to one of my assistant plant managers. I asked him to activate the emergency system by calling the rest of the staff. I dressed as hastily as I could, and I was on the way to the plant within no more than two minutes.

Of course, all during this time you begin to think of the things that could have happened and not having any real information at all, you always assume the worst. How many people were hurt? Did we have any fatalities? Would we have more explosions? Knowing about the dangers of Substance 17, I could visualize extensive damage, injuries, and potential fatalities.

I was not more than two miles from my home when the announcer on the radio came over the system saying that they had numerous phone calls about an explosion and they were trying to check its location. That was interesting to me because the radio station was in the city, an additional ten miles further away from the plant than my house. We must have really made some noise.

As I proceeded down the highway, I tried not to go more than about ten miles over the speed limit. All I needed was to get stopped for a speeding ticket or get in an accident—that would really top things off. I was about two-thirds of the way to the plant when I was passed like I was standing still by one of our employees responding with the Emergency Squad. It was a clear, cool night and I guess I really expected to see some glow in the sky from fires at the plant, but there was none. It didn't occur to me at the time that this might have been an encouraging sign.

As I neared the plant, I also expected that there would be Police Department blockades, and I was almost reaching for my wallet in anticipation of a request for my identification. The thoughts that continued to cross my mind during the entire trip from my home were: How many people were hurt, is there anyone dead, are the Emergency Squad members well protected? It's really quite a frightening thing not to know and, of course, human nature makes you always assume the worst and ask the worst questions.

As I got very close to the plant, I still could see no large fires, and I guess by that time I was beginning to get just a bit encouraged. I turned into the plant property itself and drove to the first guard house and was stopped by the security guard on duty. I pulled my car to the side, left my emergency flasher lights on, and asked the guard what had happened. The guard said that there had just been a big explosion, a lot of fire in the Unit A area, and didn't know anymore than that. I asked her had she allowed anyone to enter the plant; she said, "No," that I had been the first one to arrive. With that, I instructed her to begin taking names and the places to which people were going as I allowed them to enter the plant. Since this was an off-duty period for most employees coming in, we needed to be able to account for every person in the plant in case there were more troubles, fires, or explosions.

I had been at this gate ten minutes when the guard received a call from the main guard house that the Emergency Director had called in to say that all employees had been accounted for and that there were no injuries. Truly a miracle—I could hardly believe it! One takes a whole different set of actions—much less severe—knowing that there have been no multiple injuries or fatalities.

For about the first hour after arriving at the gate, the main thing I did was to authorize people to go into the plant. Using the guidelines in the Disaster Plan Manual, as each assistant plant manager came in, I assigned him to a specific responsibility: The assistant plant manager over the affected area was assigned to go to that area to be sure that the emergency was being properly handled and the unit personnel were all right. The assistant plant manager that was responsible for the shift organization was sent to the emergency command post to be sure that the emergency was being coordinated properly. Another assistant plant manager was sent to a different unit to coordinate the shutdown and preparations for startup of the plant once the emergency was under control. The fourth assistant plant manager was sent in to find the production planning people so that as soon as power was restored in the plant they could begin to implement a system for bringing the plant back on-stream step-by-step as quickly as possible. The fire protection specialists that we have in the plant were allowed to go in; the department heads were allowed in, some to help with the resolution of the electrical problems within the plant and the rest to coordinate the shutdown and prepare for subsequent startup. Many other off-duty personnel who came to help out were allowed in if they had emergency or fire-fighting training. The plant dispensary was opened up and staffed by volunteer emergency medical technicians.

During this first hour period, two camerapeople from the local television station came and demanded entrance into the plant to get close-up pictures. Of course, with the emergency still going on, fires, and the Emergency Squad still having their share of problems, there was no way I could let them in. It is interesting to note that this pair of individuals went around to the back of the plant, trespassed on our private property, took some pictures, and presented a story to the public that evening which was completely without fact. I had encouraged them to wait until the "all clear" had sounded, but they wouldn't.

A reporter for the local newspaper was also on the scene about five minutes after I was, and he stuck to my side like glue wanting to get every detail, wanting to hear everything that I heard, wanting to participate in every decision that I made.

Our Employee Relations manager had arrived 20 minutes after me, and it was his principal responsibility to set up the communications center to handle newspaper people, radio, television, and also the internal and external telephone communications. It was a big help to have him on board because we were able to send people directly to him. (We chose our Employee Relations manager to be our communications focal point because he is nontechnical and could not be expected to answer technical questions, thereby minimizing the possibility of being misquoted or misrepresented.)

I was relieved at the security gate by one of the assistant plant managers and went to my office about 2 A.M. By this time the electricity to the plant had been restored; and as a result, we all were feeling better. Working in the dark with flashlights can be quite unsettling. The "all clear" was sounded at 2:30 a.m., but the fire was still burning as we were allowing it to burn itself out.

Once in my office, the first thing that I did was to jot down about three or four sentences for a press release. It said something like this:

At approximately 11:55 P.M., June 12, 1978, there was a fire and explosion in the Unit A of the Dover plant. There were no injuries. Plant electrical power was out, and the entire plant shut down. The cause and extent of the damage is not known at this time.

We also contacted other area plants, law enforcement agencies, environmental and other governmental groups, local municipal leaders, and various news media to report the accident.

The young news reporter had stayed around me and was asking for a personal interview for the radio station as well as for his newspaper. I invited him into my office along with his tape recorder and his pad of questions. Before 2:30 A.M., I had given this man a personal interview taped for airing on the radio and for being quoted in the newspaper. It is amazing to see how the overall interest of the news media wanes as they find there are no fatalities, no multiple injuries, and little fire to be seen.

I was able to hold a meeting with my staff at approximately 3:30 A.M., when permanent assignments were made. One manager was responsible for the emergency control, one was responsible for the continued operation of the plant, one was responsible for the rebuild and handling of the insurance matters. The latter two managers went home to rest so they could return later in the day to provide continued management attention while the others rested.

At 5 A.M. I notified my boss in Nashville about the accident. A little later I went to the control room to meet the shift personnel on duty at the time of the explosion to reassure them. It was an emotional meeting. Everyone was still very excited and some were visibly shaken. They were testifying and writing down everything that they could recall because it doesn't take very long for those things to slip your mind. The same was being done with the Emergency Squad that responded to the emergency. They were being debriefed, asked questions—they were writing down everything that they could possibly remember. Many of these details became very useful later in the investigation. We prepared a detailed insurance statement for the insurance people when they arrived and made very detailed preparations for press tours.

By 9:30 A.M. we had a complete, very accurate plant status report of every operating unit and all utility systems. By 2:30 P.M. we also had a plan for detailed inspection, investigation, clean-up, and rebuilding of the unit. The rebuilding, of course, was very sketchy, but even at that time we estimated five months to rebuild at a cost of $4 million. These estimates proved to be extremely accurate.

In the afternoon two of our local television stations (neither one of them were the people who came out that night) asked to come out for a review of the explosion area and we were very glad to take them through, let them get right up to it, ask all the questions they wanted, interview our people and our managers; and these two television stations did what I considered to be a very good job in their news coverage that evening.

Somewhere about mid-afternoon I went to the dispensary and lay down for about an hour to rest. It was about this time that the two assistant plant managers who had gone home to rest came back and relieved the two that had been on duty during the day with me. I went home about 10 P.M. Tuesday night. The production units in the plant were still down, but all utilities had been restored.

On Wednesday morning, June 14, we had a staff meeting to discuss the entire event. There was a need to squelch rumors because despite our efforts not to draw any conclusion publicly, rumors were out about the cause of the accident and who was responsible for it. We talked about communications and the type of report we wanted to make to the Nashville office, and about the status of the damaged unit and the status of the entire plant. We were very fortunate that there were no major mechanical problems in units that were shut down. So the plant was coming back on-line in good shape. The investigation was well organized and had started already. The emergency report was being compiled, both from the Emergency Squad and the shift personnel on duty.

One of the things we continued to do for two days was to issue status reports to the plant employees about every six hours to keep them informed. We were also trying to provide information about the status of the investigation but were trying to be very careful not to draw any conclusions. At the same time, we were carefully protecting all data and all the equipment we could find that could be useful in pinpointing the cause.

My boss came to Dover on June 14. I showed him around and supplied all the information he needed to make a presentation to an executive committee which included the president of Xenon. On June 16 we sent a preliminary report to Nashville, which contained the news release, photographs, considerable plant information, and other process descriptions. Meanwhile, the insurance people had been on-site and had collected their information, so the clean-up was already starting under the watchful eyes of the investigation team. Our plant controller had already set up about ten different account numbers for the clean-up, repair, and rebuild so that all the costs associated with the explosion and fire could be captured and billed properly for insurance credit.

The work that we had done in preparing our emergency manual proved to be quite useful, since it provided a check list as we reviewed the events of that fateful night. Still, we went over the manual again, examined all sections in detail, and revised it.

REPORT OF THE INVESTIGATION TEAM

The morning after the accident, Ken Davis asked Jim Lawson, the assistant plant manager, who was responsible for Unit A, to put together a team for investigating the accident. Lawson contacted various people, and soon submitted the names of the following persons to Davis for approval:

Jim Lawson: Assistant Plant Manager

Alan Pierce: Head, Substance 17 Department

Roy Gardner: Plant Safety Director

Wayne Hay: From the R&D office in Nashville. (Hay was consid-
               ered to be the "father of Substance 17 process de-
               sign.")

Sam Jones: From the Safety Office in Nashville. (Jones had exten-
               sive first-hand experience in manufacturing pro-
               cesses.)

Investigative work began the same day. Great care was taken to preserve all evidence, such as log sheets, instrument charts, and pieces of
equipment that had been blown up. Cleanup operation could not proceed
until the investigative team authorized it. The investigation was carried
out by the team members and other consultants called in from outside.
Experiments were conducted to verify certain conjectures.

The investigative team issued preliminary reports starting two weeks
after the incident. Its final report was submitted to Davis on August 11,
1978. He, in turn, submitted it to appropriate people in the divisional
headquarters.

The report, which was over 60 pages long, covered the description of
the systems involved in the explosion, the incident itself, the extent of
damages, the basic technological hazards involved in the manufacturing
of Substance 17, and the team's conclusions regarding the causes of the
explosion and recommendations for preventing its recurrence.

Some excerpts of this report are as follows:

The June 12 explosion in Unit A started in a recovery column, which was being
reinitiated after being down for 12 days for scheduled routine preventive maintenance. Up to almost an hour before the incident the column appeared to be
responding as expected. But at 11:04 P.M. there was a 36 percent decrease in
the output from the reactor. This decrease occurred in two and one-half minutes
as the result of a decrease in the flow of [one of the input chemicals].[1] The
imbalance that was thereby created in the column caused [certain chemical
reactions to increase] which increased the heat within the column. The additional heat aggravated the chemical reactions further, thereby increasing the
imbalance. Correspondingly, the output rate gradually dropped further. By
the time of the explosion the output rate was almost half of what it was before
11:04 P.M. During this 45 and 1/2-minute period, the column temperature increased clearly beyond the safe range.

   The team's exhaustive investigation confirmed that the incident was caused
by a slowly developing condition which was monitored and recorded by visible
instruments and which could have been corrected by a simple and well-estab-
lished operator response. The temperature profile of the column was rising, the
column's output had dropped off. . . . Under this set of circumstances, the
action routinely taken by the process operator is to increase the [input flows of

---

[1]Technical parts of the report are not included in this case.

several substances] to bring the column temperature back in line. The instrument charts for this instance do not show that this action was taken in an adequate and timely fashion.

The team was unable to uncover any unusual circumstances bearing on the incident. . . . No major new elements of risk in the column's technology were discovered. There is no concern on the part of the team about the basic safety of the process. The team has no reluctance to rebuild the unit essentially as it was, with minor modifications explained under recommendations.

The team's recommendations consisted of 11 points which emphasized the necessity for adding extra alarms and automatic disaster prevention controls, performing in-depth engineering safety reviews, redesigning certain parts in the column and its protective concrete bunker walls, as well as examining and improving certain "operational aspects." Included in the latter were items such as "training, accountability, management control/audit of performance, [establishment of] performance standards, [and] clarification of responsibilities."

In mid-August, 1978, Davis was reflecting on many specific actions that he had taken already and those that he had to take soon. Dover was a big plant; its design was supposedly fail-safe. Employees had been continually trained to handle emergencies. Yet, dangerous explosions such as the one in June could occur at any time. He knew they had been lucky in June because there were no injuries. Nevertheless, one unit was to be shut down for five months, and required $4 million out-of-pocket costs for repairs. The $15 million profit budgeted for 1978 appeared hopelessly out of reach. In fact, he felt Dover would do well to break even in 1978.

Managing Disasters

Kasia Ferdows

A fire in a manufacturing plant, an explosion in a chemical complex, an inadvertent release of a toxic material to the environment, a hurricane, or other man-made or natural emergencies create important responsibilities for the manager in charge. The way he or she handles these situations often has far-reaching implications for the safety of the people in the work unit involved, the welfare of the community, the long-term profitability of the company, his or her own success as a manager, and his or her own self-esteem as a person.

Yet little has been said about this important aspect of a manager's job.[1] There appears to be an implicit belief that this is an area which should be left to the individual manager's intuition or that the events should dictate

decisions. There are, however, specific choices—sometimes among conflicting alternatives—which a manager has to make in preparing for potential disasters, handling of the crises when one strikes, and living with the aftermath. Making these choices explicit beforehand—as opposed to simply letting them emerge naturally—is an important step towards preventing and minimizing the ill-effects of disasters. The purpose of this note is to describe such choices.

Prevention and Preparation

No one disputes the desirability of being well prepared for disasters—with elaborate safety systems, contingency plans, disaster control manuals, emergency squads, and training programs. Questions of an economic nature arise both because a limit has to be set on the amount of resources to be allocated to these measures and because the allocated resources can be employed at varying degrees of efficiency. Implicitly, a compromise between the potential cost of a disaster and the incurred cost of preparedness has to be reached.

In spite of government, industry, corporate, and other institutional regulations and guidelines, the decision regarding what is the appropriate level of preparation for disasters is ultimately influenced by the *personal* values of the manager in charge. He or she[2] is almost

---

This note was prepared by Professor Kasra Ferdows as the basis for class discussion rather than to illustrate either effective or ineffective handling of an administrative situation.

[1]The episode in the Three Mile Island nuclear plant has attracted attention to this subject. Most of the available literature addressing disasters in industrial units, however, consists of descriptions of the investigations into the *causes* of these disasters and are concerned primarily with regulatory and safety measures. (See, for example, Barry A. Turner, "The Organization and Interorganizational Development of Disasters," *Administrative Science Quarterly*, Volume 21 [September 1976] pp. 378–97.) The focus in this note, on the other hand, is on the choices of the manager-in-charge who is confronting a disaster. The author has been unable to find any reference which views the problem from this viewpoint.

---

[2]Henceforth, to avoid the repetitive and space-consuming use of the term "he or she" and its variety of compound derivatives, we will utilize masculine descriptors to represent managers of both sexes.

always held partly responsible (often implicitly) for disasters caused by human error, even sometimes for natural disasters. He may *feel* responsible even if no one points a finger at him. On the other hand, expenditures for preparation usually come from his budget, hence indirectly affecting evaluation of his performance. And the responsibility for gaining the maximum benefit from every dollar spent on preparation for disasters is ultimately his. He must compromise.

But how? Tradeoffs involved are hard to quantify; hence, quantitative analysis is of limited usefulness. The risks and gravity of the adverse consequences of a disaster vary from plant to plant and from industry to industry; therefore "industry averages," even when available, require substantial modification before they are useful guidelines. Trial-and-error is impractical because it would be enormously expensive.

Yet the decision cannot be ignored, postponed, or treated lightly. Letting it emerge or evolve in fragments is likely to produce poor results. The natural tendency both in the absence and presence of disasters is overreaction; the natural forces, therefore do not lead to an equilibrium. Yet a conscious, explicit, and studied decision should be better than an unconscious and implicit one which has emerged by itself.

To improve the quality of his decisions in this area, therefore, the manager first has to go through a long and energy-consuming process of *learning*. He can learn a great deal from others' experiences. Results of the investigation of the causes and consequences of disasters in similar plants—if available—may point out neglected precautions or heretofore unforeseen complications. Analysis of the disaster control manuals, organizations, and training programs in the industries with traditionally advanced precautionary systems (such as nuclear power, chemical, and oil), or companies reputed for elaborate disaster control systems may provide useful insights into improving the efficiency of one's own precautions. For example, a chemical plant could provide recommendations regarding how to organize for disaster control. Here is one plant's system:

Emergencies during each shift are handled by a Primary and, if needed, a Backup Emergency Squad. These employees are pre-selected among regular full-time employees throughout the plant, taken through special training courses, and familiarized with different sections of the plant. The members of the Primary Squad leave their regular jobs as soon as an alarm is set off.

Both Squads work under the supervision of an Emergency Director, a senior supervisor regularly in charge of the Environmental Protection Department.

In this plant it has been possible to combine the jobs in such a way that no employee had to be assigned to emergency control full-time. In other plants this may not be possible. There are, of course, limits to the applicability of others' experiences to one's own case.

To learn more about one's own operation, disaster conditions can be simulated. Substantial data for analysis can be generated thereby, but even under the most realistic simulation of a disaster—which incidentally can be quite expensive to arrange—some of the facets of a real disaster are absent. As one experienced manager put it, "You don't really know how your people react until you go through a *real* emergency—not just a drill."

A third source of learning is history of events in one's own plant. Analysis of major, minor, and near-emergencies provide most directly applicable knowledge. In particular, the *near*-emergencies offer the least costly, yet still realistic, learning opportunities. However, unless conscious and concerted efforts are undertaken these opportunities are lost. Near-emergencies often pass unreported, known only to a few operators. They sometimes even have a dysfunctional effect: They lead operators (as well as observers) to become callous about the potential dangers involved. Knowledge of these incidents would not only help a manager *learn* but would also provide an opportunity to prevent specific dangers.

Gathering such data demands tact and discipline. Requests for elaborate reports of each seemingly minor incident can easily create resentment; random investigative blitzes can frequently be interpreted by workers as signs of management's dissatisfaction, disapproval, or mistrust; roving inspectors can cause misplacement of responsibilities; inside informers are usually ineffective in the long run; and mechanical or electronic monitoring devices are expensive. Excessive reliance on any of these modes is unhealthy. It seems that a preferred approach is to focus on instilling a safety-conscious culture within the plant. Gradually, but steadily, formal and informal organizational norms can be introduced and nurtured so that no one person or group of persons feels threatened or singled out in reporting and participating in the analysis of the near-emergency situations.

Onset of Disaster

Once a disaster strikes, the manager in charge is confronted with a different dilemma: Should he get to the front line, roll up his sleeves and take direct charge? Or should he stay back, be an overseer, and let others take direct charge?

Unless a manager makes an explicit choice *beforehand,* he is unlikely to resolve this dilemma rationally in the heat of an emergency. Experienced managers appear to favor the overseeing role. As one commented:

> I let the system that has been designed and developed to handle emergencies do the job. I would not jump in. By keeping myself away from the excitement and the drama of the scene of the accident, I have a better chance of keeping my cool, and of monitoring the system. Of course, this demands a great deal of discipline. One has to train oneself to tolerate a lot of anxiety.

Another one put it this way:

> Mere presence of senior managers directly on the scene of the accident [when the emergency is still on] takes the initiative away from those people who are usually better trained and more knowledgeable for dealing with the specific problems at hand. Exposing oneself to bodily dangers may be gallant but it isn't sensible. The senior manager should make himself available for decision-making *only* when the people on the scene want him to do so; otherwise, he should leave decisions to them, and be ready to accept responsibility for their decisions later.

Of course unless contingency plans for disaster control are drawn up beforehand, with responsibilities clearly specified, the manager in charge has little choice but to take direct control himself. A manager who finds himself in such a situation runs a much higher risk of handling the disaster poorly. A preferred approach is one which allows him maximum flexibility during the onset of an emergency. This not only means that detailed contingency plans are to be drawn up beforehand, but that the plans call for minimum routine intervention by the manager in charge. There are always unforeseen events which may need his attention. He should be available to attend to them.

He should also monitor the physical and mental conditions of the key subordinates assigned to the "front line." During a long emergency, it is important to schedule their work in such a way that not all are overtaken by fatigue at the same time. Attending to the psychological needs of his subordinates is among the important duties of the manager-in-charge during an emergency.

Aftermath

After a major emergency, particularly if there have been deaths, injuries, or extensive property damage, the manager-in-charge faces a new set of immediate tasks. There are legal reports, reports to superiors, insurance reports, investigations into the causes, audits of the damages (including the cleaning, repair, and restart of operations) and many other tasks for which the manager is solely or jointly responsible. Some of these tasks demand his attention even while the emergency is still going on. For example, some of the legal reports (such as those relating to loss of life) should be filed immediately. If the disaster attracts public attention, particularly if fire or explosion is involved, communication with the news media becomes both important and sensitive; if the area surrounding the plant is threatened by the events in the plant, authorities should be notified, etc. Procedures in the disaster control plan, however, can be specified so that the manager-in-charge need confine himself merely to monitoring others who are responsible for accomplishing these tasks.

But there are other tasks, especially when the disaster is suspected to have been caused by human error, which he cannot delegate. A man-made disaster often leaves the organization shaken, sheepish, and demoralized. The effects linger long after the initial trauma is gone. Those directly or even indirectly involved feel that a devastating black mark has been added to their records. The supervisors generally feel that they cannot afford, career-wise, to be associated with another accident for a long time.

It is the responsibility of the chief operating manager to restore normalcy. Yet he may be in the same psychological condition himself. He is "on the line," not only because a man-made disaster happened in his unit (for which he has to take responsibility), but also because he will be judged on how well he manages the aftermath of the disaster. Despite his best efforts in the future, the shutdown, repairs, and other expenses related to the disaster may already have wiped out an otherwise excellent financial performance for the year. He is indeed in a difficult situation.

Investigation of the causes of the disaster and the events during the disaster can become an issue. If he has a choice, should an "internal" team be assigned, or should "outsiders" be invited? Relatedly, should the team's reports be dispatched through him or go directly? Investigation has to start immediately, before evidence is lost. This is a time when the organization is still in trauma. It is a period during which rumors flourish, people anticipate drastic moves, and positions become insecure. Preliminary reports may reduce anxiety. Premature conclusions tend to aggravate it.

Then there is the dilemma of disciplining negligent employees. Some of them may be in the hospital. Others may be the very persons who are best equipped to restore the operations. Still others may be otherwise excellent workers or managers who have been slated for early promotion.

There are of course no formulas for resolving these dilemmas. Each case has to be handled on its own merits. One can only expect that some tough decisions and unhappy moments are to be faced by the manager-in-charge.

Summary and Conclusions

The chief operating manager carries heavy responsibilities for controlling disasters which may strike his work unit. His responsibilities include instituting precautions and contingency plans, handling the emergency after the onset of a disaster, and dealing with its aftermath. There are specific choices to be made for each. What he chooses to do often significantly affects his subordinates, the work unit and its environment, his company, and himself. Yet these responsibilities—and choices involved—are seldom scrutinized systematically. This note is a step in that direction.

Black Diamond Coal Company (B)

After three months in his new position as mine superintendent, Bill Cutler anticipated that his most pressing problems would be concerned with productivity and output at Gemstone #3. With productivity still lagging far behind designed levels, Cutler had recently been brought in to see if a new approach might help. However, on this Saturday morning in June, Bill Cutler faced a most unusual labor problem: His section foremen had gone out on strike.

Like most mines throughout the coal industry, the operations at Gemstone #3 were frequently disrupted by wildcat strikes. Eighteen days had been lost at Gemstone #3 in the first four months of 1975 due to such strikes. However, this particular work stoppage was different than the normal stoppage with potentially long lasting consequences if not handled properly. Instead of the usual picket lines involving members of the United Mine Workers (UMW), Cutler expected to see a picket line Monday morning composed of 14 to 16 of his section foremen at the mine. The four section foremen scheduled for overtime had failed to report for work that Saturday morning and had made it known that they would be on strike on Monday morning, June 5. The implied purpose of the strike was their desire to be recognized by management as a bargaining unit.

This case was prepared by Professor Steven C. Wheelwright as the basis for class discussion rather than to illustrate either effective or ineffective handling of an administrative situation.

Black Diamond Coal Company (A) 9-676-056 describes the operating tasks of the mine and the actions that affect output and productivity.

While they had presented no demands, it was clear that they wanted some changes made and that forcing a close to the mine's operations was their first step in achieving those changes.

PROBLEMS AND ALTERNATIVES

As newly appointed mine superintendent, Cutler recognized that his handling of the present situation might well determine his future in the West Virginia coal industry as well as at Black Diamond. At 27 years of age, Cutler had accumulated only three years of experience in the coal business, all at a level below that of superintendent. Since graduating from West Virginia University with a Master's degree in mining engineering in 1972, Cutler had worked as resident engineer at another Black Diamond mine. While he had done extremely well at the university and also as a resident engineer, he recognized that his recent promotion was somewhat of an experiment on the part of Black Diamond's president to see if new blood in the top operating position at a mine could help improve productivity and efficiency. Thus he knew that the top management was hoping that he would be able to bring new ideas and methods to Gemstone #3.

By Monday morning Cutler had to formulate his own approach to dealing with the striking section foremen. Since these foremen were first level supervisors not long out of the hourly union ranks, he knew that the entire mine would honor their pickets and refuse to work. Such a work stoppage was not only costly to Black Diamond but if he allowed the section foremen to "win," he knew that it would only make his job more difficult in the future.

From his discussions earlier in the day with some of his supervisory personnel, Cutler concluded that there were several alternative approaches he might take in dealing with this situation. At one extreme he could take action aimed at minimizing the chances of a head-on confrontation by choosing to deal with the problems raised by the section foremen as quickly as possible. This might involve sitting down with them and working out a better definition of their responsibility and authority and developing plans for improving their working conditions. In its extreme form, this option might simply choose to ignore the strike and focus on the problems bothering the section foremen.

At the other end of the spectrum, Cutler realized he could choose to make a real issue of the strike by refusing to negotiate until the section foremen were back at work. This would stress the point that the foremen were part of the management and that striking was not appropriate for them. An even more extreme option along these lines would be to simply state that they were not allowed to strike and fire them for doing so. If he did this, Cutler felt that he might then try to deal with them individually, getting as many of them as possible back to work and hiring or appointing new section foremen where needed.

UNDERGROUND MINING OPERATIONS

In underground mines, also called deep mines, highly specialized machines had replaced the traditional pick and shovel by the early 1970s. The oldest and most complex of the mechanized deep mining processes, conventional mining, had five major production steps. At the face of the coal seam—a vertical wall at the end of a wide corridor—a rubber-tired cutting machine rolled in and cut a deep slot in the coal with a long blade like an eleven-foot chainsaw. A mobile drill then bored into the coal for spark-proof explosives or cylinders of compressed air which ruptured with explosive force.

The cutting machine was then withdrawn, the miners moved to another portion of the mine and the blast was set off to loosen the coal. A loading machine was then moved into place which with steel arms swept the coal onto a conveyor running through the center of the loading machine. At the rear of the machine, the conveyor dropped the coal onto mobile conveyor belts or shuttle cars—low slung, electric trucks—which carried it a short distance to a permanent conveyor system or to underground rail cars which took it out of the mine.

After the shattered coal was removed from the working face, another machine would drill holes in the roof for long steel expansion bolts which would bind overlying rock layers together and help support the roof. Then the process was repeated. Since the machines of the conventional mining process worked in rotation, modern mines kept several working faces in the same section going simultaneously for the most efficient use of labor and equipment. Depending upon the depth of the coal seam and other geological factors, an experienced crew could produce about 125 tons per shift.

Recently, continuous mining machines had taken over much of the conventional underground mining process. A continuous miner in a single operation could tear the coal from the seam with revolving steel teeth and load it for movement out of the mine. Thus these machines eliminated the cutting, drilling, and blasting steps of conventional mining. However, the roof bolting operation was still required. The cycle with continuous miners was to mine a certain number of feet into a face, load the coal, and to then remove the continuous mining machine, move in the roof bolt machine, bolt the roof and then repeat the cycle. Again depending upon geological factors, a mining crew could produce 200 to 400 tons per shift with this process.

Both conventional and continuous mining followed the room and pillar mining plan. They removed about half the coal from the seam by carving out intersecting tunnels like city streets 14 to 20 feet side. Between the tunnels, large blocks of coal called pillars were left standing to support the mine roof. When the pillars were no longer needed (for example, when closing a section of the mine), they were sometimes removed to recover additional coal.

WILDCAT STRIKES AND THE COAL INDUSTRY

Though mine workers' labor contracts provided that any and all differences between the miners and operators should be settled through a five-step grievance procedure, leading ultimately—if necessary—to a final and binding decision by a neutral umpire, the provision was often ignored in practice. The industry had been plagued by innumerable strikes over local grievances which legally should have been settled by arbitration.

The magnitude of the problem of wildcat strikes could be seen from actual past experience. During the first nine months of 1973, there were an estimated 229,000 labor days of work lost in the West Virginia coal fields due to "wildcat" strikes. The cost to the industry was 2.7 million tons of coal, equivalent to revenues of $40 million (1973 prices). These were not losses due to traditional strikes which occur in connection with labor contract renewals. These were work stoppages resulting from innumerable local and even personal problems (Exhibit 1).

As one labor attorney stated:

Although many coal strike disputes involve substantial issues, a great many do not. No dispute or cause of discontent is too trivial or frivolous to result in a strike in a coal mine. Mines have gone on strike because the access roads were too bumpy; because someone stopped up the bathhouse toilets; because the water in the bathhouse showers was too hot or too cold; because someone stole a miner's hard hat, even though the company offered him a new one free of charge; because the company made an error in an individual miner's paycheck, even though the company immediately agreed to correct the error . . . Sometimes mines are even picketed by mistake. The wife of a discharged miner recently set out to picket the mine from which her husband was fired. She made a mistake and picketed the mine of an unrelated company and shut it down for twenty-four hours.

The cause of local walkouts seemed to stem from the independence and action orientation of the miners. Rather than simply debate an issue, the miners were prone to seek immediate action. The result, of course, was the sporadic shutdown of mine operations throughout the industry.

Since there were tremendous economic pressures on management to "give in" when strikes arose at the mines, beleaguered operators fre-

EXHIBIT 1   Work Stoppages—Bituminous Coal Industry 1950–1973

		Days Idle		
Year	Number of Work Stoppages	Number of Days (Thousands)	Percent of Total Working Time	Average Days Lost per Worker Involved	Renegotiation of National Agreements
1950	430	9,320.0	9.9	56.5	x
1951	549	887.0	.9	4.2	x
1952	560	2,760.0	3.3	5.8	x
1953	392	418.0	.6	3.2	—
1954	208	344.0	.6	4.2	—
1955	292	273.0	.5	3.5	x
1956	266	377.0	.6	4.4	x
1957	161	136.0	.2	2.9	—
1958	136	102.0	.2	3.4	x
1959	146	1,560.0	3.4	24.4	—
Mean	314	1,617.7	2.4	11.9	
1960	120	137.0	.3	3.7	—
1961	117	90.7	.2	3.6	—
1962	121	191.0	.5	5.6	—
1963	131	234.0	.7	6.2	—
1964	111	340.0	.9	5.9	x
1965	145	258.0	.8	4.1	—
1966	160	629.0	1.9	7.1	x
1967	207	158.0	.5	2.5	—
1968	266	956.6	2.9	4.6	x
1969	457	900.6	2.7	4.4	—
Mean	184	389.5	1.1	4.8	
1970	500	627.0	1.8	3.2	—
1971	606	4,215.1	12.4	12.0	x
1972	963	562.4	1.6	2.2	—
1973	1,039	556.8	1.4	1.9	—
Mean	777	1,490.3	4.1	5.4	

Source: From the Bureau of National Affairs, Inc., Washington, D.C.

quently succumbed to the strikers' demands. As a result, operators attempted to discover *potential* problems within the mines and to reach agreement when possible *before* the actual strikes occurred.

GEMSTONE #3

Opened in 1969 in West Virginia, Gemstone #3 was planned to be Black Diamond's largest steam coal mine with an eventual operating capacity in excess of 1.3 million tons annually. Covering over 20,000 acres at a depth of 1200 feet, the mine contained 185 million tons of low sulphur coal reserves and had an expected life of 50 years. The total invested capital in the mine was in excess of $20 million.[1] Gemstone #3 was essentially a captive mine with all its production under long-term contract to a major Midwest electric utility for approximately $20 per ton.

The development plan for the mine was based on the use of continuous miners and automated roof bolting equipment and a room and pillar layout. Above ground the mine had its own processing plant, maintenance shop, and normal support activities.

After four years of development, Gemstone #3 was placed on a production status in 1973 and by 1975 had been developed to the point where eight section crews could be working simultaneously in various portions of the mine. Each of the crews working on a face actually dealt with three to seven separate entries in order to make optimum use of the equipment. These individual entries were sometimes as far apart as 100 feet.

The mining crews that actually produced the coal at Gemstone #3 were typically comprised of nine workers: a continuous miner operator, a loader operator, a roof bolt operator and helper, two shuttle car operators, a brattice person,[2] a boom person and a mechanic. Generally, one mining section and its complete set of equipment were left unscheduled so that if a major breakdown occurred on another section, the crew could be moved to the alternate section rather than have to be idle while repairs were made. Most of the routine maintenance, not handled by the maintenance crew on each shift, was scheduled for a weekend and was handled directly by employees in the maintenance department.

For all underground hourly workers, the pay rates were established by the national contract between the coal industry and the United Mine Workers. Shift premiums were given for afternoon shift (B shift) and night shift (C shift). While the difference in day pay for trainees and the most

---

[1] In 1975 the coal industry used as a rule of thumb that for each ton of annual capacity in a new underground mine, an investment of $30 was required.

[2] The brattice person sets up temporary partitions in order to provide adequate ventilation to each of the entries being worked by the section crew.

skilled workers (grade 5) was almost 25 percent, the miners themselves felt that there was even a greater difference than that indicated by the pay in the attractiveness of the various jobs. Thus miners tended to bid for higher grade jobs whenever possible not only because of pay but because the physical requirements of the higher grade jobs tended to be less and they involved more prestige.

SUPERINTENDENT'S POSITION

In regard to control of the organization, Black Diamond's president felt that the mine superintendent was the most critical position. It was the superintendent who determined the work environment at the mine and the level of effort expended by the employees. Based on past history, the company had determined that a good mine superintendent could have a substantial impact on the profitability and total output of a mine. According to Bob Ericson, the President:

> The key person in the organization is the mine superintendent. Each mine is its own self-contained operating unit and the mine superintendent runs it. Once the engineering plans for the mine are approved, the superintendent is given full authority to operate the mine as he sees fit. As president, it is my function, with the help of the regional vice presidents, to monitor and oversee the progress being made by these superintendents.

Each year the mine superintendent was required to submit a five-year engineering plan for his mine. This plan included a physical description of areas in the coal seam to be mined (and when) as well as an annual operating budget and a capital expenditure budget. These budgets were reviewed by the regional vice president and the president and finally agreed upon by all parties concerned. Performance would then be measured against these levels for each mine superintendent. To assist corporate management and the superintendent in monitoring each mine's progress, daily and monthly operating summaries of output by working section and cost category were collected. In addition, monthly plans were submitted toward the end of each month which detailed the next month's production scheme.

One of the president's chief concerns centered on the problems and challenges that the nation's new perspective on coal might create for his operating organization. The long, lean years of the early seventies had largely restricted the possible expansion of Black Diamond's management ranks and had resulted in a shortage of young managerial talent. In an effort to overcome this problem, Ericson had begun to appoint several younger managers to key positions. Cutler's recent promotion had been one such move. The president's hope was that these individuals would soon develop into a source of new concepts that would help make his entire organization more responsive to its new environment.

ORGANIZATION OF GEMSTONE #3

Of the 590 employees on the payroll in May of 1975, 495 of those were members of the UMW bargaining unit with the remaining being management and clerical personnel. As shown in Exhibit 4, the management organization of Gemstone #3 was divided into support functions and underground operations. Reporting to the mine superintendent were the department heads of each of the support areas as well as the underground foreman. The underground foreman (sometimes referred to as a general mine foreman) supervised four assistant foremen, one for each of the three shifts and a fourth in charge of auxiliary crews. Each assistant shift foreman had reporting to him up to eight section foremen. One of the sections reporting to each shift foreman was a maintenance crew while the balance were actual mining crews assigned to work at the face of the coal seam.

The main responsibility of the management group at the mine was to achieve the desired (budgeted) levels of output for the mine at a minimum cost. Much of the work performed by the engineering staff set the general constraints within which these objectives could be achieved. For example, it was this staff that dealt with the mine development which in turn affected the availability of transportation to the various coal faces being worked, reduced the frequency of required equipment moves, and facilitated getting adequate ventilation to each working face.

Underground, the accomplishment of the mine's output and cost objectives was determined largely by the actions of the section foremen. Each foreman was responsible for a crew of nine miners working at the face. During the shift for which a section foreman's crew was assigned to a certain working area, that section foreman was responsible for all activity, equipment, and production taking place there. Since the crew's pay was not tied to the amount of coal produced, the foreman had to rely on the authority position he had built up with the crew and his own personal capabilities at motivating in order to achieve desired performance standards. Given the long standing distrust of management by the miners, this was a particularly challenging task.

Typically, the section foreman made a wide range of decisions while on the job each day. These included such things as calling in other crews— maintenance people, electricians and mechanics—performing safety checks and determining when operations should be shut down in order to take time to set them up in a more efficient manner. Rebalancing was always needed as the continuous miner moved further down an entry and transportation equipment and conveyors had to be brought into place.

Finally the section foreman was responsible for setting the pace at which his crew worked and the cycle time that they would spend in each phase of the operation. He also trained new workers (called "red-hats") for their first six months underground) and on occasion became involved in grievance procedures as a representative for management.

Section foremen were paid in the range of $18,000 per year and were the first level of management in the mine. Most foremen took on the job in order to achieve greater financial rewards as well as an opportunity for advancement. Many miners did not seek such positions and thus a natural selection process took place. Although most section foremen had experience at performing most of the underground tasks, the UMW contract clearly prohibited them from performing such work:

Section (c) Supervisors Shall Not Perform Classified Work

Supervisory employees shall perform no classified work covered by this Agreement except in emergencies and except if such work is necessary for the purpose of training or instructing classified Employees. When a dispute arises under this section, it shall be adjudicated through the grievance machinery and in such proceedings the following rule will apply: The burden is on the Employer to prove that classified work has not been performed by supervisory personnel.[3]

PROBLEMS AT GEMSTONE #3

In one sense, Gemstone #3 was a typical coal mine confronted with the problems plaguing the entire industry. (See Exhibit 2.) Beset by frequent wildcat strikes, a lack of trained labor and 8 to 22 percent absenteeism, output seemed to be continuing its decline. The available productive days had been reduced from 240 (the number stated in the UMW contract) to an anticipated level of about 200 days for 1975.

It was Bill Cutler's observation that this mine was also plagued by an assortment of problems not commonly found in most mines. These were of a much more serious nature and in aggregate they had led to the dismissal of the former mine superintendent and Cutler's recent appointment. Of particular concern to Cutler were the technical problems which had sharply restricted the output of the mine. Due to the partial collapse of one of the mine's air shafts, production at three of the mine's eight operating sections had recently been curtailed for a month until adequate ventilation could be developed.

In addition, portions of the haulage routes had started to buckle which reduced the clearance below the minimum height necessary for the passage of the coal haulage equipment. The buckling was caused by the "hard pan" materials located directly beneath the coal seam which had not been removed before the haulage track was laid in place. During the development of the haulage routes, these materials were usually broken up and removed, a time-consuming and tedious job. It appeared to Cutler

---

[3]National Bituminous Coal Wage Agreement of 1974, page 4.

EXHIBIT 2
No Easy Out: Coal Industry's Woes Becloud 'Best Answer' to the Energy Crisis*

They Involve Labor, Safety, Pollution, and Inefficiency; Can Output Triple by '85?

By Bob Arnold
Staff Reporter of the *Wall Street Journal*

HERNDON, W.Va.—Beneath the rocky slopes of Appalachia, the stubbly hills of southern Illinois and the mesas of the West lie what many believe to be the best, perhaps the only, answer to the nation's energy crisis—at least 300 billion tons of recoverable coal.

But, at the moment, it seems far from certain that enough coal can be brought to the surface each year to fulfill that promise.

Consider the problems faced by Fred Hill, superintendent of Eastern Associated Coal Corp.'s Keystone No. 2 mine here. On an average day, one in five of his 560 workers is absent or injured. Through September, he had lost 8,488 man-days—equal to more than three weeks' production—to wildcat strikes. His mine's productivity has plummeted to between six and seven tons per man per day from a high of 11 in 1969, and output so far this year is down 32%. Keystone No. 2 has incurred a loss of about $700,000 since the first of the year.

Laws and Inefficiency

Talks with coal-company executives and officials of the U.S. Bureau of Mines indicate that Mr. Hill's problems are typical of the industry's—especially in underground mines, where two-thirds of the nation's coal miners work. Because of ever-tightening federal health and safety rules, rampant wildcat strikes that the industry calculates will cost it 540,000 man-days this year, inefficient company management and stagnant technology, the industry's productivity, or output per man-day, has dropped 17% since 1969. Underground, where 52% of the nation's coal was mined last year, productivity has dropped about 25%.

One result, industry observers have said, is that the country may already face a coal crisis. Production this year is expected to fall five million tons short of last year's 595 million tons, but consumption is expected to rise 6% to 609 million tons. (The difference will come from dwindling stockpiles.) Coal operators and brokers say a shortage may already exist for this winter and spring.

Perhaps more important, it is an open question whether the industry will be able to triple its production to the 1.5 billion tons that energy analysts say the country will need annually by 1985 to supply proposed coal gasification and liquefaction plants and to meet the nation's growing

EXHIBIT 2 (continued)

electricity needs. Bureau of Mines officials say that new safety rules yet to be implemented could cut underground productivity by as much as another 10%. And while energy analysts believe that any stepped-up coal production will have to rely heavily on strip mining, industry executives say a tough federal strip-mine bill expected to be passed by Congress early next session could substantially slow development of surface mines.

Sulphur and Stacks

Moreover, current and proposed air-pollution laws clamp strict limits on sulphur emissions from coal-burning electric utilities. But machinery to remove sulphur from smokestack gases won't be available on a wide scale for perhaps three to five years. The impact of this, says John Corcoran, president of Consolidation Coal Co., is that virtually all of his company's production east of the Mississippi—46 million tons last year—could be "outlawed" by 1977.

Those problems—combined with price controls, which governed even long-term contracts until early this fall—have depressed earnings to the point where few companies are willing, or able, to risk expansion. (Most operations are divisions of other companies, so profit-and-loss figures aren't reported, but bankers and industry sources say that with few exceptions coal companies are teetering between profit and loss). Openings of underground mines in the predominantly high-sulphur seams of the Eastern and Midwestern coal fields have slowed to a trickle.

"We're heading toward a shortage and we're not making the moves that will keep us from being in a deep hole 10 to 15 years from now," warns William N. Poundstone, executive vice president of Consolidation Coal. But if the future is problematic, the present is clear: The industry has myriad problems that aren't going to be solved easily.

Labor's Allegiance

The most obvious and least predictable among them is labor. Company officials complain that they are steadily losing control over their employees and that for the most part miners today feel allegiance only to their union. "We don't have the power to make them work," says Keystone No. 2's Mr. Hill. Coal companies, he adds, "need to instill pride in the men, and we aren't doing it yet."

A major handicap, many observers agree, is the miners' style of life, especially in West Virginia. "He likes hunting and fishing. He lives very practically—doesn't want a big, fancy house. He can afford to miss a day now and then and still live in the style he wants" on union wages ranging between $42 and $50 a day, says John Higgins, Eastern Associated's vice president for production.

Moreover, "a few years in the mines has its effects physically," says Louis Antal, president of United Mine Workers District 5. "Some days you just don't feel like going underground." Whatever the causes,

EXHIBIT 2 (continued)

industry absentee rates are estimated at 10% to 15%, with resulting annual production losses of several million tons.

Turnover is similarly high. No estimates exist for the industry, but at Keystone No. 2, about 570 men, a number equal to the entire work force, have been replaced in the last four years. This is especially significant because underground miners work in seven-man teams called section crews. The longer a crew works together, the greater its efficiency.

In some cases, turnover can be attributed to such human desires as a wish to work in a newer, safer mine or in a mine closer to home that may offer a chance to work days instead of nights. Be that as it may, mine operators say the fact that a miner's pension is set by the length of his union membership, rather than by the length of his company service, doesn't help in maintaining a stable work force.

The most vexing labor problem, however, is wildcat strikes, which will cost the industry between seven million and 10 million tons of production this year. This has especially been a problem in West Virginia. Through the first seven months of this year, the state, which produces about a third of the nation's deep-mined coal, has accounted for 57% of the industry's wildcat strikes.

"They'll go on strike just about any time," says Keystone No. 2's Fred Hill. Last month, for instance, picketers from the mine union's welfare and retirement fund, who were striking against the union, showed up at Keystone No. 2 and Mr. Hill's miners walked out for one shift in sympathy.

Of course, miners also walk out for much more serious reasons. After a fellow miner is injured or killed, they frequently strike to protest against unsafe working conditions or against a foreman who they feel contributed to the accident.

A Generation Gap

There are other, more subtle production impediments from a labor standpoint. During the late 50s and early 60s, mechanization and sagging coal demand combined to interrupt the flow of new workers into the industry. Employment has since picked up, but many operators say the resulting generation gap has brought with it tensions between young and old miners and slowed the transfer of knowledge necessary to maintain peak productivity. And traditionally, middle-aged miners have been the industry's most productive workers.

But if labor difficulties hamper output, so does anemic company management. For instance the rail sidings at Keystone No. 2, where loaded cars are put until a motor is free to pull them to the surface, were designed to accommodate only five cars, fewer than half the number modern motors can pull. And the sidings are so cramped that a fully loaded car can't squeeze under the mine's roof.

A faded schedule on Mr. Hill's wall reveals that maintenance on some

EXHIBIT 2   (continued)

machinery at Keystone No. 2 is behind schedule. As a result, his miners say, at least some equipment is operating less efficiently than it should. Furthermore, the cars in which the men ride into the mine are kept outside—uncovered. When it rains, the miners must ride two or more miles to their underground stations with their feet in water. It is a situation that could easily provoke a wildcat strike, Mr. Hill feels, and one that he plans to remedy.

Also contributing to the industry's troubles are the indirect effects of the federal Coal Mine Health and Safety Act of 1969. Although regulations passed under it have helped to sharply reduce mine deaths, they have also taken a heavy toll in productivity.

Numbers in Safety

Today a superintendent can't make a change in his mine-ventilation system, move a high-voltage cable or begin installing roof supports in mining areas without prior federal approval. The law also provides for periodic inspections that can result in time-consuming repair work. For example, Mr. Hill displays a three-inch stack of violations—104 in all— from one recent inspection of Keystone No. 2's electrical system.

While even the most crusty mine operators concede that the law has resulted in safer operations, they complain about the paperwork the law requires and the time the superintendent must spend escorting inspectors around the mines. U.S. Steel Corp. estimates that its superintendents and foremen spent 4,000 man-hours last year guiding inspectors—and its mines are recognized by the Bureau of Mines as among the safest in the industry.

Another industry complaint: The need for expert inspectors under the health and safety act has siphoned off a number of its experienced foremen and superintendents at a time when the industry badly needs additional managerial talent. Consolidation Coal says it has lost 600 foremen to the ranks of federal inspectors.

There are a few glimmers of optimism in the industry: Price controls on long-term contracts were lifted this fall; energy analysts say that if the strip-mine law Congress approves turns out to be less restrictive than the bill the Senate has already passed, surface mining could boost production quickly and significantly; Arnold Miller, the new president of the mine union, is willing to meet with operators in an attempt to smooth relations at problem mines; and Eastern Associates, which owns Keystone No. 2, is giving foremen week-long courses in what it calls "labor relations, work planning and how to handle people."

Stress on Technology

There is also a new, albeit still minimal, emphasis on technology. Sophisticated machines currently in use underground can chew coal out of the earth at 600 tons an hour. But the companion tasks of keeping the roof

EXHIBIT 2 (continued)

from falling and of hauling coal away from the automatic miners are so time-consuming that 600 tons is an entire section crew's production for an eight-hour shift—under perfect conditions.

"In effect, we're still mining with a pick and shovel," says Consolidation's Mr. Corcoran. "We've improved both the pick and the shovel," but the method remains the same.

Consolidation spends between $2 million and $3 million a year on research in mining technology, a figure Mr. Corcoran says is equal to that spent by the rest of the industry. The federal government, which increasingly is looking toward coal as an important energy source, has allocated $7 million for mining technology for 1974 (the funds haven't been released yet). And the recommendations for a five-year, $10 billion energy program that are scheduled to be submitted to President Nixon on Dec. 1 by Dixie Lee Ray, the chairman of the Atomic Energy Commission, who is coordinating several presidential task forces on energy, are expected to include $25 million to $50 million a year for coal-mining technology.

Even at that rate, however, industry experts say it will be at least five years before new technology will increase coal-mine productivity.

that in an effort to accelerate development of Gemstone #3, this operation had been deferred by the former mine superintendent to be completed after the mine was in full operation. It was Cutler's estimate that 70 to 80 percent of the track would have to be upgraded or replaced in the next two years because of this problem.

Another special set of problems which concerned Cutler centered on the deterioration of worker attitudes at the mine. Productivity was running far below budgeted levels for 1975 for all operating sections and there seemed to be a complete lack of cooperation between the different shifts on the working faces. During his initial tours of the mine to acquaint himself with the layout and conditions, several of the section foremen complained to Cutler that excessive time was needed to prepare for each shift. Cutler was aware that in most mines the actual mining was completely mechanized but that the preparation was all manual. Thus maintaining productivity was very dependent on coordination between shifts working on the same section. In place of the smooth, efficient teamwork requisite for high productivity, Cutler repeatedly had observed substantial idle equipment and idle workers throughout the mine.

A third concern at Gemstone #3 was the substandard working conditions of the sections. If Cutler allowed these conditions to persist, it not only presented a risk to the safety of the employees but there was a chance that federal and state inspectors would begin to levy fines for non-compliance and also the possibility of a strike being called by the local workers' committee dealing with mine health and safety. (Such strikes were legitimate under the terms of the contract whenever the committee determined that a state of "imminent danger" existed.) Given the prevailing labor environment, Cutler felt it was extremely unwise to allow the committee any legitimate opportunity to disrupt production.

In reflecting on these problem areas, Cutler recalled that he had been surprised at the management style of the previous superintendent. During his first two days on the job, the former superintendent at Gemstone #3 had introduced him to the other management personnel and given him several tours of the mine. Even though that former superintendent was only to spend two days in acquainting Cutler with the mine's operations, Cutler had noted that he still took an active part in running the underground operations of the mine. This former superintendent had grown up in coal mines and Cutler imagined that it was difficult for him to relinquish control. While in the mine, that superintendent had tried literally to notice every possible problem in each area they visited and to bark a series of orders at the appropriate section foreman and his workers. On a couple of occasions, it had been clear to Cutler that the superintendent was countermanding directions that had previously been given (or at least agreed upon) by the section foreman. In fact, as Cutler reflected on the advice the former superintendent had given him, he recalled the superintendent had said:

> The hardest part of the job is getting the people to do what you want done. You have to stay on top of everything. It's tough but it's the only way that things get done right. During the past two years since the mine was turned over from development, output per workers as well as total output for the mine has declined. It's only by keeping a constant watch on the crews and section foremen that I've been able to make 70 percent of my targeted output.[4]

THE SECTION FOREMEN'S STRIKE

The strike by the section foremen had been precipitated when Cutler discharged one of the section foremen on the second shift on Friday afternoon, June 2. Cutler had taken the action because the section foreman had allowed a condition to exist in his operating section that was extremely hazardous and endangered the entire crew. In just the short time that Cutler had been in charge of the mine, he had repeatedly stressed the

---

[4]For the first four months of 1975, productivity at Gemstone #3 was 8.1 tons per worker-shift (compared to a budgeted figure of 10.5 tons) at the cost of $14/ton (as compared to a budget figure of $11.50/ton).

need to upgrade safety practices and to eliminate hazardous procedures. It had already been necessary to admonish this particular foreman on three other occasions and therefore when the Friday afternoon incident was reported to him and verified by his underground foreman, Cutler felt that he had no alternative but to discharge the section foreman responsible. Cutler hoped that firm, disciplinary action might provide the necessary shock for the other section foremen to recognize the need for changes in their own behavior regarding mine safety.

Just before leaving for home on Friday evening, Cutler had been informed by the underground foreman that the section foremen of Gemstone #3 were planning a strike at the mine to be initiated on Monday morning, June 5. Apparently the dismissal of the foreman on the second shift had brought to a head the dissatisfaction felt by the section foremen as a group. (See Exhibit 3.) Although the underground foreman was unable to obtain specific details with regard to the section foremen's plans or motivations, he felt that the implied purpose of their walkout was recognition as a bargaining unit and the desire for additional job security.

Early Saturday morning, Cutler had driven to the mine to be on hand when three working sections of the day shift and the maintenance crew were scheduled to arrive for some overtime work. When none of the section foremen scheduled for work showed up, he had sent the crews home. (No overtime had been scheduled this particular Saturday on the second or third shifts.)

Throughout the day on Saturday, Cutler tried to contact each of the section foremen to better determine the facts surrounding their dissatisfaction and to resolve the issue before the strike fully materialized. However, regardless of the number of times he tried, each of the section foremen was either away from home or unable to come to the phone. Cutler presumed that the foremen had made an agreement not to answer their phones during the weekend. He was able to talk with a number of the senior people on the section crews who were fairly close to some of the section foremen and also to talk with his assistant foremen in charge of each of the shifts and the auxiliary crews.

Through these conversations, Cutler was able to acquire some additional information concerning the pending strike. Apparently most of the mine's 22 section foremen were actively involved in the planned work stoppage. However, none of the assistant shift foremen were so involved. Even those section foremen opposed to the strike were planning to honor the picket line and did not intend to show up for work on Monday morning. The leaders of the striking group were two of the younger section foremen on the mine's second shift. According to the underground foreman, both individuals had a previous record of attempts to organize the section foremen group. When asked what they thought caused the strike, most of the employees contacted by Cutler explained that the section foremen were tired of working with a constant fear of losing their jobs.

EXHIBIT 3
The Coal Boss: Foreman's Job Grows Harder as the Miners, Technology Change*

'Red' Goss Knows His Work and Men; A Happy Crew Means Fewer Stoppages

By Bob Arnold
Staff Reporter of *The Wall Street Journal*

McALPIN, W.Va.—"Red" Goss, section boss, works quickly in the cool darkness, seemingly alone. The only sign that his 14-man crew is working within 50 yards of him is the distant-sounding hum of electrical machinery. The sound is muffled by the pillars of coal that separate the miners' work areas.

A shuttle car, its headlights momentarily lighting foreman Goss's blackened face, rumbles by. It disappears in the direction of a conveyor that will start its cargo of freshly mined coal on a two-mile trip to the surface of the McAlpin Mine. Around a nearby corner, two miners maneuver their loader's clawlike arms toward piles of shattered coal: the machine soon will be filling a second car.

Working by the light from the lamp on his hard hat, Mr. Goss continues his task of the moment—wedging six-inch-thick timbers upright between the coal-strewn floor and the sandstone roof only 40 inches above. Suddenly, both roof and floor quiver from the impact of an explosion, blasting loose more coal in a parallel passageway. The pungent odor of explosives fills the section of the mine known as "Five Panel Right."

Part of a Pattern

To a casual observer, the activity seems random. But to David J. Goss, 35, it is part of a predictable pattern. "I just know the cycle," he says. To prove his point, he draws a crude map on the earthen floor showing the location, at that moment, of each piece of machinery in his section.

Keeping such maps in his head is a small part of Mr. Goss's job. "A production foreman (the modern term for section boss) has the roughest job in the mines," says Reid Daniel, superintendent of the McAlpin Mine which is owned by Philadelphia-based Westmoreland Coal Co. "He's the link between me and the men. If there's no production, we're down his throat. And he's got to keep you out of Dutch" with state and federal safety inspectors.

The coal industry is being counted on to play an expanded role in helping the nation meet its energy needs. As the industry thus grows in importance, so does the foreman. In an industry long plagued by stormy labor-management relations, he has always been crucial. Now his job has become more complex than it was two decades ago, when his main

*Reprinted by permission of *The Wall Street Journal*, © Dow Jones & Company, Inc., November 6, 1975. All rights reserved.

EXHIBIT 3 (continued)

concern was responding to distant superiors' demands for increased production.

Today's increasingly sophisticated equipment requires that a "boss" be more technically expert. He has to be more expert at handling people, too; miners, who are younger than ever and have more rights than ever under protective new federal laws and improved union contracts, reward his mistakes with slowdowns, wildcat strikes or demands for his dismissal. And safety inspectors, enforcing recently enacted laws, can close his section instantly, throttling production.

Some Adapt Slowly

Some foremen are adapting slowly to the new realities of the job, coal executives say. In part, they concede, that's because many companies haven't trained their foremen well. Mr. Goss's preparation, like that of many of his peers, consisted of four years as a union miner, a passing grade on the West Virginia certification test and his own store of common sense.

Yet such a background does produce some highly competent foremen. Three days with the boss of Five Panel Right suggest he is one of them. Both his subordinates and his superiors give him high marks.

"The men respect him," says miner Ben Bane, partly because Mr. Goss was once an hourly-paid miner himself and handled a tough roof-bolting job. And mine superintendent Daniel finds Mr. Goss's approach to typically independent and suspicious miners more sensitive than that of most bosses. "Red knows how to talk to the men," he says.

Five Panel Right is a grid of passages and coal pillars about 675 feet underground. Five parallel tunnels, called entries, are carved at 75-foot intervals through a solid layer of coal. To ensure stability of the mine's roof and ribs, or walls, each entry is by law no wider than 20 feet. For reasons of economy, each is cut only high enough to get the coal—40 inches in Five Panel Right.

A Months-Long Task

It takes Mr. Goss and his men several months to carve out these entries and the connecting crosscuts, 20-foot-wide passages that intersect the entries at 90-degree angles every 75 feet. As the miners inch forward, they send the coal to the surface. Once they have driven all five entries as far as the mining plan provides, they back out of the panel, removing as much of each pillar as they can—as much as 90% in some cases—without permitting the mine's roof to cave in on them. This backing-out process takes several weeks.

Whereas many mines use machines that keep clawing away the coal, the McAlpin Mine uses an older method. A crew's first step in opening up a panel is to make an undercut, all along the working face of an entry with a chainsaw-like blade that extends 11 feet into the coal. The coal thus loosened is shattered by a shot of explosives, then is loaded into two

EXHIBIT 3 (continued)

buggies, or shuttle cars. Two pinner teams finish the cycle by taking long, half-inch-diameter steel bolts and riveting the mine's roof to the solid rock three to six feet above it, lest the intervening, less solid rock fall through. Meanwhile, the coal goes to the "top"—actually the trip to the surface is a lot more horizontal than vertical—by belt and rail, the same way the men get there. Then the miners tackle the next 11-foot-deep section of the entry.

On a typical workday running from 7 a.m. to 8 p.m., Mr. Goss continually travels his maze on all fours, his knees protected by inch-thick rubber pads, looking after production. A key task: directing traffic in the darkness so that the crew will lose little time. Equipment breakdowns are another major headache. A buggy breakdown on a recent Monday took five hours to repair, cutting the Goss crew's daily production by about 25%.

Mr. Goss often lends a hand. Frequently, for instance, so that machines in motion won't cut them, he rearranges his equipment's high-voltage power cables and the reinforced rubber lines that carry the equipment's spray-as-you-go water for suppressing lung-damaging coal dust. At other times Mr. Goss, who is 5 feet, 10 inches tall and weighs about 200 pounds, figures that his main job at the moment is just to stay visible. "If the boss is around," he says, "the men speed up a little."

In most situations, Mr. Goss's approach is relaxed and friendly. "Tim, why don't you go up and do that rock dusting?" he suggests, referring to a routine safety procedure. It is best to ask, not order, men to do things, he believes. The miners agree: "If a man starts throwing his weight around," says Chester Grubb, the section's electrician, "he'll get six cuts (the swaths of the chainsaw-like blade) a day"—about half normal production.

Mr. Goss says his basic aim is to treat the miners "like people, like trained men." His empathy for miners, he indicates, was shaped both by the mistreatment he feels he suffered as a Navy enlisted man and by the fact that his father and father-in-law helped organize the United Mine Workers in the 1930s. Thus, he lets James Mahan, the crew's loader operator, work through the 30-minute lunch break with pay, an accommodation the contract doesn't require. Each day, he brings two extra bags of chewing tobacco for his crew. His reason isn't entirely selfless: "If I give them all of mine," he says, "I won't have any."

It Seems to Work

His low-key approach appears to pay off. His men accept discipline; they may mutter obscenities, but they don't strike. Recently, a veteran miner whom Mr. Goss had rebuked for stretching his lunch break to an hour grew angry but continued to work at his normal pace. More important, Mr. Goss's men often go out of their way to be helpful: He points to a more-than-routine maneuver that his cutting crew has performed—even though he hadn't marked it, as the men might have legitimately demanded.

EXHIBIT 3   (continued)

Partly as a result of such good relations, Mr. Goss's crew routinely turns in the McAlpin Mine's highest production—150 to 200 tons a day, measured after rock and other impurities have been thrown out—and production is one of the foreman's two major concerns.

His other major concern is safety. Only once, four years ago when he "first started bossing," have inspectors closed his section. The violation: leaving too much coal dust on the mine's floor. Such an experience so soon after becoming a foreman "was about enough to make me quit," he says. "It shakes you up."

Coal dust is explosive, and a fire could set off the kind of death-dealing chain reaction that has helped make coal mining the country's most dangerous industry. So besides keeping a clean floor, Mr. Goss must check the "working face" at each entry's head four times a day for methane gas, which is liberated during mining. He measures the flow of methane-dispersing air, sucked into the mine by huge surface fans and channeled to working faces by movable plastic curtains, to make sure that it's above specified levels.

Twice in a recent two-week period, Mr. Goss withdrew his men from the section when loggers cut fan power lines on the mountain above. As his men inch forward, the foreman sees that they cover his panel's roof, floor and ribs with rock dust—ground limestone that inhibits coal dust's explosive tendencies.

Since "it doesn't take much rock to kill a man," boss Goss makes sure that roof bolts are no more than four feet apart, exactly as specified in the mine's roof plan. Occasionally, he raps suspicious-looking spots with a hammer. A sharp "clink" means the roof is sound; a hollow "clunk" means it isn't. "I never had anybody hurt yet by a fall," he says.

Safety is nevertheless Mr. Goss's most persistent headache. The combination of close quarters and heavy machinery makes serious accidents a constant threat. Last winter, Mr. Mahan suffered four cracked ribs when his loader bolted out of control and pinned him against a wall, stopping just inches short of killing him.

Yet the miners regularly ignore rules they think aren't critical to their safety, although the company would be fined if they were caught by an inspector. Mr. Goss generally says nothing, for example, as his miners follow their normal practice of getting on and off conveyor belts without first stopping the belts as the law requires.

Section boss is a job most miners wouldn't have, says Bud Alderman, a buggy operator. "I don't want the responsibility myself," he says. More to the point, Mr. Goss's base salary of $15,450 a year is a mere $700 above the top annual wage for union miners. (Counting overtime, he made $18,000 last year.) And Mr. Goss gets only 17 days off each year, about half what the average miner gets.

The pay gap between foremen and other miners was wider in late 1971. That was one reason Mr. Goss then left his roof-bolting job, and the union, to become a boss. Another reason: The union, which doesn't have a strike fund, had just conducted a six-week contract strike of the sort

---

EXHIBIT 3   (continued)

that has become almost a triennial affair. Mr. Goss felt that with a wife
and three small children to support, he wanted a salary's security. (Mrs.
Goss now works as a nurse in a hospital in nearby Beckley. The family—
still with three children—lives in a comfortable frame house a 20-minute
drive from the mine.)

Difficult Transition

Most foremen don't make the transition from miner to boss as smoothly
as Mr. Goss did. Many coal companies—belatedly, in the eyes of industry
critics—are therefore making their first attempts to teach foremen
technical and labor-relations skills. But many foremen think such
programs still lack a lot. Westmoreland Coal Co.'s course "wasn't worth a
damn," a recent graduate says. "The safety's good, but they don't teach
you anything about traffic control or pulling pillars."

In fact, the informality of Mr. Goss's own preparation is still evident at
times. Mr. Grubb, the electrician, who has mined for 41 years, feels that
Mr. Goss is "green on pillars" compared with other bosses because he
"doesn't pursue getting all the coal."

Nonetheless, Mr. Goss is as satisfied as he has ever been. Recalling
various blue-collar jobs he held between two years at Marshall University
in Huntington, W. Va., and his mid-1960s Navy stunt, he says:

"Before a job was a bore. Now I enjoy it. I run coal, kid around with
the men, and we make money for the company and ourselves. My old
man swore none of us would work in the mine, but I don't feel that way.
It's a respectable occupation."

EXHIBIT 4   Gemstone #3 Organization

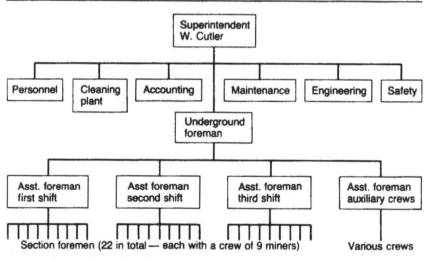

Section foremen (22 in total — each with a crew of 9 miners)   Various crews

This puzzled Cutler for the records indicated that the section foreman he had dismissed on Friday afternoon was the first discharge of a management employee (other than the former superintendent) in over a year.

PLANNING A RESPONSE

In planning his own response to the situation, Cutler had tentatively decided that at this point he should not request that the president of Black Diamond Coal Company or the regional vice president to whom he reported become involved. He thought it would be much better to formulate his own response first and to then perhaps double check with higher levels of management before implementing that plan. Even then he was not sure that it might be better to try to resolve the strike before getting top management actively involved. Since they were used to various work stoppages, he was certain that they would feel it appropriate for him to handle it himself for a couple of days before calling on help from senior people.

In formulating his response to the strike, Cutler thought it would be important to bear in mind what he had identified as some basic economics of Gemstone #3. The contract price for the mine's coal was presently approximately $20 per ton. The records indicated that before tax and before charges for corporate overhead, it had been costing approximately $13.52 per ton to produce coal ready for shipment at the mine. Of these costs, approximately $11.38 per ton were variable with the remaining $2.14 being fixed.

During early 1975, the mine had been running 22 section shifts per day (19 mining crews and 3 maintenance crews) and had averaged an output of 240 tons per mining crew per shift. While all employees of the mine, with the exception of management people, traditionally honored any picket or work stoppage, Cutler was certain that the miners would not be particularly pleased at the prospect of staying away from the mine for very long due to a strike by the section foremen. Cutler's checking of the UMW contract indicated that the average miner received fringe benefits plus $55 per day in gross wages. While the fringe benefits would continue during any period that the mine was closed due to a strike, each individual miner would lose about $40 per day in take home pay. From the corporate point of view, Cutler recognized that any lost production would have a substantial impact on the contribution from Gemstone #3 to Black Diamond's profitability.

In polling the assistant foreman in charge of each shift, Cutler had found a range of opinions as to what action would be most appropriate. The first person he talked to was Mike Rucker, assistant foreman on the third shift. Rucker had expressed his feelings as follows:

Since it's only been three months since I was promoted from the position of section foreman to assistant foreman, I think I can identify fairly well with some of the feelings and frustrations that the section foreman have. The position of section foreman is tough enough given the fact that it's the buffer between management and the union workers. However, under the former mine superintendent, it was even harder because of the emphasis he placed on productivity and output.

I think this strike of the section foremen should be handled very carefully so that they don't misunderstand your attitudes toward their position and their problems. In my view, I think the best thing to do would be to use this as an opportunity to sit down and talk with them about their work environment, their management responsibilities, and some of their problems and frustrations. I think you could do this on Monday morning with either all of the section foremen or with the three or four leaders of the group.

You might even choose to ignore the strike itself as a major issue and simply focus on what's been upsetting the section foremen. I think as long as you come out of that kind of discussion with a definite plan of action that everyone agrees will help to resolve some of their concerns, that things will be back to normal in a couple of days.[5]

One of the other assistant foremen whom Cutler had talked to about the strike situation was John Meredith. Meredith had been an assistant foreman on the second shift since the mine had moved from development into production status in 1973. Prior to that time, Meredith had been assistant foreman for five years at another mine owned by Black Diamond and had fifteen years of underground experience before that as an hourly worker and as a first level foreman. Meredith had expressed some of his feelings as follows:

Just before you took over three months ago, I told the former superintendent that I thought we should get rid of a couple of those rabble rousing section foremen on my shift. I don't think we'd have any problem at all right now if we'd have done that immediately. Some of the newer section foremen just don't have it through their heads that they're no longer part of the union but are now part of management and must accept that responsibility.

I think the only justifiable response is to demand that the section foremen return to work before any discussion of their concerns and problems can be initiated. Such action would demonstrate your insistence on being in charge and on running the mine as it should be. I guess I'd have to say that I don't believe in that new fangled participative management. We all know what our task is here and that's to mine coal. The former superintendent really made everyone toe the line in that regard and I think we should continue that policy.

A third option that Cutler had formulated as a result of his discussions with Steve Sims, assistant foreman on the first shift, was that of simply firing all of the section foremen who claimed to be on strike and then dealing with them individually concerning their rehire and the terms of their employment in management positions. While Cutler thought this

---

[5]Exhibit 5 suggests some of the actions that might be taken if Cutler were to follow Rucker's line of reasoning.

EXHIBIT 5
The Foreman: Most Misused Person in Industry*

By John A. Patton

The foreman of three to four decades ago was the man who ran the show. He was assigned and accepted broad authoritative powers and commensurate responsibilities. His primary job was to see that he met a certain schedule of production—often set by himself, and within certain quality specifications—set by himself. In most cases, he did his own hiring and firing. From the standpoint of management and the economy of the era, this was a good policy.

The foreman of today, however, is being held responsible for functions over which he no longer has any real authority or control. For some time he has not been able to hire, fire, or set production standards. He cannot transfer employees; adjust the wage inequities of his people; promote deserving people; develop better machines, methods, and processes; or plan the work of his department with anything approaching complete freedom of action. All the matters for which he is completely or partially responsible have now become involved with other persons or groups, or they have become matters of company policy and union agreement. He is hedged in on all sides with cost standards, production standards, standard methods and procedures, specifications, policies, laws, contracts, and agreements, *most of which are formulated without benefit of his participation.*

Although we have had many noticeable changes in industry over the years, the one subtle and gradual change that has been largely ignored is the erosion of the role of the foreman in the organizational structure. Today's foreman is convinced he is less rather than more effective, less rather than more secure, less rather than more important; he has received less rather than more recognition. And management has been willing neither to recognize the seriousness of this condition nor to accept it.

As a consultant with 30 years of experience in this field, I would like to outline the factors that I feel show why this sad state of today's foreman has continued to grow worse. I would also like to suggest how this trend can be reversed.

Factor 1—Income

In over 60 percent of companies today, the skilled employees in a department take home pay equal to or more than that of the foreman. Also, in too many companies the management has a policy of tying the compensation differential of the foreman to the unionized personnel by maintaining a 15 to 20 percent increment. This is hardly a way to stimulate the foreman's loyalty to his company; frankly, it makes him feel that his

*Condensed from "The Foreman—Most Misused, Accused, and Abused Man in Industry," *Industrial Management*, April 1974. © 1974 by the Industrial Management Society.

EXHIBIT 5   (continued)

well-being is tied in with the union rather than with management. If the foreman is going to have a favorable attitude toward management, his income should be governed by formal salary administration procedures with planned action by management.

Factor 2—Staff Specialists

The evolution of staff specialists has hindered the foreman rather than helped him do a more effective job. Personnel managers, industrial engineers, and other staff personnel should channel their activities so that they continually work with and through the foreman and help him increase his effectiveness.

Factor 3—Links to Management

I have never found a company where all the foremen felt they were fully informed and really a part of management. In order to remedy this situation, management should see that the foreman is kept informed on changes in policy, organization, or conditions that affect his department, receives the backing and support of higher management, and gets direct answers from above with reasonable speed.

Factor 4—Training

Although management will spend over half a billion dollars this year on foreman training, much of it will be wasted, mainly because many of the programs are canned. Foreman training programs should be tailored to fit actual shop conditions and problems as they exist on the job. Foremen themselves should have a part in working up such programs, and time-study and other staff personnel should participate in them.

Factor 5—Selection

The criteria for selecting foremen in most companies today are entirely unrelated to the criteria used for the selection of effective supervisors. In four out of five companies today, the present foreman was at one time the best assembler or machine operator, or he was the best senior man in the department. Choosing the senior man ruffles the fewest feathers. I am sure you will agree that any correlation between a good producer and a capable foreman is purely coincidental. Successful foreman selection begins with an inventory of supervisory skills and analysis of strengths and weaknesses in order to develop a valid set of criteria for selection.

The effective supervisor, in terms of both production and morale, is the one who tends to see his job primarily in terms of human problems—the management and support of people—rather than in terms of rules, procedures, technical efficiency, and direct pressure for productivity. If management can convince its foremen that it has their interests at heart, and mean it, the result will be more effective supervision than any incentive could create.

might be somewhat risky if in fact the majority of section foremen chose to seek employment elsewhere, he felt there was a good chance that only four or five would choose to do so and that those were the four or five he did not want anyway. Thus he thought he might be able to strengthen his own management position, resolve the strike and start out on the right foot with the remaining section foremen all at the same time if he were to pursue this action and it succeeded.

Regardless of the option selected, Cutler knew that the foremen were not protected under existing labor legislation nor under labor contracts of any sort. Their strike could be interpreted by management as illegal but the repercussions of a hard-line stand could not be predicted.

Cutler realized that any action he chose to pursue would have its own particular chances of success or failure. He thought it would be important to detail what the possible outcomes might be of each of these options and also to do some contingency planning as to how he would follow up in the event that the chosen option did not prove to be immediately successful. Since Cutler saw his long-term future in the coal mining industry, he thought he should be somewhat cautious about the action chosen since, if he were unsuccessful, he might put himself into a position where it would be tough to find a superintendent's job.

The Development of Manufacturing Managers for the Future

Janice A. Klein
Kevin Davis

During the early 1980s, foreign competition heightened American industry's awareness of the need for changes in manufacturing. Many corporations responded with an increased emphasis on new technology and management information systems, which in turn, created a need for a new breed of manufacturing manager. According to Wickham Skinner,[1] these new manufacturing managers need the following:

1. Comfort-level knowledge of mechanical, electronic, and management science technology, and all business functions—production, marketing, finance, control, personnel.
2. Skills that include system design—logistical, information, human systems, planning, coordination; project management and team building; interpersonal communications, handling large groups, and labor relations; general management, for example, integration of business and technical functions—in contrast to specialized technical skills.
3. Attitudes that are positive toward change, restless unless making progress in change, and that reflect security in professional life.
4. The ability to conceptualize combined with the ability to work from specifics systematically and analytically rather than intuitively.
5. Premises/assumptions that the objectives of a manufacturing system are multidimensional. Rather than always focusing the production system primarily on cost and efficiency, the manager must set objectives based on the strategic needs of the corporation. These objectives will establish both priorities among and trade-offs between:

   * cost, efficiency, and productivity
   * customer response time (delivery lead times)
   * reliability of delivery promises
   * product quality
   * investment in equipment and inventory
   * flexibility for product changes
   * flexibility for volume changes.

Manufacturing companies have begun to step up to the challenge Skinner describes in a number of ways. This reading examines the efforts of two companies, General Electric and Procter & Gamble, long considered outstanding developers of manufacturing management.

Manufacturing Management Development at General Electric

Begun in 1878 as Edison Electric Light, the General Electric Company (GE) became one of the most diversified manufacturing opera-

[1]Wickham Skinner, "Wanted: Managers for the Factory of the Future," *Annals of the American Academy of Political and Social Sciences*, 470, November 1983.

tions in the world. GE's manufacturing operations included aircraft engines, medical diagnostic systems, engineered materials, washing machines, locomotives, robots, and televisions. In 1985, General Electric operated plants and laboratories in over 200 locations in 34 states and Puerto Rico. In addition, the company had 135 manufacturing plants in 25 other countries.

Traditionally, manufacturing was seen as a dirty job that did not require brains or managers as adept as those in GE's engineering, marketing, or finance organizations. Shortly after WWII three forces pushed manufacturing into a greater role in the company. First, GE needed professional management within its factories to combat worker strikes which, according to the 1946 GE annual report, had reached an all-time high. Second, consumer demand was intense after the war, and GE forecasts showed that the only way to meet it was with extremely high productivity gains in production. Finally, competitors were beginning to use the manufacturing function as a competitive weapon. These changes required manufacturing to be elevated in status so talented individuals could be recruited. By 1950, when Ralph J. Cordiner became CEO, the need for developing key manufacturing talent had become well-recognized, which led to the formation of professional manufacturing training programs.

The precursor of the professional management programs was the Apprentice Training program, begun in the 1900s. High school graduates identified as having good technical ability were trained in advanced factory skills. This four-year program (later shortened to three) also enabled some apprentices to obtain college degrees on a special work-study program.

In 1952, in recognition of the need for more highly qualified manufacturing personnel, the Management Training Program (MTP) for manufacturing was begun. Under the direction of Art Vinson, vice president of manufacturing services, MTP was started with administrators at six company locations. Its recruiting effort aimed at college graduates who had finished in the top 50 percent of their class—the goal in engineering at that time had been the top 25 percent. Of the 112 who entered the program in its first year, half were recruited directly from college while the rest came from within the company. The program called for six six-month assignments over three years, with a move to a different location for each. This was soon modified to allow participants to spend a full year at each location. Participants specialized in one of five areas: materials, manufacturing engineering, plant utilities, quality control, or shop operations. In addition they attended evening classes on manufacturing-related subjects. By 1955 over 200 college graduates, including some MBAs, were being recruited each year for the program.

In 1972 environmental demands led to several major changes in the program's structure. First, the program was shortened to two years and four assignments. Recruiters were finding the three years under the old program difficult to sell, and administrators felt they could provide the required training in less time. To upgrade the quality of the recruits further, program entry was limited to those with a college degree. (Previously, a small number had been accepted without a degree.) Finally, the name of the program was changed to the Manufacturing Management Program (MMP).

Under the leadership of CEO John F. Welch, GE further emphasized manufacturing. The Manufacturing Leadership Curriculum (MLC), a series of nine courses covering topics such as manufacturing operations, supervisory skills, manufacturing planning, and effective communication, was added as a requirement for all MMP graduates. In addition, in 1981, a course of study not directly connected to the MMP, the Advanced Course in Manufacturing, was set up to allow engineers to attain graduate degrees while focusing on manufacturing problems at GE. (This was comparable to the Advanced Course in Engineering, established in 1923.)

The Manufacturing Management Program

The Manufacturing Management Program was a two-year rotational work and education experience consisting of four six-month assignments: the first two at one plant location; the remaining two at a different location. To underscore the program's diversity, MMPs typically had an assignment in a high-volume, low-technology environment and a complementary assignment in a low-volume, high-technology operation. Each MMP member had assignments in at least three of the following areas: manufacturing engineering, quality control, materials management, or shop operations. This rotation through different work areas assured broad technical exposure, supervisory experience, and functional understanding. Assignments were geared to the temperment, interest, and demonstrated ability of the individual, consistent with the needs of the business.

To qualify for the Manufacturing Management Program, a person had to have a Bachelor's or Master's degree in engineering, preferably in electrical, mechanical, industrial, or computer science. Beyond the basic educational requirements, GE also took into consideration extracurricular activities and grades. After the initial campus interview, those applicants most closely reflecting the kind of person believed to do well in manufacturing at GE had a structured interview at the GE facility closest to the university. Because of the high quality GE sought and the high level of competition from other firms, only one of every five applicants invited for a structured interview went to work for GE.

The curriculum of the Manufacturing Management Program spanned many manufacturing areas. It began with a one-week Manufacturing Concepts Course which introduced MMPs to GE, the Manufacturing Management Program, and the manufacturing function. The course also enabled program members to set up a network with other MMPs and managers throughout GE's manufacturing organization. During the first year of the program, participants also attended the Manufacturing Leadership Curriculum (MLC), a nine-course distributed training package. These video and personal computer-based courses, monitored by operations MMP administrations and course moderators, supplemented their daily work assignments.

At the beginning of the second year, MMPs were required to take Manufacturing Technology Overview, a one-week course in state-of-the-art manufacturing technologies. The speakers, experts in areas such as robotics, CAD/CAM, and computer applications, focused on applications and methods for introducing new technologies into the manufacturing environment.

There were five stated criteria for graduation from the Manufacturing Management Program: 1) at least one assignment on the shop floor; 2) assignments in at least three of the four subfunctions; 3) at least one supervisory assignment (preferably of hourly employees); 4) at least two technically-oriented assignments (preferably one of which involved introducing some new technology); and 5) one project assignment. Examples of project assignments included being a manufacturing engineer for the installation of a new piece of equipment, or a quality engineer finding the solution to a major quality problem.

John Evans, current Manufacturing Management Program administrator, stated that the program had two primary objectives: 1) to increase the technical competence of manufacturing; and 2) to turn out subsection manager potential within 10 to 15 years. (A sub-section manager was at the third or fourth layer of management.) The promotion record showed that MMP graduates typically reached levels higher than subsection manager in 10 to 15 years, with many attaining executive positions. The program had produced eight vice presidents and 25 general managers since it began as the Manufacturing Training Program and was viewed as an important first step in the promotion sequence. As one former MMP stated:

> Performance in the Manufacturing Management Program was crucial to later selections for long-term assignments. Top plant managers kept track of the MMPs through word-of-mouth; reputations established while on the program significantly influenced placement after graduation.

The program had been so successful that GE faced problems with other companies' recruiting its MMP graduates. Although Evans noted that GE's retention compared favorably with other large manufacturing companies, five years after graduation approximately 50 percent of program members had left the company; after ten years the attrition virtually ceased, with retention levels settling in just under 40 percent.

In 1984, 170 new college graduates were hired, 92 percent in the upper quartile of their class. According to company publications, these were by far the highest quality, most technically sophisticated people the program had ever attracted. At least part of manufacturing's ability to attract so many of the brightest engineers was attributed to new automation and favorable media exposure of the manufacturing field. A newly hired program member gave the following reasons for entering manufacturing management:

> Why did I choose the MMP? My engineering professors told us to go into design engineering. At the companies where I recruited I saw 25-year-old design engineers sitting next to 50-year-old design engineers. They sometimes work on a project for two to three years and then it doesn't get built because marketing doesn't like it. Manufacturing offered diversity and the excitement of automation. In manufacturing there is a different challenge every day.

Unfortunately there was also a down side to all the media attention on manufacturing. After being exposed to the latest state-of-the art capabilities of the new technology, new recruits were often impatient with the rate the advanced technologies were being implemented. This problem was coupled with the inherent strain caused between those managers taught under the old apprentice system and those brought in under MMP. The following evaluation of the new MMPs by an "up through the ranks" manager hinted at the strains.

> The MMPs are bright and want to advance but have several major weaknesses:
>
> 1. They lack training on labor relations; because they lack understanding of grievance procedures, they often make commitments they can't keep.
> 2. They come in too green; they don't know where to go to get help or solve problems.
> 3. They let the line workers slack up; why get into a hassle with the workers when you'll be gone in six months anyway?

To expose as many potential applicants to the MMP program's existence as possible, GE maintained contact with engineering schools across the United States. Of the approximately 100 engineering schools where GE recruited, 35 had Key Engineering School Liaison Executives who provided support in return for a healthy recruiting relationship. For instance, an executive might line up speakers from GE to speak at the college on technical subjects which, in turn, enhanced the school's curriculum. For the other 65 schools, GE had several geographically-based recruiters who maintained a close relationship with the placement office.

The Advanced Course in Manufacturing

Following the educational curriculum of the MMP, manufacturing engineers were encouraged to enroll in the Advanced Course in Manufacturing (ACM). The broad goal of this course was to bridge the gap between college and industry. The ACM program lasted three years, with participants spending four hours each week in the classroom on company time, plus 20 hours on homework each week on their own time.

During the first year, students took the Advanced Course in Manufacturing core: applying mathematics and engineering technology to the solution of manufacturing problems. Topics included application of probability and statistical inference to quality control, robotics, computer-aided manufacturing, and interactive graphics. For the second and third years, students selected a series of graduate-level courses at a participating university and completed an engineering project or thesis. GE shared the teaching job for this program with numerous universities depending on the location of the students. The program culminated with the participant receiving a graduate degree in manufacturing from the college attended and certification from GE for completion of the ACM curriculum.

Manufacturing Management Development at Proctor & Gamble

In 1837 William Procter and James Gamble combined their talents in the manufacture of candles and soap and formed Procter & Gamble (P&G), which evolved into a major manufacturer of consumer packaged goods. Its product line included packaged foods, paper products, soft drinks, citrus products, and pharmaceuticals, as well as institutional, industrial, and cellulose pulp products. Entering the 1980s the company had 40 plants in the United States and major operations in 24 foreign nations.

P&G had a strict practice of promotion from within. As the company grew, the technical challenges in developing new products, new packaging technologies, and new manufacturing processes for existing products created increasing demands on the company to recruit and train the best engineering talent available. William Cooper Procter, grandson of William Procter and manager of the Ivorydale manufacturing operation, began recruiting for P&G on college campuses in 1905, two years before he became CEO.

P&G's recruiting effort continued to grow, and by 1947 was highly decentralized with each plant manager being responsible for recruiting at specific schools. A three-member corporate-level recruiting staff was responsible for identifying long-term corporate personnel needs and providing general recruiting procedures to plant-level personnel who conducted the on-campus interviews. The staff spent several days with the interviewers training them in techniques used to identify desired traits quickly. The recruiting staff also recommended corporatewide starting salaries for college graduates.

Once identified, potential recruits went through two interviews. The first, generally on campus, lasted approximately 30 minutes, during which the recruiter assessed at least four traits: 1) goal-setting ability; 2) leadership potential; 3) communication ability; and 4) energy level. If candidates were judged to possess the right mixture of these traits they were invited for an in-depth interview either on campus or at the hiring facility. The plant manager made the final hiring decision. Recruiting at the plant instead of at the staff level accomplished two corporate objectives. First, the individual plant managers were able to construct a team built around the specific needs of their plants. Second, since the new hires were directly hired by their immediate supervisor and the plant manager, they were rapidly accepted by others at the plant.

Individual plant managers also kept in close touch with placement personnel and professors at their assigned universities to identify promising individuals. Most of the manufacturing management recruits possessed a degree in chemical, mechanical, electrical, or industrial engineering. A few nontechnical people were hired, usually whenever engineers were in short supply. In addition, although grades were considered in a broad context, those hired generally were in the top third of their classes.

The recession of 1982 led the company to scrutinize its recruiting efforts, and after studying retention and promotion records of recruits from individual schools, the corporate recruiting staff decided to limit its efforts to the best 75 out of the 150 schools where they had been recruiting. The current manager of corporate recruiting noted that they were not only able to continue to recruit the necessary number of people, but the concentration of effort led to an improvement in the overall quality of recruits.

The "Join-Up Process"

Although first assignments varied with plants, starting procedures, which P&G referred to as the "join-up process," were fairly consistent. New recruits typically remained in one plant for three to seven years and had two or three of the following assignments during that time:

1. Team Manager/Supervisor—reported to a department manager for management of a work-team (as many as 30 people). Responsibilities included evaluating operational effectiveness, identifying problems, optimizing production, and minimizing costs.
2. Area or Staff Industrial Engineer—responsible for identifying the need for and participating in design of projects to reduce costs.
3. Project Engineer—responsible for the efficient completion of new processes or the installation of new machines including costs, equipment, human resources, and time schedules.
4. Mechanical/Maintenance Engineer—responsible for planning and managing the maintenance of packaging equipment, processing equipment, and other plant machinery and equipment. The position typically entailed the supervision of 5 to 20 skilled mechanics.
5. Process/Chemical Engineer—responsible for solving all problems related to the quality of a product or process.

Several weeks prior to their official hiring date, recruits were assigned a sponsor or "buddy." The sponsor acted as a peer source of information on both the company and the local area. To decrease first-day anxiety, the sponsor and the recruit's immediate supervisor, typically a department manager, met with the recruit for dinner before the first day of work.

Although assigned a specific job right from the start, most of a new hire's time in the first few weeks was spent on general company orientation and basic manufacturing training, both in the classroom and on-the-job. During the training, in addition to becoming proficient in all areas of their own job responsibilities, recruits met with functional leaders throughout the plant and were exposed to all phases of the overall operation. These meetings permitted management trainees to learn what the functional leaders did and what they, in turn, would expect from the trainees. During the first few weeks a new recruit also spent time working with line operators on the factory floor. (People with equipment and mechanical skills who operated the equipment or line were generally referred to at Procter & Gamble as technicians; the technicians on a particular process were called a work team.) The time spent working on the line varied considerably at plants: In some locations new managers spent several months as technicians; in other plants they spent only a day or two.

During the remainder of the first three to four months, new hires prepared for "final takeover" of their specific management positions. Trainees were allowed whatever time was necessary to learn their job, with the person the trainee was to replace as the teacher. When trainees felt ready to assume responsibility for their position they were allowed to manage the team for one or two weeks while the immediate supervisor observed. This was known as "interim takeover." At the conclusion of the "interim takeover," performance was reviewed by the manager. A former management trainee described his initial training experience:

> The first month was peripheral training in employee relations, contract administration, safety, and benefits. This was done as a group of four in the conference room. The second and third months were concentrated on learning the area. This process was through two steps. One was training by my soon-to-be subordinates. In this, I worked side-by-side with the operators in the area. The second part was ongoing training by the person whom I would replace as the supervisor. This person was my primary trainer.
>
> The main emphasis was getting to know the work performance of each individual that I would be supervising. In addition, my manager provided some training on issues such as budgets. In the second to the last week of the three-month period, I was put on as an interim supervisor for one week. This was a takeover period, where the supervisor I was replacing stepped back and tried not to spend any time on the shop floor, but was available to me as a resource. At the end of this week, I then had an opportunity to say what additional training I felt I needed. The following week was devoted to this additional training. After that, I was left on my own as the new supervisor in the area.

After the "takeover" process, which typically lasted three to six weeks (the time varied from plant to plant), new managers underwent a "final qualification" to satisfy the plant man-

agement team's concern that they possessed the needed skills and qualities to be a manager. The new managers met with their department manager, staff technical engineers, operations manager, and plant manager; at each level they had to be prepared for a battery of questions that covered a wide range of subjects including corporate structure, labor relations, and technical processes within the plant.

After "final qualification" new managers continued training in those areas they or their supervisors thought needed more work. This training, both on-the-job people skills training and more formalized classroom instruction, was tailored to each manager. Although each plant typically maintained syllabuses for courses in the training program, they were tools more than absolutes and were used to maintain continuity between instructors. Course instructors decided what issues to emphasize based on their knowledge of the strengths of the students enrolled. At the completion of the "join up process" (typically two to four years), the individual was considered fully functional in supervisory and general technical abilities.

Each time a manager was assigned to a new position the "takeover" process was repeated, and a "takeover plan" was developed. A manufacturing manager described the process as follows:

> At the beginning of each assignment, including the first, we are given a training manual which describes the area the assignment is in, the objectives of the job, the knowledge required to perform it, the projects to be completed, the resources available, and the qualification requirements. We then get together with our new manager and put together a training package or process to achieve the requirements of the new job. Finally, we have to review a "take over" plan with the plant manager.

People developed and trained as manufacturing managers could generally expect promotion to the operation manager level in 5 to 10 years and to plant manager in 7 to 15 years.

A few courses were required of all new managers. One was the Management Training Exercise (MATRIX), a three-day simulation program during which the trainee assumed the plant manager position for two weeks. As the simulation progressed, the trainee made decisions on a range of problems and contended with the results of those decisions; the trainee also had to solve problems within the limitations of corporate policies and procedures. The objective of the exercise was to broaden the participants' views beyond what they had been exposed to in their first assignment. Other training courses, such as fire safety, union contracts, cost analysis, and plant technical training, were by necessity more formal in content. Beyond these required core courses the rest of the training curriculum was composed of specialized units taken as needed by manufacturing personnel.

Because of P&G's promotion from within policy, training was crucial for long-term corporate survival. Managers had therefore taken it upon themselves to be trainers. A prime example of the cultural commitment among managers to training was the System Management Development Workshop (SMDW). Set up to train younger managers for the demands of mid-level management, this program was developed by a group of mid-level managers who felt the training should not be left to outside consultants. SMDW was a one-week program run four to six times a year.

"On Boarding" Task Force

In 1983 the vice president of manufacturing commissioned a task force to study ways plants could improve how they brought new manufacturing managers "on board." The task force was prompted by two pressing concerns: 1) What should the role of the manager be within the innovative, high-technology factories of the near future? and 2) What needed to be done to improve manager-technician relations in the factory environment? The task force, comprising key managers form each corporate division, sent surveys to each plant asking for an evaluation of the strengths and weaknesses of its "on boarding" procedure. In addition, six plants were selected for on-site studies for in-depth analysis and interviews.

Based on its survey analysis, the task force made seven recommendations to the plant managers:

1. New hires for manufacturing management positions needed a sufficient time for hands-on technical training on the line. By working as an operator, reporting to a team manager, and rotating on shifts, the new hire would gain a better appreciation of the environment and culture of the plant.
2. Plants needed to spend more training time on negotiation and people skills. This area should be a prime factor in the final qualification of manufacturing managers. The consensus of the task force was that while technical training was strong across all plants, the social system training needed to be strengthened.
3. Plants needed a formal plan for training after the initial takeover. Some managers tended to view the "final qualification" as the end of training.
4. Line technician involvement needed to be increased in the "on boarding" process. Technicians should be included in interviews with potential manufacturing management recruits; they should conduct plant tours for the recruits; and their feedback on the performance of new manufacturing managers should be solicited.
5. Managers needed to lay out expectations for new hires as clearly and quickly as possible. They needed to let the new managers know what was expected from them in 6 months, 12 months, etc, right on through completion, in 2 to 4 years, of an extended "join up process."

6. Each division needed a trainer competence program. Selecting trainers with both technical and interpersonal competence was critical to program success.
7. Plants needed periodic self-audits to evaluate their "on boarding" process. At one time, the corporate personnel had a training audit team that monitored each plant's training effort. During the 1970s this was phased out; it was felt the centralized audit was inappropriate for the wide variety of work systems throughout the plants. The task force felt that the audit was important but should originate within each plant.

The recommendations were released in the summer of 1983, and by 1984 plants were working within these guidelines to strengthen their programs. In describing the future direction of manufacturing management at Procter & Gamble, the current corporate industrial relations manager noted that the role of the manager was moving toward being a visionary leader capable of team development and personal coaching versus that of an authoritarian decision-maker. While line personnel would be taking on more responsibility for functional technical knowledge and team leadership, he believed it would be critical for new manufacturing team developers to have sharp technical skills to maintain credibility with the line personnel, and to understand the process being managed. In addition, he noted, future managers had to bring energy, drive, and leadership to their management position. To find individuals with such qualities, Procter & Gamble increasingly turned toward hiring solely off college campuses for its entry-level manufacturing management positions.

The Development of Air Force Maintenance Officers

Janice A. Klein
Kevin Davis

Because of its enormous size, the Armed Forces have had many years' experience in developing a diverse mix of operations managers; in fact the military itself has long been a model for organizational design and structure. Recently, corporate America has shown renewed interest in recruiting former military officers for management positions (see Exhibit 1). This reading will describe the selection and training procedures used by the military to develop their managers. It concentrates on only one service, the Air Force, and one management job, the maintenance officer. Of the approximately 2.1 million men and women in the United States Armed Forces in 1985, approximately 600,000 were in the Air Force. Like the Army and Navy, the Air Force was composed of two different rank structures—officers and enlisted.

One-sixth of the Air Force uniformed personnel were officers. To enter the officer rank structure an individual had to possess a college degree (before 1975 outstanding enlisted personnel could be commissioned without a degree) and be commissioned. To be commissioned, recruits had to complete one of three officer training programs—United States Air Force Academy (USAFA), Reserve Officer Training Corps (ROTC), or Officer Training

School (OTS). An individual from the enlisted ranks could become an officer by fulfilling two requirements or, as was far more common, spend an entire career in the enlisted ranks.

The enlisted structure required only that recruits be at least 17 years old and receive a satisfactory score on an Armed Forces Qualifications Test, a basic aptitude test covering word knowledge, arithmetic reasoning, and space perception. The responsibilities of the enlisted ranks ranged from those of a new recruit responsible only for learning a skill to those of sergeants, the Air Force's noncommissioned officers (NCOs), who handled almost all of the first-line supervisory duties. All enlisted personnel, from the first-day recruit to the 30-year chief master sergeant, were outranked by any commissioned officer. Although roughly analogous to line technicians or nonexempt workers in industry, the parallel was somewhat tenuous: Enlisted workers were not permitted to strike or join a union and were required by law to serve for the specific period for which they enlisted. Since they had voluntarily made a long-term commitment, most considered themselves professionals.

Officer Recruiting

The first step in becoming an Air Force maintenance officer was to be commissioned, and each of the commissioning programs recruited differently. USAFA had a network of liaison officers who contacted high school juniors and seniors at school career information events and through counselors. All members of Con-

EXHIBIT 1

Fall Out: Military Officers Resign to Join Ranks of Business World*

By Helen Rogan
Staff Reporter of *The Wall Street Journal*

When E. Michael Moone was manager of sales recruiting at Procter &
Gamble Co. in the early 1970s, he hired a number of young ex-Army
officers. Now that he's president of the Beringer Winery in California,
he's still hiring them. "I prefer officers to M.B.A.s," he says. "M.B.A.s
may be smart, but they can't park their bikes straight."

Edward Haynes, an executive recruiter in Richardson, Texas, feels the
same way. Until last year he was employment manager for the equipment
group for Texas Instruments Corp., where he ignored pressure to hire
more M.B.A.s. "I'd just hire a captain with an accounting degree," he
says. "The M.B.A.s didn't have the sticking power, and you had to wait
seven to eight years for them to grow up."

There have always been retired officers working at second careers in the
defense industry. And a small number of luminaries from the upper
echelons of the military have found prestigious jobs after retiring:
Alexander Haig at United Technologies Corp., for example, or Peter
Dawkins at Shearson Lehman Brothers Inc.

But these days young officers with five to 10 years of experience are
much in demand in all sorts of industries, particularly if they have a West
Point background. The 1980 edition of the West Point Register of
Graduates lists 16 members of the class of '71 as having joined P&G,
along with 16 members of the class of '72. Ten men of the class of '71
joined Ford Motor Co., as did seven from the class of '72. Names like
Corning Glass Works, Honeywell Inc., Johnson & Johnson crop up
regularly.

The New Army

The Army that these men leave is very different from the way it once
was: its officers are primed to succeed in business. A general gives his
aides "In Search of Excellence" to read. Commanders have to learn to
balance their budgets and build a team. Recruits are called trainees. It was
the Army, after all, that invented management by objectives.

Yet still, the decision to leave is a difficult one, a careful weighing of
patriotism against pay. Although military pride has bounced back from the
low days of the Vietnam War, a married captain with eight years service
takes home around $3,000 a month in pay and allowances before taxes—
less than many of them could make in the business world.

Arlene Crook, the wife of a Fort Bragg army pilot, looks at her rocking

EXHIBIT 1   (continued)

chair and says, "It's only 10 years old but it looks terrible." That's because her husband, a 33-year-old captain, has moved eight times in 10 years. She says his friends are talking about resumes. Somebody's bought "The Hundred Best Companies to Work For." They are thinking about stability and, says Ms. Crook, "more of a family life." And the ambitious captains know that however well they perform at work, they will still mark time until they get promoted alongside less ambitious colleagues.

And so they leave, pushed by the drawbacks of military life and pulled by the network of soldiers who came before them.

Thomas C. Barren (West Point class of '65) is vice president of the West Point Society of New York, an alumni group. Mr. Barren, a management consultant at Ayers, Whitmore & Co., has the characteristics of many West Point graduates: a neat desk, a fit and youthful look, a methodical way of speaking and a class ring.

Mr. Barren says that New York members of the society are committed to helping other West Pointers. Through its Career Advisory Board, the society helps alumni with resumes, interviews, advice and, of course, contacts. Says Mr. Barren, "I've tried to help any and all graduates in my class as much as I could."

The graduates are a mixed breed. Some of them are entrepreneurs, the highfliers who become impatient with the service despite their success and good prospects. They become venture capitalists, investment bankers, management consultants and headhunters. They're also found at planning and development departments in corporations.

Daniel M. Collier Jr. is one. He left the Army in 1971, after service in Vietnam and Africa, to become an investment banker at Carl Marks & Co., in New York. "The sun was setting on the U.S. Army world-wide and rising on the U.S. banking system world-wide," says Mr. Collier, who remains a colonel in the Reserves and spends two weeks a year at the Pentagon. He relishes the fact that any particular employer "is not the only game in town."

The less exceptional soldiers fit smoothly into entry-level positions in the big corporations. John Begley is responsible for placing many of them. A former soldier (class of '50), he runs Career Seminars, a San Francisco-based placement agency that specializes in finding jobs for junior military officers. Mr. Begley says he's placed about 7,000 of them in business.

There's a good reason why soldiers are so much at home in these companies. Raymond J. Klemmer, who spent nine years in the Air Force and is now a partner at the executive recruiting firm Nordeman Grimm Inc. In New York, says it's the "comfort factor" in large corporations that makes them appealing to soldiers. "It's like the womb," he says.

Mr. Klemmer adds that to survive in the military "you learn to go along." That training, he says, "helps a lot in large corporations. So I advise my classmates to use what they have been trained to do—go to the large companies and operate by the book."

EXHIBIT 1   (continued)

Thomas-Shaw, corporate manager of professional employment at Crown Zellerbach Corp., a San Francisco-based forest-products concern, agrees that military men come well-prepared for business. "Most are very mature," he says, "because the military will give you responsibility faster than civilian organizations. And they know how to work in close quarters, where you depend on the people around you—which happens in a corporate environment."

Know the Enemy

The entrepreneurs are very conscious of the value of their military experience. They talk about their work in terms of tactics and strategy. James R. Gardner (class of '66), director of corporate planning at Pfizer Inc., says that he often thinks "in military constructions and logic, especially in planning." Mr. Gardner likens knowing your enemy with competitive intelligence.

Other entrepreneurs talk about the importance of setting an example, communicating objectives and motivating troops as basic tenets of military thinking that can be directly translated into the business world. William L. Hauser (class of '51) thinks perseverance is "the single biggest asset" he brought to Pfizer in 1979 after 25 years in the Army. "If you are given a job to do," he says, "whatever obstacles are in your way, you keep at it until you overcome them." Mr. Hauser is director of career development.

Still, there are likely to be initial problems, particularly with officers whose Army days have been cloistered in out-of-the-way places. Mr. Begley describes them as showing up for their first interview in a "nice plaid jacket." The technical types may sport white socks. All are likely to start out with uncommonly short hair and a tendency to say "Yessir" or "Nosir."

Even the highfliers find it hard at first. It can be more basic than simply adjusting to the profit motive. They have to learn to commute, to pick the right tie. Some mention the need to learn where the real power is in an environment where everyone wears suits and uses first names.

Mr. Gardner talks about trying to keep a balance in his life between ambition and overwork, between home and career. He says that "When things get hectic, I tend to respond, and I say, 'Boy, this is great, this is as close to combat as I'll ever get in this job.'"

Still, says Mr. Gardner, "Civilian life can be a bit of a comedown. You work in a large organization in a small office, with no life and death decisions or experiences. So you fall back on your prior experiences for strength. And you take everything with studied detachment—because the other stuff is for real—but this, however much you enjoy it," and he pauses, "is a way of earning a living."

gress control an equal number of USAFA appointments, and people interested in the program were advised to contact their congressional representatives for nomination. (There were also appointments available through presidential or vice presidential nomination.) Congressional nominations usually involved interviews by panels of local leaders and educators set up by the congressional representatives. From this pool of nominees, USAFA admissions personnel made the final selection.

USAFA received 12,000 to 16,000 applications each year for 1,450 slots. Of those chosen to attend, typically two-thirds were in the top fifth of their class academically, and virtually all recruits had participated in student government, athletics and/or other extracurricular activities. USAFA offered all who were accepted a full scholarship plus room and board and a salary (approximately $500 a month in 1985) from which recruits paid for uniforms, books, and personal needs. A USAFA graduate then incurred a five-year Air Force commitment. Fewer than 15 percent of the officers went this route.

Reserve Officer Training Corps, conducted at 150 colleges across the United States, was composed of both scholarship and nonscholarship students. ROTC detachments (units) recruited individuals enrolled at their college, looking for the same general academic and extracurricular background as did USAFA. The Air Force received more than 20,000 applications each year for approximately 1500 scholarships. The scholarships offered two to four years of education, including tuition, fees, books, and $100 a month for living expenses. In 1985, approximately 60 percent of those enrolled in ROTC were on scholarships. After receiving a commission, a ROTC officer served a four-year commitment.

Officer Training School, a three-month program for college graduates, offered the Air Force a way to fill unforeseen shortages quickly in any career field. In 1985, more than 3,000 trainees, or approximately 45 percent of the total number commissioned, received a commission through OTS.

Because of the demands of increasingly complex military hardware, Air Force policymakers tightened recruiting standards and increased the percentage of officers coming from scholarship programs between 1970 and 1985. By 1985, almost 40 percent of newly commissioned officers were from scholarship programs, in comparison to fewer than 25 percent in 1970. The ability of the Air Force to offer more scholarships, in light of increased tuition costs, had significantly improved the average academic quality of recruits.

More sophisticated technology also led recruiters to pursue engineers and other technical majors to the near exclusion of nontechnical majors. The Air Force sought people for specific vacancies before they began school and matched academic programs with specific career fields. Similarly, OTS personnel were selected on the basis of available Air Force jobs, nearly all of which were technical. According to Colonel Scivoletto, professor of aerospace science at MIT, the quota of ROTC scholarships to be awarded to engineering majors in 1985 was 80 percent, with an additional 18 percent for other technical majors (e.g., mathematics and physics). Those receiving scholarships were under contract to complete a specific major. People enrolled in ROTC but not on a scholarship could not receive a commission without an identified slot and, according to ROTC personnel at MIT, "slots for nontechnical degrees are almost nonexistent." Although USAFA gave far more freedom in choosing a major, all USAFA cadets were required to take extensive engineering and mathematics courses. An engineering or technical degree was listed as being a preferred background for maintenance officers.

Officer Training

The philosophy of all three commissioning programs was that one cannot lead until one has learned to follow. Therefore, initial training cast recruits in the role of enlisted personnel. Training also stressed military indoctrination or socialization, instilling recruits with the institutional values of responsibility, cooperation, teamwork, and discipline.

Before being allowed to develop their own style, recruits were exposed to numerous leadership styles. They were then assigned first-line supervisory duties with the responsibility for training several new recruits. Recruits were then assigned to the second level, where they were responsible for several first-line trainers. Finally, they were placed in charge of the entire detachment for several months. Although there were commissioned officers on staff at each unit, the detachment was basically run by recruits.

In ROTC the recruit spent the first two college years in the General Military course (GMC), attending classes on military traditions, history, organization structures, and operations. Those in GMC were considered the "enlisted ranks" of their ROTC detachment and were taught to conform strictly to dress and uniform standards. At the end of the second year of the GMC, cadets were evaluated by their commanding officers in the detachment. The officer in charge of the ROTC detachment, the professor of aerospace science, had the final decision on whether GMC members could enter the final two years of ROTC, the Professional Officer Course (POC). The decision was based primarily on a subjective evaluation of the ability of the recruit to function within a military environment. During POC cadets ran the detachment using a rank structure and organization similar to that of the Air Force. They were also responsible for training personnel in the GMC.

USAFA and OTS training formats differed slightly from ROTC's—primarily in intensity and time—but the philosophy of training and the training itself were nearly identical. At the USAFA the recruits lived in a completely military environment. OTS, on the other hand, had only a 12-week period to complete its cycle of indoctrination and leadership/followership training. To compensate, the OTS program was conducted 10 hours a day with the time spent exclusively on military training.

Once in the Air Force, officers were expected to continue professional training in residence at one of many professional military education (PME) facilities and/or through correspondence courses from these schools. Failure to complete these voluntary training programs suggested a less than professional attitude and often resulted in nonpromotion.

The first of the PME courses was Squadron Officer School (SOS). This nine-week exercise in group dynamics was generally completed some time in the first six years of service. Each year thousands of young officers participated in this course, taken in residence at Maxwell Air Force Base in Alabama. The course's objective was for the officers to learn how to build and be constructive members of a team. On arrival at Maxwell, the officers were put into groups of 12 to 14 members. Over the next nine weeks the groups (approximately 64 groups in each SOS class) competed in academics (e.g., military history, logistics and supply systems, political science), athletics, and other team activities. During the program groups learned how their performance compared with the other groups. At the end of the program groups were rank ordered. Because virtually all officers in the program held the same rank, SOS was effective in bringing out nonauthoritarian leadership styles and underlined the need to lead without resorting to rank. Officers also spent many hours giving speeches and writing short papers in an effort to improve those skills.

The Air Force also expected officers to

complete two other PME courses: Air Command and Staff College (ACSC) and Air War College (AWC). Both concentrated on the leadership qualities and techniques needed to lead or manage at senior levels of command. ACSC was typically completed at eight to ten years of service while AWC was completed at 12 to 15 years' service. Since each course lasted over nine months, the majority of officers took them by correspondence, with fewer than 10 percent of the officer corps completing the courses in residence at Maxwell. The correspondence version of the PME courses concentrated on the academic side of professional development and included readings on military history, political issues, and writing and speaking skills.

Training Specific to Maintenance Personnel

After receiving a commission from one of the three sources, all potential maintenance officers underwent training at Chanute Air Force Base in Illinois for 22 weeks. According to course administrators, the course attempted to prepare recruits for general management of maintenance resources. The course of study was broken into 11 modules, each lasting two weeks. Courses included an introduction to Air Force maintenance organizational structures, principles of management, use of management information systems, logistics systems, and numerous courses on aircraft systems. The final module was an applied maintenance management simulation where the trainees tested their decision-making skills.

Because course graduates went on to become maintenance officers for different aircraft systems (e.g., fighter, bomber, or transport aircraft), the Chanute program was tailored to provide qualifications deemed necessary for all aircraft types. Administrators from Chanute met yearly with maintenance representatives from the major Air Force commands (MAJCOMs),[1] who then set training objectives for Chanute. The sum of these objectives, referred to as the "contract," was translated into course outlines by program administrators. Tests given to the trainees throughout the program determined how well the terms of the "contract" were being met. In addition, the school sent out questionnaires each year to air base "customers" (similar to the plant level in industry) asking maintenance commanders to evaluate the training level of their new officers and requesting recommendations on how Chanute could better serve air base aircraft maintenance training needs. Graduates were also sent questionnaires requesting a rating of the applicability of their training at Chanute and recommendations of areas where more training might have been useful.

After arriving at an assigned air base, the new maintenance officer generally spent two or three weeks in orientation. During this time the officer visited all the agencies on the base involved with aircraft maintenance, learning the responsibilities, locations, and personnel of each. (See Exhibit 2 for the agencies contained in a "standard" maintenance organization.)

The next step generally was a two- to five-month training period by enlisted technicians. This training took place on the flight line so that the officer gained a sense of the process flow by watching the different maintenance branches respond to aircraft needs. Potential maintenance officers were charged with integrating and coordinating the work of the dif-

---

[1]The major commands, or MAJCOMs, were roughly analogous to corporate divisions. A MAJCOM was assigned a major segment of the USAF mission. Strategic Air Command (SAC), for example, was the MAJCOM responsible for intercontinental or strategic air power including intercontinental missiles, bombers, and reconnaisance.

EXHIBIT 2   A Typical Maintenance Organization Within a MAJCOM

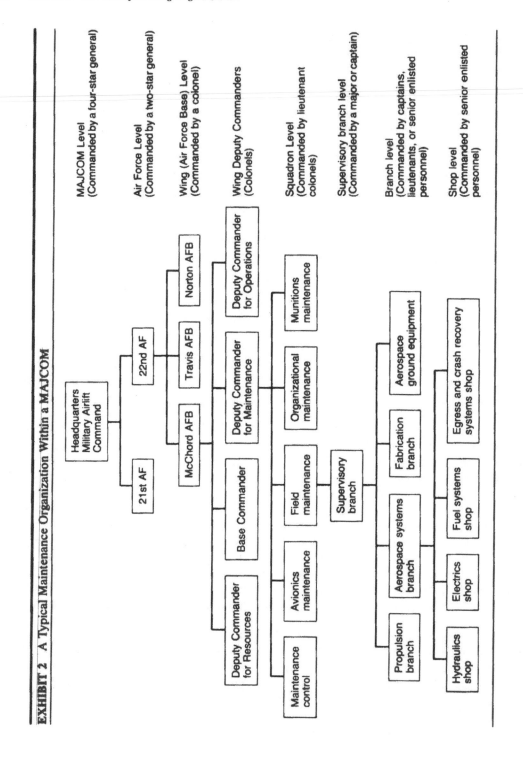

ferent maintenance branches (e.g., hydraulics, electrics, fuel systems, etc.).

At the end of the flight line training, maintenance officers were given command of a maintenance branch. As a branch chief, the officer generally supervised more than 100 enlisted maintenance technicians. (Within a branch there were typically several shops, each with 40 to 50 people and an enlisted supervisor who reported to the branch chief.) Since each branch of maintenance directly handled only a small portion of the total maintenance effort (e.g., the propulsion branch handled only aircraft engine requirements), maintenance officers typically were moved yearly to command a different branch to broaden their knowledge of maintenance.

After three or four years at an air base, maintenance officers generally received an assignment at their MAJCOM headquarters, which allowed the officer to assimilate a broader sense of "corporate" goals. Following several years at the MAJCOM assignment the officer could expect assignment to a new base as a maintenance supervisor, to be responsible for the production of three to five branches and the supervision of 350 to 600 people.

Very little, if any, time was devoted to academic training at the individual bases; instead, the new officers were put right on the line to learn by doing while under supervision. The rationale was that the technicians knew the nuts and bolts of maintenance, and the job of the maintenance officer was managing people. According to a branch chief at McChord Air Force Base in Washington:

> Interpersonal skills are critical. While the maintenance officer needs some knowledge about the maintenance process, it is generally recognized that the technicians are the systems experts. Maintenance officers are supposed to be the people experts. The Air Force has moved away from authoritarian management. There is now clear

recognition that a leader must make his people feel that they are a part of the mission. We used to ask if a manager was "mission-oriented." Now we agree that there is no mission if the people are not motivated. Younger airmen look to their leaders for sincere concern. If you don't take care of one of your people, word-of-mouth will kill you. You couldn't begin to do the job without the support of your first-line supervisors and technicians.

After 12 to 15 years, maintenance officers could be given a squadron command. However, at the squadron level and above maintenance officers had to compete with pilots and navigators for command positions. Not unlike industry, where finance and marketing dominated the command roles, pilots dominated decision-making jobs in the Air Force. In 1984, while pilots constituted fewer than 25 percent of the officer corps, approximately 75 percent of all Air Force generals were pilots.

Training of Enlisted Personnel

In contrast to officer training, enlisted force training was system-specific. Initial enlisted training occurred at one of several training bases, depending on the maintenance specialty of the recruit. Recruits learned the basics of maintenance, from how to read technical orders[2] to how to maintain a safe work environment, and were then assigned to a base to complete on-the-job training.

On base, enlisted personnel began as apprentices, being trained for approximately one year, in the numerous "tasks" specific to their maintenance specialty. This training, conducted by a qualified specialist, was supplemented by manuals and monthly tests covering academic portions, which were called

---

[2]Technical orders were books containing all the information on a particular maintenance system including troubleshooting procedures.

Career Development Courses (CDCs) for their maintenance specialty. After enlisted personnel had passed CDC exams, been signed-off by their trainer as being able to accomplish all the necessary "tasks," and approved by their supervisors, they were promoted to the skilled level. At this level, work did not need to be cross-checked and they might be assigned as a trainer.

After completing more CDCs, and satisfying those in their chain of command as to their competence, enlisted personnel could be promoted to the supervisor level. As supervisors, they were responsible for varying numbers of skilled-level people. Finally, by proving to be competent as supervisors, enlisted personnel could be promoted to the management level, eligible to serve as branch chiefs or as staff advisors to the deputy commander for maintenance. In this capacity, however, noncommissioned officers could supervise only other enlisted personnel.

Enlisted personnel could not progress above branch level supervisor—the second level of management—without being commissioned as officers through one of several programs; for example, the Airman Scholarship and Commissioning Program (ASCP) and the Airman Early-Release Commissioning Program (AECP). In 1984, out of the 85 enlisted personnel who applied for officer training programs, 46 were accepted. According to a maintenance NCO at McChord: "Whenever we feel an enlisted recruit has officer potential, we counsel the individual on available programs. The Air Force encourages the enlisted to become officers and clearly advertises the means for them to be accepted into officer programs."

Implementing Workplace Change

Traditional manufacturing organizations, like those introduced in the previous module, have limited ability to adapt to new competitive environments quickly. Now that competition is truly worldwide, any manufacturing organization that hopes to compete in this global marketplace must recognize that its manufacturing facilities are primary competitive weapons. Not only must these organizations introduce change, their operations must be pushed for *continual improvement*; maintaining the status quo or only meeting current customer demand is no longer sufficient. The goal must be never-ending revitalization, and the entire manufacturing process should be constantly reevaluated for ongoing improvement and learning.

This imperative of continual improvement demands a better understanding of how the people, machines, and materials interconnect to make the product (or service). Increasing the operations knowledge base is essential, not only so immediate problems can be fixed but so they won't recur. The previous module on the manufacturing operation investigated where knowledge typically resides in traditional operations; this module explores how that knowledge can be enhanced and better used.

LEVELS OF KNOWLEDGE

Managing knowledge continues its critical importance in implementing workplace change, particularly when the desired outcome of that change is continuous improvement. To improve an operation, operations managers need first to determine how much they know about the process; only then can they marshal their resources to learn more about that process. Thus, an operations manager needs to know not only the location of knowledge about the current product and/or process, he or she must identify the *level* of that knowledge. Only when this is understood is it possible to determine whether a contemplated change is an appropriate fit.

Knowledge about a product and/or process typically varies along a continuum from very crude to very sophisticated (Bohn, 1986; Jaikumar and Bohn, 1986):

0. Total ignorance—Cannot tell good product from bad.
1. Can tell good product from bad, but do not know why—Pure "art"; knowledge as to what makes a good product is in the head of an expert, who is unable to articulate process variables/characteristics.
2. Have tentative "feel"—Good output often seems to happen at random, but one can list possible relevant variables/characteristics that generate good output.
3. Can recognize which variables are more important than others—Patterns begin to emerge by noting correlation with good output, but experts still disagree on relevant variables/characteristics.
4. Able to measure variables—Beginning to move toward "science"; some variables can be articulated.
5. Able to control variables—A notion of "procedure" (scientific method) begins to emerge; can develop repeatable recipe which produces acceptable output with some degree of repeatability.
6. Understand contingencies—Understand direction and strength of effects of changes in control and environmental variables on output characteristics; procedure is documented and possess some knowledge of contingencies.
7. Able to control contingencies—Can develop precise quantitative model that relates environment and controls to output characteristics; automation becomes possible, with frequent human intervention.
8. Complete procedural knowledge—Process and environment are so well understood that one can head off any problems in advance, by monitoring environment and adjusting controls; fully automated, unattended operation is feasible.

In general, most managers would like to move up the continuum towards greater knowledge, but it is not clear that all aspects of an operation should strive to the highest levels; new products or processes often make higher levels of knowledge obsolete. Keeping a mix of people who can operate at different levels is important. Nonetheless, operations managers, once having determined the level of their operations, must decide where they would then like to move. The key is maintaining consistency between the level of knowledge and the style of management and production methods. For example, if management and the work force do not understand the process, then process controls, rigid time standards, or automation will not work very well.

Moving between levels, typically toward enhanced knowledge, requires introducing mechanisms to bring processes under better control, which thereby leads to improved operations performance. Quality, inventory, or information systems are generally viewed as prime vehicles for improved process controls. New technology or changes in organizational structures can also make way for increased learning. All three methods are explored in this module.

MODELS FOR CONTINUOUS IMPROVEMENT

Achieving continuous improvement in traditional manufacturing organizations requires breaking down functional barriers and working towards common goals; organizational energies must be focused on some mutually understood target. The target comes in various forms: An abundant number of "three-letter" programs (e.g., CIM, SPC, JIT) are available, each addressing a different segment of the operation. If designed and implemented correctly, each of these programs leads to more efficient and flexible operations. By itself, each is beneficial; but taken together, they have complementary and often synergetic effects. These programs or models can be organized around three themes—new technology, new systems, and new structures.

New Technology.   Models here include computer-aided design (CAD), computer-aided manufacturing (CAM), computer-aided engineering (CAE), computer-integrated manufacturing (CIM), flexible machining systems (FMS), and many more. The cases addressing these programs presume a basic understanding of the technology; the goal is to look at its impact on the workplace and organization.

New Systems.   Models in this group include total quality control (TQC), statistical process control (SPC), just-in-time (JIT) inventory, and group technology (GT). The cases in this group also assume the reader has been exposed to these systems; they apply that general understanding and examine what is needed to implement the programs.

New Structures.   Measurement systems, organizational structure, and job design are the models in this group. The cases dealing with them focus on introducing changes to existing structures in an effort to align the organization's human resource policies with its operational goals of greater flexibility and continuous improvement.

MANAGING THE CHANGE PROCESS

Change is not brought about by magic. An operations manager must plan and manage the change process. One key, as noted above, is to assure that the contemplated change is an appropriate fit with the current state of knowledge about the operation. Change also creates commotion within an operation: Resources must be gathered, training conducted, and contingency plans put in place.

There are emotional as well as physical impacts of change. When introducing new technology, systems, or structures, product flow and workplace layout are often altered; job descriptions and daily tasks are affected. Knowledge invariably shifts, which leads to power shifts between functions or across layers of the organization. Each model for change has its unique affects and requires specific expertise to implement, but as will become apparent by the end of this module, there are generic implementation issues that operations managers encounter regardless of the targeted change.

One such issue is resistance to change. Operations managers must always be alert to the causes of resistance to change; what looks like foot dragging may be something deeper. For example, with the growing crisis in workplace literacy, an apparent unwillingness to change may reflect an underlying inability to understand the basics. An estimated 45 million adults holding jobs today are either functional or marginal illiterates (Goddard, 1987), that is, unable to read, write, calculate, or solve problems at a level of coping with even the simplest tasks. Clearly, introducing new technology, systems, and structures requires greater skills, particularly the ability to handle statistics and to be computer literate. One study found that statistics and computers were the two areas that middle managers and first-line supervisors felt most ill-equipped to handle (Bittle & Ramsey, 1982); such fears are undoubtedly greater within the nonexempt work force.

OVERVIEW OF MODULE

The module begins with a background reading on managing change: "The Effective Management of Organizational Change." The remaining readings and cases examine individual workplace changes in each of three major categories—technology, systems, and structures. After several types of changes have been investigated, the impact of multiple changes within the same operation will be explored.

New Technology. This group of cases focuses on the technical and social factors associated with introducing new equipment, processes, and technology (EPT), ranging from process changes, to new stand-alone equipment, to fully computer-integrated manufacturing (CIM) operations. The *Owens-Illinois: 10 Quad* case series compares the start-up of a new machine on a new product application in two facilities: One was extremely successful, the other encountered numerous difficulties. Technological advancements opened the door at *AT&T Communications: Connecticut Custom Services District* for the remote testing of long-distance telephone lines. Rather than simply changing the equipment, management chose to relocate employees to a new site to "break old habits." The case, therefore, describes the site selection, work station design, relocation processes, and technical and social difficulties encountered. At *Allen-Bradley,* management opted to simultaneously design a new product and process. The outcome was a highly successful CIM pilot. While the (A) case traces the chronology of the design and start-up of the CIM line, the (B) case investigates the skill requirements of the first-line manager of such an operation. Two readings, "Manufacturing by Design" and "Toward the Factory of the Future," provide background on the design and implementation processes for new technology.

New Systems.   This group of cases looks at the introduction of new quality (statistical process control), inventory (just-in-time), and product/ process groupings (group technology and modular manufacturing) programs. In the *Detroit Tool Industries (A)* case, the president of a small machine tool company looks to statistical process control (SPC) as the solution to all his operational ills. In addition to highlighting the importance of understanding the current operation before trying to introduce change, the case examines how various organizational levels and functions view change. At *Consolidated Transformer Company (A),* process controls were successfully implemented but a change in raw materials has led to a high level of in-process rejects, even though the product still meets customer specifications. The manager must decide whether to change internal standards. A reading on "Constructing and Using Process Control Charts for Statistical Process Control" is provided for background information on SPC.

Issues concerning interdisciplinary coordination needs, compensation, and measurement systems surface in the introduction of just-in-time (JIT) inventory systems at *General Electric Thermocouple (A).* Similar issues arise at *EG&G Sealol (A)* in the introduction of group technology. (Two articles, "What's Your Excuse for Not Using JIT?" and "Group Technology and Productivity," provide background information on implementation issues surrounding the use of JIT and group technology.) Finally, the question of dedicating production lines to particular products is investigated in *Rio Bravo Electricos, General Motors Corp.*

New Structures.   This group of cases examines organizational needs in refocusing manufacturing strategic direction, including measurement systems, job design, and supervisory/managerial resistance to change. At *Dulaney Sound Systems,* a department manager must decide which parts, if any, to outsource in the face of capacity shortfalls. His task is complicated, however, by ambiguous signals concerning changes in key measurements.

Andreas Stihl Maschinenfabrik had the opportunity to experiment with alternative work structure within the same facility. The case presents preliminary performance results in productivity, quality, cost, and employee satisfaction. Finally, resistance to change, particularly at the first-line supervisory position, is explored in *Century Paper Corporation* during job redesign. The readings "Why Supervisors Resist Employee Involvement" and "Traditional versus New Work Systems Supervision: Is There a Difference?" summarize recent research on the subject.

The module ends by looking at the impact of multiple changes within an operation. In *Sedalia Revisited,* JIT and SPC play havoc with a high commitment work system (described in *Sedalia Engine Plant (A)*); the process controls are viewed as counter to the autonomy granted to the work force under the human resource policies. Finally, a new plant manager at *The Council Bluffs Plant* is faced with introducing a new product while transforming a "traditional" operation into a world-class manufacturing (WCM) model plant. WCM, as defined by corporate management, includes new technology, total quality systems (including SPC), JIT, and employee involvement.

REFERENCES

Bittle, Lester, R., and Ramsey, Jackson, E., "The Limited, Traditional World of Supervisors," *Harvard Business Review,* July–August 1982, pp. 26–37.

Bohn, Roger E., "An Informal Note on Knowledge and How to Use It," Harvard Business School Case #9-686–132, 1986.

Goddard, Robert W., "The Crisis in Workplace Literacy," *Personnel Journal,* December 1987, pp. 73–81.

Jaikumar, Ramchandran, and Bohn, Roger E., "The Development of Intelligent Systems for Industrial Use: A Conceptual Framework," *Research on Technological Innovation, Management and Policy,* Volume 3, 1986.

The Effective Management of Organizational Change

David A. Nadler

Consistently effective organizations are those that appropriately position themselves in their environment. They have determined what they need to do differently in order to take advantage of shifts in the markets, industries, technologies, or localities in which they operate. This repositioning may be forced by sudden or unexpected environmental shifts; it may be planned in anticipation of environmental moves; or it may be calculated to precipitate movement in the environment. The classic cases of failure (such as Penn Central, W. T. Grant, or Braniff) are organizations that were not able to reposition themselves in the face of environmental change.

Repositioning an organization involves modifying the way it functions or "does business." Strategies, formal structures, processes, cultures, operating styles, and people all may have to be changed. Thus a critical issue for organizations operating in dynamic and uncertain environments is the management of change.

In recent years, the management of change in organizations has become a more critical concern. Organizations face tremendous uncertainty and instability in the current environment. Technological change has been accelerating; deregulation has disrupted market patterns in traditionally stable industries such as communications, banking, and transportation; multinational competition is increasingly

important; and a host of destabilizing general macroeconomic and political forces are at work. In addition, organizations face challenges as a result of their own success: greater size, scope, and complexity also necessitate change.

Organizations have responded to these challenges with new business strategies, new ways of structuring the work, new technologies, the addition of new elements (through acquisition or merger), the casting off of elements (divestitures or splits), new cultures or operating styles, relocation to new settings, or the appointment of new management teams. All of these responses raise questions of change management. Experience has indicated that the effectiveness of any response depends as much on *how* the change is implemented as on *what* it is. The task of implementing new aspects or elements of organization is fundamentally a problem of change management.

This chapter presents an approach to change management based on both systematic research (Nadler 1981) and observation, over a number of years, of managers who have been responsible for implementing change. Because it is critical to have some view of how an organization functions before considering how to change it, the chapter begins with a brief outline of a model or approach for thinking about organizations. Next we present a way of thinking about changes and discuss some classic patterns of change management that have characterized large U.S.-based organizations. Then the issues in managing change are considered, as well as the techniques that effective change managers seem to

Reprinted with permission from J. Lorsch, ed., *Handbook of Organizational Behavior* (Englewood Cliffs, N.J.: Prentice-Hall, 1987), pp. 358–369.

employ. The chapter concludes by focusing on the specific case of managing change to uncertain future states. The objective is to provide a framework for thinking about change as well as a structure for planning effective management of significant organizational change.

A View of Organizations

There are many different ways of thinking about organizations and the patterns of behavior that occur within them. During the past three decades, there has emerged a view of organizations as complex open social systems (Katz and Kahn 1966), mechanisms that take input from the larger environment and subject it to various transformation processes, which result in output.

As systems, organizations are seen as composed of interdependent parts. Change in one element of the system will result in changes in other parts of the system. Similarly, organizations have the property of equilibrium: the system will generate energy to move toward a state of balance. Finally, as open systems, organizations need to maintain favorable transactions of input and output with the environment in order to survive.

While the systems perspective is useful, systems theory by itself may be too abstract a concept to be useful as a managerial tool. Thus, various organizational theorists have attempted to develop more pragmatic theories or models based on the system paradigm. The particular approach used here, known as "congruence model of organizational behavior" (Nadler and Tushman 1977, 1979, 1982), is based on the general systems model. In this framework, the major inputs to the system of organization behavior are the environment, which provides constraints, demands, and opportunities; the resources available to the organization; the history of the organization; and, perhaps the most crucial, the organization's strategy. Strategy is the set of key decisions about the match of the organization's resources to the opportunities, constraints, and demands in the environment within the context of history.

The output of the system is, in general, the effectiveness of the organization's performance in meeting the goals of strategy. Specifically, the output includes *organizational performance*, as well as *group performance* and *individual behavior and affect*; the latter two, of course, contribute to organizational performance.

The basic framework thus views the organization as a mechanism that takes inputs (strategy and resources in the context of history and environment) and transforms them into outputs (patterns of individual, group, and organizational behavior).

The major focus of organizational analysis is therefore the transformation process. The model conceives of the organization as composed of four major components: (1) the *task* of the organization—the work to be done and its critical characteristics; (2) the *individuals* who are to perform organizational tasks; (3) the *formal organizational arrangements*, including various structures, processes, and systems designed to motivate and help individuals in the performance of organizational tasks; and (4) *informal organizational arrangements*, which characterize how an organization actually functions. These patterns of communication, power, influence, values, and norms usually evolve without planning or explicit formulation.

A critical question is the degree of consistency, congruence, or "fit" among those four components. For example, are the demands of the organizational task consistent with the skills and abilities of the individuals available to perform it? Are the rewards that the work provides well-suited to the needs and desires of the individuals? All in all, six possible relationships among components need to be examined for congruence (see Exhibit 1). The

EXHIBIT 1   A Congruence Model of Organizations

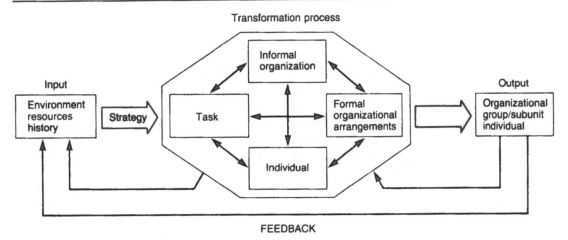

basic hypothesis of the model is that *organizations will be most effective when their major components are congruent with each other.* Problems of effectiveness due to management and organizational factors will stem from poor fit, or lack of congruence, among organizational components.

This is a contingency approach to organizations. The model recognizes that individuals, tasks, strategies, and environments may differ greatly from organization to organization, and hence different patterns of organization and management will be appropriate. There is no one best organization design or style of management or method of working.

A Change Scenario

The organization model provides us with a road map for thinking about how changes occur. A good example is the very dramatic changes that have occurred in the American

Telephone & Telegraph Company (AT&T). (See Nadler 1982 for more detail on this case.)

For many years, the AT&T organization exemplified good fit. In response to the environment, the company developed an effective strategy of maintaining its monopoly through high-quality and low-cost universal phone service. Because this environment was relatively stable and manageable, AT&T could define a stable and predictable set of tasks to performance—"the end-to-end provision of universal telephone service." A highly formalized organizational structure and procedures fit this task well. Over time the company recruited, developed, and rewarded individuals who functioned well in this structured environment and could identify with the value of high-quality service to the public. Finally, the informal organization, including the culture and leadership style, stressed high-quality service, adherence to rules and procedures, and compliance with the highly formal organizational structure. With an extremely high level of congruence, AT&T performed extremely well for many years.

What changed all this? As in most major organizational transitions, the impetus for change came from the outside. Starting in the late 1960s and through the 1970s, the regulatory environment posed increasingly serious challenges to the Bell monopoly. Concurrently, new and powerful competition arrived on the scene. Technological advances, such as satellite communications and private data-transmission networks, also challenged the Bell System's dominance in the communications market.

Over time AT&T began to modify its strategy to accept a redefinition of the monopoly and to become more market-driven, responsive, and, in some lines of business, competitively oriented. As the strategies of the organization changed, so did the nature of the task. Work became more unpredictable and faster paced, requiring greater coordination among the different units serving specific customer needs.

In recognition of this new set of strategic and work demands, AT&T began to reshape its formal organizational arrangements. Major reorganizations were undertaken in 1974, 1978, 1980, and again in 1982.

Events at AT&T fit the pattern of the classic change scenario shown in Exhibit 2. A series of catalytic events in the environment lead to an adjustment in strategy that requires a redefinition of the nature of the organization's work. Key organizational arrangements are then modified to permit implementation of the strategy and performance of the new tasks. This sequence of events is typical of large, complex organizations.

Several points about this change scenario are worth underlining. First, most change is brought about by events or forces outside the organization. As open systems, organizations try to adapt to significant environmental

EXHIBIT 2   The Classic "Change" Scenario

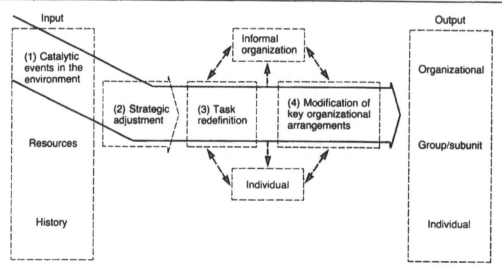

changes. While these changes may present new opportunities, they are more frequently seen as threats to the organization's existence or success.

Second, the pace of the sequence will depend on the situation. The first three steps shown in Exhibit 2 could occur within a few months, but required ten years in the case of AT&T.

Third, the sequence is usually less linear than suggested by Exhibit 2. In reality, change is very much tied up with the political life of the organization and, as such, is usually the result of various political processes: conflicts, bargains, agreements, and so on.

Finally, even if the organization successfully moves through these phases, it still faces several unresolved problems. The organizational arrangements will not have been fully adjusted for the new strategy and task, particularly the measures, rewards, and coordinating devices. In addition, the informal organization or culture may not fit well with the new task and organizational arrangements. And individuals may not fit the new task or organizational arrangements.

The Task of Implementing Change

Implementing a change involves moving an organization to some desired future state. It is useful to think of changes in terms of transitions (Exhibit 3). The effective management of change involves developing an understanding of the current state (A) and an image of a desired future state (B), and moving the organization from A through a transition period (C) to B (Beckhard and Harris 1977).

Major transitions usually occur in response to anticipation of organizational input (environmental or strategic shifts) or outputs (problems of performance). In terms of the congruence model, a change occurs when managers determine that the configuration of the components in the current state is not effective and the organization must be reshaped.

Organizational change is effectively managed when:

1. The organization is moved from the current state to the future state.
2. The functioning of the organization in the future state meets expectations, that is, it works as planned.
3. The transition is accomplished without undue cost to the organization.
4. The transition is accomplished without undue cost to individual organizational members.

Of course, not every organizational change can be expected to meet these criteria, but they do represent an appropriate planning target. The question is how to manage the implementation process so as to maximize the chances that the change will be effective.

EXHIBIT 3   Organizational Change as a Transition State

R. Beckhard and R. Harris, *Organizational Transitions* © 1977, Addison-Wesley, Reading, Massachusetts. Adapted material. Reprinted with permission.

Problems of Change

Effective change requires an understanding of both *what* the change should be and *how* it should be implemented. Clearly, managers must diagnose and understand organizational problems and causes before developing solutions. Otherwise problem solving becomes an expensive trial-and-error process.

We focus here on a second question—how the changes are implemented. Three types of problem seem to be encountered in some form whenever a significant organizational change is attempted. The first is *power*. Any organization is a political system made up of different individuals, groups, and coalitions competing for power (Tushman 1977; Salancik and Pfeffer 1977). Political behavior is thus a natural and expected feature of organizations, in both current and future states. In transition, however, these dynamics become even more intense as the old order is dismantled and a new order emerges. Any significant change poses the possibility of upsetting or modifying the balance of power among groups. The uncertainty associated with change creates ambiguity, which in turn tends to increase the probability of political activity (Thompson and Tuden 1959). Individuals and groups may take action based on their perception of how the change will affect their power in the organization. They will be concerned about how the conflict of the transition period will affect the balance of power in the future state and will maintain or improve their own position. Individuals and groups may also engage in political action for ideological reasons: the change may be inconsistent with their shared values or image of the organization (Pettigrew 1972).

The second problem is individual *anxiety*. Change involves moving from something known toward something unknown. Individuals naturally wonder whether they will be needed in the future state, whether their skills will be valued, how they will cope with the new situation, and so on. These concerns are summarized in the question so frequently voiced during a major organizational change: What's going to happen to me? To the extent that this question cannot be fully answered (and frequently it cannot), individuals feel stress and anxiety. High levels of stress have some harmful behavioral effects, leading to difficulty in integrating information or hearing things, resistance to the changes, or, in the extreme, irrational and self-destructive acts. The resistance response is well documented (Watson 1969; Zaltman and Duncan 1977). Frequently, however, instead of actively resisting a change, people react in a variety of ways that objectively do not appear to be constructive for either the individual or the organization.

A third problem is that of organizational *control*. Change disrupts the normal course of events within an organization and undermines existing systems of management control. Change may make formal control systems irrelevant and/or inappropriate; thus it may be easy to lose control of the organization during a change. As goals, structures, and people shift, it becomes difficult to monitor performance and make corrections as in normal control processes.

A related problem is that most formal organizational arrangements are designed for stable states, not transition states. Managers focus on designing the most effective organizational arrangements for the future, and think of the change from A to B as simply a mechanical or procedural detail. They ignore the special needs of the transition state, which are ordinarily not well met by the steady state management systems and structures developed for A and B.

Implications for Change Management

The preceding discussion of problems in implementing change suggests some relatively straightforward implications for management. First, if a change may present significant problems of power, it requires the management of the organization's political system to shape the political dynamics associated with the change, preferably before the actual implementation. Second, if change creates anxiety and the associated patterns of dysfunctional behavior, it is critical to motivate individuals through communications and rewards to react to the change in a constructive manner. Finally, if a change presents significant problems of control, it is important to pay attention to the management of the transition state.

Shaping Political Dynamics

Most significant changes involve some modification of the political system, thus raising issues of power. As shown in Table 1, steps can be taken in four specific action areas to *shape and manage the political dynamics* before and during the transition period.

First, it is important to *get the support of key power groups* within the organization in order to build a critical mass of favor for the change. The organization is a political system with competing groups, cliques, coalitions,

TABLE 1 Shaping Political Dynamics

Action	Purpose	Technique
Get support of key power groups	Build internal critical mass of support for change	Identify power relationships 　Key players 　Stakeholders 　Influence relationships Use strategies for building support 　Participation 　Bargaining, deals 　Isolation 　Removal
Demonstrate leadership support of the change	Shape the power distribution and influence the patterns of behavior	Leaders model behavior to promote identification with them Articulate vision of future state Use reward system Provide support and resources Remove roadblocks Maintain momentum Send signals through informal organization
Use symbols	Create identification with the change and appearance of a critical mass of support	Communicate with: 　Names and graphics 　Language systems 　Symbolic acts 　Small signals
Build in stability	Reduce excess anxiety, defensive reactions, and conflicts	Allow time to prepare for change Send consistent messages Maintain points of stability Communicate what will not change

and interests, each with its own views on any particular change. Change cannot succeed unless there is a critical mass of support. The first step in building the necessary support is identifying the power relationships as a basis for planning a political strategy. This step may involve identifying the key players in the organization or the individual and/or group stakeholders—those who have a positive, negative, or neutral stake in the change. In thinking about these relationships, it may be useful to draw a diagram or create a stakeholder or influence map. This map should show the relationships among the various stakeholders, who influences whom, and what the stakes are for each individual.

There are several possible approaches to building support. The first is participation, which has long been recognized as a tool for reducing resistance to change and gaining support. As individuals or groups become involved in the change, they tend to see it as their own rather than something imposed upon them. Participation may not be feasible or wise in all situations, however. In some cases, it merely increases the power of opposing groups to forestall the change. Thus, another approach may be bargaining with groups—winning support by providing some incentive. A third method is isolation. Some members of the organization may resist participation or bargaining and persist in attempting to undermine the change. Their impact on the organization can be minimized by assigning them to a position outside the mainstream. In an extreme case, individuals who cannot be isolated or brought into constructive roles may have to be removed from the scene through transfer or firing. Clearly, one would prefer to rely entirely on the first two methods, participation and bargaining. But it would be naïve to assume that they will be successful in all cases.

Leader behavior can also have a powerful influence in support of change. Leaders can shape the power distribution and influence patterns in an organization and can create a sense of political momentum by sending out signals, providing support, and dispensing rewards.

Both through explicit statements and through their behavior, leaders provide a vision of the future state and a source of identification for different groups within the organization. Secondly, in addition, they can play a crucial role by rewarding key individuals and specific types of behavior. Leaders can provide support (or remove roadblocks) through political influence and needed resources and can use their public statements to maintain momentum. Finally, leaders can send important signals through the informal organization. During times of uncertainty and change, people throughout the organization tend to look to leaders for indications of appropriate behavior and the direction of movement in the organization. Often actions of little apparent consequence, such as patterns of attendance at meetings or the words used in public statements, send potent signals. By careful attention to these subtly significant actions, leaders can greatly influence the perceptions of others.

The third action area involves the *use of symbols* associated with the change. Language, pictures, and symbolic acts can create a focus for identification and the appearance of a critical mass within the political system of the organization. Widely employed in public and social movements, symbols are equally useful in dealing with the political system within the organization. A variety of devices can be used, such as names and related graphics that clearly identify events, activities, or organizational units. Language is another type of symbol; it can communicate a different way of doing business. People who want to function effectively need to learn new terms of expression; by doing so, they create the perception of broad-scale support. And important signals can be sent through symbolic acts: a particular promotion, a firing, a moving of an office, or an open door.

Finally, managers can help implement change by *building in stability*. Too much uncertainty can create excess anxiety, defensive reactions, and political conflict. The organization must provide certain "anchors" to create a sense of stability during the transition. If organization members are prepared for the change by being given information in advance, they will be somewhat protected against uncertainty. Moreover, some stability can be preserved if managers are careful not to send inconsistent or conflicting messages to the organization during the period of change. It may also be important to maintain certain very visible aspects of the business, for example, by preserving certain units, organizational names, management processes, or staffing patterns, or by keeping people in the same physical locations. Finally, it may help to communicate specifically what will *not* be different—to assuage the fears that everything is changing or that the change will be much greater than what actually is planned.

Motivating Constructive Behavior

Most people are made anxious by the uncertainty associated with significant organizational change. They may react with withdrawal, panic, or active resistance. The task of management is somehow to relieve that anxiety and motivate constructive behavior. Four areas for management action are outlined in Table 2 and discussed here.

First, managers can expose or create *dissatisfaction with the current state*. People may be psychologically attached to the current state, which is comfortable and known in comparison with the uncertainty associated with the change. They can be helped to see how unrealistic it is to assume that the current state has been completely good, is still good, and will always remain good. The goal is to "unfreeze" people from their inertia and create some willingness to explore the possibility of change. Their anxiety is based in part on fantasies that the future state will create problems, as well as fantasies about how wonderful the current state is.

A good approach to this problem is to supply organization members with specific information, for example, about the events in the environment that have created the need for change. In addition, it is useful to help people understand the economic and business consequences of *not* changing. It may be helpful to identify and emphasize discrepancies between what the situation is and what it should be. In critical cases, it may be necessary to paint a disaster scenario to show people what will happen if the current state continued unchanged. For example, one manager remarked that if the division did not become successful within eighteen months, "they'll pull buses up to the door, close the plant, and cart away the workers and the machinery."

An alternative to management's presenting this kind of information may be to get organization members involved in collecting and presenting their own perceptions. The information may seem more believable if it comes from peers in the work force.

Here and in other action areas, it is important that managers overcommunicate. Above some threshold, anxiety impairs normal functioning; thus, people may be unable to hear and integrate messages when they first receive them. It may be necessary to communicate key messages two, three, four, and even five times through different media and methods.

Participation in planning and implementing change can also provide motivation. Employees' participation in the change process tends to capture people's excitement. It may result in better decisions because of employee input, and it may create more direct communications through personal involvement. On the other hand, participation also has some costs. It

TABLE 2   Motivating Constructive Behavior

Action	Purpose	Technique
Surface or create dissatisfaction with the current state	Unfreeze from the present state; provide motivation to move away from the present situation	Present information on: Environmental impact Economic impact Goal discrepancies How change affects people Have organization members: Collect and present information
Obtain the appropriate levels of participation in planning and implementing change	Obtain the benefits of participation (motivation, better decisions, communication); control the costs of participation (time, control, conflict, ambiguity)	Create opportunities for participation Diagnosis Design Implementation planning Implementation evaluation Use a variety of participation methods Direct and indirect Information vs. input vs. decision making Broad vs. narrow scope Expertise vs. representation
Reward desired behavior in transition to future state	Shape behavior to support the future state	Give formal rewards Measures Pay Promotion Give informal rewards Recognition and praise Feedback Assignments
Provide time and opportunity to disengage from current state	Help people deal with their attachment and loss associated with change	Allow enough time Create opportunity to vent emotions Have farewell ceremonies

takes time, involves some concession of management control, and may create conflict and increase ambiguity. The question, then, is where, how, and when to build in participation. People may participate at different times, depending upon their skills and expertise, the information they have, and their acceptance and ownership of the change. Participation can be direct and widespread, or indirect through representatives. Representatives may be chosen by position, level, or expertise. Some form of participation is usually desirable in view of the costs of no involvement at all.

Managers can also enhance motivation by *rewarding the desired behavior in both the future and the transition state*. People tend to do what they perceive they will be rewarded for. If their behavior seems likely to lead to rewards or outcomes they value, they will be motivated to perform as expected.

During change implementation, the old reward system frequently loses potency before new rewards are established. Sometimes the existing measurement system punishes people for doing things that are required to make the change successful. Management needs to pay special attention to the indicators of performance, to the dispensation of pay or other tangible rewards, and to promotion during the transition. In addition, informal rewards, such as recognition, praise, feedback, or the assignment of different roles should be used to support constructive behavior. And an appropriate reward system for the future state must be clearly reestablished.

Finally, managers can alleviate individual anxiety by *providing time and opportunity to disengage from the current state*. People feel a sense of loss associated with having to change, and predictably go through a process of "letting go" of the old structure. Management should allow the appropriate time for this essential "mourning" period, while giving people enough information and preparation to work through their detachment from the current state. Another technique may be to organize small group discussions in which people are encouraged to talk about their feelings concerning the organizational change. Some may object that such sessions are likely to promote resistance rather than facilitate change. But people will undoubtedly talk about these issues, either formally or informally. If management recognizes the concerns and encourages people to express their feelings, they may be better able to let go of them and move into constructive action. It may also be useful to create ceremony, ritual, or symbols, such as farewell or closing-day ceremonies, to help give people some psychological closure of the old organization.

Managing the Transition

Transitions are frequently characterized by high uncertainty and problems of control because the current state is being disassembled before the future state is fully operational. Managers need to devote as much care, resources, and skill to managing the transition as they would to any other major project. Four specific action areas are relevant (see Table 3).

The first is to *develop and communicate a clear image of the future state*. It is difficult to manage toward something when people do not know what that something is. No one is sure what is appropriate, helpful, or constructive behavior. Without a clear direction, the organization develops "transition paralysis," and activity grinds to a halt. Managers can help in several specific ways. First, they can articulate a vision of the future state and develop as complete a design as possible. It may also be useful to draw up an impact statement describing the effect the change will have on different parts of the organization and on people. And it is important to maintain a stable vision and to avoid unnecessary or extreme modifications or conflicting views of that vision during the transition.

Finally, there is a need to communicate, repeatedly and through multiple channels—whether video, small-group discussions, large-group meetings, or written memos. It is critical to think of this communication as both a telling and a selling activity. People need to be informed, but it is also important to convince them that the change is important. This may necessitate repeated explanations of the rationale for the change, the nature of the future state, and its advantages. The future state needs to be made real, visible, and concrete. Communications should include information on future decision-making and operating procedures. How this is communicated can help shape the vision of the future. For example, one company provided clear and memorable

TABLE 3 Management of Transition

Action	Purpose	Technique

Develop and communicate a clear image of the future state	Provide direction for management of transition: reduce ambiguity	Develop as complete a design as possible
		Generate impact statements
		Communicate
		  Repeatedly
		  Multiple channels
		  Tell and sell
		Describe how things will operate
		Communicate clear, stable image or vision of the future

Use multiple and consistent leverage points	Recognize the systemic nature of changes, and reduce potential for creating new problems during transition	Use all four organizational components Anticipate poor fits Sequence changes appropriately
Use transition devices	Create organizational arrangements specifically to manage the transition state	Appoint a transition manager Provide transition resources Design specific transition devices (dual systems, backup) Develop a transition plan

Obtain feedback about the transition state; evaluate success	Determine the progress of the transition; reduce dependence on traditional feedback processes	Use formal methods
		  Interviews
		  Focus groups
		  Surveys and samples
		Use informal channels
		Use participation

images of the specific types of customer service it was attempting to provide by showing television commercials both inside and outside its organization.

Managers should *use multiple and consistent leverage points for changing behavior.* During a transition, when certain aspects of the organization are being changed, problems may arise from a poor fit between the component elements of the organization. Managers need to use all of these levers for change, modifying the work, the individuals, the formal structure, and the informal arrangements as appropriate. It is also important to monitor and/or predict some of the poor fits that may occur when any of the organizational components are changed. Changes should be planned so as to minimize the mixed messages or the inconsistencies created among elements of the organization.

Certain *transition devices* may prove useful. Because the transition state is different from the current and future states, special organizational arrangements may be needed to manage it. These may include (1) the appointment of a transition manager; (2) the allocation of specific transition resources, including budget, time, and staff; (3) the use of specific transition structures, such as dual-management systems and backup support; and (4) the development of a transition plan. All of these devices can be helpful in bringing needed management attention to the transition itself.

Finally, managers should *build feedback and evaluation of the transition state*. The feedback devices managers normally use to collect information about how the organization is running often break down during a transition. This is particularly serious during a period of change, when anxiety may run high and people may be hesitant to deliver bad news. Therefore, it becomes critical to build in different channels to obtain feedback. Formal methods may include individual interviews, various types of focus-group data collection, surveys used globally or with select samples, or the gathering of feedback during a normal business meeting. Informal channels include senior managers meeting with individuals, having breakfast with groups, informal contacts, or field trips. Finally, feedback is more readily obtained when representatives of key groups participate directly in planning, monitoring, or implementing the change.

The Problem of Uncertain Future States

The current state/transition state/future state paradigm is a useful means of conceptualizing change and structuring the problems of change. Many organizational changes, however, do not fall into this neat framework. Often, for example, a particular transition is just one of a long and continuous series of changes. The future state is not a stable state but merely a platform for the next transition.

Perhaps even more complex is the transition toward a highly uncertain future. This occurs when an organization moves into a new area of activity about which little is known or in which it will be subject to powerful forces beyond its control. In these cases, managers must plan changes involving movement toward uncertain or unknown future states.

Several special problems are associated with this type of transition. First, the power problem is greatly intensified, because the ambiguity of the unknown future state adds significantly to the uncertainty of the transition period. Because there are so many unknowns, political activity increases as individuals and groups try to position themselves to benefit from real, probable, or imagined events in the future. With an uncertain future state, the transition period is frequently extended—lengthening the time during which political activity is likely.

Second, the level of anxiety is greater. Managers are often unable to describe a specific future state and answer concretely the question, What's going to happen to me? The lack of a clear future state and a clear role for each individual makes it difficult for people to focus on the future and direct their energy at working toward the future. Instead, the free-floating anxiety is left to feed on itself.

Third, the problem of control becomes immensely more difficult if there is no clear future state toward which to manage. The whole concept of transition management seems at first glance to depend on the existence of a defined future.

Finally, situations with uncertain future states are by definition turbulent and potentially threatening to the organization. As a result, senior management is frequently preoccupied with environmental and strategic challenges with little time to attend to the human organization during the transition.

Clearly, then, the uncertain future state poses serious challenges to the basic concepts of implementation management. While it is not yet entirely clear how the problems can be solved, observation indicates several promising approaches for managing the uncertainty. As we become more experienced and collect more data, we should be able to test the validity of these suggested action steps. Table 4 summarizes possible managerial actions related to each of the three problems described previously.

Shaping Political Dynamics

Several approaches may be useful in managing the political dynamics of transition toward an uncertain future. The first is stabilization through the formation of a small, cohesive planning group at the senior level of the organization to monitor movement toward the future on a regular basis. This group would be responsible for integrating information from different sources and responding to new environmental changes. It actually might be composed of the same people as the senior policy-making or decision-making group, but it would meet regularly with an agenda different from the normal—monitoring and guiding the transition. By staying informal, the group could develop a consensus and a set of relationships that would facilitate quick reactions and decisions when needed.

For such a process to work, the bases of reward in the organization might have to change. What is needed is a systemwide perspective in which each key manager thinks about what moves are best for the organization as a whole, not just for his or her piece of it. If this perspective is to prevail, both formal and informal rewards must be explicitly and significantly tied to performance as a team member rather than tied to one's performance in one's own individual domain.

Finally, leaders must be much more visible to the entire organization during this period. Given the increased levels of anxiety and political activity, the visible actions and words of organizational leaders demonstrating confidence and displaying consistency can help to reduce the possible perceptions of "drift" or lack of leadership that frequently occur and in turn lead to increased political maneuvering.

TABLE 4   Implications for Change Management Toward Uncertain
Future States

Shaping political dynamics
1. Create a senior planning group
2. Design special rewards for senior-management collaboration
3. Increase leader visibility

Motivating constructive behavior
4. Create a vision of the future
5. Prepare people for uncertainty
6. Define the future state as made up of transitions

Managing the transition
7. Define a series of short, incremental transitions to alternative futures
8. Maintain tight linkage between planning and transition management
9. Create increased two-way communication flows

Motivating Constructive Behavior

While there is no concrete future state on which to focus, some picture of the future can be constructed and communicated. In this case, it is more likely to be a vision or a set of principles or guidelines for doing business than a concrete structure or set of organizational arrangements. It may simply be a statement of "what we will be and what we won't be," or a description of "why we are where we are, and where [in general terms] we're headed."

At the same time, it is important to prepare people to deal with the uncertainty. This may involve education or information that can help employees recognize the sources of the uncertainty and the necessity of living with that uncertainty for some time.

People may want to believe that the uncertainty and turbulence of the transition will someday cease and stability will be restored. In many cases, however, the tranquility of the past will never return. (Frequently the past is only retrospectively perceived as tranquil—things were not really that stable.) Thus, individuals may have to be prepared to move to a future state that is made up of successive and continuous transition states.

Transition Management

Transition management in an uncertain context must be refined into a series of shorter, smaller transitions toward a hypothesized future state or a set of alternative scenarios of the future. Thus, instead of one transition to a fixed future state, we might envision five transitions toward a set of different, possible, ultimate future states, with the first transition being concrete and feasible for all of the possible ultimate alternatives, the second being less concrete, the third even less so, and so on. Change is thus managed incrementally toward an evolving future.

During this process of incremental change management, those responsible for planning (including environmental analysis, strategic planning, and organizational planning) must work very closely with those who manage the transition. When uncertainty is high, the need to coordinate these two functions is critical, and information must move freely, in both directions, between them.

Finally, the management of these transitions requires greatly increased communication flows within the organization. Given the possible intensity and speed of activity, information must flow out to the organization regularly and effectively, and feedback data must be collected to monitor how people are perceiving and dealing with the change.

It is important that senior management balance the time and attention spent on strategy and implementation. The temptation is to focus entirely on strategic issues—to work on resolving the causes of the uncertainty. But the problems of power, anxiety, and control will persist even as senior management tries to find stability or determine directions. If these problems are ignored, the implementation process may become unmanageable. The key implication for managing transition to an uncertain future, then, is to balance the attention given to strategic and implementation concerns: they are concurrent, not sequential.

Summary

The effect of implementation of organizational change is a critical task for managers. Change management requires an understanding of how organizations function as well as an appreciation of the peculiar dynamics of transitions. The view of change presented in this chapter is based on observation over a period of years, of how managers actually handle organizational transitions. Combining the empirical base with the insights of an organizational model, I have tried to provide concepts that can be helpful to managers in understanding and managing complex changes in demanding and uncertain times.

REFERENCES

Beckhard, R., and R. Harris. *Organizational Transitions*. Reading, Mass.: Addison-Wesley, 1977.

Katz, D., and R. L. Kahn. *The Social Psychology of Organizations*. New York: Wiley, 1966.

Nadler, D. A. "Managing Organizational Change: An Integrative Perspective." *Journal of Applied Behavioral Science* 17, no. 2:191–211, 1981.

Nadler, D. A. "Managing Transitions to Uncertain Future States." *Organizational Dynamics*, Summer 1982, pp. 39–45.

Nadler, D. A., and M. L. Tushman. "A Diagnostic Model of Organizational Behavior." In *Perspectives on Behavior in Organizations*, ed. J. R. Hackman, E. E. Lawler, and L. W. Porter. New York: McGraw-Hill, 1977, pp. 35–47.

Nadler, D. A., and M. L. Tushman. "A Congruence Model for Diagnosing Organizational Behavior." In *Organizational Psychology: A Book of Readings*, ed. D. Kolb, I. Rubin, and J. McIntyre. 3d ed. Englewood Cliffs, N.J.: Prentice-Hall, 1979.

Nadler, D. A., and M. L. Tushman. "A Model for Diagnosing Organizational Behavior: Applying a Congruence Perspective." In *Managing Organizations*, ed. D. A. Nadler, M. L. Tushman, and N. G. Hatvany. Boston: Little, Brown, 1982.

Pettigrew, A. *The Politics of Organizational Decision-Making*. London: Tavistock, 1972.

Salancik, G. R., and J. Pfeffer. "Who Gets Power and How They Hold onto It: A Strategic-Contingency Model of Power." *Organizational Dynamics*, Winter 1977, pp. 3–21.

Thompson, J. D., and A. Tuden. "Strategies, Structures and Processes of Organizational Decision." In *Comparative Studies in Administration*, ed. J. D. Thompson et al. Pittsburgh: University of Pittsburgh Press, 1959, pp. 195–216.

Tushman, M. L. "A Political Approach to Organizations: A Review and Rationale." *Academy of Management Review*, no. 2:206–16, 1977.

Watson, G. "Resistance to Change." In *The Planning of Change*, ed. W. G. Bennis, K. F. Benne, and R. Chin. New York: Holt, Rinehart and Winston, 1969, pp. 449–93.

Zaltman, G., and R. Duncan. *Strategies for Planned Change*. New York: Wiley, 1977.

Owens-Illinois: 10 Quad

People have been making glass in one form or another for over 14,000 years with the world's first glass containers being produced in Egypt around 2,000 B.C. Although glassmaking has undergone changes and improvements throughout the ages, interestingly enough, by the end of the nineteenth century, glass bottles and jars were still being made the way they had been for 2,000 years. Not until 1903 was the first fully automated bottle-making machine invented by Michael Owens, paving the way for mass production in the glass container industry.

In 1903, Owens and three associates organized the Owens Bottle Company in Toledo, Ohio. In 1929, the firm merged with the Illinois Glass Company to become Owens-Illinois (O-I). By 1987, O-I had become one of the nation's largest corporations with annual sales exceeding $3.5 billion and worldwide employment of more than 46,000 people. It operated nearly 100 plants in the United States, and an equal number of facilities in Latin America, the Caribbean, Africa, Europe, and the Far East/Pacific region. O-I was an international manufacturer of packaging products of glass, plastic, and paper; a producer of technical and consumer products for diversified markets; and a provider of health care and financial services. In March 1987, the corporation was acquired by Kohlberg, Kravis, and Roberts, a group of private investors, for $3.64 billion.

O-I was known for being a progressive company whose engineering was done in house at a technology center in Toledo. (See Exhibit 1 for an organization chart.) The center's role was to design new state-of-the-art technology and equipment, encourage the plants to adopt the latest ad-

---

EXHIBIT 1   Owens-Illinois Organization Chart

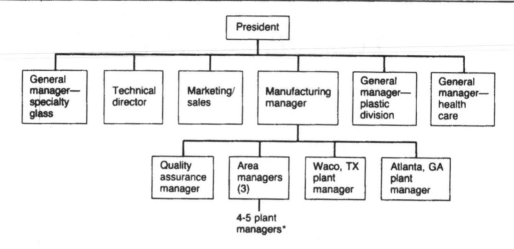

*Streator, IL, plant manager reported to one of the area managers.

---

vances, and aid in their introduction. The tech center was also available for doing diagnostic testing and problem solving to help the plants optimize their operations. They were aided by an on-line central computer system which could monitor various process parameters on equipment at all O-I facilities (the computer was not, however, used to "audit" the plants.)

One of the most recent projects the tech center had been working on was the 10 quad, a glass-forming machine which was capable of producing 33 percent more bottles per minute than the industry's next best machine. The quad was being used for a new application, the 16-ounce single service bottle, in two operations: Streator, Illinois, and Atlanta, Georgia.[1] Although the plants had started manufacturing the bottles within six months of each other, they had experienced different results in introducing the quads. The tech center's computer showed that variations in process parameters across the two facilities, however, were minimal and could not account for the differences in each plant's performance. Ken Lemke, senior vice president of technology for Owens-Illinois, suspected the differences stemmed, in part, from the plants' ability to handle the hot end of the quad's bottle-making process (described below). He knew the Atlanta plant had had prior experience making 10-ounce bottles on a quad

---

[1] 16-ounce bottles had previously been made on another bottle-forming machine called the 10 triple.

but believed there were other factors contributing to Streator and Atlanta's dissimilar experiences with the 16-ounce bottle. Lemke pondered how to better understand these differences so he could use that knowledge to improve performance across all 14 of O-I's domestic bottle-forming operations.

GLASS CONTAINER MAKING

Owens-Illinois' oldest and most basic business, the production of glass containers, had grown in sophistication since the early 1900s. At O-I's glass plants, glass container production began with the arrival of raw materials including silica (ordinary sand), which made up approximately 70 percent of a glass container, limestone, and soda ash (sodium carbonate). Recycled bottles, which were ground up into crushed glass, or cullet, were also frequently used as raw material.

The bottle-making process was separated into the hot end, which included the furnaces and forming machines, and the cold end, which included the lehr (annealing), inspection, labeling, and packaging. (See Exhibit 2 for a summary of the glass container manufacturing process.) The hot end process began in the batch house where raw materials were weighed to make sure the ingredients were properly proportioned and then mixed into a batch. Using dump buckets on overhead rails, the batch was transferred and automatically fed into a furnace, a huge brick room with slits in the walls from which the batch entered. Currents, caused by differences in temperature within the tank, made the molten glass circulate inside the furnace, which was being heated at approximately 2,700°F. (The furnace operated continuously seven days a week, shutting down only for certain holidays or repair periods.) Each furnace typically fed molten glass to two to four bottle-forming machines; the number of machines varied depending on their size and the volume of glass that each needed to pull from the furnace. From the furnace, molten glass flowed through a throat or skimmer into a chamber to be refined. Next, it flowed into the forehearth, which was designed to ensure proper consistency and uniform temperature as glass flowed to the feeder of the bottle-making machine.

The first step in forming the bottle was the cutting of a glass gob, basically an elongated cylinder of molten glass. The feeder controlled the size and shape of the gob, proportioned for the bottle or jar to be produced. Shearing blades at the bottom of the feeder, adjustable for both speed and height, cut the gob at just the precise moment. The cut gob dropped onto a scoop, slid down a deflector and into a funnel atop a blank mold. Immediately after the gob dropped into the mold, the mold was covered with a baffle, and air was injected from the bottom to transform the gob into

EXHIBIT 2   Glass Container Manufacturing

1. The manufacture of glass containers begins in the batch house where raw materials—sand, soda ash and limestone—are accurately measured and combined. The mixed ingredients are then transported through a conveyor to the furnaces where they are heated at 2700 degrees farenheit until they become molten or "liquid" glass.

2. From the furnace, the molten glass flows through refining troughs (feeders) to an orifice, where it pours through the openings and is sheared off into "gobs" and delivered by chutes to the bottle molds on the forming machine.

3. In the forming machine, the gob is preformed into molds in preparation for the final "blow" which gives the container its actual shape and form.

4. The formed bottles are then released from special hangers to a conveyor.

5. The conveyor takes the bottles to a special "cooling down" oven known as the annealing lehr (pronounced leer).

6. As the bottles pass through the lehr, they are gradually cooled. It takes about an hour in the "cooler" before the bottles are cool to touch. The annealing process strengthens the containers and enhances their desirability.

7. After cooling, each container is subjected to a series of visual and electronic checks. Containers not to specifications are removed and recycled back into the manufacturing process.

8. After inspection, the containers are then packed into cartons, or in bulk, and stored in the warehouse for shipment to customers.

an intermediary shape called a parison. From the blank mold, the parison was transferred into the finishing mold where it was again inflated with air to form the final dimensions of the bottle. (See Exhibit 3 for a process diagram.) The forming machine was designed so that new parisons could be formed in the blank molds at the same time as the previously transferred ones were converted to bottles in the finishing mold.

Red hot when they came out of the finishing molds in the forming machine, the freshly blown bottles moved over 25 feet along a conveyor belt to the lehr, a large oven-like structure where the annealing process took place. Annealing tempered the glass through reheating and then gradual cooling. Depending on the speed of the conveyor within the lehr and the size of the bottle, annealing could take anywhere from 35 minutes to one hour. During the annealing process, special protective coatings were applied to the bottles to make them more durable and to protect them from scratches or abrasions.

After annealing, bottles were subjected to electronic, mechanical, and visual inspections to ensure high level quality. Each bottle was checked for a variety of critical tolerances such as the inside and outside circumference of the bottle's finish, or top, its sealing surface, bottle diameter, and height. In addition, the bottles' ability to withstand shock and rapid temperature changes was checked, and rejected bottles were sent to a collection area where they were reduced to cullet and remelted. After inspection, the bottles proceeded to the labeling and packing areas to be either bulk packed or packed in corrugated shippers.

PRODUCTIVITY IMPROVEMENTS

Over the years, O-I had continually worked to develop new machines that could manufacture more glass bottles than their predecessors. The volume of glass containers produced by a glassmaking machine was determined by the number of sections operated and the number of cavities, or gobs, formed in each section. For instance, an 8 section triple gob machine meant each of the machine's eight sections, which operated independently, produced three bottles at once. The earliest glass container-making machines were one section single gob units. Throughout the years, the number of sections on the machines had increased; in 1987, the largest machines had 10 sections. The number of cavities per machine had also increased over time. For instance, double gob machines became popular in the 1920s and 1930s, and in the early 1970s, O-I began using the first triple gob units (8 sections).

Shortly after the introduction of the triple gob machine, the glass container industry began to be threatened by plastics. The lowest production cost for plastic bottles with carton and label was $64 per thousand and that figure was expected to drop to $56 per thousand. Glass manufactur-

EXHIBIT 3   Bottle-Forming Process

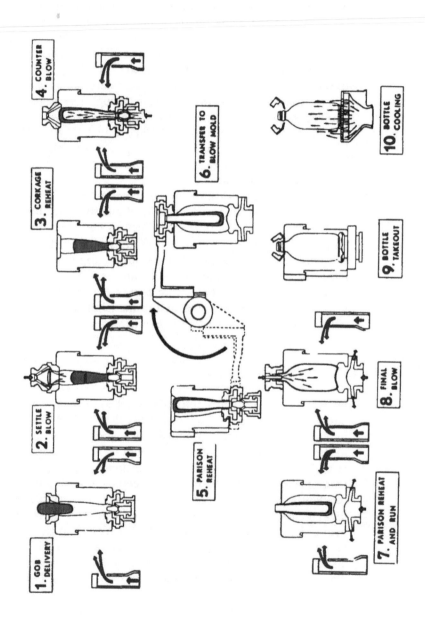

ing costs at that time were $59 per thousand bottles. In response, O-I decided it needed to increase productivity. In 1978, the 10-section triple gob was introduced. Shortly thereafter, Ken Lemke conceived the idea for a 10 quad machine. This unit would be capable of producing over 530 bottles per minute, which was 33 percent more bottles than the next best machine, and would make bottles at a cost of $50 per thousand.

Lemke, affectionately referred to as "The Quad Father" throughout the company, assigned his engineers the difficult task of designing the 10 quad. The major difference between the quad and single, double, or triple machines was the number of mold cavities within each section box. The quad had to have the same 21-inch size section boxes as the 10 triple in order to be cost effective; this meant that the quad would be the same width and length as the older unit, although it would contain four mold cavities rather than three. As a result, the maximum bottle size that could be produced in the quad would be approximately 16 ounces. Realizing that the 16-ounce bottle would be near the limits of the machine, the first quads were used to produce 12-ounce beer and 10-ounce single service soft drink bottles; a beer bottle had a diameter of 2⅜ inches and a 10-ounce bottle had a diameter of 2 inches. Moreover, bottle-handling on the quad could be potentially difficult because of the large increase in the number of bottles that would have to be moved along the conveyor.

DEVELOPMENT AND INTRODUCTION OF THE 10 QUAD

With the goal of having the first quad installed and operating before 1983, development began in 1980. In addition to the tech center's design and development team, production managers from four of O-I's beer and soft drink bottlemaking plants were invited to the tech center to discuss the project (Winston-Salem, North Carolina; Waco, Texas; Volney, New York; and Atlanta, Georgia.) Each plant was a potential first site for the quad. The machine itself would be assembled and tested in O-I's Godfrey, Illinois machine shop before being disassembled and sent to a plant.

The first 10 quad unit went on line in Winston-Salem in November 1982. Winston-Salem was chosen as the site of the first quad for three main reasons. Serving the customer's needs was the top priority. The plan was to take advantage of consistently long runs, such as those occurring in the production of beer bottles, because the most efficient use of the quad required infrequent changes in the size of the bottle being produced. Second, the facility itself was well suited to the introduction of the quad. The metal line (the distance from the supply of glass to the bottom of the machine), which was the benchmark to determine what size machine would fit underneath a furnace, was high enough for a quad unit. A quad required a higher metal line than other machines because of loading: it needed a strong force of gravity to feed the molten gob into the blank, especially in the outer sections, and this was provided by a high metal line. Also, the annealing process was the correct size (i.e., the lehr capacity was sufficient to hold the larger volumes of bottles that would be produced by the quad). Finally, the Winston-Salem facility had the reputation of being a successful operation with a good atmosphere, good management style, good work ethic, and the right attitude, that is, the employees were said to be talented and receptive to the development of new processes.

"Winston-Salem was an outstanding project, a sounding ground, a great start-up and success from day one, which really encouraged us to go forward with the next project," explained a tech center engineer. By May 1983, the second quad, which would be making 10-ounce soda bottles, had been installed in the Atlanta plant. However, no new quads were set up until three years later when a plant began making 12-ounce beer bottles on a unit with improved electronics. The three-year hiatus had been due to a number of developmental factors. However, in 1985 O-I again demonstrated its commitment to the glass business by installing additional quads.

In May 1986, one month after the beer plant quad installation, the fourth quad was introduced at the Streator, Illinois, plant. The decision to put a quad into Streator was again based on customer demand; there was a large demand for 16-ounce soda bottles in the Midwest and so the quad was set up to run that size container. There was also a significant 16-ounce bottle demand in the South and Southeast United States in the mid–1980s, which led to the decision to put the fifth 10 quad machine in Atlanta in January 1987. Additional quads were installed following the Atlanta installation.

KEYS TO PERFORMANCE

Through years of continuous experimentation and improvements, the process of manufacturing glass containers was being transformed from an art form into a science. Computer simulations had been developed to determine the optimal mixture of raw materials and to provide better understanding of the interrelationships of the various process parameters (e.g., glass temperature, mold designs, etc.). Using this knowledge in an effort to maintain uniformity of product, the tech center prescribed the majority of process parameters to be used by the plants. Inherent variation in the raw materials and day-to-day variations within each operation (particularly ambient temperature and humidity), however, required each operation to fine tune these parameters continually. Therefore, the ultimate performance of a particular plant, measured by its 24-hour production output and its pack-to-melt ratio (tons of glass packed divided by tons melted), was influenced to a large extent by each plant's ability to manage the molding and bottle-handling processes.

An important factor in the molding process was glass composition, which affected the viscosity of the glass. Although the tech center set the various percentages of raw materials for each type of glass, the plants could vary the amount of cullet used in each furnace, within a range of 10 percent to 40 percent and higher. In general, the advantage of using cullet was that it lowered the melting temperature of the glass, thus saving on energy costs.

Another factor in the molding process, which itself was affected by glass composition, was the set time of the glass, that is, the time it took for the molten glass to solidify into its shape. A quick set time, therefore, meant the machine could run faster, but it also meant there was less margin for error in other parameters, especially cooling temperature and time. Thus, a quick set time could require more precise control of the process. The set time was affected by the quantities and characteristics of the soda ash and limestone used; soda ash and limestone in different parts of the country tended to vary a great deal.

Besides the set time, the temperature of the molten glass itself also affected the molding process because of the need to balance the glass temperature with the mold temperature and the cooling. The molds, made mostly out of iron, were typically cooled by jets of air. The amount of air needed and the direction and timing of the jet were influenced by the amount of metal in the mold and its contour. If a mold were either too cold or too hot, the bottles would not hold their shape. Because the molding process required finding the right combination of glass temperature, mold temperature, and cooling, many considered the process to be something of an art.

Once the bottle was formed, it had to be transported to the lehr. Although the glass had already set, the bottle was still extremely hot as it started to be handled, especially as it was automatically swept onto the conveyor, the beginning of the bottle-handling process. Bottle handling was difficult because, with a high volume of bottles, the timing of all the different mechanical devices that moved the bottles along had to be in total synchronization. (See Exhibit 4 for a diagram of the bottle-handling mechanisms.) For instance, the sweepouts (mechanisms that pushed bottles onto the conveyor) had to be timed with the conveyor belt, and at the curve where bottles were transferred from one conveyor to another, the two conveyors also had to be synchronized. Furthermore, the lehr loader arm (which pushed bottles into the lehr) had to move in parallel with the bottles as it pushed them so the bottles did not collapse or collide. As such, bottle handling was also considered to be an art. As one tech center engineer explained, "The measure of many glass processes is just the ability to handle that bottle-handling process."

Typically, the expertise for various key performance measures lay with people in the plant. For example, there was a mold repair supervisor responsible for overseeing the maintenance of mold equipment. The bottle-handling expert at a plant was the machine repair supervisor, while the

EXHIBIT 4   Typical Layout of Bottle-Handling Process

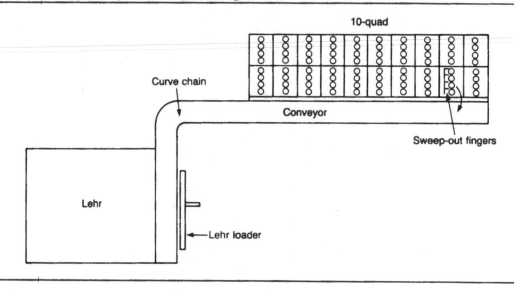

forming supervisor specialized in keeping the overall machine running smoothly. The forming supervisor had to work closely with the head of the batch and furnace department since so much of what went on in the batch house affected the molding process. The day specialist was the plant's hot end technical expert, typically responsible for problem solving. The production manager coordinated all activities relating to the machine. (See Exhibit 5 for an organization chart of a typical bottle-making plant.)

PERFORMANCE MONITORING

The tracking of machine speed, line output, and pack-to-melt figures was considered so critical at all O-I glass plants that these measures were displayed throughout the facilities on television monitors. The figures for each machine were updated every 20 minutes and were scrutinized by managers, all of whom had monitors in their offices. The ongoing tracking put constant pressure on machine operators and supervisors to assure that proper speed and pack-to-melt levels were maintained. If one of these measures fell unexpectedly, management would be quick to question the cause.

The machines and furnaces at every O-I glass plant were also on-line to a central computer at the corporate tech center. The plants could there-

EXHIBIT 5   Typical Glass Plant Organization Chart

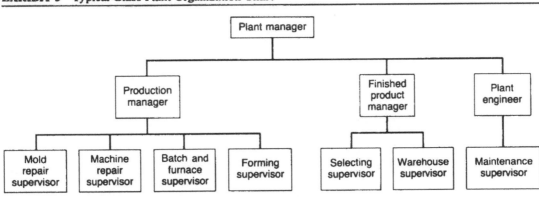

fore request special diagnostic testing from the tech center if they needed help with a particular problem.

Since coordination between functions and performance monitoring were so strongly stressed, the plants typically conducted daily production meetings to review productivity, quality, and any mechanical breakdown problems. These were 20 minute morning meetings held Monday through Friday, which were usually attended by the plant manager, production manager, forming supervisor, mold repair supervisor, machine repair supervisor, batch and furnace supervisor, plant engineer, and representatives from the quality department. In some plants, afternoon meetings were also held daily; in others, afternoon meetings were scheduled only when there were pressing issues to discuss.

THE 16-OUNCE BOTTLE

The 16-ounce bottle being produced at both the Streator and Atlanta plants was a challenge to manufacture on the 10 quad, especially when compared to the 10-ounce soda bottle and 12-ounce beer bottle that other quads were making. This was because the 16-ounce was a taller and wider bottle (the 16-ounce was 6⅞ inches tall and 2⅝ inches wide whereas the 10-ounce was 5¾ inches tall and 2 inches wide.) In addition, 10-ounce and 16-ounce bottles had different glass-to-capacity ratios (the weight of the glass in the bottle versus the bottle's capacity). A bottle with a 10-ounce capacity had a weight of 5.5 ounces whereas a 16-ounce bottle weighed 7.5 ounces. All of these factors meant bottle forming, handling, and cooling on the 16-ounce had to be especially precise as they learned to run this item on the quad for the first time.

EXHIBIT 6   10 Quad Performance at Atlanta and Streator

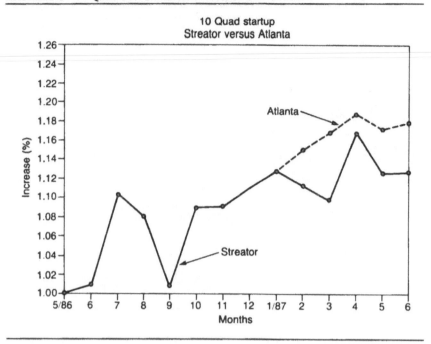

Curiously, the Streator and Atlanta plants were experiencing different results producing the 16-ounce bottle on their quads. (See Exhibit 6 for a comparison of the performance results.) In mid–1987, as Lemke started thinking about future quad installations, he contemplated reasons for the differences between the two facilities. Were they due to the plant personnel's technical skills or attitudes, the facilities' physical layout, the composition of the glass being used, tech center engineering support, or other factors?

Owens-Illinois: Streator 10 Quad

The 16-ounce is taxing the limitation of the machine. It was developed for 10-ounce soda and 12-ounce beer bottles, and I think the 12-ounce capacity should be the limit. The 16-ounce creates too many bottle handling, forming, and cooling problems.

John Valduga, plant manager of Owens-Illinois' Streator, Illinois plant thus stated his view of the 10 quad bottle-making machine that had been introduced into the plant in May 1986. "The monster," as Streator personnel referred to the quad, had been a source of continual problems since its startup. First had come a series of electrical malfunctions followed by bottlehandling and mold trouble. Even now, in July 1987, tech center personnel felt the quad was not running at its full potential. (See Exhibit 1 for monthly production statistics.)

PLANT HISTORY

Built in the early 1900s by the American Bottling Company and later purchased by O-I, the Streator plant had once been the largest glassmaking plant in the world. Spread across 68 acres, it was a 47-building complex comprising 1.9 MM square feet of floor space and which included 1.2 MM square feet of warehouse space. The plant's peak years had been in the

EXHIBIT 1   I-2 Monthly Production Statistics

	Speed*	Job Efficiency†	Average 24-Hour Production‡
5/16/86 Start-up	475.2	86.7%	4000
6/86	474.0	87.4	4038
7/86	473.6	95.7	4407
8/86	473.5	93.4	4317
9/86	477.2	87.2	4030
10/86	479.6	93.2	4359
11/86	484.3	92.4	4362
12/86	485.8	94.3	4438
1/87	489.8	95.1	4513
2/87	497.0	91.8	4451
3/87	503.5	89.4	4392
4/87	504.1	95.0	4670
5/87	505.9	91.3	4501
6/87	506.8	91.9	4538
7/87	507.3	95.2	4712

*Bottles/minute
†Bottles packed/theoretical
‡Gross bottles

Note: In March 1987, the quad switched from running flint (white) to running green glass; the quad switched back to flint in June.

late 1950s and early 1960s when there were approximately 2,500 employees and 10 furnaces with 33 bottle machines in operation. However, a shrinkage of the glass container industry in the early 1970s led to massive layoffs. By 1987, the plant's work force had been reduced to 77 salaried and 675 hourly employees, the latter represented by the Glass, Pottery, Plastics Allied Workers (GPPAW).

The work force, which viewed itself as a close-knit family, felt especially proud that their older plant had survived the turbulent 1970s when many new glassmaking facilities had been closed. In fact, in the early 1980s Streator was made into an international showplace for Owens-Illinois to demonstrate what could be done with an old facility. Tours were often held more than once a week for customers and representatives of O-I and its affiliates. The visitors were able to see several one-of-a-kind glassmaking machines such as an 8-section narrow-neck press and blow unit, a major technological breakthrough whose method of manufacturing was predicted to replace the current molding process. Streator also operated a 10 quad and four additional glassmaking units, utilizing three furnaces. (A fourth furnace had been dismantled in November 1986 and was to be rebuilt in 1988.)

Although the hot end working environment was less than optimal and

EXHIBIT 2   Furnace and Forming Machines*

	G Furnace	
G–1	*G–2*	*G–3*
10 double	8 triple (narrow-neck press and blow)	8 double
Age: 4½ years	Age: 1½ years	Age: 8 years
Changeover frequency: 3–4 per year	No changeovers	Changeover frequency: once per week

	I Furnace
I–1	*I–2*
10 triple	10 quad
Age: 1 year	Age: 1 year
Changeover frequency: every three weeks	No changeovers

	J Furnace
J–1	*J–2*
8 triple	10 double
Age: 15 years	Age: 8 years
Changeover frequency: twice per year	Changeover frequency: 2–3 times per week
Conversions: three times per year	

*There were 40 job changes and 7 machine conversions between January and May 1987.

the work demanding, the layout was orderly. The furnaces ran down one side of the building and the machines lined up perpendicular to them. The machines were crowded, however, with the 10 quad located in the middle of the plant. The equipment was shut down every five to six weeks for preventive maintenance; in addition, several of the units underwent frequent changeovers, and a few were sometimes converted from double to triple gobs or vice versa. (See Exhibit 2 for details on furnaces, machines, and changeovers.)

The glass containers produced on the various machines were for customers in the soft drink, liquor, food, and pharmaceutical industries.[1] Typically Streator furnaces ran cullet levels in the 15 to 50 percent range. At 103.8 seconds, Streator's set time for flint (white) glass was faster than other plants. The set time for green glass was 105 seconds.

----

[1]The plant also made corrugated boxes for shipping some of its containers.

PLANT MANAGEMENT

In early 1985, John Valduga took over as plant manager at Streator; his predecessor, Jack Allison, had been transferred to O-I's Atlanta operation. The two managers had sharply contrasting styles: Allison was considered to be classic theory X, while Valduga was labeled classic theory Y. A Streator employee commented:

> It was a big change when John came. When Jack left, some of the key people let down their guard and relaxed too much, which wasn't really right. It was like some of the people took a deep breath and enjoyed the cloud moving somewhere else. John is very laid back and nice; he's very caring and fits in well with the small-town, family-like environment here. Initially, there was a let down, and we had to get people back together.

Having to deal with this difficult personnel situation was just one in a series of rapid-fire challenges Valduga faced upon arriving at Streator. First, the facility was in the midst of rebuilding the J furnace. Then, soon after the completion of the rebuild, the narrow neck press and blow unit was installed. While Streator personnel were attempting to adjust to this totally new process, they also had to contend with a transition on the H furnace from flint to champagne green glass for a line of ware they had never run before. In addition, the forming supervisor, a long-time plant employee, was acting as production manager since there had been no production manager at the plant for over two years. According to one tech center engineer, "The lack of a production manager created a lack of leadership in the forming, mold repair, and machine repair areas."

Valduga described the scene at the plant during this time:

> With all the changes being made we really were mixed up, and when things go badly there's a tendency for people to start blaming each other. It was my job not to let that infection set in; I had to build morale. My philosophy is to give people full responsibility and authority to do the job with my guidance. I believe people should work as a team. If you don't have the people working together, I don't care what kind of widget you put out there, the people will find a way to make it not work. But there was so much going on here, I had to address the nuts and bolts before I could address management styles.
>
> For 60 to 90 days it was extremely difficult. We were running every day hand-to-mouth. We couldn't even get the place settled down after the narrow neck arrived, then came the H tank, and then came the quad.

INTRODUCTION OF THE 10 QUAD

The quad arrived in April 1986, a few months after the startup of the narrow neck press and blow machine. Because the plant had known for about nine months that it would be getting the new equipment, it had ample time to rebuild the I furnace, including raising the metal line up 15 inches. Unlike most other O-I plants, Streator customarily did all of its own contracting for furnace rebuilds; for example, they hired outside bricklayers and steelworkers who were put on the plant's payroll and supervised by O-I's facilities supervisor.

In preparation for receiving the 10 quad at Streator, the forming supervisor, the machine supervisors from each of the four shifts, four machine operators, and the day specialist visited the Winston-Salem and Atlanta plants to view their quads. According to one of the supervisors, "Going to Atlanta helped a whole lot. The guys down there were really good and answered all our questions. We spent two full days, and we could have used more time."

A plant employee explained how the operators were selected:

> You couldn't be a rookie starting on this machine or a senior person. All the senior people were asked if they wanted the quad, but most didn't want it because it was something new, and everyone knew it would be a bummer to start up. They know it's more work even when it's running well; it keeps you going all the time. Even though the quad pays more, the pay is not high enough to make it worthwhile, especially for the senior workers.[2] You don't get too many who want to work on the quad.

Machine operators were responsible for keeping the machine clean, sampling bottles several times per shift, and swabbing, a method of lubricating the molds so the gob would slip easily into the metal mold. In addition, every three to five months they were responsible for installing new molds; since each section could be shut down independently, mold changes were typically done while the machine was running.

On April 22, when Joe Smith, the tech center's forming specialist, who was responsible for helping plants install new equipment and training local personnel, arrived at Streator, the machine had already been unloaded and set in place. He spent the next few weeks assisting in the installation, a task made easier by the high level of expertise in the maintenance operation, whose members were experienced in setting up new equipment. During this time, the forming specialist worked closely with Streator's forming supervisor rather than with the production manager, as was typical, since the latter only started at the plant on May 15.[3]

On May 26, Streator personnel began dry cycling (running without glass) the quad and on May 29, it began to run with glass. The tech center

---

[2] Quad operators earned $.32 more an hour than operators of the 10 triple, which was the next highest paying machine. In addition, there was a corporatewide bonus system which gave the workers incentive for increasing output.

[3] According to one tech center employee, the forming supervisor was a gruff, hard-nosed manager with better technical than interpersonal skills who was discontent that he did not receive the production manager position. Valduga, who had postponed hiring a production manager until he became accustomed to the plant, finally hired someone whom the tech center claimed was more interpersonally than technically skilled.

forming specialist left that day because the quad appeared to be running well; the installation and startup had gone smoothly, and plant personnel, who viewed the quad as just another new machine, were not expecting any trouble. Problems, however, soon arose. First, in the beginning of June, trouble developed with the lehr loader. The loader was similar to those used on other machines and had been used for almost five years with no trouble, but because the quad ran at such high speeds, the loader was cycling too fast, and the drive belts kept jumping off the tracks. The Streator machine repair group recommended a cam modification, but the tech center computer simulation said it was not feasible. Streator personnel decided to try it out anyway and it worked. During this same time, the quad's conveyor belt was also changed. The machine had come equipped with a newly designed one-inch pitch[4] conveyor, but after two weeks of problems, the machine repair supervisor decided to replace it with an older style half-inch pitch conveyor used on other Streator machines.

ELECTRICAL PROBLEMS AFFECTING THE 10 QUAD

A couple of weeks after the lehr loader and conveyor problems had been taken care of, the quad began shutting down about once a week for no apparent reason. The quad's scoops stopped oscillating which in turn caused all the gobs to load into one section creating a huge pile of molten glass. At first, plant personnel attempted to locate the source of the problem on their own; for instance, the electrical supervisor brought in his own meter to check the power lines. Valduga was hesitant to call Toledo for help. He explained:

> At first we tried to find the problem ourselves. I don't scream unless I have a serious problem, whereas a lot of guys like to have their hands held, so they call right away. We had 14 engineers in here with the narrow neck press and blow; I had to chase them out. There's such a thing as too much help. In fact, it can be worse than too little.

Valduga did not have to worry about being inundated with tech center engineers when it came to the quad, however. The area manufacturing manager explained why:

> From my perspective, we'd built such a good machine that you just needed to plug it in and it would work. The technical support people didn't feel the need to give as much support to Streator as they had to the other quad sites. By the time we got to Streator, we thought it was a plug-in and startup procedure. Besides, Streator had a tendency to be more self-sufficient trying to troubleshoot their problems. They were less willing to ask for tech center help.

---

[4]Pitch was the distance between teeth on the sprockets.

However, over the next few months, the frequency of the quad's shutdowns increased until by the beginning of August the unit was going down once a shift. Finally, Valduga lost patience: "I got personally involved. I called the VP of engineering. When that machine stops, it's a pretty big problem," he exclaimed. Several tech center engineers came to the plant for a day and then returned to Toledo to try to simulate the problem on paper. On August 5, a vendor for the multiple inverter computer control system (MICC), a part of the drive system which turned the quad's motors in synchronization, installed a diagnostic board on the machine's feeder drive to see whether the shutdowns were being caused by drops in the power line attached to the quad. The board showed that they were, and after further investigation, technicians were able to pinpoint the cause. The quad's drive system was tied to the same power line as its electric heating system and a new label machine being installed. When two of the three units were in use and the third turned on, the resultant spike (sudden drop and return) of the line voltage caused the quad to shut down.

The internal wiring problems were fixed, and the quad ran smoothly for the next two weeks; but then the shutdowns began again. This time, Streator's power engineer met with a representative from the electric company to investigate the plant's external power lines. They discovered a bad capacitor bank which was not holding the proper voltage and thus causing spikes on the incoming line. The electric company quickly rectified the problem. Meanwhile, summer electrical storms in the area were causing periodic power outages which also led to quad shutdowns.

For the next few months the quad ran trouble-free, and the plant began to feel confident that its electrical problems were gone. However, in late October 1986, the entire system began stopping periodically. On October 21, a tech center engineer came to the plant with representatives from the MICC vendor. Using a diagnostic board, they found that the quad's 5480 inverter (part of the MICC package) was too small for the feeder drive and was tripping out because of overloading. (Streator's quad was the first one equipped with the MICC; the 8 triple unit had an MICC but it did not carry as heavy a load as the quad.) On October 30, in an effort to solve the problem, a separate isolation transformer was put on the quad to filter spikes in the incoming power.

After realizing the problem was still not solved, a tech center engineer and a vendor representative returned to Streator on November 5 to replace the 5480 with a 5305 stand-alone inverter. Although this new inverter had more than enough power (7½ horsepower), it was not compatible with the quad's timing system. The 5305 was a Silicon Controlled Rectifier (SCR) inverter, which acted like a transistor for higher voltages and wattages but which was slightly slower than a regular transistor. Because of the speed differences, the machine kept getting out of synchronization; on November 11, for instance, the 5305 became five degrees out of sync in less than 45 minutes.

Finally, on November 13, plant personnel, assisted by a tech center engineer, installed a 290 inverter which was compatible with the quad's other electrical equipment. This was a 20 horsepower, stand-alone inverter which was smaller than the 5305 but larger than the 5480. According to Valduga:

> Our plant engineer and his guys wanted the 290, and I forced the issue. The thing had been going on too long and I lost patience, so quite frankly, I forced the issue. Toledo didn't want to do it. It cost the plant $12,000, and I said, "I'll absorb the cost." We were convinced it would solve the problem 99 percent.

Because of its size, the 290 would not fit inside the cabinet where the previous inverter had been; instead, it was wired remotely from an entirely separate cabinet that was not located directly near the quad. According to the plant, tech center personnel felt the 290 was like a "huge appendage" on the quad and were unhappy that the quad no longer looked like a "neat package" with all the equipment in one place. They wanted to redesign the system to eliminate the need for the 290, but Valduga was opposed. He told them, "If it works, don't mess around with it. That's not a research center out there, it's a production area."[5]

After the 290 was installed, plant personnel increased the speed of the quad to 486 bottles per minute. It was at this time that the conveyor started shutting down. The electricians discovered that some defective wires were shorting out, so during the Thanksgiving shutdown, the wires were replaced. From that time on, there were no more electrical problems on the quad.

Yet, the continual electrical problems that plagued the quad had taken their toll on Streator's personnel, especially the operators. As one described:

> When you have a jam up down there like that it's really difficult. The glass just keeps running out for a few minutes until you can shut it off. The startup period was really demoralizing—as soon as we got the machine running pretty good, it went down. Once the machine goes down, it's hard to start back up at the level you leave off at, especially if the machine gets cold.

MOLD CHANGES

In early 1987, the tech center sent a new type finishing mold to Streator to replace their second set of molds, which had worn out (molds typically had to be replaced every three to five months). Although the tech center had already been at work designing the new mold, it was the Atlanta plant which gave them the impetus to finish the project quickly. Atlanta had

---

[5]During the plant's Christmas shutdown, tech center engineers came to study ways to change back to a smaller horsepower inverter, but they were unsuccessful.

started up O-I's second 16-ounce quad in January, and even before they had received the machine, they had been calling Streator to find out how their quad was running. Atlanta had learned that Streator was having cooling problems making 16-ounce bottles on the quad and so had requested that their second unit (which arrived in November 1986) come equipped with improved molds.

Unlike the original molds used by Streator, the new mold, referred to as the singleback, had improved cooling characteristics thus cooling the mold quicker and more evenly. Another unique aspect of the singleback mold was that it contained copper cooling rods to speed up heat removal. After using the molds for a few months, Streator personnel found they had to modify the size of the copper rods in an effort to improve their cooling. As the mold repair supervisor stated, "Singleback molds are what everyone's using, so I guess that's what we'll keep using."

Once Atlanta's new quad was operating, Streator personnel began periodically calling Atlanta to learn what they were doing on their quad. Streator was having breakage problems due to the pressure value of the bottles and trouble with light shoulders in their bottles. They found that Atlanta was using a different set of blanks (the molds that made the parisons). Atlanta's blanks put more glass into the shoulder area because the dimensions of the base were wider which enabled the "gob" to be loaded deeper in the blank, thus making the shoulder heavier. Streator ran blanks similar to Atlanta's for one week, but could not get the gob to load properly. Therefore, the molding department redesigned the blank, using the original opening from their own blank mold but keeping the same neck dimensions as Atlanta's. After three weeks, the forming supervisor pulled the modified blank mold and replaced it with Streator's original design. The mold repair supervisor explained, "He said he was getting light shoulders and didn't like the overall appearance of the bottle. He just didn't like it."

Though Streator had adopted the new singleback molds being used by Atlanta and had attempted to use its blank mold design, the plant had opted not to switch to the plate cooling system being used there. Streator used two cooling hoses for each mold, with the hoses blowing air onto the molds. (Atlanta, on the other hand, used a plate cooling system in which air was put directly into the molds; this eliminated the need for two of the four cooling hoses.) Though there were advantages to the system used at Atlanta (e.g., it was easier for operators to change molds without having hoses in the way), Streator felt there were also major drawbacks. Specifically, they felt that using plates rather than hoses allowed the air going into the molds to get warmed up, thereby slowing down cooling. According to Streator's mold repair supervisor, "When the quad was initially brought here, they wanted plate-type cooling. But the molds weren't fit for that type of system, so we went with what we're using now and never changed."

BOTTLE-HANDLING PROBLEMS

In May 1987, tech center engineers brought in a video camera to take slow motion films of the quad's electronic sweepout section. The plant had been noticing problems with the sweepouts when they began increasing the quad's speed from 486 bottles/minute to 506 bottles/minute. The films showed that the sweepout fingers were hitting the hot bottle at the wrong angle; the problems were exacerbated by the quad's increased speed level. By July, Toledo engineers had made some adjustments to correct the problem. Commenting on the latest 10 quad change, the machine repair supervisor explained, "People realize it takes about a year to debug and redesign equipment. Things level out after a year of hard work, and our people understand this."

As in all O-I glassmaking plants, 24 hour production and pack-to-melt measures were continually tracked at Streator. Typically, the hot end had the major influence on output, but sometimes a reduction in the productivity numbers was caused by a bottleneck in the cold end (where bottles were label-wrapped, inspected, and packaged); the production count could temporarily drop by as much as 15 gross of ware in five minutes because of cold end backups. As a result, although operators were supposed to look for defects in the bottles when the production count dropped, they tended to be less concerned about momentary dips in production because they assumed they were caused by the cold end. Bottlenecks in the cold end were often blamed on the poor layout; as Valduga explained, "Cold end handling on the quad has been difficult since the day we started."

PERFORMANCE IMPROVEMENT

Streator's conversations with the Atlanta plant and their subsequent changes on the quad enabled them to increase the quad's speed up to 506 bottles per minute. (See Exhibit 3 for a chronology of changes affecting the quad.) According to the forming supervisor, the plant had recently been able to start increasing speed for three main reasons. As he explained:

> First, we worked on wind build up and got better cooling on the molds. We also worked on glass conditioning, maintaining a lower glass temperature so it would be more uniform. And, we changed the blank cavity design; we opened it up to allow the gob to load deeper. I think we could get the speed up to 512, but it would be a bottle-handling nightmare. Even though that's machine repair's problem to worry about getting the bottles into the lehr, I wouldn't try to increase the speed now. I very rarely try to increase speed on a hot July day like this when the temperature outside is close to 100 degrees.

EXHIBIT 3 Chronology of Events

Month	Electrical Changes	Material Handling Changes	Mold Changes
May '86		Lehr loader; cam modified Conveyor replaced with older style	
August '86	Power line problems (internal and external)		
October '86	Inverter problems		
November '86	Installed 290 inverter		
February '87			Changed to singleback mold
March '87			Tried new blank design but switched back to original mold
July '87		Electronic sweepouts adjusted	

Speed was something that the machine repair department did not look forward to regardless of outside temperature. As the machine repair supervisor stated:

Speed is the enemy now when we're trying to deliver bottles into the lehr. It makes them more unstable. The object is to run it as slow as possible because slowness means stability. We could crank the machine up 10 or 20 bottles, but then we would have to fight bottle handling. We try to do what's best for us here at Streator.

Owens-Illinois: Atlanta 10 Quad

It was a tremendous effort by people in the plant. When you challenge people sometimes they'll rise to the occasion, and I think that's what happened. People worked seven days a week 24 hours a day. It wasn't like the world was going to end if we missed the date, but it became a challenge. It would have been the loss of production and a customer inconvenience, which is hard to put a dollar value on.

Jack Allison, plant manager at Owens-Illinois' Atlanta, Georgia, plant thus described the January 1987 installation of the facility's second 10 quad machine. The installation of the quad, and the furnace rebuild that preceded it, had proved to be extremely tough projects; but once the quad started up, it ran smoothly. Operators had pushed for increased performance and by July 1987 had the machine running at a speed of over 500 bottles per minute; they were continuing to look for ways to increase the quad's speed. (See Exhibit 1 for monthly production statistics.)

PLANT HISTORY

Built in 1957, by its thirtieth anniversary the Atlanta plant had become O-I's largest glassmaking facility producing bottles at a rate of over 30,000 gross per day. As of July 1987, total employment was 1,000 people, 70 of

EXHIBIT 1　D–1 Monthly Production Statistics

	Speed*	Job Efficiency†	Average 24-Hour Production‡
1/6/87 Startup	500	95.0%	4500
2/87	440.3 (Flint)	94.9	4458
	504.7 (Green)	93.0	4604
3/87	505.9	96.4	4674
4/87	502.5	98.0	4751
5/87	504.9	96.7	4689
6/87	501.0	97.7	4716
7/87	501.8	97.4	4694

*Bottles/minute
†Bottles packed/theoretical
‡Gross bottles
Note: On March 23, 1987, D–1 switched from running flint (white) to running green glass.

whom were salaried employees.[1] The plant covered 84 acres and had 1.3MM square feet total under roof which included 900M square feet of warehouse. The plant had gone through numerous plant managers and supervisors over the years and was viewed as a training ground for management. In fact, many corporate executives, including the glass container division's current vice president/manufacturing manager, had spent time as plant managers there.

In addition to its two 10 quad units, Atlanta also operated nine other bottle-making machines utilizing four furnaces. As the operation grew and each new furnace was built, the building was expanded; thus, the layout appeared a bit haphazard. The plant's second quad, D1 or "Dog 1" as workers affectionately called it, was the only machine connected to the D furnace. According to Tom Waller, production manager:

> It's a nice advantage to have one machine on one furnace because of glass conditioning; you can work to get what you want for a particular job temperaturewise, colorwise, furnacewise. It doesn't affect the speed, but it minimizes your upsets.

All machines were shut down every 60 to 90 days for preventive maintenance; in addition, several of the units underwent frequent changeovers, and a few were sometimes converted from double to triple gobs or

---

[1] Hourly employees were represented by the Glass, Pottery, Plastics Allied Workers (GPPAW).

EXHIBIT 2 Furnace and Forming Machines*

A Furnace

A–1	A–2	A–3
8 triple Age: 1 year No changeovers	10 double Age: 6 months No changeovers	10 quad Age: 4 years No changeovers

B Furnace

B–1	B–2	B–3

10 triple
Age: 9 months
Changeover frequency: Twice
   per year
	8 double
	Age: 9 months
	Changeover frequency: Twice
	   per month
		10 double
		Age: 8 years
		Changeover frequency:
		   Three times per month

D Furnace

D–1
10 quad Age: 7 months No changeovers

E Furnace

E–1	E–2	E–3	E–4

10 triple
Age: 10 years
Changeover frequency:
   Once per year
	8 triple
	Age: 1½ years
	Changeover frequency:
	   Twice per month
		10 triple
		Age: 6 years (2 months
		   in Atlanta)
		Changeover frequency:
		   Four times per year
			8 double
			Age: 6 years (4 yrs. in
			   Atlanta)
			Changeover frequency:
			   4–5 times per month

*There were 56 job changes and one conversion between January and May 1987.

vice versa. (See Exhibit 2 for details on furnaces, machines, and change-overs.) The glass containers produced on the various machines were for customers in the soft drink, liquor, and food industries. Typically, the machines used glass that contained cullet in the range of 10 to 40 percent. Furthermore, Atlanta's set time was 105 seconds for flint (white) glass and 103 seconds for green glass.

Because glass consistency was crucial, viscosity and set time were monitored daily by the batch and furnace operation. In addition, Atlanta's batch and furnace personnel measured glass density, which was typically measured by the quality department.

PLANT MANAGEMENT

In early 1985, a new plant manager, Jack Allison, arrived at Atlanta. Allison had been transferred from the Streator operation where he had been for three years; before that, he had worked in a number of other O-I facilities. Allison had the reputation of being a very demanding manager who placed extremely high standards on his management staff. As one worker described, "Jack's patience only goes so far with you, and when he runs out of patience, he's done with you. He measures people on results, not how hard you work or how many hours you put in." After Allison had been at the Atlanta plant for one year, its pack-to-melt percentages rose 2.4 percent. This was a significant increase, especially because the plant's pack-to-melt figure had been at 2 to 3 percent less in the three years prior to his arrival.

Allison looked to Waller, an experienced glassmaker, to oversee the daily technical running of the hot end. Waller had a similar management style to Allison's. As Waller described:

> I probably demand too much from my people. My people tell me that. They say I'm never satisfied. I have very high goals and ideals. Jack does too. Maybe that's why we've always gotten along. We both believe in dealing with problems immediately, and getting results. I don't look at problems and say "It's not that bad." I say, "It's not good enough." I like to be involved in everything, and I believe strongly in the chain of command. I give my people a half-hour to solve problems on any machine, and if they can't, they have to call me.

THE FIRST QUAD

Even before O-I's first 10 quad was installed at the Winston-Salem facility in November 1982, Atlanta had already been notified it would receive the second one. Although Atlanta had been targeted as a potential site for the first quad, Waller had told the tech center the plant was not ready for it. This was because in late 1982, Waller, who had come to Atlanta especially to introduce the 10 triple to the plant, was still in the process of "selling" triples to the workers, many of whom were reluctant to embrace the new technology.

Since Waller knew Atlanta would ultimately be getting its own 10 quad, he had kept up to date on the machine's development. In September 1982, for instance, he and his machine repair supervisor visited Godfrey to examine O-I's first quad when it began dry cycling.[2] Later, before Atlanta's

---

[2]"Dry cycling" was a method of running the machine without any glass to check for mechanical or electrical problems.

own quad started being built, Waller suggested changes for the new unit.

In April 1983, Atlanta's quad arrived, and by May it was running three shifts per day staffed by one operator per shift. Initially, there had been some concern that two operators would be required for the quad. However, experience had shown the machine could be handled with one operator. The quad's operators had been handpicked by management. An assistant supervisor explained how this was done:

> We made an agreement with the union and told them about the quad. We said, "It's going to scare the hell out of you when you first see it; it's big and it's fast. We really feel we have something here, though." So we were able to handpick the operators through a team effort with the union. It was the first time we did anything like that. We did that on the second quad too.

As part of their training, the operators assigned to the quad traveled to Winston-Salem, along with a shift supervisor, forming supervisor, and day shift specialist to see O-I's first quad in operation.

Shortly after starting up the new quad, Waller began trying to increase its speed. He was able to increase the speed by 35 bottles per minute, but the quad experienced sweepout motor problems when he tried to go faster. The tech center designed a new profile (computer software) which enabled Atlanta to speed up the motors, and at the same time, new sweepout fingers were designed in house by the machine repair group.

As Atlanta personnel continued to test the quad's limits, the curve chain cam and then the lehr loader became the next roadblocks to increasing the machine's speed. Throughout 1986, Atlanta consistently asked for help, but since no other plants were complaining about the problem, the tech center did not view the problem as a high priority. Finally, in late 1986, Waller wrote to Lemke (with a copy to the vice president of manufacturing), because as he put it:

> I thought they were dragging their feet trying to help us. They said, "We have higher priorities and don't have time to work on your bottlehandling problems. Do it yourself." I had pushed my machine repair supervisor to his limit, and he didn't know what else to do. So I wrote telling them how I felt. It went high enough up that we finally started getting some attention.

On December 1, 1986, new curve chain and lehr loader cams that had been designed by tech center engineers were installed on the quad. After using them for a short time and noticing an improvement, Waller ordered additional cams for the second quad, which had just arrived and was scheduled to begin production by January 5.

THE INSTALLATION OF ATLANTA'S 16-OUNCE QUAD

In May 1986, corporate management recommended putting another 10 quad unit in the Atlanta plant because of market demand for 16-ounce bottles in the South and Southeast. They had initially considered reopening a plant in Louisiana or putting equipment into their Florida facility but eventually decided that the best ROI and least cost could be obtained by rebuilding a furnace that had been idle for four years at the Atlanta operation. On June 12, 1986, O-I's board of directors approved the request.

In preparation for the new equipment, a contractor was hired on October 29, 1986, to rebuild the D furnace, which included raising the metal line three feet. It soon became apparent that the contractor was having problems with the project; at one point things got so bad that the contractor's own superintendent quit. The tech center sent a furnace rebuild engineer and a refiner and forehearth specialist to Atlanta to advise the contractor.

Due to the delays caused by the contractor, the furnace had to go through an "accelerated" rebuild in order to be ready by January 5, 1987, the date Allison and his manager selected for the start up of D–1. As Allison noted, "We needed to get into production as soon as possible, especially since the A furnace was going to be down. We also needed the production from the quad to meet the forecasted business demands."

On December 8, Joe Smith, the tech center's forming specialist, arrived at Atlanta. His role was to work with the plant in installing new equipment and training local personnel. When he got to Atlanta, the quad was waiting on the loading dock. Before moving it into position, he worked with plant personnel to make corrective changes on the quad's guide air system (a means of handling hot bottles without contacting them by using low-pressure compressed air that blew on the bottles) because Atlanta had learned that Streator was having problems with that particular system on their quad.

As the installation program got underway, Waller ordered all change-overs to be postponed so that he, Gene Marshall (the forming supervisor) and the day specialist could focus solely on the new machine. In mid-December, electrical wiring of the quad began. The electrical contractor was having problems with the project, however, partly due to inadequate electrical prints; according to one engineer, the tech center had not chosen the best locations on the machine for the electrical equipment and the plant never had time to review them. In addition, the electrical job typically entailed rewiring an existing machine, but since the furnace had been idle for four years, they were starting from scratch. Wayne Leidy, the tech center electrical engineer assisting in the installation, took over and began instructing the contractor's workers about what they needed to do. Technicians had to work seven days a week, 12 hours per day to complete the project. (After D–1's startup, Allison urged Leidy to stay on as the plant's electrical supervisor, which he agreed to do.)

The lengthy wiring project led to delays in checking the machine's electrical and mechanical functions. On December 29, they finally started programming the quad. On December 30, the remaining functions were programmed, and for the next two days they were debugged; however, technicians were not able to test the machine completely because not all of the wiring had been finished.

The quad began to be dry cycled on January 2 and 3 but only for a few hours each day since the wiring and the air lines were still being hooked up. On Sunday, January 4, glass ran in the quad for the first time. Waller, Marshall, and the day specialist worked through that night preparing for the machine's first production run on Monday.

Waller was impressed with the plant personnel's efforts throughout the installation. He noted:

> Everybody we had on the quad told us the problems instead of just throwing their hands up; there was good communication among people. They had good attitude, and I think that was key. We didn't have time to trouble shoot the machine before running with hot glass. But I can understand Jack's position—we had to work quickly.

There was indeed an underlying sense of urgency surrounding the quad's installation, but plant employees were used to dealing with pressure. As one employee declared, "We had a date to meet and we made it because you don't miss dates around here."

The Atlanta plant had been aided in its mission by extensive help from the tech center. According to one tech center manager:

> Atlanta had more engineering support on the quad project than any other project of that magnitude throughout O-I. We all realized the importance of a good startup, and Jack Allison was not about to let us forget it.

D–1'S STARTUP

Once the quad was up and running, it was initially staffed by operators from the first quad, A3. The A furnace had gone down for a rebuild during Christmas and was not going to be restarted until February 9. In mid-January, double staffing on the quad began with new workers receiving training from the experienced employees.

The new operators chosen to work on the second quad were, according to one supervisor, "People who had good attitudes. They looked at the quad as a challenge rather than a job. They saw it as a progressive machine." They were responsible for keeping the machine clean, sampling bottles several times per shift, and swabbing, a method of lubricating the molds so the gob would slip easily into the metal mold. In addition, every three to five months, the operators were responsible for installing new molds; since each section could be shut down independently, molds were typically changed while the machine was running. The quad's operators earned $.32 more an hour than operators of the 10 triple, which was the next highest paying machine. In addition, there was a corporatewide bonus system which gave the workers incentive for increasing output.

During the first days of production, the quad periodically shut down for no apparent reason. It was immediately restarted. After a week and a half of continued problems, Atlanta called the tech center for help. Soon after arriving at the plant, tech center engineers found the source of the trouble. The quad and the batch house mixer were tied to the same power line as a sand elevator; when the elevator was running and the mixer kicked on, it caused a drop in the power level which affected the quad. The engineers had been able to detect and solve the problem quickly because they had gone through a similar experience at the Streator plant.

Shortly after the electrical problems had been solved and the quad began running consistently, the machine repair supervisor changed the machine's conveyor belt to a smoother belt with a smaller pitch,[3] which was identical to the one being used on the first quad.

MOLD CHANGES

Once Atlanta heard they were getting a new quad to produce 16-ounce bottles, they had called Streator to discuss their experience with the machine. This led them to use Streator's timing values as a basic starting point as they began production. As a result, the quad started operating at a speed of 486 bottles per minute, but the bottles were "out of round" because the molds were cooling too quickly. The molds that Atlanta used on D–1 were a new design not used on A–3 or other quads. This was because Atlanta had learned from Streator that they had been having cooling problems making the 16-ounce bottle and that the tech center was already at work designing a new mold. Atlanta pushed the tech center to finish the new design so that the molds would be ready for use on D–1. "The squeaky wheel gets the grease. I guess Atlanta was the squeaky wheel," Waller commented.

Unlike older style molds, the new mold, referred to as the singleback, had improved cooling characteristics, thus cooling the mold quicker and more evenly. Another unique aspect of the singleback was that it contained copper cooling rods to speed up heat removal. Ironically, the singleback design was so effective it was cooling the molds too quickly. In an effort to solve the problem, Waller decided to increase the quad's ma-

---

[3]Pitch was the distance between teeth on the sprockets.

chine speed to over 500 bottles per minute, thereby running more hot glass through the molds to keep the iron hotter. The cooling problems were thus solved.

OTHER FORMING CHANGES

In April, the plant switched from using cooling hoses, which blew air onto the molds, to using plate cooling, which piped air directly through the molds. According to the mold repair supervisor:

> Our first equipment duplicated Streator's equipment because we felt they must have solved some of the bugs if not all of them. But, we decided to change the equipment when we saw it hooked up with all that garbage [hoses] on it.

The hoses had had to be continually replaced because of the twisting and fraying they experienced each time the mold cavities opened. Plate cooling made it possible to eliminate two of the four cooling hoses on each mold which made it easier for operators to work on the machine, especially when changing molds. Although the change to plate cooling did not directly increase the quad's speed, it reduced mold changeover time and downtime.

Although Atlanta had initially opted to try Streator's cooling system, they had decided against using the blanks used by them. The blanks that Atlanta used on D–1 were the same design as those on A–3 and all other machines in the facility. Atlanta had examined the Streator quad's blank, which was different from their own, when they found out they would be getting D1, but they felt it would be too difficult to load the gob into the blank, and they wanted to keep the same blank throughout the entire operation.

BOTTLE-HANDLING CHANGES

In late April, Atlanta began fine tuning the quad's bottle-handling process. For example, the machine repair supervisor worked on increasing the lehr loader's performance. He first tried to get it to load 42 bottles across, but the gauging equipment would not work. He increased the number to 44 bottles, but then the curve chain was not in synchronization. He kept adjusting the machine and eventually changed the shift cam and curve chain so that he could find the perfect mix: 46 bottles per loader bar sweep. Also during this time, an operator suggested putting sweepout fingers like those used on A–3 onto D–1 since he felt they might work better, and indeed they did. A tech center manager commented on Atlanta's bottle-handling finesse: "Atlanta is a forerunner in bottle-handling. They know all the tricks because they've gone through it with the first

EXHIBIT 3　Chronology of Events

Month	Electrical Changes	Material Handling Changes	Mold Changes	Cooling System Changes
Oct '86			Ordered new singleback molds for second quad	
Dec '86		Ordered new curve chain and lehr loader cams for second quad		
Jan '87	Changed wiring of internal power line	Replaced conveyor belt		
April '87		Built new sweepout fingers Changed shift cam and curve chain		Changed from hose to plate cooling

quad. Also, they've got several key people there who really understand the process." (See Exhibit 3 for a chronology of changes affecting the quad.)

WALLER'S PROMOTION

In July 1987, Waller received word that he was being promoted to the position of plant manager at the Winston-Salem operation. His knowledge of glassmaking and his performance throughout his career (especially while at the Atlanta plant) had earned him the promotion. Marshall, the forming supervisor, had been chosen to replace Waller as Atlanta's new production manager. Marshall had been a key figure throughout D1's installation and had a reputation for being highly skilled. A tech center engineer described him by saying, "Gene used to work at the tech center and so has done a lot of installations. He's technically sharp and is used to handling some difficult startups. He also knows how to handle people."

AT&T Communications: Connecticut Custom Services District

You have to balance as best you can the skills and needs in this office with the needs and skills of people at the new office, with an underlying sensitivity to individual people's needs and preferences.

That was what Lynn Curtis, operations manager of the Hartford Serving Test Center (STC), considered his challenge to be for the five-year transition from local to total remote testing in the Connecticut District of AT&T Communcations' Custom Services Unit. In May 1985 AT&T Communications (AT&T-C) faced an environment and a technology that were totally different from what Lynn Curtis had experienced when he started with AT&T 26 years before. Even in the five years he had managed the STC in Hartford, the squeeze of competing with other long-distance phone companies had made it increasingly important to take full advantage of the new technologies at hand.

Long-distance line testing had been revolutionized by two processes: digital transmission (rather than analog) and remote testing by computer (rather than jack testing on-site). Digital transmission was a significant improvement over analog because the signals could be transmitted more cleanly for truer reproduction, which when combined with better con-

---

ducting mediums like fiber optics, provided even clearer, higher quality transmission. The Remote Test System made it possible to test a variety of complex special circuit connections end-to-end from a remote location. A centralized computer (that controlled a network of minicomputers) could sectionalize and locate faults on circuits. Then, a technician at a remote site could activate a terminal that would test circuits from that end and relay the information back to the central site. Problems could thereby be diagnosed without having people actually on-site. The system would ultimately be able to locate faults automatically. The company's long term objective was to convert to digital circuitry completely. Until all circuits were converted to digital, there was software available to remote test analog circuits which was more cost-effective than jack testing on site.

With its potential for reduced costs and improved efficiency of testing procedures, this new technology could not be ignored. The Connecticut District had decided to consolidate the testing for the entire state in a Special Services Center (SSC) in Wallingford, Connecticut. Frank Falciano, district manager of Connecticut Custom Services, had responsibility for planning and managing the implementation of the new SSC. Within his district, five facilities were affected: Hartford, Stamford, Bridgeport, New London, and New Haven. (See Exhibits 1 and 2 for a map of the facilities and an organizational chart.)

Curtis had been a member of the transition team that had worked for over a year to plan for moving to Wallingford. During that time, he had become aware of the changes that would occur: The SSC would have an impact on every level of service delivery and would affect the nature of line jobs from clerk through manager. He also was aware that once the implementation process got underway, problems that the team had not anticipated could arise and jeopardize the improvements they had carefully planned.

AT&T COMMUNICATIONS

American Telephone & Telegraph, which operated as a regulated monopoly until 1984, was broken into intrastate and interstate portions. Both local Bell operating companies and a number of independent telephone companies, not part of the Bell system, provided intrastate service; interstate service was provided by the AT&T Communications (formerly called Long Lines) division, which managed a nationwide telecommunications network connecting the local operating companies.

The AT&T-C network could be viewed as an interstate highway system, with cars, buses, and trucks all traveling to certain destinations, getting on and off the highway at various connection points, moving at dif-

EXHIBIT 1   Location within Connecticut of Wallingford SSC and STCs

SSC	STCs
Wallingford	Bridgeport
	Hartford
	New Haven
	New London
	Stamford

ferent speeds in different lanes. When well functioning, the system could handle whatever kind of transport was required and route it efficiently. For AT&T-C, the "transport" was voice or data communication.

A typical interstate call involved three different entities. The calling telephone subscriber's local phone company provided the initial link, Point (1) in Exhibit 3. Then the message was sent by the local carrier (2) to an AT&T-C STC (3). Next, it was sent over appropriate interstate facilities (AT&T-owned equipment) to the general destination area. Finally through the same process in reverse, the message reached the receiving telephone over the lines of the local company at the receiving end. If there were a failure in the connection, at any point, it was up to AT&T-C to find it and have it fixed.

Responsibility for maintaining the AT&T-C system was divided into regions, each of which had two divisions: *Custom Services* (customized services for business accounts) and *Network Services* (the standard long distance telephone service). Before divestiture, local phone companies

EXHIBIT 2   Organization Chart as of May 1985

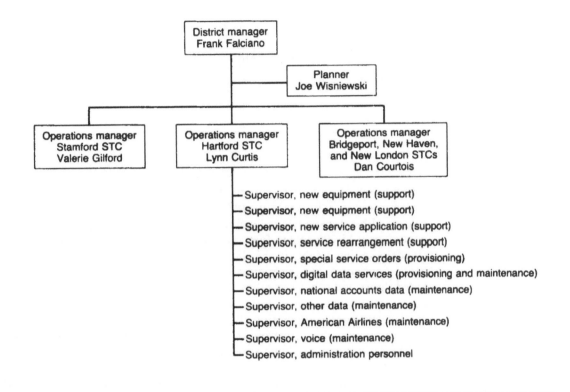

District manager
Frank Falciano

Planner
Joe Wisniewski

Operations manager
Stamford STC
Valerie Gilford

Operations manager
Hartford STC
Lynn Curtis

Operations manager
Bridgeport, New Haven,
and New London STCs
Dan Courtois

— Supervisor, new equipment (support)
— Supervisor, new equipment (support)
— Supervisor, new service application (support)
— Supervisor, service rearrangement (support)
— Supervisor, special service orders (provisioning)
— Supervisor, digital data services (provisioning and maintenance)
— Supervisor, national accounts data (maintenance)
— Supervisor, other data (maintenance)
— Supervisor, American Airlines (maintenance)
— Supervisor, voice (maintenance)
— Supervisor, administration personnel

had been intimately involved with long distance transmission; now these local companies, like Southern New England Telephone (SNET) in Connecticut, became "suppliers" since they were no longer directly in AT&T-C's system.

THE CONVERSION TO REMOTE TESTING

For the Connecticut District, implementing the SSC was a response to competitive cost pressures and growth; on the single factor of 24-hour coverage, for example, a major savings would be achieved by consolidating the service for several STCs into one SSC. (See Exhibit 3 for a summary of the cost benefits.) Another major savings would be realized

EXHIBIT 3   Carriers Involved in Long Distance Transmission

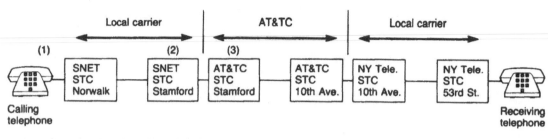

through reduced space requirements.[1] (Hartford had already been projected to outgrow its space by 1989.) An STC required one acre of floor space, with intensive floor-loading capacity and high ceilings; an SSC needed only 15,000 square feet, with standard floor-loading capacity and standard ceiling height. In the SSC, space would no longer be filled with heavy equipment and the rooms would be cleaner and quieter, with carpeted floors and other amenities. (Even in STCs that were converting to digital equipment, space costs were reduced because equipment was smaller and lighter—requiring altogether one-twentieth of the space.) It was estimated that Hartford could save $654,000 per year, while $1,354,000 per year could be saved in rental costs for the entire Connecticut district.

From a technical standpoint, only one or two remote testing centers would be necessary for the entire North East Region (Maine to New Jersey). However, because Connecticut had 17 national accounts, along with numerous smaller accounts, from Customer Service's perspective, the state should have at least one SSC to maintain service quality. The optimal range of serving links for an SSC was between 25,000 SL (serving links)[2] and 35,000 SL, and the projected 1989 need in the district fell within that range. An SSC would enable the district to handle the growth without a corresponding increase in personnel: Frank Falciano, district manager of Connecticut Custom Services, hoped for a 10–15 percent annual increase in growth with a 0 percent increase in people over the next

---

[1]Space costs per square foot at the Hartford analog location were $75/SF and $105/SF for the digital location; space at the SSC would cost $19/SF.

[2]A serving link was one customer line that ran from an AT&T-C office to a customer's premise.

EXHIBIT 3   Projected Savings of SSC

	1984	1989 Projections
Number of technicians	130	140
Number serving links	24,000	40,000

STC Test (Nondiscounted)

Year	Capital	Expenses
1985	$8,353,100	$36,248,200
1986	544,600	38,712,100
1987	1,055,200	42,625,600
1988	1,108,900	46,771,800
1989	1,143,800	50,549,800

SSC Test (Nondiscounted)

Year	Capital	Expenses
1985	$9,452,200	$21,763,100
1986	1,228,100	20,678,300
1987	1,481,100	22,632,100
1988	757,500	24,892,400
1989	1,039,800	27,081,400

---

five years. He also believed a new SSC facility, with its state-of-the-art technology and modern architecture, could be an attractive marketing tool.

Other benefits would accrue from an SSC. Interface with the district's major supplier, Southern New England Telephone, also located in the state, could be more easily managed. Employees would not have to relocate (at a cost of $45,000 each). And finally, there would be a reduction in "call-outs." Currently, if service were down, technicians would not know if equipment had failed in Hartford or New York City, and they would have to call the offices to check equipment and sectionalize the trouble. With the remote test system at an SSC, all the testing could be done from one site.

To have these benefits realized and to compete effectively, implementing the conversion had to be done quickly. Not only could capital cost benefits be maximized, but current employees would be more likely to support the new technology if they realized their jobs were secure.

THE WALLINGFORD SSC

The decision to go ahead with an SSC was made at a district staff meeting in June 1983, and the task of researching the particulars of the Connecticut SSC was assigned to Joe Wisniewski, a 12-year AT&T veteran who had joined the district staff as a first-line planner in December 1982. His first six months were spent planning the generic size of the SSC (the number of people required, the number of serving links which could be accommodated, the number of circuits to be involved) and the optimum geographic location in Connecticut. During that time, he visited an SSC in Denver and also investigated similar technologies at other companies.

In December 1983, the decision to establish an SSC in Connecticut was formally communicated to all first-line supervision in the district. This was followed by another presentation in January 1984 to the local union representatives of the Communication Workers of America (CWA). In the spring of 1984, the three (second-line) operations managers of the key facilities joined Falciano and Wisniewski as members of the transition team, which met monthly. These managers were: Lynn Curtis of Hartford; Valerie Gilford of Stamford (manager since January of 1984); and Dan Courtois who managed New Haven, Bridgeport, and New London. Courtois had been with the company for 30 years and had served as an operations manager for ten years. Gilford, who had seven years with AT&T, was on a management development track to become a district manager.

There was some disagreement about and resistance to the idea of a centralized SSC, but once the lease was signed for a building in December of 1984 and a formal "transition book" was assembled, any skepticism as to whether the SSC would be a reality disappeared. The company objective in planning the new facility was to involve technicians and clerks to the greatest extent possible. Executive vice president of the CWA, John Trainer, joined the team in January 1985. Joe Barone, a first-line supervisor from the support group at Bridgeport, joined the next month; and in March, Steve Johnson, a first-line supervisor from provisioning in Hartford, joined to develop methods and procedures for work flow at the new facility.

SITE AND BUILDING SELECTION

In April 1983, in anticipation of the SSC, Wisniewski researched potential locations. He first focused on where employees currently lived, believing that if a majority of the work force could easily commute to the new site, relocation would be minimized and those workers outside the reasonable-commute distance could work at one of the three other SSCs planned for

---

EXHIBIT 4   Site Study

Total Number of People within 33 Miles of Proposed SSC Sites

Proposed Sites	Number of People	Percent of Total Connecticut Force (185)
Bridgeport	103	56
Hartford	97	52
Meriden	120	65
New Haven	96	51

Commuting Distance (Currently)

STC	Miles—Average One Way
Bridgeport	14
Hartford	18
New Haven	9
New London	17
Stamford	18
Average	15

Commuting Distance (to Proposed SSC Site)

Proposed Sites	Miles—Average One Way
Bridgeport	18
Hartford	21
Meriden	16
New Haven	12

New England. He took a 30-mile radius from the proposed sites, which was later expanded to a 35-mile radius. Two areas became obviously workable: Greater New Haven and Hartford. Meriden fell in the middle. (See Exhibit 4 for excerpts from the site study.)

One potential site was an existing building in Bridgeport, jointly owned by AT&T-C and Southern New England Telephone, but it would not have available space until 1987. In addition, it was an equipment building constructed with no windows and was in a bad section of town where workers' cars had been vandalized. Frank Falciano also wanted to get out of the "trap" of locating in buildings owned or leased by AT&T-C and create a fresh start in a brand-new environment. The other SSCs in the country had been put in existing buildings, and the only real changes had been in the name of the facility and work assignments.

From March through May 1984 meetings were held with supervisors, technicians, and clerks to ascertain their wishes for the new facility.[3] Several concerns were raised. Technicians wanted windows at their work stations and to be near banks, stores, restaurants, etc. They also wanted ample parking and were concerned about traffic patterns for commuting.

In May 1984, the search for a facility began. An appropriate one-story modern brick building was found available for lease in Wallingford, near Meriden. The facility satisfied most of the concerns: It was a 20-minute drive from both Hartford and New Haven and was only three miles away from a major artery (Rte. 91) and close to another (Merrit Parkway). In June, the decision was made to locate the SSC in Wallingford.

HARTFORD: A TYPICAL STC

The environment of the Serving Test Center at Hartford was more like a living museum than an office or a factory. Ceiling-high metal racks (bays) supported intricate ribbons of wires leading into large cables that then went through the floor and out of the building. The facility employed 91 people and housed equipment that served over 500 customers, making it the largest STC in Connecticut.

Most of the technology was analog (much of it outdated electromechanical) so that on one floor the clatter of relays from the equipment that had been installed in the 1960s and 1970s was audible. There was even older equipment that was heavier and larger. Altogether the equipment occupied three floors in two buildings.

On average, 20 days were needed from customer request to service delivery. Seven to eight hundred orders were run per month, about 20 percent of which were new lines while the remainder was service rearrangement (changes in existing lines). For example, in the previous five years, a major insurance company in Hartford had changed service requirements five times.

A typical order was initiated by a salesperson who took the necessary information from the customer, found out from a "cookbook" exactly what parts were required to fill it, and wrote it up for engineering. There the order was interpreted and translated into electronic terms. Then, at the STC, circuit assignments were made, supplies were checked, and coordination was arranged with other AT&T-C offices. Customers could receive either voice or data service transmitted by fiber optics, radio, or coaxial cable.

---

[3]In 1980, AT&T had negotiated a joint labor-management Quality of Life Work Program with the CWA which had steering committees on all levels of the organization. Managers and technicians within the Connecticut district had established problem-solving teams to work on issues of workplace change.

Lynn Curtis was responsible for the equipment and maintenance of 10,000 SL of AT&T-C's Custom Services for the Hartford region. His organization was divided into three groups—support, provisioning, and maintenance. The *support* function (which would be staying in Hartford) was responsible for the installation and wiring of the equipment. *Provisioning* was the process of creating equipment assignments to hook up a new service or, more likely, change an existing one. *Maintenance* was the function of testing and trouble shooting existing service. The last two groups would be transferred to the Wallingford SSC. Both provisioning and maintenance required a high degree of customer involvement. Clerks, technicians, and managers used an interactive software system to track all information and critical dates for the orders and to respond to customer questions about implementation and timing. The work flow, as described below, is diagrammed in Exhibit 5.

Eleven supervisors reported to Curtis. Some were responsible for specific products (such as data or voice services) while others were assigned major customer segments. To maximize their utilization, technicians were part of a pool and were organized by the function they performed, such as support, provisioning, or maintenance.

Support. Support technicians, organized into four groups, serviced the provisioning and maintenance groups and had little direct customer contact. Two groups could be considered as the "hardware" people: one group worked on newer digital equipment and one on the other analog circuits. These technicians received new equipment, tested it, installed it, and maintained it. The other two groups were responsible for the "software" part of the orders.

EXHIBIT 5 Hartford STC Work Flow

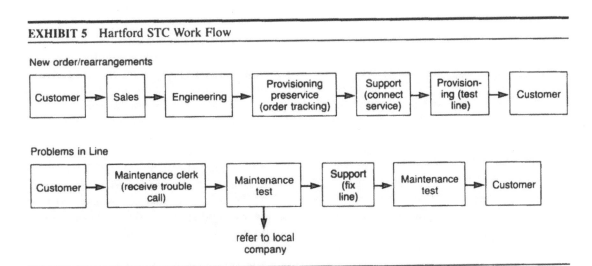

New order/rearrangements

Customer → Sales → Engineering → Provisioning preservice (order tracking) → Support (connect service) → Provisioning (test line) → Customer

Problems in Line

Customer → Maintenance clerk (receive trouble call) → Maintenance test → Support (fix line) → Maintenance test → Customer

Maintenance test → refer to local company

When new lines were wired in, a support technician who did the installation would find the assignment for the wire via the computer support system that maintained the inventory. Then the technician physically translated that inventory into the wiring pattern, connecting it to the appropriate local carrier and interstate facilities.

The individual wires were electronically combined first in groups of 12; then five groups of 12 were combined for a group of 60; then ten of those were finally combined for a group of 600 which left the building as a complex single signal on a coaxial cable, microwave system, or fiberoptics. This was the physical mechanism for the "interstate highway system"—different types of transmissions traveling at different rates with different destinations through these facilities at the same time. As the groupings grew bigger, the skills to maintain, install, or test became more complex.

To serve Hartford region customers, lines left the building and went to their destinations via a local telephone office (e.g., Southern New England Telephone) where there were sets of bays similar to the Hartford arrangement; then the wires were cabled to a specific destination (such as American Airlines). The whole transmission process was a layering of a simple single circuit application: one customer phone, one customer voice or data. To add lines, new wires were added (physically soldered in).

Provisioning.   The "preservice" function received new orders or service rearrangement requests. The process began with an accounting packet (paper documentation), made up by clerks, containing a wiring spec that designated what type of wires were needed and where they needed to be run. This information was then sent to the support technicians who did the physical installation of the line.

After a new line was wired in, a provisioning technician tested the circuit by physically plugging into it with a jack attached to a feedback mechanism. The task was to determine if the line was ready for customer use. At least one technician was on duty 24 hours a day.

Maintenance.   Maintenance technicians were responsible for maintaining quality service for the customer. When a trouble report was called in by the customer, a clerk or technician entered the customer circuit number in a computerized tracking system and a serial number was automatically assigned to the problem, which was then recorded on a "ticket."

The maintenance technicians picked up the tickets by entering into the computer system to judge (by date, time, and service requirements) which reports were priority. The tube also indicated which jobs had already been referred out of the office, for example, to Boston, for someone else to repair. (Problems were tested locally before another city was called.)

Technicians first looked at the wiring diagram in the packet to understand where in the circuit the problem would most likely be. The testing technique used by maintenance technicians was similar to that used by testers in provisioning. It basically answered two questions: Is the circuit functioning properly? If not, where is the break? The technician then transferred the problem "diagnosis" to the support groups for the physical repair process. Once the repair had been completed, the technician once again tested the line before notifying the customer that the problem had been corrected.

WORK FLOW AT SSC

The new SSC would provide not only a new environment (see Exhibit 6) but an opportunity to improve the work flow. The transition team's first objective was to standardize both the procedures between the five STCs and the individual clerk and technician tasks to take advantage of the economies of scale of having the volume from all the STCs in one office. At the same time, the team wanted to increase the clerks' and technicians' identification with the customer. The resulting layout placed eight technicians and/or clerks in a "pod." Two pods (one for provisioning and one for maintenance) were assigned to the existing geographic designations under each of the three operations managers, making six pods in total. Two preservice pods were allocated for the initial customer contact (one for order receiving, the other for receiving trouble calls). See Exhibit 7

EXHIBIT 6 Views of the Hartford STC and Wallingford SSC

Hartford STC

Wallingford SSC

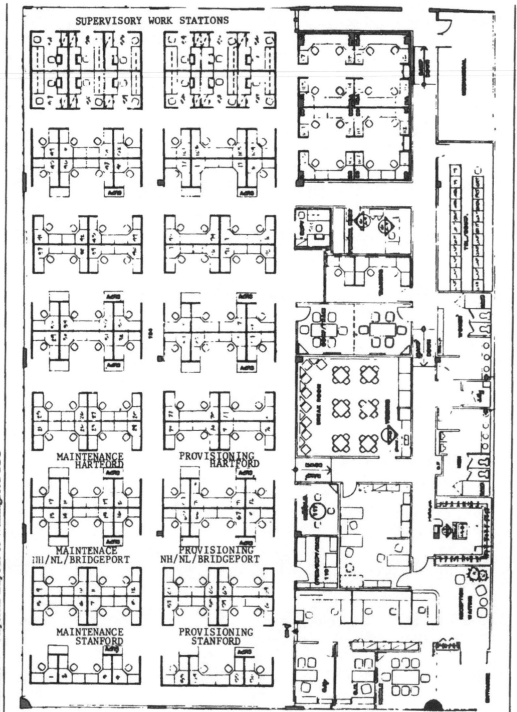

EXHIBIT 7   Physical Layout of Wallingford SSC

EXHIBIT 8   Wallingford SSC Proposed Work Flow

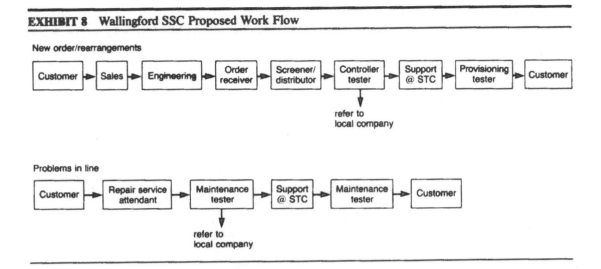

New order/rearrangements

Customer → Sales → Engineering → Order receiver → Screener/distributor → Controller tester → Support @ STC → Provisioning tester → Customer

refer to local company

Problems in line

Customer → Repair service attendant → Maintenance tester → Support @ STC → Maintenance tester → Customer

refer to local company

for the entire layout (the four additional pods were for two other functions being transferred to Wallingford).

Rather than having a pool of technicians and clerks to assign as needed, each clerk and technician would be given a more specialized task at the SSC. The proposed work flow is diagrammed in Exhibit 8. To improve the efficiency and quality of order handling, new orders or service rearrangements were to be received by a clerk, designated as an "order receiver." Similarly, in an effort to improve customer service, trouble calls would be taken by a "repair service attendant." The service difference, in comparison with the STC where potentially hurried technicians handled multiple tasks, would be analogous to calling an airline for a question about a flight and having a worker in the hangar pick up the phone and mumble "yeah?" instead of having a customer-oriented service representative as the first company contact.

For new orders or service rearrangements a technician, called a "screener/distributor," checked the engineering documentation and ensured that everything was ready to be installed. This was a totally new role. (In the past, technicians would discover errors only as the installation was being done.) Up to this point, the order remained in the preservice pods. Within these pods, there were also installation and maintenance analysts—experienced technicians who would be available for problem solving and other special assignments.

The order would then proceed to a provisioning pod where another technician, or "controller tester," would coordinate the process of the order completion. This person would work with the support groups remaining at the STC or with other local exchange companies and STCs outside AT&T-C's Connecticut district. Similarly, a tester within a maintenance pod would handle trouble calls. After the order or trouble call was completed/corrected by the support technicians, a group of testing technicians would check the circuits and notify the customer that the line was in service.

Clerks and technicians had been asked whether the supervisory work stations should be in or outside of the pods. They decided they wanted the supervisors close by to help out with problems but not "hovering." Managers and supervisors wanted to have an area away from the work so they could have a quiet desk, but they still wanted to "keep a hand in it." It was jointly decided that there would be a supervisory work station with a table, phone, and an in-basket in each of the work areas. However, when things were quiet or when they were on special projects, the supervisors could go back into a supervisory area (12 cubicles on the far side of the facility where they would have a desk, file cabinet, and a CRT tube to tune into activity in the work area). It would be up to each individual supervisor to determine how much time would be spent directly within or away from the pod. Management hoped that this arrangement would encourage supervisors to avoid the traditional "overseer" role.

IMPLEMENTATION ISSUES

As he contemplated the move to Wallingford, Lynn Curtis was concerned about his ability to continue to deliver the same high level of performance in service during the transition. As an operations manager, he was evaluated against four measures: (1) on-time preservice delivery, (2) minimum service interruption on trouble calls, (3) a high customer satisfaction index, and (4) per-unit cost (measured in dollars per serving link). Ultimately, he was accountable for the maintenance of that service through all changes affecting his customers' lines, including the upcoming transition. That transition was complicated by a number of factors:

Line Conversion

The conversion from analog to digital circuits coupled with the change from on-site to remote testing meant that at some time customer service would be interrupted. The goal was that Hartford, being the largest STC, would begin the conversion to digitalization at a rate of 20 percent of their lines per year with the aim that all five STCs would be completed by 1989. The process of converting analog lines interrupted service for one to two hours so the conversion had to take place at off-hours, most often in the middle of the night, to minimize the negative impact on the customer.

In addition, during the transition period, there was the risk of a fragmentation of service: A major customer could have some lines on analog and some on digital. Until the SSC got on-line there would be the added confusion of how and to whom (the SSC or the STC) the client would report trouble. The plan was to have the SSC handle all customer contact via one "800" number.

Equipment Delays

The SSC being implemented at Wallingford was to employ an updated version of a remote test technology. After divestiture, there had been a last-minute change to an AT&T Technologies software package. The Remote Test System software was still being modified and as of November 1984 was not available. The developers of the software promised it would be available in January 1986, but until then, testing would have to be done at the STCs.

To address this problem, a transition plan, consisting of two implementation phases, was developed: The first involved the movement of all non-testing operations (i.e., preservice coordination and centralized trouble reporting); the second would involve testing functions that would commence in the first quarter of 1986. For the first phase, Bridgeport was the first to transfer in May. New London and New Haven were next, and Stamford was scheduled for July. Hartford's preservice group was scheduled to move August 26.

Relocation of Personnel

Although Joe Wisniewski had done a thorough analysis of the optimal site to minimize the disruption on employees' lives, the physical relocation was not expected to go smoothly. The union wanted the transfers based on seniority, but the company was determined to use multiple factors. To assess employee preference of work location, a "dream sheet" was mailed to all district personnel. This questionnaire asked employees to specify their desire for assignment at either the SSC or one of the five existing STCs. Employees were also asked to specify a first and second choice.

Curtis had also personally asked each of his employees whether they wanted to go to the SSC, stay in Hartford, or be transferred to some other site. He tried to take into account each individual's preference along with the skills needed at the SSC, the skill-needs remaining at the STC, seniority, and individual skill levels. Curtis was very careful not to rank the five criteria by priority. For Curtis, who knew every member of the Hartford plant as an individual, the question of "who goes, who stays" was a personal decision. (See Exhibits 9 and 10 for staffing estimates for the SSC and STCs.)

EXHIBIT 9   Technicians Needed to Staff Analog STCs*

		1984	1985	1986	1987	1988	1989
Bridgeport	SL	2,146	1,717	1,373	1,098	549	0
	Force	6	5	4	3	2	0
Hartford	SL	7,440	5,952	3,761	2,008	267	0
	Force	24	20	16	8	2	0
New Haven	SL	1,768	1,414	1,131	904	452	0
	Force	5	4	3	2	2	0
New London	SL	762	808	0	0	0	0
	Force	2	2	0	0	0	0
Stamford	SL	3,243	2,594	2,075	1,660	830	0
	Force	7	6	5	4	2	0
Total	SL	15,359	12,485	8,340	5,670	2,098	0
	Force	44	37	28	17	8	0

*Maintenance only.

Communicating Continual Change

The rumor mill had been active from the time an SSC for Connecticut was first considered. The transition team had aimed to keep everyone informed, but day-to-day changes made total communications impossible. The technician meetings had helped, but all technicians and clerks within the district had not been involved. The initial surveys and meetings lost their impact as time passed. As one technician noted, "At first we had a lot of meetings, but we would ask a lot of specific questions that management didn't have answers for. After that, the meetings tapered off."

The coordination of five offices created a scheduling nightmare. Hartford, being the final office to transfer was the last office to identify and notify employees who would be involved in the transfer. This played havoc with people's personal lives. For those who planned to relocate, the delayed notification complicated purchasing new homes and moving families.

Impact on Jobs

AT&T migration from an analog technology to a digital- and software-controlled network was rapidly changing the manner in which circuits were provisioned and tested. In addition, much of the physical manipulation performed by test technicians in the STCs would be automatically performed by computer command. Thus, it was felt that technically adept

EXHIBIT 10   Wallingford SSC

	Year Ending 1985	Year Ending 1986	Year Ending 1987	Year Ending 1988	Year Ending 1989
SL SSC	14,322	21,278	27,100	34,207	40,271
SL STC	12,485	8,340	5,670	2,098	0
Maintenance tester job (technician)	21	31	41	45	48
Controller tester job (technician)	23	25	27	30	34
Records analyst administrator job (technician)	5	5	6	7	8
Screener/distributor job (technician)	4	4	4	4	5
Installation analyst job (technician)	1	1	1	1	1
Maintenance analyst job (technician)	1	1	1	2	2
Total technicians	55	67	80	89	98
Repair service attendant job (clerk)	5	5	5	5	5
Order receiver job (clerk)	2	2	2	2	2
Total clerks	7	7	7	7	7
Total nonmanagement	62	74	87	96	105
Total supervisors	9	10	12	13	14
Total employees	71	84	99	109	119

technicians might feel that their jobs would become less challenging. To address this concern, the transition team's plan was to enhance the technicians' jobs with increased customer-focused responsibilities, including customer premise evaluations and assigning specific customers to individual technicians to encourage technician-customer ownership. In addition, to assist employees in their adjustment to the change, management developed a two-day transition seminar for employees to provide up-to-date information and support mechanisms to help with the stress and change.

Another issue was the evaluation of the skill level of these "changed" jobs. All technicians were paid the same dollar amount after five years on the job, allowing for differences in hiring pay rates due to experience. But standardizing the work flow and tasks meant clerks could now take on "nontraditional" clerk tasks, such as interpreting (not just recording) what the customer would say, checking the CRT and notifying the local phone company, knowing what information to ask for, etc. These tasks typically had been performed by technicians. About 10 percent of the technician's job had always been clerical, and the company argued that by taking those percentages and adding them up, a clerk job would be more cost-effective—saving about $200/employee/week.[4] The software embodied in the Remote Test System made the line between technicians and clerks very fine. The union had already filed several grievances disputing the new job assignments.

Supervisors' lives would also be transformed. Rather than having a pool of technicians to call upon when problems arose, they would now have a pod of eight people assigned to them at all times. Because technicians would have increased customer contact, supervisors would have more time to be problem solvers. For those supervisors whose style was more hands-on, this presented a potential problem.

Finally, as employees began working at the SSC issues concerning separation would become apparent. Technicians would be physically isolated from equipment they knew intimately and from people who were part of their team. Technicians would have to call the equipment centers and give directions to other technicians who were not their subordinates. Clerks who were once at the center of activity in the STC would be isolated from technicians' work and the opportunity to learn from it. There was a fear of losing expertise which had taken several years to accumulate. Two to three years had been necessary to become proficient at the traditional testing job at an STC; a technician could now learn the new menu-driven remote testing commands in something like three days and become pro-

---

[4]A TG-4 clerk (pay scales for clerks ranged from TG-3 to TG-6) made $20K, whereas a tech made $30K.

ficient in the total job in less than half that time. There was also concern that employees who remained at the STC might feel left out and left behind.

Work Flow Balancing

Each pod of eight technicians and one supervisor would be responsible for approximately 25 (large and small) customers. This was totally different from the demand system in effect at the STCs, where work was assigned as it came in. Several questions remained unanswered: If one pod had no work, would they assist technicians in the other pods? How would they do so? If there were a major crisis (storm, power outage, etc.) affecting one geographical area, how would that pod enlist others to help? Further, since pods were customer-specific, one pod would have a particular knowledge of a customer base that the other pods did not have. If they shared work, how would they maintain that accountability without losing productivity?

MANAGEMENT AT WALLINGFORD

Although preservice had begun to go on-line in Wallingford in May, as yet no one manager had been designated to run the SSC. The operations managers would each be managing their home STC and their own people at the new facility, which was quickly picking up the name, "Wally World." Meanwhile, Joe Wisniewski continued to handle the day-to-day coordination of the SSC facility needs.

Frank Falciano and the transition team had discussed various staffing alternatives. The most likely organization, once the SSC was fully operational, would be three managers: one SSC manager responsible for northern Connecticut, another SSC manager for the southern portion, and an STC manager responsible for the support people remaining at all five equipment centers. Although Curtis was not worried about his own job security, he did wonder what his job would ultimately be. In addition, he wondered who would decide on priorities at Wallingford and how much control he would have over the physical and personnel resources there.

Meanwhile, Curtis was concerned about the transitional period. Coordinating activities at two locations would be difficult. To complicate matters a budget freeze restricted his ability to add personnel or incur additional costs to cover the transition. This meant that the job responsibilities of many of his clerks, technicians, and supervisors had to be expanded to cover both locations.

Allen-Bradley (A)

We did everything right, right from the start. Our production people were under a lot of pressure to trim costs and reduce waste—yet maintain our high standards for quality. As we began to look at how we could achieve such a mandate, it was obvious that time-tested, traditional methods would have to be scrapped. So we threw out the rules and started looking at production methods worldwide. It was a challenge, but one we accepted because we knew success would strengthen the worldwide presence we were fostering. Our number-one objective was to automate all of our processes into the most efficient, cost-effective system possible.

This was how Larry Yost, vice president of operations for the Industrial Control Group at the Allen-Bradley Company, described the factors that led to the creation of World Contactor One (WCI), Allen-Bradley's computer-integrated manufacturing (CIM) facility located on the eighth floor of the company's Milwaukee, Wisconsin headquarters. The facility, which began manufacturing IEC (International Electrotechnical Commission) contactors and relays in April 1985, was capable of producing four styles of contactors and three styles of relays with up to 1,025 different customer specifications in lot sizes of one, within 24 hours of the order placement and with no direct labor and zero defects. Moreover, the 150 × 300 foot "factory within a factory" served as a showcase for customers and visitors. As of October 1986, the line was still considered a pilot operation but was scheduled to go to full production in December 1986, which meant a hand-off from New Product Industrialization (development manufacturing engineering) to Manufacturing and a return to the bargaining unit for the hourly workers.

ALLEN-BRADLEY

The Allen-Bradley Company traced its origin to 1893, when 15-year-old Lynde Bradley created a homemade motor controller which he attached to a toy lathe in the cellar workshop of his family home in Milwaukee. In 1903, with the financial backing of his friend Dr. Stanton Allen, Bradley incorporated the Compression Rheostat Company (renamed the Allen-Bradley Company in 1909) to manufacture his latest invention, a unique type of crane controller. Harry Bradley, Lynde's 18-year-old brother, joined the firm soon thereafter; for the remainder of their lives, the two brothers ran the business together.

The brothers' philosophy permeated every part of the organization: They both had an intense dedication to quality not only in products and customer service, but also in dealings with suppliers and all others who interacted with the firm. Over the years they worked hard to foster employee satisfaction and to encourage strong relationships among workers, in part, by instituting various social and educational programs. The Bradley brothers were fond of emphasizing that their company was really the "Allen-Bradley Family."

By 1986 Allen-Bradley had become an international firm employing over 14,000 people in 20 plants in the United States, along with factories in Latin America, Europe, Australia, and Asia, and manufacturing affiliates in nine nations. Though it had been founded as a manufacturer of electric motor starters and controllers, Allen-Bradley had become a leading international supplier of electronic and electrical controls and systems for industrial automation as well as a manufacturer of electronic and magnetic components used in consumer, industrial, and commercial products. In 1984 the company had sales of $942 million.

On February 20, 1985, the Allen-Bradley Company was acquired by Rockwell International Corporation at a price of $1.65 billion in cash. Rockwell, one of the largest multi-industry corporations in North America with 1985 sales of $11.3 billion, was a leading defense contractor, involved in aerospace, automotive, general industries, and industrial automation businesses. In discussing how Rockwell dealt with its new acquisition, one Allen-Bradley manager commented, "Rockwell leaves us alone. The management of Allen-Bradley is identical to what it was before. Rockwell only looks at the books."

ALLEN-BRADLEY'S DECISION TO ENTER THE IEC MARKET

Historically, the Allen-Bradley Company had focused on the domestic industrial controls market, with only 3 percent of its 1975 industrial control business coming from outside North America. In the late 1970s, the firm's market research revealed that a global market was emerging due to an influx of foreign industrial machinery, most notably from Japan and Europe, which created the need for European-style devices meeting standards set by the IEC. IEC-rated products totaled to an approximate $100 million per year domestic market, but outside the United States the market for IEC contactors and relays was about $700 million per year.

Contactors and control relays were mechanical devices used to open and close electrical circuits. Contactors were switches (motor starters) for heavy-duty, high voltage power circuits; relays were used for lower current applications where relay logic was employed. The first computers used thousands of control relays to process information; currently they were used in logic switching of simplistic machines. The main difference between the contactor and the relay was the composition of the contact points on the switch: Points in a contactor were 90 percent silver and 10 percent tin, while the relay points were 99 percent silver.

Previously, Allen-Bradley had built products that only met the standards of the National Electrical Manufacturers Association (NEMA). NEMA devices were standard and useful for many applications, whereas IEC products came in over 125 configurations and numerous sizes and were application-specific. IEC devices could be one-third the size, weight, and price of the NEMA product but would start the same rated horsepower motor. Markets for the mechanisms included the machine tools, petrochemical, automotive and mining industries, forest products, heavy equipment, and more recently, industrial garage door openers and tanning beds.

DESIGN OF WORLD CONTACTOR ONE

The idea for the WCI Facility was conceived in late 1981 in response to market changes and Allen-Bradley's strategic decision to become more international; because one of the firm's goals was to become a world-class manufacturer by the mid–1980s, corporate management was committed to the project from the outset. The company knew its product had to be cost competitive in all markets and compete against overseas manufacturers, some of which were subsidized by their countries (one French company, for example, in 1980 was selling contactors for $4.50, even though the materials cost nearly $5.25). Allen-Bradley, therefore, began focusing on reducing the labor content of the product. High quality and fast turnaround would also affect the line's success. Yost, who along with other executives and engineers had traveled extensively in Europe and Japan to learn about automation, and who was considered by many to be the "father of CIM" at Allen-Bradley, explained:

> We decided to make labor a non-issue. Our concern was having the most automated facility in the world that was building a product that Engineering wanted and Marketing could sell.

Unit cost, as opposed to return on investment, became the driving factor. Marketing came up with a target cost; if Allen-Bradley's manufacturing cost were equal or lower, it could match its toughest competitor. Initial estimates showed product cost to be about 50 percent material, 40 percent overhead, and 10 percent direct labor. Subsequently, with the first rough conceptualization of the automated system, the direct labor was eliminated, but the 10 percent reduction was insufficient. To reduce cost further, overhead costs had to be dropped too. Management turned to a CIM system to eliminate the need for material handlers, inspectors, group leaders, supervisors, and a warehouse. The corresponding decrease in overhead would make the target cost attainable. In total, the cost of a typical contactor could be reduced from $13.25 to $8.35 (see Exhibit 1). In 1986 the sales price of a contactor ranged from $10 in Australia to $25 in the United States. (After the decision had been made to proceed, it was found that the inventory savings equaled the cost of the capital investment.)

When Allen-Bradley management looked at the business and manufacturing objectives, it became clear that the product and the automated line should be designed for each other. Moreover, market research data had indicated that the new product had to be introduced within two years, which meant that many disciplines would have to work simultaneously on the project. As a result, a series of task forces were established to jointly design the product and the process. (See Exhibit 2 for the organization chart.)

One task force, the Project Management Team, was charged by senior management with developing a contactor and relay that would be competitive worldwide. The group consisted of a manufacturing representative, Jim Heimler, at that time manager of Production Engineering, a Marketing manager, and a Development manager. Shortly after the team began working on the design of the product, an Automation Equipment Engineering Task Force was established with the responsibility for designing and ultimately building most of the equipment for WCI. Members included a project manager responsible for special equipment design; a project engineer in charge of designing the electrical system for the machinery; a supervisor of machine builders who oversaw the equipment construction; a project supervisor who did equipment planning; a project

---

EXHIBIT 1   Projected Cost Estimates

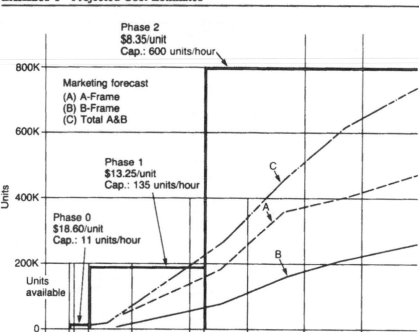

engineer who created the facility layout; and a maintenance superinten-dent in charge of facilities.

In January 1983, Tracy O'Rourke, president of Allen-Bradley, gave the go-ahead to invest $15 million in the new CIM line. Shortly thereafter, another task force, spearheaded by Yost, began to meet weekly (and often twice a week) to consider ideal features for the line. Yost's team consisted of 25 people: engineers, technicians, accountants, and other specialists from all the departments that would be affected by the project, such as Finance, Marketing, Quality, Information Systems, Special Equipment, and Development Engineering.

In November 1983, John Rothwell was appointed department manager for WCI. Rothwell had worked at Allen-Bradley since 1972 primarily in the Special Equipment Design Group. Before becoming department man-ager of the new facility, he had been working on a project in the Special Equipment Group that involved using a database to schedule long-term projects, which included keeping track of the WCI project.[1] Yost com-

---

[1]See Allen-Bradley (B) for a more detailed description of Rothwell's background.

EXHIBIT 2 Organization Chart, October 1986

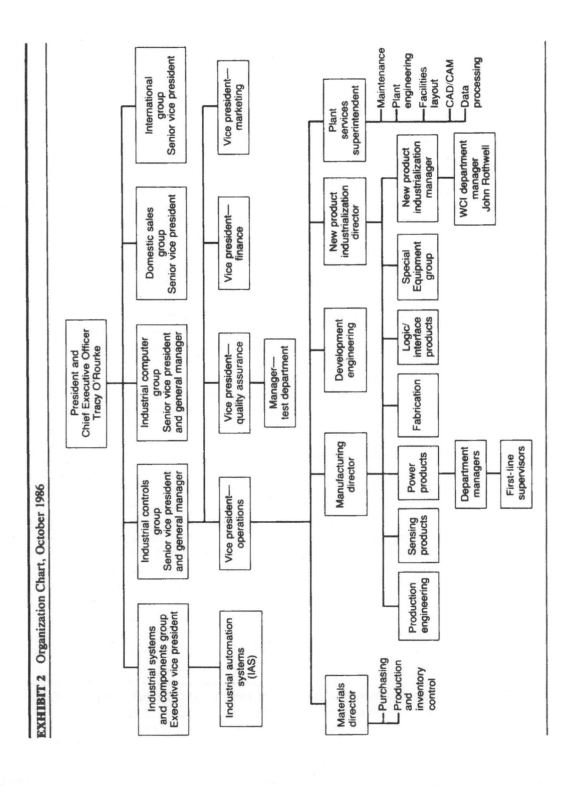

mented on the timing of Rothwell's appointment: "Whenever you're being a pioneer in a new area, you need to get the person or team that will ultimately be responsible for living with it involved as early as possible. You need a person who will be schooled in everything eventually. Why not make that person a champion?" Heimler elaborated, "John was chosen and put in so he could help nursemaid the computer system all the way through, and at the same time, he could debug the mechanical system. It was crucial that we got him in there early."

Once Rothwell was assigned to his new role, he began to learn about the make-up of the parts of the product, and he started to work on getting the product released. With the aid of two pilot lab workers, he began putting the contactor together by hand, completing the first unit on January 20, 1984. (The first pilot units were sold in June 1984.)

After the contactor had been developed, Rothwell began to focus on the automated line. On March 16, 1984, Rothwell led the first meeting of the Implementation Task Force, a group that replaced and carried on the work of Yost's task team. One member of Rothwell's group was a representative from Allen-Bradley's Industrial Automation Systems Group in Cleveland who had come from Yost's previous team and was project manager of the software team that wrote the functional specification for the automated line. Other members included representatives from Test, Planning and Inventory Control, and Manufacturing, plus four employees from the Facilities and Equipment area. This group took the "wish list" generated by the larger 25-member team and began to condense and prioritize it. Rothwell's group was also in charge of laying out the facility, coordinating incoming equipment, and working on equipment design. Its efforts were aimed at ensuring the area would be ready by December 1, 1984, the date marketing research had estimated would be necessary for Allen-Bradley to enter the IEC market ahead of other domestic competition.

Before the equipment was ordered, upper management decided to make the facility into a showcase, which had not been part of the original plan. The operation was being designed based on a concept Allen-Bradley referred to as the "productivity pyramid" (see Exhibit 3), and the Sales Group needed a running model of the pyramid. This was a system architecture which linked the machine level to the order procurement process, using various Allen-Bradley products. When they began to refurbish the area where WCI was to be located, Rothwell explained, "Management realized that for a few more dollars they could get a tangible asset out of this. It could be valuable both as a manufacturing facility and a sales showcase."

Over 90 percent of the equipment was designed by Allen-Bradley and about 60 percent was built by them (in the Special Equipment Group). Ninety-five percent of the equipment was on the floor by December 1984, with the exception of the base and housing assembly machine which was

EXHIBIT 3 Allen Bradley's Productivity Pyramid

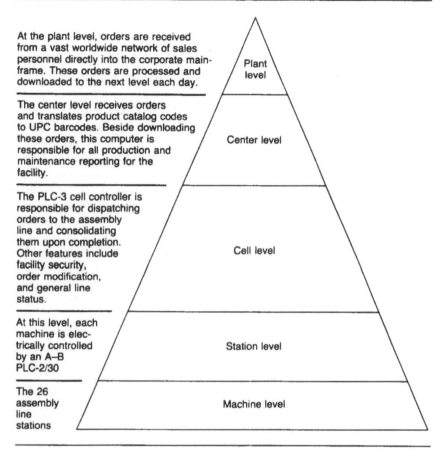

At the plant level, orders are received from a vast worldwide network of sales personnel directly into the corporate mainframe. These orders are processed and downloaded to the next level each day.

Plant level

The center level receives orders and translates product catalog codes to UPC barcodes. Beside downloading these orders, this computer is responsible for all production and maintenance reporting for the facility.

Center level

The PLC-3 cell controller is responsible for dispatching orders to the assembly line and consolidating them upon completion. Other features include facility security, order modification, and general line status.

Cell level

At this level, each machine is electrically controlled by an A–B PLC-2/30

Station level

The 26 assembly line stations

Machine level

built by an outside machine builder. In January 1985 debugging of the line started, and on April 15, 1985, two weeks after Rockwell's first board meeting at Allen-Bradley, WCI began operation.

WCI'S MANUFACTURING PROCESS

The production cycle started when staff at an Allen-Bradley distributor or field sales office entered an order into their computer terminal which interfaced with Allen-Bradley's mainframe computer. All 80 United States sales offices and about 100 of the 500 distributors worldwide were connected by telecommunications to Milwaukee's mainframe, which received sales orders and integrated them into manufacturing, sales, and accounting data.

Each day at 5:00 A.M. the IBM mainframe downloaded the previous day's orders for the product to another, smaller computer (a VAX 11/780 Area Controller) located in the WCI control room. The area controller translated those orders into specific production requirements and downloaded the information into an Allen-Bradley master controller (PLC-3) at the cell level. This master controller transmitted instructions to 26 smaller Allen-Bradley controllers (PLC-2/30) on the factory floor. The PLC-2/30s instructed each machine on the asssembly floor what to do next. Two local area networks (communication links) were used to transmit information about the assembly process between the PLC-3 controllers in the control room and the PLC-2/30s.

WCI was designed to produce seven styles of IEC products: three relays and four contactors. The contactors consisted of two type "A" (4 and 5.5 kilowatts) and two type "B" (7.5 and 11 kilowatts). Although the line was set to run at a rated capacity of 600 units per hour, it also had the ability to run at either 500 or 700 units per hour. The line utilized all DC motors, which enabled it to have variable speeds. As of October 1986 the line was producing approximately 10,000 units per week.

In addition to the control room, which was typically unattended, the WCI facility comprised a plastic molding cell, a contact fabrication cell, and 26 machines that constituted the CIM assembly line. Material such as steel, brass, molding powder, coils, and springs entered through a side door that also served as the exit for finished contactors and relays. (See Exhibit 4 for a process flow diagram.)

The CIM line began at the base subassembly machine where a bar code identifying the specifics of the particular unit (packaging, electrical configuration, product markings, etc.) was applied to the bottom of the contactor or relay base. A springed latch was also assembled at this station. At the next station the yoke was inserted. The yoke comprised 42 thin layers of laminated steel that had been fabricated in the punch press department (part of the production operation). The layers were riveted together at the yoke assembly machine with a force of two tons and staked with copper coil and then ground to a tolerance of $\pm .0004$ inches. A laser inspected the height of the yoke and provided automatic feedback to the grinding wheel for height adjustments.

Electrical coils were received from an outside vendor in stacks of 30 and loaded into the coil dispensing machine, which had a capacity of over 5,000 coils and had been designed to be reloaded prior to the start of each shift. At the coil insertion machine, coils were positioned into the base, checked via a probe to assure that the voltage was in accordance with the bar code, and then tested with 800 volts to check for open circuits in the coils.

EXHIBIT 4   WCI Process Flow

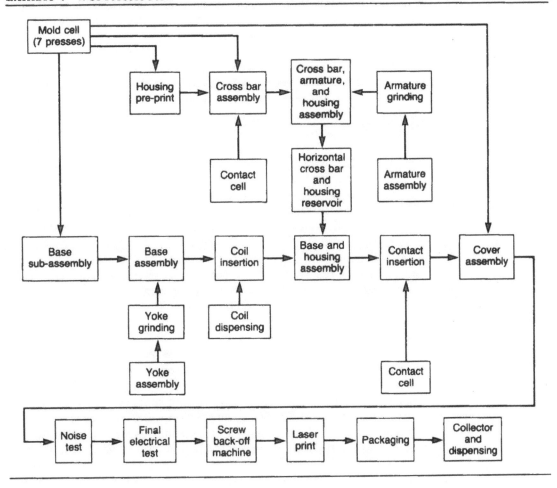

Housings, which were formed in the molding cell, entered the line at the housing preprint machine. Here, the housing was labeled and a vision system camera checked the printing. The housing then proceeded to another assembly machine where the crossbar was inserted. Once again, a camera checked the presence of the component parts. Simultaneously, the armature was being ground and sprayed with a rustproofing lubricant. Once the crossbar, armature, and housing were assembled, they proceeded to a horizontal crossbar and housing reservoir.

Two automatic screw driving machines assembled the base to the housing, and the torques were automatically checked before a 50-pound force fit was used to insert the stationary contacts (contact points). At the cover assembly machine, the cover was attached with three screws, and a paper label and plastic clip were inserted on the top. Three amps of current were then passed across the contact points to remove any oxidation, and the unit proceeded through a noise and final electrical test where 100 percent of all electrical parameters were checked and SPC (statistical process control) data recorded on the Allen-Bradley 7890 computer system.

The unit then advanced to the screw back-off machine, which had been designed in-house and built at a cost of $200,000. During the design stage, Marketing had discovered that a desired product feature was to have the screws backed off for ease of installation. A side benefit, subsequently found, was that backing off the screws would uncover defective threads.

Finally, the unit proceeded through a laser that printed the catalog number, the Allen-Bradley logo, and date code, and then on to the packaging machine which boxed the units and labeled the box via an ink jet printer. At the end of the line, the unit was loaded into the collector and dispensing machine that automatically sorted the units into customer orders.

All molded plastic parts such as base frames, housings, crossbars, covers, and spring seats used in WCI products were made in automatic molding machines in the molding cell located in a room adjacent to the primary assembly area. There was typically a two-shift inventory buffer between the mold cell and the assembly line. The cell also provided components for another department and was staffed on a two-shift basis.

Of the spanners and terminals used on the WCI line, 95 percent were made in the contact cell. For spanners, a bronze reel was fed into a series of presses to punch holes, drop a weld of silver, and then punch out the spanners. Typically, the cell had a one-week supply of raw materials and ran batches of 80,000–100,000 spanners as the line required them.

Allen-Bradley's Total Quality Management System, including statistical process control, was incorporated into the production process. There were more than 3,500 data collection points and 350 assembly test points checking each component and the final product as it moved from station to station. Any component that failed to meet predetermined quality parameters was rejected, and the system automatically began to build a new one so that at the end of the day the order was complete.

Several times each shift the line operators checked the "reject drawer" at each machine (defective parts moving through the line were automatically shunted off to the drawer) and manually repaired any units that could be reworked. If a product with a fault was repaired and returned to the line, it was sent to "Orphan Alley," a holding lane from which unscheduled assemblies were retrieved automatically when needed. In addition, the operators would examine several products on the line each day, and if they discovered a reject, they would go back and look at every item that had gone through the line, since a reject could conceivably pass all the automatic checks.

A variety of multilevel production diagnostics were also built into the system. Every machine was monitored by a three-light system and sound alarm which alerted operators when there was a problem. A blue light signified that a bulk feeder was low; a yellow indicated a part jam; red meant there was a machine malfunction. When an assembly station malfunctioned, the system activated an automatic shutdown of all preceding processes. A color graphics terminal in the control room could also monitor diagnostics. In total, the system generated over 80 different manufacturing reports, ranging from maintenance to output totals.

THE CURRENT OPERATION

Originally, management planned to hire eight operators, but Rothwell had felt six would be enough to oversee the assembly line operation. (If the product were assembled manually, 100 or more workers would have been needed.) Rothwell began looking for operators in the spring of 1984. During several general meetings with groups of 20 to 30 people who had been recommended by other managers or who had heard about the new job openings through the Allen-Bradley grapevine, applicants filled out requirements sheets, listing their mechanical assembly experience and any related schooling. Rothwell intensively interviewed 15 applicants and ultimately narrowed the choice to seven operators (six line and one contact cell operator). (See Exhibit 5 for an organization chart.) In describing those he selected, Rothwell remarked:

> I have individuals out here who are self-starters, individuals who are all leaders in a sense. They are all very conscientious people who want to do better and who can see that this is the wave of the future. The backgrounds of the people were all very similar, but their attitude toward automation got me from 15 down to seven.
>
> These people had been recommended to me by department managers. I know all the assembly managers and I'm on good terms with them, and I took some of their best people away. Their bosses told them to do it because we needed the best for CIM. From talking to people on the floor, I know I got the best.

The line operators all worked during the day shift (during the second shift the only employees in the area were two mold cell workers), and their pay rate was based on their skill level, which was related to their knowledge of the machines on the line. After an initial six-month assignment to gain a basic knowledge of the machines, they rotated stations (groupings of machines on the line) every three months. One operator commented:

> We've been here 18 months and we don't know it all yet. It would take about two years for average learners to reach the point where they'd feel they could

EXHIBIT 5   WCI Organization

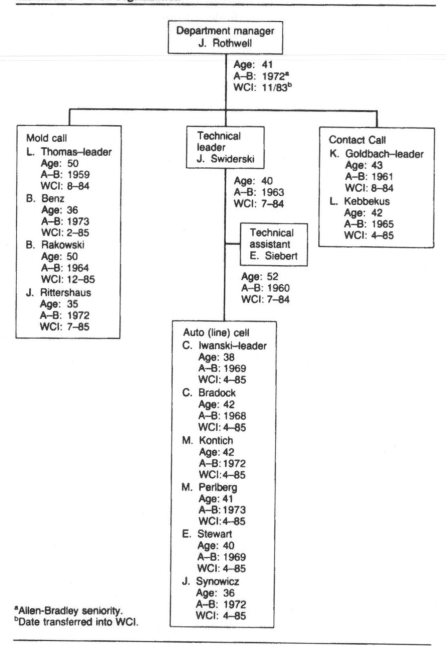

Department manager
J. Rothwell

Age: 41
A–B: 1972[a]
WCI: 11/83[b]

Mold call
L. Thomas–leader
   Age: 50
   A–B: 1959
   WCI: 8–84
B. Benz
   Age: 36
   A–B: 1973
   WCI: 2–85
B. Rakowski
   Age: 50
   A–B: 1964
   WCI: 12–85
J. Rittershaus
   Age: 35
   A–B: 1972
   WCI: 7–85

Technical
leader
J. Swiderski

Age: 40
A–B: 1963
WCI: 7–84

Technical
assistant
E. Siebert

Age: 52
A–B: 1960
WCI: 7–84

Contact Call
K. Goldbach–leader
   Age: 43
   A–B: 1961
   WCI: 8–84
L. Kebbekus
   Age: 42
   A–B: 1965
   WCI: 4–85

Auto (line) cell
C. Iwanski–leader
   Age: 38
   A–B: 1969
   WCI: 4–85
C. Bradock
   Age: 42
   A–B: 1968
   WCI: 4–85
M. Kontich
   Age: 42
   A–B: 1972
   WCI: 4–85
M. Perlberg
   Age: 41
   A–B: 1973
   WCI: 4–85
E. Stewart
   Age: 40
   A–B: 1969
   WCI: 4–85
J. Synowicz
   Age: 36
   A–B: 1972
   WCI: 4–85

[a]Allen-Bradley seniority.
[b]Date transferred into WCI.

muddle through, not even necessarily be comfortable with it. I don't even think any of us would say we were completely familiar with the equipment at the stations we've left.

The operators were expected to complete the sixth rotation by early 1987, when they would advance to a higher rate of pay. Thereafter they would continue to rotate their positions. Each operator also had "co-lateral duties" in addition to regular assignments. These tasks included filling out the labor and attendance slips, handling hazardous materials, working on SPC, and keeping track of inventory.

Lunch time, quitting time, and breaks varied daily depending on the timing of visitors' tours through the facility. For example, if a tour came at lunch time, the workers would postpone eating. If a tour came through at the end of the day, they might have to stay an extra half hour or more (for which they received overtime pay). Employees were entitled to one break a day, but according to one operator, "Nine times out of ten we seldom take a break because we're too busy."

Line operators covered for one another. When one person was away on vacation, for instance, a fellow employee had to handle several machines and move around to different areas of the line. In addition, even if it were not his or her primary area of responsibility, if an alarm went off at a machine and no one else responded, the operator who noticed the alarm would answer it.

Rothwell liked to have at least five people working on the line at once, but there had been times when it had operated with only four. According to Rothwell, the line could run with as few as two workers if a mechanized material handling system were put in place. It was doubtful that the line would ever run completely unattended, however, because of problems with incoming product quality (nonconforming parts caused line jams and subsequent machine shutdowns) and because of the design of the hoppers (feeder bowls) which continually caused parts to jam. If these flaws were corrected, the line could run unattended, but as Rothwell explained, "It could probably be done in a year if we applied ourselves, but it doesn't really seem necessary right now. We need to have the volume (millions of contactors per year) to make it worthwhile."

To keep in close touch with the day-to-day operation of the line, Rothwell spent a great deal of time on the floor. Said one operator, "Most supervisors usually sit behind a desk, but John likes to get involved. He'll come out on the floor and pitch in, which is great." Rothwell's duties also involved keeping abreast of current changes in products and manufacturing techniques affecting the fields of automated assembly and robotics and applying those that could achieve cost savings and improved performance to the World Contactor line.

The second in command at WCI was Joe Swiderski, who was the technical leader in the department. Swiderski had been involved with the facility since the beginning when the machines were being built. His role as technical leader involved debugging and repairing the equipment, dealing with part problems, and handling operator complaints. He also filled in for Rothwell when the latter was away from the area.

The third link in the chain of command was Chris Iwanski, the assembly cell leader. Rothwell described Iwanski's role by explaining, "He's the focal point for people's problems. He tries to get answers. He's in charge of the line." His daily activities included going to other departments to get stock, finding an engineer to help out with a problem, and reassigning the operators to cover for someone who was sick or on vacation. If a machine went down for an extended period, Iwanski also had the authority to decide to go to batch process.

MAINTENANCE AT WORLD CONTRACTOR ONE

During the pilot stage, two maintenance mechanics were permanently assigned to WCI. Heimler explained:

> We felt from the beginning, in order to keep ourselves as self-sufficient as possible, we needed to have mechanical and electrical maintenance people reporting in to us. We didn't want to have to wait for help when problems came. When building a product from beginning to end in 24 hours, you can't afford a shutdown.

The line operators also did minor maintenance work. Each operator had a tool box and was permitted to do limited repairs such as replacing drives and clearing jams on the lines. According to both Swiderski and Rothwell, the lack of quality of the incoming parts was the biggest problem on the line because out-of-tolerance parts caused jams in the feeder bowls.

Rothwell felt that the maintenance system in his department gave the operators "freedom to do more of their jobs. It gives them freedom of control in their area, which is an incentive for working here." The maintenance personnel did not seem to mind that the operators performed maintenance tasks. As one operator explained:

> The maintenance people will teach the operators new things if we are capable of understanding. They prefer if we learn because they are very busy doing complicated tasks, like those that require taking apart machines. This is not like an assembly line where maintenance people are afraid you'll take their jobs. Everyone here wants to learn more. I feel personally if you're comfortable enough and you know what you're doing, go ahead and fix it. We need to keep the line running up here because if one part goes down, the whole line goes down.

When problems occurred in a machine, operators had to insert their identification badge in a magnetic card reader before correcting the problem. This enabled the computer to monitor the length of time it took the operators to get the line up and running again. One operator, who complained of feeling "computer-monitored" and of having no sense of privacy due to the card reader system, commented, "I wonder if they'll think I'm slower than another worker. I don't like it because I might repair a jam a little differently from another operator. It's unfair to compare us."

In the past, only major machine breakdowns had been documented; the mechanics had felt documentation of minor problems would be a waste of time, mainly because the line was still in a start-up stage. Yet, Rothwell had recently begun instructing the operators to keep track of the time, date, and nature of all changes or repairs made to the machines. The documentation would serve as the groundwork for a preventive maintenance system which Rothwell felt would soon be needed. His plans to change the maintenance system had been prompted, in part, by a recent machine malfunction caused by a loose screw. He realized that other equipment might also have loose parts, and he planned to have an operator check all of the machines on either an off shift or on a Saturday.

PILOT TO FULL PRODUCTION

In 1981, Local #1111 of the United Electrical Workers, the union representing Allen-Bradley's production employees at the Milwaukee facility, had agreed to a contract provision whereby employees assigned to the development of new products, technology, and manufacturing methods, including automation, would be considered outside the bargaining unit. If, after the product had been developed and the new manufacturing method established, Allen-Bradley decided it would be feasible to manufacture the new product in Milwaukee, the work would transfer into full production, and the jobs would fall within the bargaining unit with wage rates and other terms and conditions of employment to be negotiated with the union.

Early on it was decided that WCI would be established in the Milwaukee facility rather than at a greenfield site because, as Yost noted, "It made sense to keep it as close to Engineering and Marketing as possible. Also, we wanted to keep it close to our greatest pool of manufacturing resources." This was important to the union because, by 1986, hourly employment had dropped to 2,500 from a peak of 5,500 in 1975, a result primarily due to the relocation of major divisions such as Electronic Components, which moved to Texas and North Carolina, and Magnetics, which went to Oklahoma.

December 1986 was targeted as the date for the move into full production, and that meant a return to the bargaining unit. The transition also meant that a new job classification system would have to be developed. One operator discussed the initial steps taken to classify jobs on the assembly line:

> It's very hard to compare us to other departments. Personnel has been up here analyzing the group. They originally tried to classify each job, but you can't do that because they're all connected. They will have to come up with a title for all of us. I wouldn't want their job. It's tough. It's going to be a mess.

Rothwell believed that although the assembly line job sounded like an unskilled job, the operators needed to be multiskilled and flexible to be able to do both typical assembly line tasks and troubleshooting, which required mechanical ability. As a result, the company had proposed to the union that the operators on the line be put under one job classification and that they be made the highest skilled production workers at the company, aside from the skilled trades. In addition to the single job classification for the assembly line, the company was also planning to propose two other job classifications for the plastic molding cell and the contact fabrication cell.

The company's proposal to the union also included a new approach for selecting future department workers. A local college was developing a series of skills and abilities tests that would be validated by current employees and later used to choose new workers. Hence, future operators at the CIM facility would be selected based first on skill, and then on seniority, which was the reverse of the procedure used in the other departments. Rothwell noted, "In the future, the test will be used instead, and we'll have to hope for good attitude."

If the proposals were approved by the union, the operators would receive an increase in their pay rate which would put them near the top of the Allen-Bradley wage structure. Yet, even though they would be the highest skilled production workers, during layoffs, skilled tradespeople could still "bump" into the operators' positions if they could pass the necessary skills and abilities tests. As a result, Rothwell felt that protecting operators in the new job classification system would be a "sticky issue." While in pilot production, the line operators' jobs had been "grandfathered" so that they did not lose their seniority (their hire date at Allen-Bradley), but most still had short service relative to the entire work force. The average length of service for Allen-Bradley employees was about 18 years, with the lowest level of employee seniority being approximately 12 years; some had as much as 30 to 40 years of service at the company. In a recent reduction of 150 production employees, two of the operators would have been bumped by the layoff.

Another issue raised by the upcoming shift to union status was whether or not the department would be able to continue operating as flexibly as it had been. Several people felt that changes were bound to take place:

Rothwell:

"I'm a little worried because I run things more flexibly than most out there. When it does go into the union, I may be restricted as to what I can do for the people. I'll be afraid of setting precedents. I can still be flexible, but with some people I may have to play by the book."

An operator:

"The group up here and John all stick together. Now, there is give and take and it's like a family situation. I went through it before in another department. It became a different area completely. You had to go by union rules, and you lost your freedom. For example, you couldn't leave early. A lot of little things will change. We'll lose the coffee pot up here. There's a problem with setting precedents. The people outside the door would complain."

A mold cell worker:

"Right now we're pretty free and easy. When problems come up we go downstairs to another department and get them fixed. The union slows things down. We'll have to go through the proper channels."

An operator:

"Right now there's no strict line between what operators and maintenance people do. They work on anything and that's a nice situation. But, that will change when we move into the bargaining unit. Then, people will work on tasks defined by their job classification. If people start filing grievances, that will be a problem."

HAND-OFF BETWEEN ORGANIZATIONS

Besides the switch in union status, another major change that would take place when the facility moved from pilot to full production was a hand-off between the Industrialization and Manufacturing organizations. Since initially World Contactor was a pilot area with a new product and process, it had been designed and brought on line by the Industrialization Group. With the transfer, Rothwell would begin reporting to Tom Retzer, the production manager of Power Products. Retzer had been indirectly involved with WCI in a previous role as a test engineer, where he had evaluated the prototypes that became the World Contactor and had been involved in the concept of the relay. In addition, in his current role he received copies of the minutes of weekly WCI update meetings, which covered the hot issues in the department, e.g., problematic machines.

With the hand-off between organizations fast approaching, management had begun working on a manufacturing audit of the WCI facility that would determine whether the process was under control and the equipment in place. For example, they were investigating the total quality control programs, certification codes, and tool acceptance and specification acceptance agreement procedures. According to Retzer, the reason for the audit was "to assure us that Industrialization has done its job. The audit won't be a stumbling block to bringing the department into full production but rather will provide a 'things to do list.'" Dan Challgren, director of New Product Industrialization, noted:

> The audit will help to resolve quite a few issues even before the operation is handed over. Once we pass the acceptance, if we find that there still are problems to take care of, my automation or industrialization crew will deal with them.
>
> The number one key issue with the audit and the hand-off is trust. There's a good relationship established between organizations and a knowledge that the other person will back you up if you need it.

Accompanying the hand-off would be a change in technical support at the facility. While in the pilot status, technical support had been carried out by the New Product Industrialization Engineering Group, but after December, it would be supported by assembly engineers who reported to Retzer. By the end of October, it was planned that Joe Swiderski would be promoted to the salaried position of senior manufacturing technician, reporting to the Assembly Engineering manager and acting as WCI assembly engineer. He would be physically located at the CIM facility, however, and would continue to act as a quasi-supervisor backing up Rothwell. Swiderski had low seniority, which meant that as an hourly worker, he would not be able to stay in the department when the operation transferred into full production. Rothwell noted, "If the union wasn't coming in, I would have left this organization chart exactly as it was."

Two other technical areas would also be affected by the transition. Maintenance would be supplied by the Maintenance Department with full-time personnel assigned to WCI, at least initially. Eventually, maintenance support would revert to an on-call basis, as with other production departments.

Computer support would also become limited to a once-a-week back-up as the line became more self-sufficient. During the design and start-up phases, members of the Milwaukee facility's CAD/CAM group had programmed the IBM computers for the automated line, whereas personnel from Plant Engineering had programmed the line's 26 PLC-2/30s; personnel from Allen-Bradley's IAS Group in Cleveland had programmed the VAX and PLC-3 computers. Swiderski had been trained and had already begun to operate and program the PLC-3 and PLC-2/30s.

CHANGE IN MEASUREMENTS

Yet another change that would come about when WCI moved into full production would be a shift in the measurements upon which Rothwell, and the facility in general, were evaluated. During pilot production, Rothwell had been evaluated by Heimler on "interface and project work." "Interface" was Rothwell's interaction with his employees, other managers and executives, and the public: the company officials, customers, journalists, and other visitors who constantly came to tour the WCI showcase. "Project work" related to the company's management by objective (MBO) system. Heimler explained:

> We set a few specific objectives, about five to seven for each manager. The objectives are based on the strategic plan of the business, and you really want to get those things accomplished. In John's case, cost, quality, and delivery have been number one even though they're in the pilot stage. When you're talking about the customer, you can't fail. Our goal has been to be at 100 percent on on-time shipments, and we've been there for the last 14 months. The quality performance has been equally impressive. WCI has been running at 15 ppm on field returns, whereas the rest of Allen-Bradley averages 120 ppm.

Retzer was still uncertain as to what Rothwell's objectives would be once the area moved into full production:

> I must admit that in an area like this where computers are so critical, I don't know yet what John's objectives will look like. There will probably be objectives on machine downtime and pieces per hour, but not on employees since they're indirect labor.
>
> Costs and improvements of gross margin will be very important. I think that right now there's an awful lot of external support being charged into that area, and some of that will disappear as the area matures. Also, you need the volume to get the costs down.

Other department managers in Manufacturing were typically measured on financial goals, housekeeping, communications, improving test yield, stockless production, and cost-reduction programs. For example, supervisors were required to return 10 percent of their own salary via cost reductions while engineers had to submit cost-reduction proposals equal to their salaries. Whether a similar system would apply to WCI had not been decided, but one thing was for certain—as in other departments, cost reduction would become a central measurement used to evaluate the facility. Yost commented:

> We will start tightening up on cost. It's time to pay attention to that. There will be a different set of measurements. Every hour will have to be effectively utilized by the operators. There will be a focus on more productivity.

Rothwell, well aware of the role that cost containment would play in the future of the CIM facility, remarked, "I had an open checkbook there for a while, but it looks like I'm going to have to tighten my belt."

Allen-Bradley (B)

At 7:00 A.M. on October 15, 1986, John Rothwell, department manager of Allen-Bradley's World Contactor Facility (WCI)[1] was sitting in his office looking over the schedule of tours coming through his area that day. The governor of Wisconsin would be visiting later that morning, and one of the company's key customers would be arriving in the afternoon. For Rothwell the day would not be atypical since being a tour guide of the CIM facility was part of his daily routine. As he explained, "There are usually two to three tours per day. I give at least one of them, which can last up to two hours or more with the question-and-answer period."

Besides the tours, at 7:30 A.M. every morning Rothwell held a five-minute meeting with his employees while the line was warming up. In addition to discussing the tour schedule for the day, the group talked about the production schedule, quality issues, and any problems that had arisen on the line during the previous day.

After the meeting, Rothwell usually went back to his office to see if there were messages on his answering machine (he had no secretary) and to check his "things to do" list, which typically included writing employee evaluations, reviewing production reports from the VAX and PLC-3, doing scheduling for fill-in work that had to be manually entered into the system, working on the budget, or dealing with daily production problems. Rothwell also continuously "interfaced" with Quality Control and Engineering and as part of his daily routine, spent a lot of time out on the floor "putting out minor fires and making snap decisions." He

---

Copyright © 1987 by the President and Fellows of Harvard College Harvard Business School Case 0-687-074.

[1]See Allen-Bradley (A) for details of the World Contactor One Facility design and start-up.

described his role by saying, "I picture myself as a 'go-for': if they need something on the line, I get on the phone and get it." (See Exhibit 1 for Rothwell's job description.)

Rothwell prided himself on always trying to do the best he could for his employees. He commented: "People will make or break the line; the quality of the people and the morale of the group are essential. I spend a

---

EXHIBIT 1
Rothwell's Job Description

POSITION TITLE: Manager, World Contactor Automation Facility

POSITION SUMMARY:

To organize, manage, and initiate the start-up and development of automated manufacturing systems and techniques for new products in accordance with pre-established company and I. C. Division production objectives at optimum cost consistent with quality requirements.

Principal Duties and Responsibilities

1. Develops and implements department organization policies and procedures that are consistent with I. C. Division and Corporate goals and ensures proper adherence to them with respect to all operations within Department 260.
2. Plans, organizes, and controls assigned manufacturing operations in accordance with guidelines and schedules established by the Manager of Power Product Industrialization.
3. Develops, trains, and motivates hourly employees to acquire maximum efficiency, productivity, cooperation, and morale.
4. Maintains close surveillance over assembly operations to ensure the proper use of manpower skills, equipment, and facilities.
5. Keeps current with the changes in technology, products and/or approaches as they affect the entire field of automated assembly and robotics and applies those found to be of value in achieving cost savings and/or improved performance.
6. Coordinates and participates in general and technical programs such as cost reduction, methods, communications, staffing, employee training, employee relations, grievances, and disciplinary actions.
7. Interacts with staff support groups to expedite any changes in department operations and works with other supervisory and managerial personnel to meet production schedules and to assist in meeting cost reduction and quality improvement objectives.
8. Anticipates production problems such as potential delays, material shortages, equipment maintenance and repairs, and devises and implements procedures to prevent or minimize loss of labor hours and scheduled interruptions.

lot of time thinking about my people and how I can make their job easier, and I work hard at taking care of them and making sure they get what they need to make the job satisfying." One way Rothwell tried to increase worker satisfaction was by giving a lot of freedom to his employees. For instance, one employee was permitted to work from 7 A.M. to 3:30 P.M., instead of the regular shift, on days when he wanted to go golfing. (He had been in an Allen-Bradley golf league prior to joining WCI and asked if he could continue for the remainder of the summer.)

Rothwell also gave his workers considerable responsibility. "I think I do more with my people than other managers. I try to put myself in their shoes. I want to involve them as much as possible," he remarked. An example of the way he increased his employees' involvement was their evaluations. In the coming fiscal year, Rothwell would be asking the operators to set their own objectives, and they would be measured on how well they met them. This was the technique by which Rothwell himself was evaluated by his boss.

The workers enjoyed the freedom and responsibility Rothwell gave them. According to one operator:

> With John we get more freedom. We don't have to report to him every time we walk out the door. He trusts us. He gives us more responsibility and makes us feel like we're accomplishing something. It's not like being a robot like on the assembly line.

JOHN ROTHWELL

While in high school Rothwell began working part-time as an automotive repairer, an experience he claimed was significant, for it helped him to develop a background in mechanics. After high school, Rothwell enrolled in Milwaukee Area Technical College to study architecture. A year and a half later, he switched his major because "architecture was not mechanical enough." He realized that machine design better fit with his skills and interests. "I was cut out to be a machine designer. I really enjoy tinkering. I can spend hours drawing machine designs because I find it fascinating," Rothwell declared. In 1968, Rothwell received his associate's degree in machine design.

After serving two years in Viet Nam as an engineman in the Navy, Rothwell spent a year as a product design draftsman for a manufacturer of large fittings before working for another year at a small engineering consulting firm. There he did mechanical drafting and designed high speed automated equipment and heavy construction machinery.

His career at Allen-Bradley began in May 1972 when he was hired as a layout draftsman in the Special Equipment Design Group, which was part of the New Product Industrialization Area. He then became a machine designer, responsible for creating the concept and design of special machinery, furnishing design specification data, and preparing cost proposals. During this period, Rothwell joined a local chapter of Toast Master's International (a business club whose members were taught public speaking skills) and eventually became chapter president. In describing his participation in the club, Rothwell remarked, "I had a great time and learned a tremendous amount. It was a real asset and had a big effect on my career."

In 1979 Rothwell was promoted to supervisor of the Special Machinery Department, his first exposure to a union environment and his first supervisory role. His duties included developing prototype machines and special equipment and supervising more than 30 employees, including machine builders, machinists, and sheet metal mechanics, in both a numerically controlled and a conventional machine shop operation. Describing that role, Rothwell stated, "It enabled me to learn a tremendous amount about the machinery and the duties of various machine operators. It made me a much better designer."

In 1981 Rothwell took on the part-time job of being chairman of United Way at Allen-Bradley, which he claimed helped him "to get to know the people and to get exposure and visibility." A year later, he was transferred back to the Special Equipment Group to do long-range machine planning. He was responsible for developing and directing project and resource planning for the development, construction, and installation of special equipment and systems. Rothwell remained in that position until November 1983 when he was selected to become the department manager of World Contactor One. Also in 1983, he received a Bachelor's degree in manufacturing engineering (he had attended evening classes). In addition, he had also participated in seminars in robotics and computer applications.

There were several reasons why Rothwell was chosen to head the new CIM facility. Larry Yost, vice president of operations for the Industrial Control Division, explained:

> Automation requires a manager to understand high technology, for instance, computers. You need a new type of manager, a type of person who is technically competent and who has people skills. My feeling is you can take a person with a technical bent and teach them people skills, but you can't teach everybody technical skills.
>
> John had a broad background which was necessary. He had a smattering of all the technologies and worked well with it. He had formal training, and he had moved around the company and therefore had developed people skills. He was the only person for the job. We made it known that he was the first of a new breed of manager.

Jim Heimler, manager of New Product Industrialization and Rothwell's immediate manager, remarked:

> Ideally, you'd like to think a CIM manager would understand automation equipment. You ask, "Is he a good mechanic? Does he understand how machines operate mechanically and electronically?" John's background fell right in.

I don't think there are that many mechanical-background people who have strong computer backgrounds. They don't necessarily need to know it, but they need to understand it. I don't think John had all the attributes, but he learned them. He had a very good basic understanding of machines and needed to learn the computer aspect, so he took a lot of courses in programming and PLC-3.

REPLACING JOHN ROTHWELL

There was beginning to be talk of a World Contactor II facility, which would be even more automated than WCI and for which planning would start in 1988. Rothwell hoped to be given the opportunity of applying his knowledge and experience to this new endeavor. Since he was well regarded, he appeared to be the natural candidate to head up such a project.

This, however, raised an important question. What kind of skills would Rothwell's replacement need? Rothwell himself explained, "My replacement would need to understand the integration of the whole computerized system. It's important for a supervisor to know the software too. You don't have to be an excellent technician, you just have to be able to supervise the one you've got." Other Allen-Bradley executives also commented on the skills that Rothwell's replacement would need:

Larry Yost:

"The new manager would have to be a person with very similar skills. The technical skills that are needed won't diminish as time goes on. The people skills will still be necessary. So, the requirements won't really change. The only difference is maybe you could get by with a person who was a bit less ambitious than John, more of a caretaker type."

Jim Heimler:

"I don't think the new supervisor will have to be as refined as John. They won't need to be as much of an inventor. The next person will just have to keep it stable. This role could fall to one of our present managers, providing he or she has the same basic skills."

Tom Retzer, the production manager of Power Products:

"Rothwell's replacement should be someone from the quality area. One of the reasons for the Japanese success is because they put quality first, and at Allen-Bradley, quality is part of our name. Also, the person should be degreed in electrical engineering because of the high technology and should have over five years of experience. He will also need the right kind of personality to fit in there. That person will have to supervise the workers directly because there are no other supervisors, and will have to give tours to visitors and businesspeople."

Rothwell's potential departure from World Contactor One also raised the issue of how he would transfer all of his knowledge of the CIM facility to his replacement. Yost remarked, "When we get to that point, we'll put the person in there to train with John for a long time period. Still, there's no way he'll be able to transfer all of his knowledge. It's not going to be easy, there's no question about it."

Another aspect of the knowledge transfer issue involved the way Rothwell could help other supervisors who would be managing new CIM facilities being started by the company. Part of Allen-Bradley's long-term strategy included implementing CIM as part of several other operations, and many plants already had such facilities at various stages of planning and installation. At Milwaukee, for instance, a second CIM facility called the Solid State Project was well into the planning stage. Solid State would comprise a smaller area than World Contactor One, but its volume would be tremendous. It would also be different from WCI because neither the product nor the equipment was being specially designed.

The Solid State department manager, who had already been selected, held a Bachelor's in electrical engineering and had spent his 16 years at Allen-Bradley working in manufacturing and engineering, dealing mainly with electronics. He had spent time with Rothwell to gain insight on issues such as how to select the work force, set up an accounting system, and establish the computer hierarchy. He wanted to find out what went right and what went wrong with WCI, and how he could avoid any pitfalls that had previously been encountered.

Manufacturing by Design

Daniel E. Whitney

In many large companies, design has become a bureaucratic tangle, a process confounded by fragmentation, overspecialization, power struggles, and delays. An engineering manager responsible for designing a single part at an automobile company told me recently that the design process mandates 350 steps—not 350 engineering calculations or experiments but 350 workups requiring 350 signatures. No wonder, he said, it takes five years to design a car; that's one signature every 3½ days.

It's not as if companies don't know better. According to General Motors executives, 70 percent of the cost of manufacturing truck transmissions is determined in the design stage. A study at Rolls-Royce reveals that design determines 80 percent of the final production costs of 2,000 components.[1] Obviously, establishing a product's design calls for crucial choices—about materials made or bought, about how parts will be assembled. When senior managers put most of their efforts into analyzing current production rather than product design, they are monitoring what accounts for only about a third of total manufacturing costs—the window dressing, not the window.

Moreover, better product design has shattered old expectations for improving cost through design or redesign. If managers used to think a 5 percent improvement was good,

Reprinted with permission from *Harvard Business Review*, July–August 1988.

[1] J. Corbett, "Design for Economic Manufacture," *Annals of C.I.R.P.* vol. 35, no. 1, 1986, p. 93.

they now face competition that is reducing drastically the number of components and subassemblies for products and achieving a 50 percent or more reduction in direct cost of manufacture. And even greater reductions are coming, owing to new materials and materials-processing techniques. Direct labor, even lower cost labor, accounts for so little of the total picture that companies still focusing on this factor are misleading themselves not only about improving products but also about how foreign competitors have gained so much advantage.

In short, design is a strategic activity, whether by intention or by default. It influences flexibility of sales strategies, speed of field repair, and efficiency of manufacturing. It may well be responsible for the company's future viability. I want to focus not on the qualities of products but on development of the processes for making them.

Converting a concept into a complex, high-technology product is an involved procedure consisting of many steps of refinement. The initial idea never quite works as intended or performs as well as desired. So designers make many modifications, including increasingly subtle choices of materials, fasteners, coatings, adhesives, and electronic adjustments. Expensive analyses and experiments may be necessary to verify design choices.

In many cases, designers find that the options become more and more difficult; negotiations over technical issues, budgets, and schedules become intense. As the design evolves, the choices become interdependent, taking on the character of an interwoven, historical chain in which later decisions are conditioned by those made previously.

Imagine, then, that a production or manufacturing engineer enters such detailed negotiations late in the game and asks for changes. If the product designers accede to the requests, a large part of the design may simply unravel. Many difficult and pivotal choices will have been made for nothing. Where close calls went one way, they may now go another; new materials analyses and production experiments may be necessary.

Examples of failure abound. One research scientist I know, at a large chemical company, spent a year perfecting a new process—involving, among other things, gases—at laboratory scale. In the lab the process operated at atmospheric pressure. But when a production engineer was finally called in to scale up the process, he immediately asked for higher pressures. Atmospheric pressure is never used in production when gases are in play because maintaining it requires huge pipes, pumps, and tanks. Higher pressure reduces the volume of gases and permits the use of smaller equipment. Unfortunately, the researcher's process failed at elevated pressures, and he had to start over.

Or consider the manufacturer whose household appliance depended on close tolerances for proper operation. Edicts from the styling department prevented designs from achieving required tolerances; the designers wanted a particular shape and appearance and would not budge when they were apprised of the problems they caused to manufacturing. Nor was the machine designed in modules that could be tested before final assembly. The entire product was built from single parts on one long line. So each finished product had to be adjusted into operation—or taken apart after assembly to find out why it didn't work. No

one who understood the problem had enough authority to solve it, and no one with enough authority understood the problem until it was too late. This company is no longer in business.

Finally, there was the weapon that depended for its function on an infrared detector, the first of many parts—lenses, mirrors, motors, power supplies, etc.—that were glued and soldered together into a compact unit. To save money, the purchasing department switched to a cheaper detector, which caused an increase in final test failures. Since the construction was glue and solder, bad units had to be scrapped. Someone then suggested a redesign of the unit with reversible fasteners to permit disassembly. But this time more reasonable voices prevailed. Reversible fasteners would have actually increased the weapon's cost and served no purpose other than to facilitate factory rework. Disassembly would not have been advisable because the unit was too complex for field repair. It was a single-use weapon—with a shelf life of five years and a useful life of ten seconds. It simply had to work the first time.

Manufacturers can avoid problems like this. Let's look at a success story. One company I know wanted to be able to respond in 24 hours to worldwide orders for its electronic products line—a large variety of features in small-order batches. Engineers decided to redesign the products in modules, with different features in each module. All the modules are plug compatible, electrically and mechanically. All versions of each module are identical on the outside where assembly machines handle them. The company can now make up an order for any set of features by selecting the correct modules and assembling them, all of this without any human intervention, from electronic order receipt to the boxing of final assemblies.

In another company, a high-pressure machine for supplying cutting oil to machine tools requires once-a-day cleaning. Designers recently reconfigured the machine so that normal cleanout and ordinary repairs can be accomplished without any tools, thus solving some bothersome union work-rule problems.

There are no guarantees, of course, but the experiences of these companies illustrate how design decisions should be integrated, informed, and balanced, and how important it is to involve manufacturing engineers, repair engineers, purchasing agents, and other knowledgeable people early in the process. The product designer asks, "What good is it if it doesn't work?" The salesperson asks, "What good is it if it doesn't sell?" The finance person asks, "What good is it if it isn't profitable?" The manufacturing engineer asks, "What good is it if I can't make it?" The team's success is measured by how well these questions are answered.

The Design Team and Its Task

Multifunctional teams are currently the most effective way known to cut through barriers to good design. Teams can be surprisingly small—as small as 4 members, though 20 members is typical in large projects—and they usually include every specialty in the company. Top executives should make their support and interest clear. Various names have been given to this team approach, like "simul-

---

Designing for Predictability: New MCAE Tools

A well-designed product is a predictable product. Managers particularly need to predict reliability, manufacturing costs, and manufacturability. In the past, engineers have dealt with these three issues only after engineering has completed the drawings, the near-final stage in the development cycle. But by then it may be too late. Moreover, when a product has hundreds, or even thousands of components, it's just not feasible to design, prototype, and redesign slightly different versions of each.

Two new, integrated, mechanical computer-aided engineering (MCAE) systems permit engineering teams to test before they build, so they can design for total quality with reliability, performance, and manufacturing costs in mind from the start. The first of these new systems models products not just physically but analytically: a designer can not only draw a component on the screen but prompt the computer to model and test the feasibility of its features, in the manner of a business planner using an electronic spreadsheet to play "what if" games. Engineers can vary assumptions about materials, speeds, loads, size, and other operating conditions. In this way, developers can both see the effects of hypothetical stresses and estimate product costs while making design decisions.

A company making internal combustion engines, for example, may use an integrated MCAE system to design reliability and smoothness of operation into the counterbalance for the crankshaft. The optimal counterbalance keeps forces from the piston's up-and-down motion and centrifugal forces from the unbalanced crankshaft both low and constant—to reduce vibration. The system works as follows: a desktop workstation paints an image of a cylinder with all its operating parts on the screen. The engineer then selects values for counterbalance features like angle, thickness, diameter,

taneous engineering" and "concurrent design." Different companies emphasize different strengths within the team. In many Japanese companies, teams like this have been functioning for so long that most of the employees cannot remember another way to design a product.

Establishing the team is only the beginning, of course. Teams need a step-by-step procedure that disciplines the discussion and takes members through the decisions that crop up in virtually every design. In traditional design procedures, assembly is one of the last things considered. My experience suggests that assembly should be considered much earlier. Assembly is inherently integrative. Weaving it into the design process is a powerful way to raise the level of integration in all aspects of product design.

A design team's charter should be broad. Its chief functions include:

1. Determining the character of the product, to see what it is and thus what design and production methods are appropriate.
2. Subjecting the product to a product function analysis, so that all design decisions can be made with full knowledge of how the item is supposed to work and all team members understand it well enough to contribute optimally.
3. Carrying out a design-for-producibility-and-

(continued)

etc., from menus. As choices are made, the system automatically computes the merits of the design, based on about 100 engineering equations, including compression ratio and stroke. So design variations are tried, evaluated, and discarded with near instantaneous response. This puts robustness and performance optimization into the very first counterbalance designs.

The second new system is an expert system which projects probable production costs for various part or assembly configurations and provides guidance as to their manufacturability. Another parts maker might use this system to project the manufacturability of and costs for, say, its stamped carburetor parts. Typically, the PC screen has several menus listing aspects of a part's design, like materials used and manufacturing process involved. Each menu holds progressive layers of possible choices. If the engineer selects metal from the list of materials, the system offers a choice of ferrous or nonferrous. Under ferrous metals, one can pick from carbon steel, stainless, cast iron, and so on.

There are automatic default values that the engineer might not normally specify, such as surface finish and carbon content. The system also draws on its own data base for manufacturing information like material density and base unit cost. Once the designer completes the menu sequence, the system produces an approximate part cost that includes materials, processing, and tooling expenditures.

MCAE systems don't eliminate prototyping entirely, but they can drastically reduce the number of trials and help engineers address reliability, performance, and cost problems early—and thus design for total quality from the very beginning.

—Philippe Villers

usability study to determine if these factors can be improved without impairing functioning.

4. Designing an assembly process appropriate to the product's particular character. This involves creating a suitable assembly sequence, identifying subassemblies, integrating quality control, and designing each part so that its quality is compatible with the assembly method.

5. Designing a factory system that fully involves workers in the production strategy, operates on minimal inventory, and is integrated with vendors' methods and capabilities.

The Product's Character

Clearly it is beyond the scope of this article to establish by what criteria one judges, develops, or revamps the features of products. Recently in these pages, David A. Garvin has analyzed eight fundamental dimensions of product quality; and John R. Hauser and Don Clausing have explored ways to communicate to design engineers the dimensions consumers want—in the engineers' own language.[2]

Character defines the criteria by which designers judge, develop, or revamp product features. I would only reiterate that manufacturing engineers and others should have something to say about how to ensure that the product is field repairable, how skilled users must be to employ it successfully, and whether marketability will be based on model variety or availability of future add-ons.

An essential by-product of involving manufacturing, marketing, purchasing, and other constituencies in product conception, moreover, is that diverse team members become fa-

miliar enough with the product early in order to be able to incorporate the designers' goals and constraints in their own approaches. As designers talk with manufacturing or field-service reps, for example, they can make knowledgeable corrections. ("Why not make that part out of plastic? I know a low-cost source." "Because the temperature there is 1,000°; plastic will vaporize." "Oh.")

Product Function Analysis

This used to be the exclusive province of product designers. But now it is understood that to improve a product's robustness, to "design quality in" in Genichi Taguchi's good phrase, means thoroughly understanding a product's function in relation to production methods. Product designers and manufacturing engineers used to try to understand these relations by experience and intuition. Now they have software packages for modeling and designing components to guide them through process choices—software that would have been thought fantastic a generation ago. (See the insert, "Designing for Predictability: New MCAE Tools.")

Recently I worked on a product containing delicate spinning parts that had to be dynamically balanced to high tolerances. In the original design, partial disassembly of the rotating elements after balancing was necessary before the assembly could be finished, so the final product was rarely well balanced and required a lengthy adjustment procedure. Since total redesign was not feasible, the team analyzed the reassembly procedure solely as it pertained to balance and concluded that designers needed only to tighten various tolerances and reshape mating surfaces. Simple adjustments were then sufficient to restore balance in the finished product.

Another important goal of product function analysis is to reduce the number of parts in a product. The benefits extend to purchasing (fewer vendors and transactions), manufactur-

---

[2]David A. Garvin, "Competing on the Eight Dimensions of Quality," *Harvard Business Review* November–December 1987, p. 101; and John R. Hauser and Don Clausing, "The House of Quality," *Harvard Business Review* May–June 1988, p. 63.

ing (fewer operations, material handlings, and handlers), and field service (fewer repair parts).

When a company first brings discipline to its design process, reductions in parts count are usually easy to make because the old designs are so inefficient. After catching up, though, hard, creative work is necessary to cut the parts count further. One company I know saved several million dollars a year by eliminating just one subassembly part. The product had three operating states: low, medium, and high. Analysis showed that the actions of one part in the original design always followed or imitated the actions of two others. Designers eliminated the redundant part by slightly altering the shapes of the other two parts.

This change could never have been conceived, much less executed, if the designers hadn't had deep knowledge of the product and hadn't paid attention to the actions underlying its engineering.

Design for Producibility

Recently, a company bragged to a business newsweekly about saving a mere $250,000 by designing its bottles for a new line of cosmetics to fit existing machines for filling, labeling, and capping. This plan seems so obvious, and the savings were so small as compared with what is possible, that the celebration seemed misplaced. But it's a better outcome than I remember from my first job with a drug company. It spent a fortune to have a famous industrial designer create new bottles and caps for its line. They were triangular in cross section and teardrop shaped, and they would not fit either existing machines or any new ones we tried to design. The company eventually abandoned the bottles, along with the associated marketing campaign.

Obviously, nothing is more important to manufacturing strategy than designing for the production process. In the past, this has meant designing for manufacturing and assem-

bly, and value engineering, which both strive to reduce costs. But now we have to go beyond these goals.

To take the last point first, value engineering aims chiefly to reduce manufacturing costs through astute choice of materials and methods for making parts. Does the design call for metal when a ceramic part will do? If metal, should we punch it or drill it? Value engineering usually comes into play after the design is finished, but the thoroughness we seek in design can be achieved only when decisions are made early.

Moreover, design for producibility differs from design for assembly, which typically considers parts one by one, simplifies them, combines some to reduce the parts count, or adds features like bevels around the rims of holes to make assembly easier. Valuable as it is, this process cannot achieve the most fundamental improvements because it considers the product as a collection of parts instead of something to satisfy larger goals, such as reducing costs over the product's entire life cycle.

Nippondenso's approach vividly illustrates how an overriding strategy can determine a product's parts and the production process. The Delco of Japan, Nippondenso builds such car products as generators, alternators, voltage regulators, radiators, and antiskid brake systems. Toyota is its chief customer. Nippondenso has learned to live with daily orders for thousands of items in arbitrary model mixes and quantities.

The company's response to this challenge has several components:

- The combinatorial method of meeting model-mix production requirements.
- In-house development of manufacturing technology.
- Wherever possible, manufacturing methods that don't need jigs and fixtures.

The combinatorial method, carried out by marketing and engineering team members, divides a product into generic parts or subassemblies and identifies the necessary variations of each. The product is then designed to permit any combination of variations of these basic parts to go together physically and functionally. (If there are 6 basic parts and 3 varieties of each, for example, the company can build $3^6 = 729$ different models). The in-house manufacturing team cooperates in designing the parts, so the manufacturing system can easily handle and make each variety of each part and product.

Jigless production is an important goal at this point, for obvious reasons. Materials handling, fabrication, and assembly processes usually employ jigs, fixtures, and tools to hold parts during processing and transport; the jigs and fixtures are usually designed specifically to fit each kind of part, to hold them securely. When production shifts to a different batch or model, old jigs and tools are removed and new ones installed. In mass-production environments, this changeover occurs about once a year.

In dynamic markets, however, or with just-in-time, batches are small, and shifts in production may occur hourly—even continually. It may be impossible to achieve a timely and economical batch-size-of-one production process if separate jigs are necessary for each model. Nippondenso's in-house manufacturing team responds to this problem by showing how to design the parts with common jigging features, so that one jig can hold all varieties, or by working with designers to make the product snap or otherwise hold itself together so that no clamping jigs are needed.

By cultivating an in-house team, Nippondenso also solves three difficult institutional problems. First, the company eliminates proprietary secrecy problems. Its own people are the only ones working on the design or with strategically crucial components. Second, equipment can be delivered without payment of a vendor's markup, thus reducing costs and making financial justification easier.

Third, over the years the team has learned to accommodate itself intuitively to the company's design philosophy, and individual team members have learned how to contribute to it. Designers get to know each other too, creating many informal communication networks that greatly shorten the design process. Shorter design periods mean less lead time, a clear competitive advantage. (It is worth noting that many Japanese companies follow this practice of designing much of their automation in-house, while buying many product components from outside vendors. American companies usually take the opposite tack: They make many components and buy automation from vendors.)

Nippondenso uses combinatorial design and jigless manufacturing for making radiators (see the diagram). Tubes, fins, headers, and side plates comprise the core of the radiator. These four snap together, which obviates the need for jigs, and the complete core is oven soldered. The plastic tanks are crimped on. The crimp die can be adjusted to take any tank size while the next radiator is being put in the crimper, so radiators can be processed in any model order and in any quantity. When asked how much the factory cost, the project's chief engineer replied, "Strictly speaking, you have to include the cost of designing the product." A factory isn't just a factory, he implied. It is a carefully crafted fusion of a strategically designed product and the methods for making it.

Without a guiding strategy, there is no way to tell what suggestions for improvement really support long-range goals. Some product-design techniques depend too much on rules, including rule-based systems stemming from expert systems. These are no substitute for experienced people. Volkswagen, for example, recently violated conventional ease-of-assembly rules to capture advantages the company would not otherwise have had.

In the company's remarkable Hall 54 facility in Wolfsburg, Germany, where Golfs and Jettas go through final assembly, robots or spe-

How a Raditor Is Made—The Combinatorial, Jigless Method

Inlet tank

Header

Side plate

Outlet tank

Sheet stock
|
Fins
Side plates
Headers
Tubes
|
Sheet stock

Inlet tank ┐

Core → Oven solder → Finished core → Crimp → Test

(snaps together)   No jig   Outlet tank ┘   No jig

Except for final testing, this radiator is fabricated entirely without manual labor. Radiators differ by core length, width, and depth. They are available in various sizes and offer many heat-transfer capacities. Without changing part sizes, a designer can program different shapes for fins and diameters for tubes, thus allowing the same production system to achieve new heat-transfer capacities.

cial machines perform about 25 percent of the final-stage steps. (Before Hall 54 began functioning, Volkswagen never did better than 5 percent.)[3]

[3]E.H. Hartwich, "Possibilities and Trends for the Application of Automated Handling and Assembly Systems in the Automotive Industry," International Congress for Metalworking and Automation, Hannover, West Germany, 1985, p. 126.

To get this level of automation, VW production management asked to examine every part. It won from the board of directors a year-long delay in introducing the new models. Several significant departures from conventional automotive design practices resulted, the first involving front-end configuration. Usually, designers try to reduce the number of parts. But VW engineers determined that at a cost of one *extra* frame part the front of the car could be temporarily left open for installation of the engine by hydraulic arms in one straight, upward push. Installing the engine used to take a minute or longer and involved several workers. VW now does it unmanned in 26 seconds.

Another important decision concerned the lowly screw. Purchasing agents usually accept the rule that low-cost fasteners are a competitive edge. VW engineers convinced the purchasing department to pay an additional 18 percent for screws with cone-shaped tips that go more easily into holes, even if the sheet metal or plastic parts were misaligned. Machine and robot insertion of screws thus became practical. Just two years later, so many German companies had adopted cone-pointed screws that their price had dropped to that of ordinary flat-tip screws. For once, everyone from manufacturing to purchasing was happy.

Assembly Processes

Usually assembly sequence is looked at late in the design process when industrial engineers are trying to balance the assembly line. But the choice of assembly sequence and the identification of potential subassemblies can affect or be affected by—among other factors—product-testing options, market responsiveness, and factory-floor layout. Indeed, assembly-related activities with strategic implications include: subassemblies, assembly sequence, assembly method for each step, and integration of quality control.

Imagine a product with six parts. We can build it many ways, such as bottom up, top down, or from three subassemblies of two parts each. What determines the best way? A balance of many considerations: construction needs, like access to fasteners or lubrication points; ease of assembly (some sequences may include difficult part matings that risk damage to parts); quality control matters, like the operator's ability to make crucial tests or easily replace a faulty part; process reasons,

like ability to hold pieces accurately for machine assembly; and, finally, production strategy advantages, like making subassemblies to stock that will be common to many models, or that permit assembly from commonly available parts.

Again, software now exists to help the designer with the formidable problem of listing all the possible assembly sequences—and there can be a lot, as many as 500 for an item as simple as an automobile rear axle. It would be impossible for a team to attack so complex a series of choices without a computer design aid to help, according to a preestablished hierarchy of goals like that just discussed—access to lubrication points, etc. Another virtue of this software is that it forces the team to specify choices systematically and reproducibly, for team members' own edification but also in a way that helps justify design and manufacturing choices to top management.

Consider then, automatic transmissions, complex devices made up of gears, pistons, clutches, hydraulic valves, and electronic controls. Large transmission parts can scrape metal off smaller parts during assembly, and shavings can get into the control valves, causing the transmission to fail the final test or, worse, fail in the customer's car. Either failure is unacceptable and terribly expensive. It is essential to design assembly methods and test sequences to preempt them.

With respect to assembly machines and tooling, manufacturers should consider the following questions:

- Can the product be made by adding parts from one direction, or must it be turned over one or more times? Turnovers are wasted motion and costly in fixtures.
- As parts are added in a stack, will the location for each subsequent part drift unpredictably? If so, automatic assembly machines will need expensive sensors to find the parts, or assembly will randomly fail, or

parts will scrape on each other too hard.

- Is there space for tools and grippers? If not, automatic assembly or testing aren't options.
- If a manufacturing strategy based on subassemblies seems warranted, are the subassemblies designed so they do not fall apart during reorientation, handling, or transport?

There are clear advantages to combining consideration of these assembly procedures and/or quality control strategy with design. Designers who anticipate the assembly method can avoid pitfalls that would otherwise require redesign or create problems on the factory floor. They can also design better subassemblies to meet functional specifications—specifications that will be invaluable when the time comes to decide whether to take bids from outside vendors or make the part on the company's own lines, specifications that will determine how to test the subassembly before adding it to the final product.

Designers concerned about assembly must ask:

- What is the best economic combination of machines and people to assemble a certain model-mix of parts for a product line (given each machine's or person's cost and time to do each operation, plus production-rate and economic-return targets)?
- How much time, money, production machinery, or in-process inventory can be saved if extra effort is put into design of the product, its fabrication and assembly processes, so that there are fewer quality control failures and product repairs? A process that yields only 80 percent successful assemblies on the first try may need 20 percent extra capacity and inventory—not to mention high-cost repair personnel—to meet the original production goals.
- Where in the assembly process should testing take place? Considerations include how

costly and definitive the test is, whether later stages would hide flaws detectable earlier, and how much repaired or discarded assemblies would cost.

These are generic problems; they are hard to answer, and they too are stimulating the development of new software packages. This new software enhances the ability of manufacturing people to press their points in (often heated) debates about design. Hitherto, product designers, more accustomed to using computer modeling, have had somewhat of an upper hand.

Factory System Design

Many features of good product design presuppose that machines will do the assembly. But automation is not necessary to reap the benefits of strategic design. Indeed, sometimes good design makes automatic assembly unnecessary or uneconomic by making manual assembly so easy and reliable. Regardless of the level of automation, some people will still be involved in production processes, and their role is important to the success of manufacturing.

Kosuke Ikebuchi, general manager of the General Motors-Toyota joint venture, New United Motors Manufacturing Inc. (NUMMI), believes that success came to his plant only after careful analysis of the failures of the GM operation that had preceded it: low-quality parts from suppliers, an attitude that repair and rework were to be expected, high absenteeism resulting in poor workmanship, and damage to parts and vehicles caused by transport mechanisms.[4] The assembly line suffered from low efficiency because work methods were not standardized, people could not repair

---

[4]Kosuke Ikebuchi, paper read at the Future Role of Automated Manufacturing Conference, New York University, 1986.

their own equipment, and equipment was underutilized. Excess inventory, caused by ineffective controls, was another problem. Work areas were crowded. Employees took too much time to respond to problems.

NUMMI's solutions focused on the Jidoka principle—quality comes first. According to NUMMI's factory system today, workers can stop the line if they spot a problem; the machinery itself can sense and warn of problems. Two well-known just-in-time methods of eliminating waste—the kanban system of production control and reductions in jig and fixture change times—are important to NUMMI's manufacturing operation.

But lots of other things also contribute to this plant's effectiveness: simplified job classifications, displays and signs showing just how to do each job and what to avoid, self-monitoring machines. NUMMI has obtained high-spirited involvement of the employees, first by choosing new hires for their willingness to cooperate, then by training them thoroughly and involving them in decisions about how to improve the operations.

Design Means Business

The five tasks of design bring us back to the original point. Strategic product design is a total approach to doing business. It can mean changes in the pace of design, the identity of the participants, and the sequence of decisions. It forces managers, designers, and engineers to cross old organizational boundaries, and it reverses some old power relationships. It creates difficulties because it teases out incipient conflict, but it is rewarding precisely because disagreements surface early, when they can be resolved constructively and with mutual understanding of the outcome's rationale.

Strategic design is a continual process, so it makes sense to keep design teams in place until well after product launching when the same team can then tackle a new project. Design—it must be obvious by now—is a companywide activity. Top management involvement and commitment are essential. The effort has its costs, but the costs of not making the effort are greater.

Toward the Factory of the Future

P. T. Bolwijn
T. Kumpe

The factory of the future has had a wonderful press. Automation, the media never tire of telling us, will soon reshape the face of manufacturing industry. Articles and TV programs on the future of computer-based systems often give the impression that in the fairly near future all manufactured goods will be produced in computer-run factories where all shop-floor employees will have been replaced by robots.

The reality is rather different. To be sure, continuous process industries like petroleum refining are already well advanced in automation, and computer technology has played a major role in the steel industry, most notably in Japan. In much of manufacturing industry, however, computer-integrated production has made remarkably little headway: worldwide, just two industries—aerospace and automaking—account for some 80 percent of installed robots.

Yet there are plenty of potentially profitable applications even in relatively prosaic manufacturing operations. A manufacturer of precision-machined pump cylinder blocks, for example, found that its worn-out boring equipment could no longer meet required tolerances. Instead of simply replacing the equipment at a cost of $1.3 million, management decided after careful investigation to invest $15

Excerpted and reprinted with permission from *Flexible Manufacturing: Integrating Technical and Social Innovation*, Elsevier Science Publishers B.V.

The authors wish to thank Graham Sharman of McKinsey & Company for his helpful comments and suggestions.

million in a totally computer-integrated flexible manufacturing system (FMS). As a result the company succeeded in reducing overall manufacturing costs by 30 percent and inventory by 85 percent, while achieving order-of-magnitude improvements in product quality. And since it can now make 99 percent of customer requirements directly to order, it has gained an important competitive advantage.

The Competitive Imperative

Intensifying competition gives manufacturing management a compelling reason to pursue such opportunities. It has forced electronics manufacturers, for example, to become not only more flexible and efficient but also more quality-conscious as they strive to outdo each other in tempting consumers with innovative design, sophisticated features, high style and product variety. In the past decade, product life-cycles in the North American and European consumer electronics industries have contracted under the pressure of competition from three or four years to, in some cases, a few months. The time available to develop new products, set up production facilities, and train marketing and sales people has shrunk just as radically.

To meet competition from low-cost producers, European and American manufacturers traditionally relied on the economies of scale made possible by large-scale batch manufacturing operations. Product variety was limited by the demands of efficiency. Quality, instead of being built into the manufacturing system, was achieved by weeding out defective items.

The objective was a reasonably good product, manufactured at a cost that would permit competitive pricing.

Today, efficiency is no longer the overriding imperative in manufacturing. Product offerings from the Far East have made customers much more quality-conscious and accustomed them to frequent model changes. And traditional batch-manufacturing techniques lack the flexibility required to bring out high-quality new products at yearly intervals or less. To meet the fast-changing demands of today's fashion-conscious markets without maintaining huge stocks of finished products, companies need greater flexibility, smaller batch sizes and continuous flow—in short, precisely the benefits promised by FMS.

Startling gains in operational effectiveness are indeed possible through FMS, but only if the groundwork has been properly laid. Without it, massive applications of computer power can be a waste of resources: The cost of simply automating existing manual operations in a factory built to produce yesterday's products will usually exceed the anticipated labor savings and may actually result in a loss of flexibility.

Most companies, if the truth be known, have not even begun to tap the improvement potential that they could realize by optimizing their existing manual systems. "In a lot of applications where we start out looking at robots," said Larry J. Kerr, vice-president of manufacturing at Firestone Tire & Rubber, in a recent interview, "we wind up finding another way to do the job at half the cost and a quarter of the complexity."

A recent British study suggests that a major share of the performance improvement achieved by companies that have introduced FMS often results from a thoroughgoing preliminary examination and overhaul of operations—a point underlined by Philips vice president Martin Kuilman. "Analyzing our operations in the runup to FMS really opened our eyes to two sources of waste," he says. "One was the materials wastage represented by products or components that had to be rejected for quality reasons. The other was the investment tied up in unnecessary stocks of raw materials, subassemblies and finished products."

The principal explanation for the slow progress of FMS, in our view, is not so much a shortage of technical expertise as a lack of genuine management commitment to the demanding task of laying the operational groundwork for the factory of the future. This challenge boils down to three essential requirements. The first is to instill an integrated managerial viewpoint across the entire business system. The second is to build in the flexibility—managerial, technical and motivational—that is needed to make factory automation a successful reality. The third is pragmatism in implementation. Let us look at each in turn.

An Integrated View

Most manufacturers have a tendency to approach FMS as if it were merely a matter of automating the individual segments of their business system—the chain of functional operations, beginning with product design and materials procurement and ending with physical delivery and after-sales service, by which the product is conceived, created, marketed, and put at last into the customers' hands. In other words, they have typically used automation to optimize individual links in the chain rather than refashioning the chain as a whole. As a result, they end up paying the full price for automation without capturing the benefits of system-wide optimization—which alone can justify its cost. To realize the full potential of flexible manufacturing, companies must learn to see their business systems in integrated terms.

Bringing this about will require three kinds of change:

1. *Instilling integrative skills.* In preparing for FMS, the production system needs not only to be considered in its totality, but also to be seen as an integral part of the overall business system. Thus, managers with a business-wide outlook as well as outstanding technical, operational and organizational skills will be needed to direct the integration process. To make a real success of FMS over the long term, most companies may well need to rethink their recruiting criteria and redesign their management training programs to replace functional thinking with an integrated, business-wide perspective.
2. *Redesigning products to exploit system-wide synergies.* To make the best use of existing facilities and prepare the ground for FMS, management will need to take a hard look at the existing product line. Commonality of parts among different products, often a minor consideration in past product design, becomes crucially important with FMS: The more parts are shared by different products, the lower the costs of switching production from one to another in response to changing market demand. An extensive product redesign program may be needed to maximize parts commonality.

   Individual products may also offer scope for simplification through redesign. A survey of product designs in consumer electronics revealed that the number of components in a given product could typically be reduced by a third, with a commensurate reduction in assembly times, a saving of almost 50 percent in investment in transportation and assembly systems, and a major improvement in logistics performance.
3. *Rethinking conventional wisdom.* Reluctance on management's part to reexamine accepted capital investment evaluation criteria has, we believe, seriously impeded the spread of FMS to date. Seen in terms of conventional financial decision-making criteria, a full-fledged FMS is simply harder to justify, to "sell" internally, than were the routine, one-off mechanization and automation projects of the past.

Most companies base their investment decisions on the potential returns to the specific business system segment where the new asset will be put to work and on its potential to improve profit margins. In the context of flexible manufacturing these criteria will not do. For one thing, the advantages of FMS are found not so much in increased profit margins as in higher capital turnover, product quality and customer service levels. For another, the savings may not show up in the form of lower manufacturing costs. Instead, they may surface elsewhere in the business system, in the form of lower stock levels, shorter planning horizons and lower service costs.

Another bit of conventional manufacturing wisdom that stands in the way of FMS holds that machine utilization should not drop below about 85 percent of capacity. Consistently followed throughout a factory, this results in piles of work-in-process inventory between one operation and the next—a situation to which, in the past, most manufacturers have been resigned. With FMS, on the other hand, some machines or groups of machines may be running at only 50 percent of capacity—a situation that can be justified only from the viewpoint of the business as a single integrated system. Hence the need for a new breed of manager—or at least for new management thinking.

Built-In Flexibility

More than an integrated managerial view is required if flexible manufacturing is to work. Within the factory itself, three further kinds of change are needed. First, the organization must become flexible as well as efficient; second, the production process itself will have to be redesigned for maximum flexibility; finally, operating managers and their people will have to be motivated effectively.

Organizational Flexibility

Many businesses have been so single-minded (and often so successful) in their pursuit of minimum costs that they hardly conceive of the possibility of operating in any other way. Yet the handwriting is on the wall for the long life-cycle, scale-oriented mass production operations to which their existing organization and systems are tailored. The costs of inflexibility are rising all the time. Inflexible organizations must continually be improvising to meet unforeseen (if not unforeseeable) changes and challenges. They are likely to be burdened with outdated products and obsolete stocks. They are underpriced by competitors who turn flexibility to their advantage while escaping the penalty of high cost. Market forces are inexorably forcing a shift toward shorter production runs, shorter life-cycles and higher product quality.

Under today's conditions, then, flexibility is really a competitive necessity. Like quality, it is not an element of cost but a source of value for the industrial customer as well as the producer. It enables the customer to speed the introduction of products incorporating the latest technological advances, and enables both customer and producer to improve their profitability by reducing work-in-process and finished goods inventories.

Flexibility and bureaucracy, however, will not mix. The leadership of the flexible organization must be visible and vigorous. Reporting relationships must be participative rather than authoritarian. Systems must ensure rapid communications and feedback of information throughout the organization. Simple, clear-cut, product-oriented organizational structures must be designed to facilitate the linkage between decisions and actions across the entire business system. Interfunctional barriers must be minimal or nonexistent.

All this means that small enterprises, or small, quasi-autonomous, product-based, profit-center units of larger organizations, provide a much more congenial climate for flexibility than do large, functionally oriented corporations. For a large company, a commitment to flexibility may mean abandoning functional organization structures and highly specialized production facilities. Depending on the degree of flexibility required, large-scale manufacturing operations may have to be reorganized into smaller, simpler, more flexible "factories within a factory."

The organizational shift from conventional efficiency-oriented organization to flexible factory will almost always prove a more radical change than management anticipates. As Exhibit 1 suggests, differences are to be found at every stage of the business system—most critically, perhaps, in the production process itself and in its design criteria.

Redesigned Production Processes

Companies that have routinely sacrificed flexibility on the altar of minimum variable product cost tend to show certain characteristic symptoms. These include excessive stock levels, overly complex product flows, and throughput times out of all proportion to production times. In almost any European factory where the production process has been mechanized one step at a time, anything from 95 to 99 percent of product throughput time is waiting time. Factories have become glorified warehouses.

That the traditional piecemeal approach to mechanization should have such results is not surprising. When individual processes and mechanisms are optimized without regard to how each relates to the others in the total manufacturing flow, bottlenecks are the natural result. When individual machines are run for

EXHIBIT 1 From Efficiency to Flexibility

	Efficient Organization	*Flexible Organization*
Scale	Large	Small
Division of labor	Extensive	Moderate
Organization	Functional	Product oriented
Production	Batch/mass	Continuous flow
Machine	Workload	Lead time
Discipline	Individual	Team/work group
Management	Authoritarian	Participative

maximum efficiency and optimum utilization regardless of other stages of the production process, operating cycles remain unsynchronized, stock levels build up and throughput times lengthen.

With FMS the picture changes radically. Back in 1975, it took 12 hours to manufacture a television set; today it takes no more than 90 minutes. Prior to redesigning production, factory throughput time in a typical Philips plant took six weeks; today it is down to a fortnight and still shrinking.

Worker Motivation

As we move toward the factory of the future, with its far-reaching implications for organizational systems and structure, the importance of motivating people cannot be overstressed. No one has a monopoly on FMS technology; the people controlling the machines are the key competitive variable. Partly because quality standards are so much higher with FMS, and partly because work-in-process inventories are disappearing, continuous-flow production is more easily disrupted than the old style assembly line. Without skilled, motivated people on the shop floor to keep things running smoothly and carry out basic repairs on site, costly shutdowns can play havoc with the economics of production. Moreover, by the very nature of FMS, production workers are much more closely in-

volved in the engineering aspects. For all these reasons, a company that can inspire employees with real enthusiasm for the effort has clearly gained a competitive advantage.

The greatest threat to worker motivation and morale probably comes from a one-sided technical approach on management's part, leading to a factory designed as if the unmanned factory of the future were already there. Observance of a few simple caveats can help to keep worker morale and product quality high:

- Don't assign anyone permanently to a single station.
- Don't isolate workplaces.
- Don't create groups of over 20 people.
- Don't let machine pace drive the system; let quality be the driving force.
- Don't make cycle times too short.
- Don't do testing far away from the workplace.
- To ensure optimum flexibility and to minimize system breakdowns, don't link too many workstations into a single system.

Pragmatic Implementation

To realize the full potential of FMS, as we have seen, significant changes must be made both in the manufacturing process and in the business system as a whole. These changes will demand a great deal of management thought and effort, and they will take time. Our experience at Philips indicates that the transition from conventional manufacturing to FMS should realistically be seen as a three- to five-year program comprising three discernible stages.

Stage 1: Preparation

Once developed, the plan for the shift to FMS must be vigorously communicated to the entire organization. In communicating the destination and course of the journey, clear evidence of top management's commitment and support is vital; the message must be repeated again and again. At Philips, responsible top managers chair awareness-building seminars to explain the rationale and methodology of impending changes in strategy and policies to factory managers, begin to instill an integrated managerial outlook, and introduce them to the required changes in managerial style.

Redefining job content and retraining the workforce are vital elements of the preparation stage. In a medium-sized electronic instruments factory situated in a high-unemployment region of France, management set up a massive program to retrain all its shop-floor workers to operate the new computerized equipment. "We want to disprove the notion that unskilled people are doomed to stay that way," as the CEO told the press. Five years later, once-semiskilled workers are inspecting incoming goods, analyzing and reporting quality, doing work layouts and operating computer terminals.

Stage 2: Optimization

Before the automation of processing operations can begin, everything possible must be done to optimize existing manual operations. Accordingly, factory managers should be required to analyze inefficiencies and bottlenecks in the production process and come up with new ideas and proposals for improving

factory flexibility and control, restructuring operations where necessary to simplify the work flow. A striking real-life example of the kind of simplification that can sometimes result from such analysis is shown in Exhibit 2.

At Philips, factory managers are asked to analyze their own operations systematically with the aid of a questionnaire, to generate concrete improvement ideas and to present them as proposals in open forum before top management, committing themselves to achieve specific improvements in specified performance indicators. At one assembly plant management merged Engineering and Development in a single department and succeeded by systematic effort in reducing by more than 30 percent the number of components used in its different products. At the same time, redesigning the products for automated production enabled it to increase the proportion of robot-produced components from 30 to 80 percent.

Stage 3: Installation

The actual installation of FMS should take place only when all manual operations have been optimized and all manual systems are working perfectly. At this point, all practices that could threaten the success of the program—piecemeal evaluation of specific hardware investments, inappropriate design practices, or purely efficiency-oriented management appraisal systems, and so on—will have been identified and corrected. The organization will have avoided the pitfalls of automating existing production methods and procedures, or trying to automate a poorly designed manual system. More important still, the successes scored in Stage 2 will have spurred the shift from a factory narrowly oriented to efficiency to a flexible production organization.

Following a shakedown period for debugging, balancing and refining the new system, tangible results should soon be in evidence. In one Philips plant manufacturing throughput time was reduced from a minimum of four

EXHIBIT 2 Goods Flow Before and After Analysis

Spring-loaded cord winder of a vacuum cleaner

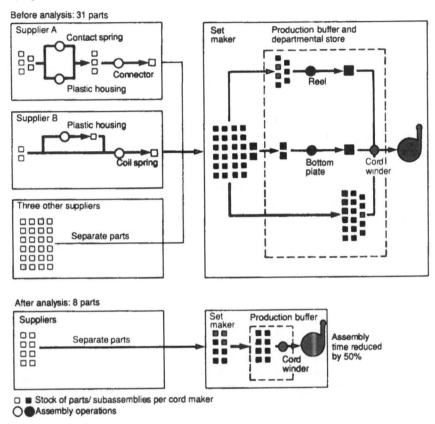

□ ■ Stock of parts/ subassemblies per cord maker
○ ● Assembly operations

weeks to a maximum of one; five months of inventories were reduced to six weeks' worth; and customer complaints dropped by two-thirds. Such results are not exceptional.

In the course of their journey from conventional manufacturing to the factory of the future, many companies will have to discover all over again that the shop floor is, in the last analysis, the heart of the business. In putting the new technology to work, there is no substitute for hands-on, experience-based understanding of physical production. The factory of the future can indeed become a profitable present-day reality—but only if the job is done from the ground up.

Detroit Tool Industries (A)

We need to implement statistical process control (SPC) for three reasons. First, many of our customers are moving quickly into SPC. They haven't yet demanded that we use it, but there is a strong hint that it will soon become a requirement if we want to have a continuing relationship with some customers. Second, we have noticed a tightening of customer quality requirements. Over the past six months we have seen an increase in customer rejects and returns. Lastly, I feel that the participative management aspects of SPC might help in getting some heads back into the job. I see it as a tool to remotivate some of our people, to get some of them back on board who had begun to look askance at what we are trying to accomplish as a company. The biggest problem is getting it into place.

This was Ed Palm, president of Detroit Tool Industries, speaking of his perception as to why SPC was critical to his company's success. Eleven years earlier, in 1974, Palm, along with several partners, had purchased the company through a leveraged buy-out; one partner had left in 1982, and in 1985 Palm retained sole operating responsibility and owned about 40 percent of the company. Exhibit 1 displays the organization structure as of March 1985.

Detroit Tool Industries was a 50-year-old company specializing in precision ground threading tools, its only business. The company had pioneered many innovations in the threading tool industry, for example, multiple-thread (crush) grinding, and had developed its own proprietary methods of generating relief in thread forms. The major industries it served included automotive, construction and farm equipment, aircraft,

---

Copyright © 1985, Revised 1988, by the President and Fellows of Harvard College, Harvard Business School Case 9-681-016.

EXHIBIT 1   Organization, March 1985

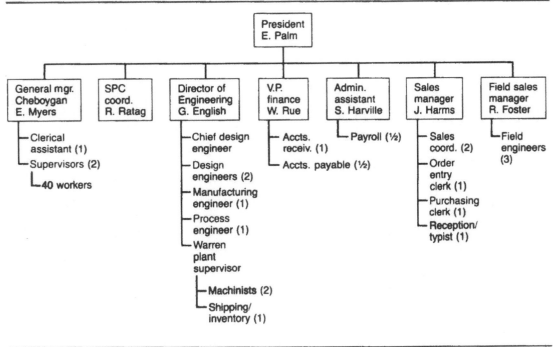

and fluid power. The machine tool industry had gone through a period of increasingly stiff competition from overseas; a number of smaller companies had closed their doors while large corporations tried to sell off many of their nonprofitable machine tool divisions. Detroit Tool Industries' major competitors included machine tool divisions of TRW, Bendix, Litton, and Gulf & Western. In 1979, its net sales had reached over $6 million, but the recession in the early 1980s and the downturn in the automotive industry had had a drastic impact. Nevertheless, although 1980 through 1981 had ended in losses, by 1983 the business turned a profit with over $3 million in sales. (See Exhibits 2 and 3 for historical trends of financial results.)

Ed Palm entered the business world at age 14; by age 16 he had started his first company, making sails for sailboats in Cleveland. He bounced in and out of three colleges, never completing a degree. After working in marketing and sales for another tool company, which he tried to purchase, he bought Detroit Tool at age 29. He was also president of the Cutting Tool Manufacturers of America and very active in the Young President's Organization.

EXHIBIT 2   Consolidated Statement of Operations Summary for Years Ended December 31

	1983	1982	1981	1980	1979
Net sales	$3,109,624	$3,269,134	$4,611,563	$4,646,770	$6,246,812
Cost of sales	1,448,617	1,965,790	3,744,469	2,951,530	3,510,026
Gross profit	1,661,007	1,303,344	867,094	1,695,240	2,736,786
As a percent of net sales	53.4%	39.9%	18.8%	36.5%	43.8%
Expenses:					
Marketing	434,420	552,587	805,058	1,004,159	1,091,172
Corporate	862,374	739,901	813,922	828,780	826,011
	1,296,794	1,292,488	1,618,980	1,832,939	1,917,183
Income (loss) from operations	364,213	10,856	(751,886)	(137,699)	819,603
Other income (expense)					
Interest expense, net of interest and other income of:	(101,675)	(171,104)	130,581	91,325	48,054
$41,735 in 1983					
$36,912 in 1982					
$33,329 in 1981					
$ 7,119 in 1980					
$22,629 in 1979					
$10,566 in 1978					
Gain on sale of assets	88,760				
Costs of manufacturing consolidation net of gain on sale of asset of $104,864				98,256	
	(12,915)			189,581	
Net income (loss) before provision for income taxes	351,298	(160,248)	(882,467)	(327,280)	771,549
Provision (credit) for income taxes	100,000	0	(454,000)	(181,000)	300,000
Net income (loss) before extraordinary item	251,298	(160,248)	(428,467)	(146,280)	471,549
Extraordinary item: tax carryforward	100,000				
Net income (loss)	351,298	(160,248)	(428,467)	(146,280)	471,549

EXHIBIT 3 Consolidated Balance Sheets for the Year Ended December 31

	1983	1982	1981	1980	1979
Assets					
Current assets					
Cash	$ 124,835	$ 137,448	$ 121,813	$ 19,840	$ 342,989
Commercial paper	308,000	0	0	0	0
Accounts receivable, less allowance for doubtful amounts of $20,000 in 1983 and 1982 $11,000 in 1981 and 1980 $16,000 in 1979	470,381	285,326	532,918	511,452	663,292
Federal income tax carryback receivable	0	0	387,114	101,764	0
Inventories—LIFO	507,547	650,142	814,815	842,542	991,133
Prepaid expenses and other assets	108,687	134,896	181,538	134,244	103,879
Total current assets	1,519,450	1,207,812	2,038,198	1,609,842	2,101,293
Investment in joint venture	88,410	0	0	0	0
Property, plant, & equipment					
Land, building, and improvements	284,162	466,601	467,725	466,602	460,531
Machinery and equipment	1,934,008	2,020,163	2,027,600	1,528,697	1,399,332
Office furniture and equipment	467,289	449,799	454,211	288,813	271,764
	2,685,459	2,936,563	2,949,536	2,284,112	2,131,627
Less—accumulated depreciation	1,537,766	1,514,936	1,381,294	1,239,739	1,173,487
Total property, plant & equipment	1,147,693	1,421,627	1,568,242	1,044,373	958,140
Total assets	$2,755,553	$2,629,439	$3,606,440	$2,654,215	$3,059,433

EXHIBIT 3 (continued)

	1983	1982	1981	1980	1979
Liabilities and Shareholders' Equity					
Current liabilities					
Notes payable to bank	0	0	198,974	200,000	0
Current portion of long-term debt	277,917	84,366	54,907	25,673	32,158
Current portion of subordinated note	0	0	0	0	145,438
Accounts payable	242,692	432,253	824,730	207,924	259,853
Accrued liabilities	453,673	294,446	297,717	294,482	369,169
Accrued income taxes	0	0	0	0	204,529
Total current liabilities	974,282	811,065	1,376,328	728,079	1,011,147
Deferred income taxes	0	0	0	75,658	54,040
Long-term credit agreement with vendor Secured by related steel inventory	0	0	150,000	0	0
Long-term debt, less current portion above	761,375	959,522	1,061,012	247,292	244,780
Subordinated promissory note	344,690	534,944	534,944	690,563	690,563
Shareholders' equity					
Common stock, no par value, 50,000 shares authorized, 10,000 shares issued and outstanding	10,000	10,000	10,000	10,000	10,000
Class B stock, $1 par value, 10,000 shares authorized, 6,346 shares issued and outstanding	6,346	6,346	6,346	6,346	6,346
	16,346	16,346	16,346	16,346	16,346
Retained earnings:					
Balance beginning of year	307,562	467,810	896,277	1,042,557	571,008
Net income (loss) for year	351,298	(160,248)	(428,467)	(146,280)	471,549
Balance end of year	658,860	307,562	467,810	896,277	1,042,557
Total shareholders' equity	675,206	323,908	484,156	912,623	1,058,903
Total liabilities and shareholders equity	$2,755,553	$2,629,439	$3,606,440	$2,654,215	$3,059,433

CHEBOYGAN PLANT

The main manufacturing facility for Detroit Tool was located in Cheboygan, Michigan, 300 miles north of the company headquarters in Warren, Michigan (a suburb of Detroit). The plant employed 40 machinists, who were represented by Local #64 of the Mechanics Educational Society of America (AFL-CIO). Cheboygan, located on Lake Huron, was in the heart of Michigan's summer resort area. The area unemployment rate was about 26 percent and seldom fell below 18 percent. Although most companies in the area had closed or moved over the past 10 years, leaving very few manufacturing operations, the majority of the workers in the plant were unconcerned because they had second incomes coming from trapping, hunting, or some sort of resort-related activity. The Cheboygan area was also heavily influenced by the UAW and its nearby Black Lake Union Conference Center.

During the first few years that he owned Detroit Tool, Palm spent a couple of days every other week at the plant. The plant manager had just retired and Palm replaced him with the plant's second-in-command, who then died of a heart attack several months later. Palm subsequently promoted Harry Heath, a supervisor, to the position, but the latter proved to be an ineffective manager and was returned, one year later, to his prior job. In 1977, Ernie Myers was hired as plant manager. Palm described him as someone with experience (over 30 years in the industry), who was mature and methodical. Palm noted, "Overall, Ernie has done a pretty good job considering the people he's got in the plant."

In the early 1970s, the plant was viewed as the "Cheboygan Country Club." The original owners, of Scandinavian descent, had a socialistic perspective; for example, all workers in the plant were paid the same wage, regardless of their job or skill level. In Palm's view, the union ran the plant:

> I didn't feel that I could come in with a hard hand because it was a leveraged buy-out, and a strike would mean that I would be out of business. However, I made a few mistakes when I first got there. The people there thought I was a relative of the former owners and a spoiled, rich kid from Grosse Pointe [a wealthy suburb of Detroit]. I really upset them the first week when I told them that we had a job to do and that I was going to do it with or without them.

Palm noted that, as time had passed, he believed he had been relatively consistent in his decisions and people had come to trust him, even if they did not like him or what he had to say. One of his first actions was eliminating a runaway bonus system and replacing it with one based on efficiency as opposed to the old one based on dollars of volume shipped. In 1982, the union agreed to 10 percent wage concessions. By 1984, the business had picked up and the first bonus was paid.

In 1980, Detroit Tool closed its Greensville, Ohio, plant and transferred the work and machines to Cheboygan. Greensville had specialized in standard taps, and the work was more production oriented than the specialty items the workers in Cheboygan were used to. Palm believed that his machinists never really wanted the more production-oriented work. The Cheboygan plant had not been competitive in products like standard taps, which required workers merely to load material on automated equipment and monitor longer runs. The machinists considered themselves craftsmen and preferred to work on special one- or two-item orders. As a result, the work and equipment, originally from Greensville, were being transferred to the Warren facility. The company headquarters had recently been relocated to a new building and space to set up an alternate manufacturing plant was available. Long term, Palm planned to transfer even more work to Warren: He ultimately hoped to set up a highly automated, nonunion machine shop. Exhibit 4 compares production efficiencies for special and standard products at the Cheboygan plant.

PRODUCTION PROCESS

Detroit Tool received, on average, between 400 and 500 orders per month. The plant viewed itself as a specialty tool shop, but a large portion of its output was what was considered standard taps or thread rolls and cutters. Exhibit 5 provides a monthly breakdown of orders. In addition to the 20 percent that were new orders, about 20 percent of the repeat orders needed some sort of redesign. The engineering department had a throughput time of two days to process repeat orders and four days for new ones. During this time they generated process drawings and sheets (Exhibit 6), entered the order in a computerized production tracking system, and then sent the order on to the manufacturing operation. The introduction of group technology[1] into the design engineering stage had helped to classify new orders quickly into product families and reduce the number of new designs needed.

The Cheboygan plant received about 300 work tickets per day, and the throughput was between two to four weeks for each order. If a blank was already in stock, delivery could be made within two weeks. As orders were received, they were manually added to a shop loading report by Ernie Myers, the general manager, who listed all the orders by due date. John Archambo, one of the two supervisors, then scheduled the shop based on his knowledge of how long it would take to produce a particular part, machine availability, and the raw bar stock on hand. The tracking

---

[1]Group technology was a method of standardizing the design and/or manufacturing process through the grouping of geometrically similar parts or parts with similar manufacturing processes.

EXHIBIT 4

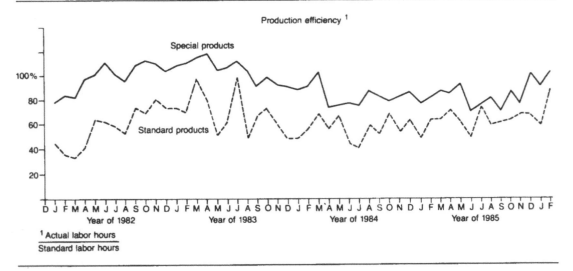

Production efficiency [1]

^{1} $\dfrac{\text{Actual labor hours}}{\text{Standard labor hours}}$

of work through the shop was done through labor vouchers; each operator was required to note the time spent on each operation. The accounting department generated an activity report daily that denoted the location and progress of each order. In addition, it compared the time vouchered for each part with the standards listed on the process sheets.

The production process was broken into three parts: 1) soft-side machining, 2) heat treat (done by an outside vendor), and 3) thread grinding and finishing.

Soft-Side Machining. The objective of the soft-side machining process was to get the bar stock into its "near net shape," i.e., to remove the majority of stock prior to heat treat in the least expensive and quickest manner. The initial step was to produce the slug, a piece of bar stock cut to the length of the tool to be produced. This could be done in one of three ways depending on the diameter of the tool and the lot size to be produced. An abrasive wheel was used for small lots of small diameter tools (under 3¼"). For larger lots of the small diameter tools, the bar stock was turned on a screw machine; a saw was used for all large diameter slugs. The slug then went to a drilling machine to be centered. After centering, the slug went through a series of operations on one of several machines, including a milling machine, broach, shaper, or tapping head, to cut the flutes, pockets, squares, or holes. Palm planned to transfer the entire soft-side of the process, which amounted to four or five jobs, to the Warren facility within the next five years.

EXHIBIT 5   Customer Orders

| | 1984 | | | | | | | | | | | | | 1985 | | |
Number of Days:	21 Jan.	21 Feb.	22 Mar.	21 Apr.	22 May	21 June	21 July	23 Aug.	19 Sept.	23 Oct.	19 Nov.	17 Dec.	250 Total	22 Jan.	20 Feb.	21 Mar.
	Orders Per Month Completed															
Standard products:																
Customer, finished goods, replacement, research	39	45	55	12	33	36	29	27	26	46	47	58	453			
Plant inventory	17	39	43	7	27	41	31	27	33	30	18	4	317			
Total standard products	56	84	98	19	60	77	60	54	59	76	65	62	770	109	113	81
Special products:																
All orders	289	332	366	222	405	390	353	401	405	391	362	351	4267	485	479	384
Total orders completed	345	416	464	241	465	467	413	455	464	467	427	413	5037	594	592	465
Total orders per day	1.9	2.1	2.5	0.6	1.5	1.7	1.4	1.2	1.4	2.0	2.5	3.4	1.8			
Average plant inventory per day	0.8	1.9	2.0	0.3	1.2	2.0	1.5	1.2	1.7	1.3	0.9	0.2	1.8			

erage standard products per day	2.7	4.0	4.5	0.9	2.7	3.7	2.9	2.3	3.1	3.3	3.4	3.6	3.1	5.0	5.7	3.9	3
erage special products per day	13.8	15.8	16.6	10.6	18.4	18.6	16.8	17.4	21.3	17.0	19.1	20.6	17.1	22.0	24.0	18.3	18
erage orders per day	16.4	19.8	21.1	11.5	21.1	22.2	19.7	19.8	24.4	20.3	22.5	24.3	20.1	27.0	29.6	22.1	21

Pieces/Orders per Month Completed

dard products: stomer, finished goods, replacement, research	130.1	204.8	152.3	780.7	283.0	193.8	250.9	604.3	319.0	277.4	67.4	143.4	230.6				
ant inventory	612.4	467.2	239.2	545.1	520.1	399.6	245.0	515.4	394.7	539.3	67.9	2401.3	424.9				
l pieces/orders	276.5	326.6	190.4	693.9	389.7	303.4	247.9	559.9	361.3	380.8	67.5	289.0	310.6	222.2	202.7	180.4	20
ial products:																	
l pieces/orders	6.3	5.1	3.9	8.8	6.8	3.1	4.6	5.7	5.3	4.8	6.1	5.9	5.4	5.0	3.3	7.5	
l pieces/orders completed	50.2	70.0	43.3	62.8	56.2	52.6	39.9	71.4	50.5	66.0	15.5	48.4	52.1	44.8	41.4	37.6	3

EXHIBIT 6   Process Sheet

SEQ-OPER	Operation Description/Operator Instructions	Setup	Process Piece	Total
001–820	Cut-off/rec	20	2.50	60
002–112	Face			
	Both ends and center			
003–114	Turn OD	30	13.24	318
	Shank end			
004–112	Face	15	3.50	84
	Face collar			
005–122	Trace OD	25	19.81	475
	O.D., Chamfer, pilot			
006–119	Recess cut	20	5.00	120
	Behind collar, turn ring			
007–119	Recess cut	20	2.00	48
	Turn u'cut behind thread portion			
008–202	Flute mill	45	36.66	880
	Use steady rest			
009–701	Soft inspection			
	Stamp to b/p			
010–700	Heat treat			
011–702	In-process inspection			
	Insp. & record rockwell, cent. lap			
	*****Straighten			
012–402	External grind	20	2.00	48
	Shank			
013–402	External grind	20	6.11	147
	Flute dia			
014–402	External grind	10	4.20	101
	Pilot dia.			
015–402	External grind	10	2.23	54
	Tap lead			
	******Straighten			
016–408	Flute	20	15.50	372
017–702	In-process inspection			
	*****Straighten			

Heat Treat.   At an outside vendor, the tools were dipped into a salt bath at approximately 2,100–2,200°F, quenched, and tempered (one to three times) to increase the hardness and ductility of the material.

Thread Grinding and Finishing.   Upon return from heat treat, the tools were rough ground to clean the surface (both inside and outside diameters) and then center lapped to ensure good concentricity. The heart of the process was the grinding operations. The sequence of operations var-

EXHIBIT 6 (continued)

SEQ-OPER	Operation Description/Operator Instructions	Setup	Process Piece	Process Total
018–520	Crusher	120	97.95	2351
019–616	Chamfer	20	10.00	240
020–612	Convolute	10	4.30	103
021–700	Heat treat			
	Nitride and diffuse			
022–702	In-process inspection			
	Center lap lightly			
	******Straighten			
023–601	External grind	10	2.00	48
	Shank			
024–601	External grind	10	4.84	116
	Pilot			
025–616	Chamfer	10	3.50	84
	Circle grind front			
026–616	Chamfer	20	4.05	97
	Circle grind chamfer on collar			
027–602	Flute	25	15.00	360
028–702	In-process inspection			
	****Straighten			
029–609	Locate	30	7.50	180
	Grd. location to b/p			
030–407	Surface	30	6.00	144
	Grd. slot thru collar			
031–407	Surface	20	4.25	102
	Grind flat on collar			
032–604	Notch	10	5.50	132
	Grd. whistle notch			
033–615	Buff/polish		2.00	48
	Stone edges of pilot			
034–703	Final inspection			
035–810	Shipping			
	Ship to customer			
	Total time	570	279.64	6712

ied with the design of a specific tool but included thread, chamfer, flute, or special surface grinding. Because of the equipments' age and condition, operators had to make many manual adjustments to maintain extremely tight tolerances. With the exception of five recent investments (a mill, ID grinder, saw, and two rebuilt thread grinders), most equipment was of 1945–1953 vintage. Several thread grinders had been upgraded from their original design, which dated to 1930. Every two to three years the thread grinders were overhauled to clean the way systems and replace the lead screws. About 30 machinists were assigned to the thread grinding and finishing operations. After a finish surface heat treatment, the tools were marked, inspected, packaged, and shipped.

Quality Control.   Three operators were assigned to inspection tasks. Once bar stock was received, there was a visual check to ensure that there were no cracks or voids in the material. On the soft-side, operators checked the work piece against a blueprint to make sure that each met the "near net shape" dimensions. After heat treat, one inspector performed a Rockwell hardness test on each lot.

There was first-piece inspection on all thread grinding operations. The operator made the first piece (a master), took it to the inspection room to have it checked against a shadow graph, and then used it as a master against which the rest of the lot was checked. For OD and ID grinding, each operator was responsible for checking every piece against a blueprint.

The finished goods inspection (done in the inspection room) was primarily a visual, cosmetic check. Rechecking the physical dimensions of each thread was extremely time consuming, and it was presumed the machinists had done it properly. Experience showed, however, that most customer returns were due to visual defects as opposed to fine dimensions.

ERNIE MYERS' VIEW OF THE PLANT

Ernie Myers noted that his first four or five years as general manager had been spent just getting the plant back in control. A turning point had been a six-week strike in 1980 over health benefits. The plant had to be operated by managers brought up from Warren, which elicited an extremely violent reaction from the union—to a point that some managers had feared for their lives. Once the strike was settled, things calmed down. Shortly thereafter, the work force was reduced from 150 to 10, a ploy to send a message to the work force. After three weeks the core of the work force was rehired.

This remaining work force was, for the most part, hard working and concerned about the company. (See Exhibit 7 for a seniority and age listing.) There were, however, nine to twelve troublemakers, whom Myers called the "dirty dozen," who basically ran the place. Although they were skilled craftsmen and did good (to exceptional) work, they were extremely negative toward the company and obstructed change in the plant. For example, several machinists who had attended an external training program returned to severe reprisals, including having had their machines greased up. Myers feared to confront the group because of its potential violence; he even believed that some would not hesitate to bring a gun into the plant.

EXHIBIT 7 Employee List—March 1985

Name	Age	Seniority
Bishop, J.	47	September 1962
Bishop, L.	61	April 1952
Block, A.	60	February 1952
Brandau, R.	42	October 1965
Brown, G.	43	February 1962
Charboneau, C.	53	August 1966
Compeau, H.	50	July 1973
Comps, J.	58	January 1952
Cool, G.	54	November 1951
Christ, R.	42	June 1973
Drake, W.	65	March 1952
Duffiney, E.	48	December 1961
Palmdy, J.	46	February 1962
Elliott, K.	56	August 1952
Friday, W.	61	February 1952
Friday, W. Jr.	34	March 1973
Fowler, J.	57	December 1962
Gahn, G.	56	December 1951
Gahn, V.	52	August 1965
Hawley, J.	52	February 1966
Hiar, V.	39	August 1966
Hull, L.	45	January 1963
Knight, R.	44	February 1962
Laveque, R.	56	April 1952
Leask, O.	57	June 1952
Modrak, C.	56	February 1952
Mulka, G.	48	October 1963
Paquin, G.	45	August 1966
Percy, J.	53	May 1952
Reimann, F.	46	August 1966
Schramm, J.	46	April 1966
Shields, D.	44	February 1965
Simmons, M.	35	June 1976
Stillwell, L.	43	November 1963
Van Slembrouck, A.	56	May 1952
Venzlaff, J.	58	April 1952
Wheaton, J.	54	January 1952
Wheelock, D.	60	June 1952
Willey, T.	38	October 1965
Zbieracz, R.	54	January 1952

One of Myers' first actions upon arriving at the plant had been to make a thorough job description analysis of all jobs. (It had never been used.) There were five levels of thread-grinding skills in the plant, but all workers were paid at the same wage rate. Contractually, there were three wage groups, but during the downturn (1980–81) all of the employees in the lower two wage groups were laid off. Over the years, informal work rules had been established whereby workers were assigned to one job or one machine. Although everyone earned the same wage, very few rotated to different jobs: Myers estimated that only 10 to 15 percent of the work force wanted, could, and were willing to rotate to different jobs. Many of the longer service workers were the least skilled.

When the business had declined, one of the first areas cut was the maintenance department. In the maintenance department were now two machinists, both assigned to do other things: One was assigned to the heat treat area and was required to take parts in and out of the oven every 30 minutes; the other was often assigned to production work to fill in on emergency jobs. Hence, only about three quarters of a full-time employee was regularly assigned to maintenance, which included repairing machines, cutting the lawn, and washing windows. Fortunately, one of the supervisors was also one of the best maintenance operators; he was able to do much of the troubleshooting and thereby expedite work that had to be done.

Myers also faced a problem with the standards and process sheets. The process sheets, used for routing parts through the plant, were mostly viewed as a joke by the supervisors and workers. The sheets had been created by a design engineer who was not trained, had no manufacturing background, and who had filled out the sheets based only on memory of what had been done. In March 1985, the previous Warren plant supervisor was reassigned to a position of process engineer to work on the process sheets. "They are a real hodge-podge and very few of them are accurate," Myers observed. "To make matters worse, several years ago the engineering department down in Warren decided to cut the standards by 20 percent across the board because the company was forced to cut prices by 20 percent."

Myers' other major problem was his two supervisors. They knew the product and processes extremely well, but "they're very independent, have done things their own way for years, won't delegate or tell their people to get back to work, and can't be told what to do," he noted in frustration. "They are definitely more union employees than management. Both have very strong socialistic union leanings." The supervisors were required to punch a time card and chose to park in the same lot as the machine operators, rather than in the management parking lot in front of the building, which Myers, the SPC coordinator, and the clerical assistant used. Recently, John Archambo, one supervisor, defied Myers' decision that both he and Harry Heath (the other supervisor) could not take vacation time on the same days. Both men had second incomes from trapping and hunting, and it so happened that the trapping season overlapped the deer season by one day. Myers had informed Archambo that he could not take the same days off as Heath, but he took two days off anyway.

Gale English, director of engineering, observed that, although the supervisors were extremely well qualified technically, they in fact did no supervision. He noted that every time he was in the plant he found people lounging around. He believed that productivity could improve by about 20 percent if people would get to work; there could be an additional 20 percent if people worked smarter.

STATISTICAL PROCESS CONTROL (SPC)

In August 1984, Ed Palm hired Tom Larson, an industrial psychologist, to interview all members of his organization to uncover possible personnel difficulties. The two had recently collaborated on a grant from the state of Michigan that provided financial support for the introduction of SPC. The state was funding educational activities to encourage Michigan companies to implement SPC, thereby becoming more competitive.

During a periodic company-union grievance meeting, Ed Palm informed the union leadership that customers were beginning to require tighter conformance and if the plant did not implement SPC, it would not receive additional business. Subsequently, Palm and Myers held two formal sessions with the union leadership to describe why their customers were requiring SPC and how it could be implemented. In the first session they showed a movie on the benefits of SPC; for the second they brought in Tom Larson as the SPC facilitator and Mike Martinuzzi, an SPC instructor funded by the state. Because everyone was concerned the work force would resist the program, Larson also met with small groups of seven to eight workers to explain why SPC was critical to Detroit Tool's future.

Formal training for the union officers and shop committee representatives began in late January of 1985 with eight hours of intensive SPC instruction. (Myers and the shop supervisors also attended this session.) This same training, conducted by Larson and Martinuzzi, had been given to the management team in Warren. Myers had also attended a seminar on SPC held by another company in Detroit. Finally, during the last two weeks of March, the entire work force at Cheboygan was trained in SPC. The group was split in half, and each half attended four hours of training on Friday afternoon and four hours on Saturday morning. (They were paid overtime for the Saturday session.)

On February 25, 1985, Palm hired Ray Ratag to be the SPC coordinator. Prior to joining Detroit Tool, Ratag had owned and operated (with a partner) a small business specializing in field equipment and remachining parts of large machines. Ratag noted that he had become interested in SPC because he felt that it would be good for his own career. His formal exposure to SPC was a one-week SPC course at the University of Illinois during his first month at Detroit Tool. He described his new job:

In theory, SPC is a rather simple idea, but implementation is the hardest thing. First I'll have to gain acceptance from the operators. My role will be to answer questions and assist them in doing the charting. Once the charts are filled out by the operators, I'll place control limits on them and analyze the results. After a few months, I will run process capacity studies of the equipment. If the equipment does not get repeatability, then I'll run a machine capacity study which will check the machine without the operator.

I'm trying to convince the operators that SPC is a good thing by telling them it is a tool a machinist can use. One problem in this plant is that many of the orders are only one or two in volume. Therefore, the charting will only be done on orders having at least a dozen pieces. Currently, the attitude of the operators is mixed. Some are willing to do the SPC charting because they're worried about their job and say, "If that's what's needed to keep our jobs, we'll do it." Others are less concerned about their job security and will do it if they see that it will make their job easier. There are some, however, who are pretty negative, and I've got to admit that they put up a good argument. They say that the company places limits on the parts they have to produce, and every part that is made is checked and made within those limits, so why should they have to do SPC charting? That's a hard argument, but I'm telling them that SPC will make their job better and that even though their work is within limits, we want every piece to be identical. From what I've heard, eventually SPC will sell itself, so what I need to do is start with the easier workers and get them to buy in.

Ed Palm told me that the SPC coordinator role is a management job with no administrative duties, such as discipline. But within the plant, I'm in charge of quality control and I report directly to Ed. When I uncover a problem from the charts, I'll first approach the supervisors to correct the variables, and then I'll tell Ernie Myers what needs to be done. Ernie will have to fix the problem or he wouldn't be doing his job. Ernie is my management connection at the plant, and I keep him filled in on the administrative aspects of SPC; but when it comes to making recommendations to management relative to my findings, management is Ed. My job is to pinpoint variables in the machines and the processes, and if the variables aren't controllable, then I'll recommend to Ed which machines need to be repaired.

VIEWS OF SPC

On March 25–26, 1985, the casewriter interviewed key people within Detroit Tool to record their views of the SPC program. Their comments follow:

Ed Palm, president:

"My view of SPC is that the operators will do the charting and then Ray [Ratag] will take their charts and analyze them. Once they've been analyzed, Ray should go back to the supervisors and the operators and say, 'Hey, what do you think about this?' and they should brainstorm ideas to solve the problems. At that point, if there is a need for an investment of significant dollars, the idea should be presented to me for approval. Maintenance is going to be a big issue. SPC is the only way to sort out whether the unrepeatability of the machines is due to poor maintenance or operator performance. I really don't believe that the operators want to have the machines repaired and up-to-date because if they are, then there won't be a need for their skill level and they'll have to do something different. The basic problem is that the whole plant is insecure; they think I'm planning on shutting the plant down. In all honesty, the decision point will be next spring when the contract expires. If I get too much hassle from the union, I'll be very tempted to shut the whole operation down. I'll move what equipment we need to Warren and sell off the rest." Palm also noted that an additional SPC benefit was that he would be able eventually to eliminate all the inspection operations. The heat treating vendor was in the process of implementing SPC and would have it in place within six to eight months. This would eliminate the need to do the Rockwell hardness test. After the steel vendor received SPC certification, he could also eliminate the raw material inspection. Finally, with all operators doing their own charting, there would be no need for any final inspection.

Bill Rue, vice president of finance:

"SPC is really our quality program, and it is critical to our survival as a company. We can't get it into place fast enough. It is critical to our reducing costs and to our keeping Detroit Tool in a competitive position." But when asked how it was being implemented, he replied, "It's horrible. We have a dictum but there is no plan in place, no objectives, no goals, and no timetables. They don't even know where to start."

Gale English, director of engineering:

"SPC will fail within one year. Ray has no background or training in SPC and he's up there without any support. The general manager is a Theory X manager and doesn't really support this whole thing. Even though both Steve Leaney [Warren plant supervisor] and I have worked with SPC at the company we just came from [both had been hired within the previous nine months], Ed has basically told me to keep my nose out of it. I was up in Cheboygan last week right after they had finished the first go-round of training. Ray was just walking around and talking with the guys because he didn't know what he was supposed to do. I sat down and helped him lay out a plan."

Ernie Myers, general manager—Cheboygan Plant:

When asked who was the driver behind SPC, Myers quickly replied: "There's no question, Ed is the person behind SPC. Although it's something new for the plant, it's not really new for the Cheboygan area. Another small company down the road with 100 employees has had SPC in place for about four years. They are a vendor for the Ford Motor Company and have a Q1 rating. The attitude in our shop toward SPC is still 'wait and see.' They don't really trust it but they haven't seen enough yet to say no. Ray Ratag's role is to teach people what they need to know in order to do the charting and to monitor the charts and make recommendations. Ray was told that he was to do anything that affects quality, but that's not very well defined. He really doesn't know what his job is. He also needs to recognize that I'm the authority in this plant and that any recommendations should be made to me rather than going directly to Ed. The first thing he needs to do is to learn what I call the four Ps—product, plant, process, and people. He's fairly well-accepted in the shop because he listens to their problems and then lists them on a clipboard and brings them to me. He's become their go-between."

John Archambo, supervisor:

"We're going to SPC because our customers say we have to do it, but I'm not convinced it will work. The operators already check every piece, sometimes as many as three times, so charting won't show us anything new or different with every piece already being mic'd (measured with a micrometer). It might be o.k. on three or four of the machines which have automatic feed. At least it will be interesting to see how good the machines are doing. It won't work on the rest of the jobs because we already have everything under control. In addition, the charts won't show all the adjustments that have to be made on the machines for every piece.

"My concern, as well as my operators', is that we will spend more time on paperwork than on doing work. In order to do the charting they will have to clean their hands to get the grease off, and that takes a lot of time. Charting on the lathe won't be much of a problem because they can do it while the machine is running. But, on the OD grinder, the operator has to feed the machine constantly and dress the wheel every so many times. Therefore, he will have to leave the machine, dry off his hands in order to keep the pencil and paper clean, and then do the charting. Besides, the charts won't show when things are oversize because if something is oversize, it goes back into the machine to be ground some more. The only time that charting will show something out of dimension will be when something is undersize, and if that happens we set those aside anyway.

"SPC won't have any change in my job. As I see it, SPC is Ray's job and if anyone comes to me with a question about SPC, I'll send him to Ray. The biggest benefit that can come from SPC may be that we'll finally get some maintenance done that we couldn't before. They really don't need SPC for that, though, because if they would only ask the supervisors, we could identify what machines need to be fixed. We've gotten lip service for so long that it's now just wait-and-see whether or not SPC is really anything different."

Union shop committee representatives:

"The union is not too concerned with SPC. We're more concerned about the future of the company and where the jobs are going to be. There's a big problem with no maintenance and maybe SPC will help that. There haven't been any complaints yet, probably because we haven't gotten into it much. Charting may make sense on a production basis but not in a specialty shop like ours. Another problem is that SPC charts the operator, not the machine. We have to make continual adjustments to the machines because of poor maintenance and the charts won't show that. They'll only show what comes out after the operator adjusts for the poor machinery. Initially, Ray Ratag was billed as nonpartisan, that is, not management, not union; but that's now changed, and when you go to him with a problem he says he has to go through the normal management chain of command. The biggest problem in the plant is Ernie. He won't listen to us and is only nice when Ed or other visitors come up. He doesn't know the equipment and the processes and won't listen to any of our ideas."

THE FUTURE OF SPC

Ed Palm wondered how he could help to move the SPC process further along. The state facilitator and instructor were beginning to indicate that things were not progressing as rapidly as they had hoped. Palm's major concern, however, was a recent letter from a division of a Big Three auto company stating it would be assessing key vendors to determine which would become primary vendors for a large new project. A major criterion would be having an active SPC program in place. To this end, the auto company had scheduled an SPC audit of the Cheboygan plant for June 21, 1985.

Constructing and Using Process Control Charts

Roger Bohn

Statistical process control (SPC) is a philosophy, a system, and a set of specific techniques for controlling and improving production and service processes. Developed in the United States in the 1930s and 1940s by W.A. Shewhart, W.E. Deming, J.M. Juran, and many others, SPC has been used effectively in some American industries and many Japanese industries for decades. The use of SPC is on the upsurge in the United States and other countries, both among industries discovering it for the first time and those industries that are rediscovering it.

One of the most important tools of statistical process control is the *process control chart*. Control charts are easy to construct and can be done on the shop floor using only paper and pencil. The charts provide a common language for people from many functions and at many levels of a company. They apply to a diversity of industries and situations, ranging from lathes to restaurants to white-collar operations.

Control charts are graphs that show how a process is performing over time. They can be used to

- detect when something is wrong with the process;

- establish what the process is inherently capable of achieving;
- help diagnose the causes of abnormal behavior of a process;
- monitor and control the process;
- tell when apparent abnormal behavior is, in fact, normal so that no remedial action is needed and the process should be left alone.

Control charts are not a panacea. This note closes with a caution on their limitations.

Processes

A *process* is any set of people, equipment, procedures, and conditions that work together to produce a result. This might be one person at a keypunch or many people and machines preparing an aircraft for departure.

The results of a process vary over time. Each succeeding product or service will be slightly different due to variations in materials, equipment, environmental conditions, and the physical and mental actions of the people who are a part of the process. If these variations are minor, they are not of concern. For example, if 99.8 percent of all airplanes left the gate within 30 seconds of the scheduled departure time, then on-time departure would probably not be a concern to airline managers. More often in modern operations, however, some processes are not performing up to par. Because competitors are continually improving their products and processes, companies must constantly look for ways to improve. In these situations, control charts often help diagnose, control, and improve the process.

This note was prepared as the basis for class discussion rather than to illustrate either effective or ineffective handling of an administrative situation.

Processes can be measured by many attributes, and control charts can be constructed for virtually any such attribute. Control charts are commonly used to manage product quality. Any important quality attribute of the product could have its own control chart. For example, a restaurant's "product" would be an enjoyable and delicious meal. Managers can construct control charts for various product attributes such as the following:

- Time the customer waited in line before being seated.
- Time between seating and when the order was taken.
- Time between ordering and arrival of the meal.
- How often food is sent back to the kitchen by a dissatisfied customer.
- Temperature of the hot food when it leaves the kitchen.

These attributes will fluctuate due to various causes—large or small, chronic or intermittent. For example, one oven may heat food differently than another; one chef may prepare the recipe differently than another; sometimes the food may sit on a counter before being served. Part of the management process is to identify and, where necessary, correct these variations.

Definition of Control Charts

A *control chart* is a graph of the performance of a process over time, arranged to emphasize the variation of the process. For example, the chart might show the temperatures of successive meals during one night. Superimposed on each graph are lines called the upper and lower *control limits* (UCL and LCL) for the process. The control limits are drawn so that if the process ever performs outside the control limits, something unusual has happened. A centerline (usually the mean of the process performance) is also drawn on the graph. If

values are consistently above or below the centerline, something has happened. Thus, the control chart quickly gives a visual answer to the question: Is this process behaving the way it usually does or has something changed? There are also various specific problems that the chart can visually highlight.

A control chart is a neutral tool that acts to identify and describe a situation objectively. Avoid using it to blame someone for a problem. In fact, companies find that putting control charts on the shop floor, where the work is being done and the measurements are being made, serves to make everyone aware of how the process is doing and gives them quick feedback about the effects of changes. This leads workers and managers alike to be much more alert to problems and more responsive to fixing and eliminating them, whether the problem is caused by a faulty machine, bad materials, or human error. Management often becomes more sensitive to providing whatever assistance is needed to keep the process in control.

The simplest use of control charts is probably to tell when a process is in or out of *statistical control*. When a process is in (statistical) control, it does not mean that variation in the process is zero; that would be impossible. Instead, it means that the variation is due to a multitude of small, purely random fluctuations. Conversely, when a process is out of control, it means that some of the process variation is due to a few large, irregular causes of variation. Out-of-control processes are hard to manage since on any given day anything can happen. Such processes also tend to perform erratically, reducing their effectiveness. Much of SPC consists of tools to detect, diagnose, and fix the causes of out-of-control situations.

It is easy to detect when a process is out of control on a control chart. Any measurement that falls outside the upper or lower control limits signals an out-of-control situation. The control limits are drawn so that such a measurement would only happen 1 time in 1,000 by chance, if the only sources of variation were small, random fluctuations. Therefore, if a measurement falls outside these limits, something else must be going on.

Constructing Control Charts

Control charts measure what a process produces over time. Because many processes produce hundreds or even thousands of parts in a day, it would be time consuming and expensive to plot every single part on a control chart. Fortunately, this is not necessary. Instead, choose a sample size, and periodically collect data in samples of that size. A typical sample size is five, so the data are collected by measuring five consecutive parts every few minutes. These five parts are then aggregated, as described below, to produce a single point on the control chart. The next five-part sample would be the next point on the control chart.[1] A sample size between two and ten is usually chosen and is referred to as n.

Two different control charts are usually constructed:

- The $\overline{X}$ (X-bar) control chart measures the mean of each sample. It reveals any tendency for the process mean to drift or jump around over time.

---

[1]Managers should choose an appropriate sampling plan. For example, one option is to measure the first five parts made each hour. What sampling plan is best will depend on what kind of problems you expect the process to have. It might be best to sample more intensively just before and just after a shift change, to quickly determine whether anything has changed due to the shift change. At other times of day, one sample of five each hour may be adequate. This subtlety is easily overlooked and may require a mix of judgment and experimenting with different sampling plans. Effective control-charting programs should evolve over time as the process evolves and the needs of management change.

- The R (range) chart measures the range of each sample (i.e., the largest member of the sample minus the smallest). It reveals any tendency of the process to behave more randomly or less randomly over time. (The standard deviation of the samples in some ways better measures variability than does their range. However, the range is easy to calculate by eye, while the standard deviation requires a calculator.)

The mechanics of constructing control charts are as follows:

1. The first step in constructing control charts is to choose the sampling plan and sample size (n).
2. The second step is to collect 20 or more sets of samples of size n.
3. Calculate the range (R) and mean ($\overline{X}$) of each sample. The range is simply the highest value in the sample minus the lowest value.
4. Calculate the average range ($\overline{R}$) of all the samples. This will be the centerline of the R chart. Do the same for the average mean of the $\overline{X}$ values. (This is called $\overline{\overline{X}}$ or X double bar: the mean of the means.) This will be the centerline of the $\overline{X}$ chart.
5. Now calculate the upper control limit for the R chart. From Exhibit 1 look up the factor D_4 in the row corresponding to the sample size n. The upper control limit is $\overline{R}$ times D_4. For example, if the sample size is 5 and the average of the sample ranges is 12.5, the centerline in the R chart is 12.5, and the upper control limit is 12.5 × 2.11 or 26.4. Any sample with a range greater than 26.4 would indicate that the process is out of control.
6. The lower control limit for the R chart is $\overline{R}$ times D_3, where D_3 comes from Exhibit 1. In this example the lower control limit is zero.
7. The upper and lower control limits for the

EXHIBIT 1 Factors for Setting Control Chart Limits*

Sample Size n	$\overline{X}$ Chart Factor A_2	R Chart Factors† Lower Limit D_3	Upper Limit D_4	Factor to Estimate σ' d_2
2	1.88	0	3.27	1.128
3	1.02	0	2.57	1.693
4	0.73	0	2.28	2.059
5	0.58	0	2.11	2.326
6	0.48	0	2.00	2.534
7	0.42	0.08	1.92	2.704
8	0.37	0.14	1.86	2.847
9	0.34	0.18	1.82	2.970
10	0.31	0.22	1.78	3.078

*Formulas for larger sample sizes are available.
†Definitions: n = Sample size
$\overline{X}$ = Mean of each sample
R = Range of each sample (largest − smallest)
$\overline{R}$ = Average of all the samples' ranges
Upper limit = Upper control limit (UCL)
= $D_4 \times \overline{R}$ for R chart
= $\overline{\overline{X}} + A_2\overline{R}$ for $\overline{X}$ chart
Standard deviation estimate = $\sigma' = \overline{R}/d_2$

Source: J. M. Juran, *Quality Control Handbook* (New York: McGraw-Hill, 1974, 3rd edition) Appendix 2, Tables Y and A. These numbers are based on three standard deviations, assuming a normal distribution.

$\overline{X}$ chart are calculated slightly differently. They are measured above and below the centerline $\overline{\overline{X}}$. The upper control limit is $\overline{\overline{X}} + (A_2 \times \overline{R})$, while the lower control limit is $\overline{\overline{X}} - (A_2 \times \overline{R})$. For example if $\overline{\overline{X}}$ is 100, with sample size 5, $A_2 = 0.58$, so $A_2 \times \overline{R} = 7.25$, and the control limits would be 107.25 (100 + 7.25) and 92.75 (100 − 7.25). $\overline{X}$ points outside these limits would signal out-of-control situations. (These control limits are three standard deviations above and below the mean. There is nothing sacred about using three standard deviations, but it works well in practice.)

8. This establishes the centerline and limits for both $\overline{X}$ and R charts. The next step is to prepare a graph for analysis and daily use. Several graphical conventions are useful:

• The $\overline{X}$ chart is drawn directly above the R chart.
• The centerlines of the two charts are shown as solid lines, and the upper and lower limits are shown as dashed lines.
• The physical distances between the upper and lower limits of both charts are nearly the same.
• The control limits are drawn at three standard deviations from the centerline.

Exhibit 3 shows the $\overline{X}$ and R charts for the example in Exhibit 2. The Appendix explains how to construct $\overline{X}$ and R charts using a personal computer spreadsheet.

EXHIBIT 2   Example of Initial Capability Study Using $\bar{X}$ and R Charts*

A restaurant's manager is concerned about the temperature of a food item being served. The manager never wanted the dish to be cooler than 147°F or warmer than 160°F when it left the kitchen to be served. The manager decided to do a capability study over 10 days. Each day, she randomly picked three servings and tested the temperature just as the dish was being served. The following are the data that she collected.

Day	Reading, °F			Range, °F	$\bar{X}$, °F
1	150	160	155	10	155.0
2	140	150	155	15	148.3
3	145	150	150	5	148.3
4	150	150	155	5	151.0
5	130	155	150	25	145.0
6	140	140	145	5	141.6
7	150	150	150	0	150.0
8	155	155	160	5	156.7
9	160	160	160	0	160.0
10	150	160	165	15	158.3
				$\bar{R}$ = 8.5°F	$\bar{\bar{X}}$ = 151.5°F

*n = -3
R *Chart Limits*
    Lower control limit = $(D_3)\bar{R}$ = 0(8.5°F) = 0°F
    Upper control limit = $(D_4)\bar{R}$ = 2.57(8.5°F) = 21.8°F
    Center control line = $\bar{R}$ = 8.5°F
$\bar{X}$ *Chart Limits*
    Lower control limit = $\bar{\bar{X}} - (A_2)\bar{R}$ = 151.5 − 1.02(8.5°F) = 142.8°F
    Upper control limit = $\bar{\bar{X}} + (A_2)\bar{R}$ = 151.5 + 1.02(8.5°F) = 160.2°F
    Center control line = $\bar{X}$ = 151.5°F

Using the Charts

The next step is to determine if the process is in a state of statistical control. Examine each chart separately. The simple visual tests for out-of-control conditions are as follows:

1. Any points above the upper control limit or below the lower control limit or
2. Eight points in a row above the centerline or eight points in a row below the centerline

Applying these conditions or "tests" to Exhibit 3, we see that both the $\bar{X}$ and R charts indicate process instability. Clearly the process is not in a state of process control. When the manager checked the records, she remembered that on day eight a new chef took over. The new chef seems to have tighter control (small range on the R chart) but tends to overheat the food.

The manager met with the new chef to stress the importance of delivering the food consistently and not overheating it. Exhibit 4 shows the data collected for the next 10 days. Applying the tests to charts based on this data, we observe that the process now seems to be in statistical control. Setting up control charts with the new values and tracking performance with the new limits should indicate the occurrence of any unnatural patterns in the future.

EXHIBIT 3 Initial Capability Study $\bar{X}$ and R Charts

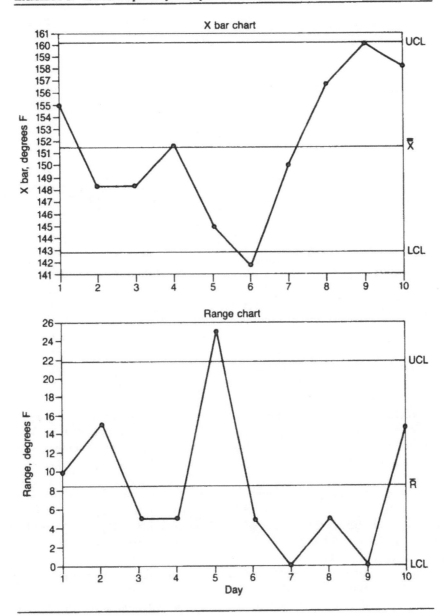

Note: ◇ signifies UCL   △ signifies LCL

EXHIBIT 4   Capability Study of Improved Process*

Day	Reading, °F			R, °F	$\overline{X}$, °F
11	151	147	150	4	149.3
12	151	150	147	4	149.3
13	150	152	151	2	151.0
14	150	156	151	6	152.3
15	150	148	152	4	150.0
16	148	151	155	7	151.3
17	152	157	149	8	152.6
18	147	151	150	4	149.3
19	150	150	156	6	152.0
20	150	155	151	5	152.0

$$\overline{R} = 5.0°F \quad \overline{\overline{X}} = 150.9°F$$

*R chart limits: UCL = 12.9°F LCL = 0°F
$\overline{X}$ chart limits: UCL = 156.0°F LCL = 145.8°F

Product Tolerances

Just because a process is in control does not necessarily mean that it is within tolerances. Tolerances refer to how the part must be made to be usable later in the process or by customers. Tolerances are a characteristic of the design for the part while control limits are characteristic of the manufacturing process. If the design tolerances are too tight, then a particular process may be physically incapable of meeting the tolerances 100 percent of the time, even if the process is statistically under control.

For the restaurant example, suppose that the manager sets the tolerances at 150 to 155 degrees. Any meal outside that range is "unacceptable." Exhibit 4 clearly shows that many of the meals are outside this limit. The process will have to be fundamentally modified to meet the tolerances. Until this is done, some of the meals will be outside the tolerance band or out of specification. Of course, if the process were not under statistical control, the problem would be much worse since even more of the meals would be out of specification.

Note that it is not enough to have the $\overline{X}$ control limits and the tolerances coincide. The control limits refer to the mean value of a sample of parts. Even if the mean is within the tolerance on the $\overline{X}$ chart, some members of the sample may not be within tolerance. Simple statistical rules are available to show how far the control limits must be inside the tolerances.

The way tolerances are set is controversial in American manufacturing. Many observers believe that some engineers routinely overspecify the tolerances, that is, they call for tighter tolerances on a part than are really needed, hoping that the actual process variability will be acceptable. This encourages the manufacturing organization to be less diligent about ensuring that all parts meet the tolerance and that all processes are capable of meeting them. Some companies have no formal procedures for specifying tolerances during design, and some tolerances may be underspecified while others are overspecified.

An early step in a quality-improvement situation, therefore, is to make specifications meaningful again. They should be no looser and no tighter than what is actually needed.

EXHIBIT 4 (continued)

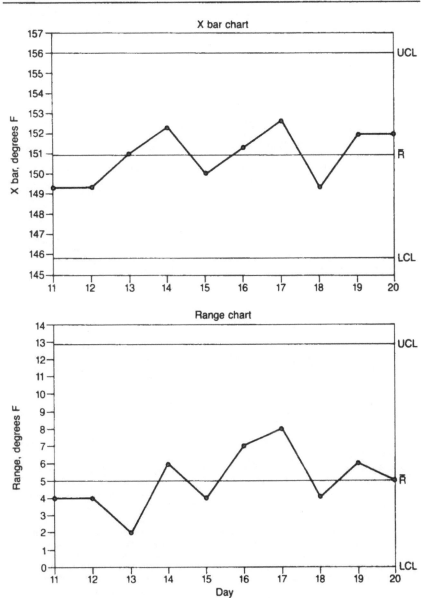

EXHIBIT 5   Diagnosis of Patterns of Process Instability

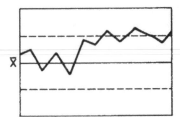

A. This abrupt upswing in X̄
suggests a changed machine setting
or new employee unfamiliar with
the specification. Also, perhaps
a new material was introduced.

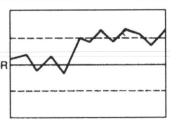

B. This abrupt change in R
suggests an untrained employee,
a machine needing maintenance,
or a new material supplier.

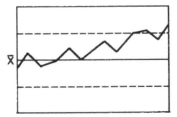

C. This gradual upward drift
in X̄ indicates a tool being
gradually worn down or a gradual
slippage in machine settings.

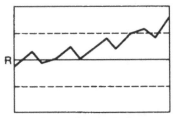

D. This gradual rise in R means
greater randomness. It might be
the deterioration of a machine,
but it can also indicate a decline
in operator discipline.

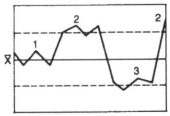

E. One might suspect we have three
different operators who set up and
operate at different levels. Also,
this might indicate a "sticky"
setting in the process.

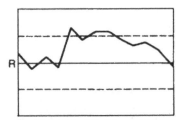

F. This looks like the case of a
new worker who eventually learns
the job and approaches normal
results. But it might be the same
worker who is distracted and is
coming into control.

EXHIBIT 5 (continued)

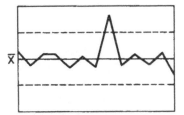

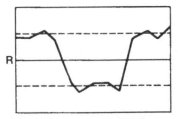

G. This looks like the case of a "freak" point. One might check into the causes of the freak, but be cautious about overreaction.

H. This pattern suggests there might be different employees with different skills and training levels.

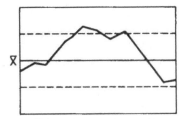

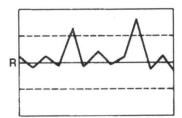

I. This suggests some cycle that interferes with the process. This might be seasonal or monthly. Watch to see if it repeats.

J. This is the chart of a process that shows failure every 5 days. The problems all occurred on Monday after a weekend shut down, so they are probably not truly freaks.

(This may take some experimentation to determine because setting tolerances is an art.) Suppose that after this some parts are still outside of specification. Then the choices available to the manufacturing operation are as follows:

1. Bring the process under statistical control.
2. If this is insufficient, modify the process to have a tighter range.
3. Assign the part to a different machine.
4. As a last resort, use inspection and sorting to cull parts that are out of specification, and either rework or discard them.

Using Control Charts for Diagnosing Process Problems

We have shown how to use control charts to detect when a process is not under statistical control. When looked at for the first time, most processes will not be under statistical control. Even after they are brought under control, new factors may intrude and disturb the process.

Control charts can also help in isolating and diagnosing process problems. Exhibit 5 shows several control chart patterns and their related diagnoses. Exhibit 5-A shows an abrupt upward shift in $\overline{X}$. If there is no associated change in R, this is probably due to a deliberate change in the process. This might result from an altered machine setting or the introduction of a skilled employee who has an upward bias. If $\overline{X}$ and R move together, as in Exhibits 5-A and 5-B combined, it is more likely due to an untrained employee or a breakdown in the equipment. Exhibit 5 provides various patterns that deserve study.

Examples of Control Charts

Most repetitive processes can be tracked by control charts. However, considerable management judgment is needed to decide what to track since it is easy to be buried in too many control charts. When examining a process for the first time, look at several characteristics of potential interest. Eventually you will find out which ones are most indicative of the proper functioning of the overall process.

Control charts and other tools of SPC have been developed more fully in manufacturing than in services. With the growth, diversity, and increasing competitiveness of the service sector, however, opportunities abound in services also. A few companies in each industry are leading the way. Examples of what to track with control charts are as follows:

1. The size of a part being machined is a candidate for charting. Individual workers and work centers may track various critical dimensions while higher management just audits the process.
2. Labor time or elapsed time per part can be charted. An out-of-control situation here might signal problems with incoming materials or with machines, requiring more time to do a proper job. It might also signal operator fatigue.
3. Quality measures for mass-service organizations such as restaurants (already discussed) and banks lend themselves to charting. Both front-room operations, such

as waiting times for tellers, and back-room operations can be charted. Speed and accuracy of information processing are often useful targets for control charting. Examples include lockbox operations for banks, handling telephone orders and inquiries for WATS line order-taking systems, and data-entry operations in any large information-processing company.
4. For airlines, various measurements can be control-charted. The entire maintenance function of some airlines is subject to sophisticated statistical process control. Late departures, lost baggage, overbooking, underbooking, and other measures of customer service can also be charted.
5. Personalized services are more challenging. Control charts may or may not be directly applicable. To use a control chart, the operation or some portions of it must follow a recurrent pattern. In some instances an out-of-control signal on a control chart is due to a shift in the environment rather than in the internal operation. However, even early warning of this can be valuable.

A Warning

Besides control charts, many other SPC tools are useful to diagnose and solve process problems. These include process-flow diagrams, scatter plots, and fishbone diagrams, as well as more sophisticated methods such as designed experiments. The unique virtue of the control chart is an attention-signaling device. It gives unambiguous evidence that something has gone wrong. The managers and workers must then use various tools and their common sense to isolate and fix the problem.

Other types of control charts exist as well. The so-called np chart is used to track defectives. The "cusum" chart measures cumulative deviations and is more useful in some situations than standard charts.

Control charts are often the most important part of an SPC program, but they should never be the only part. Companies that are starting out with SPC sometimes interpret SPC to mean maximizing the number of control charts. This is unfortunate.

See the references for further information. See also "Statistical Quality Control for Process Improvement" Harvard Business School Case No. 9-684-068 for an introduction to other tools of SPC.

References

This note gives the basics of control charts. Considerable experience about them exists and has been collected by various authors. These references also discuss other equally important topics in quality control and SPC.

1. Kaoru Ishakawa. *Guide to Quality Control*. Second Revised Edition 1982, copyright Asian Productivity Organization, 1982. Available in the United States from UNI-PUB (800) 274-4888. This book is a brief and clear introductory text for nontechnical audiences.
2. J. M. Juran, *Quality Control Handbook*, Third Edition, New York: McGraw-Hill Book Company, 1974. Huge, comprehensive, and clearly written. Chapter 23 is specifically about control charts.
3. *Statistical Quality Control Handbook*, Second Edition, Indianapolis: Western Electric Company, 1956. Ref: Select code 700-444. This classic book is available only through AT&T (800) 432-6600. It is a down-to-earth guide on the use of control charts.

Appendix: Constructing Control Charts Using Lotus 1-2-3

Control charts are deliberately designed to be simple enough to do by hand. However, with the easy availability of spreadsheets and personal computers, it is often convenient to construct one on a spreadsheet. This is particularly useful when setting one up for the first time, since it automates calculation of the control limits.

Assume that you have already arranged the data in some tabular format, where each row is a sample of size n, and there are n columns. (If the data come in some other form, such as 100 percent inspection of consecutive lots, you can easily arrange it in this form.) We will show you how to use the next two columns for the $\bar{X}$ (mean) and R (range) charts. We will use the notation of Lotus 1-2-3.

Suppose that you have 15 samples of size 5, that column A is the sample ID number or date, and that columns B through F are the data. Suppose the first row of data is row 11. (See sample at end of Appendix.) In column G we will build the $\bar{X}$ chart; in column H we will build the R chart.

In the following text what you type is **boldfaced**:

In cell G11, put the formula **@AVG(B11.F11)**. This is the mean of the first sample. For the range, use the @MIN and @MAX commands. In cell H11, put **@MAX(B11.F11)-@MIN(B1.F11)**. Now/Copy these formulas for the other samples:

/C FROM: G11.H11 TO: G12.G25. (Change G25 to the last row of your data.)

Now you need to find the control limits. At the bottom of columns G and H you will find the $\bar{X}$ and $\bar{R}$ values, which you will put in row 27. This will give you your centerlines. In cell G27, put **@AVG(G11.G25)**. Then /Copy this formula to cell H27. We will need the control chart factors from Exhibit 1 for your sample size ($n = 5$). These should be permanently stored near the top of the spreadsheet for later use. Say you have put the A_2, D_3, and D_4 factors in cells B7, C7, and D7, respectively. (If you want to get fancy, you can use the table LOOKUP feature of 1-2-3.) Put the UCLs in row 28, and the LCLs in row 29. Here the formulas are completely different for the Xbar and R charts.

For Xbar, in G28 put **+G27+($B$7*H27)**   (upper control limit).
     In cell G29 put **+G27−($B$7*H27)**   (lower control limit).
For R, in H28 put   **+$D$7*H27**   (upper control limit).
     In cell H29 put **+$C$7*H27**   (lower control limit).

Now you have all the numbers. You should type in labels to know what is what. To graph this you will need to define two graphs, one for each chart. The 1-2-3 default graphing options (such as automatic scaling) work quite well for this purpose. Type/**Graph A**data range **G11.G25.** So you know what this is, define a title: Options**Title**First *X*bar **chart for widget process**. Then **Quit**, then **View**.

The R graph is defined the same way, except substitute column H and remember to redefine the title. However, you might as well create these graphs permanently by using the /**GraphName** command: /**GNC**reate *X*bar for the first one, and /**GNC Rchart** for the range.

When you view these graphs, you can call up either one using the /**GraphNameUse** command, and print them out. (Use PrtSc if you are using 1-2-3 version 2 and do not care about the print quality.)

The easiest way to put the centerline, UCL, and LCL on the graphs is to physically draw them on the paper printouts with a pencil. However, a more elegant approach for the Xbar centerline is to define a B range with 15 equal data points, each equal to **+$G$28**. Do the same on the Xbar graph with C and D ranges for the UCL and LCL. Repeat this for the Range graph.

That is it. The second time you do this it should take less than five minutes (assuming you already have the data). When you have more samples to add, use /**Worksheet Insert Rows** to put more rows just before the last sample (Row 25). This will automatically modify all the formulas to pick up the new data.

If you want to get fancy or have to do this a lot, you can define a spreadsheet with the formulas but without data. Then use /File Import to move your data into the spreadsheet, and save it under a new name. (Or move the formulas into a blank region of your data spreadsheet.)

	A	B	C	D	E	F	G	H
1								
2								
3								
4	Sample control chart template							
5	REB 1/31/86							
6	Factor			A2	D3	D4		
7	n = 5			0.58	0	2.11		
8								
9	Sample							
10	Number						Xbar	Range
11	1						@AVG(B11..F11)	@MAX(B11..F11)
12	2						@AVG(B12..F12)	@MAX(B12..F12)
13	3						@AVG(B13..F13)	@MAX(B13..F13)
14	4						@AVG(B14..F14)	@MAX(B14..F14)
15	5			The raw data			@AVG(B15..F15)	@MAX(B15..F15)
16	6						@AVG(B16..F16)	@MAX(B16..F16)
17	7			goes in here			@AVG(B17..F17)	@MAX(B17..F17)
18	8						@AVG(B18..F18)	@MAX(B18..F18)
19	9						@AVG(B19..F19)	@MAX(B19..F19)
20	10						@AVG(B20..F20)	@MAX(B20..F20)
21	11						@AVG(B21..F21)	@MAX(B21..F21)
22	12						@AVG(B22..F22)	@MAX(B22..F22)
23	13						@AVG(B23..F23)	@MAX(B23..F23)
24	14						@AVG(B24..F24)	@MAX(B24..F24)
25	15						@AVG(B25..F25)	@MAX(B25..F25)
26								
27	Means						@AVG(G11..G25)	@AVG(H11..H25)
28	UCL						+G27+($B$7*H27)	+$D$7*H27
29	LCL						+G27−($B$7*H27)	+$C$7*H27
30								
31								
32								
33								
34								

31-Jan-86   05:54   PM

Consolidated Transformer Company (A)

Dan Keller, plant manager of the Cleveland plant of Consolidated Transformer, was reviewing the factory specifications for his main product line, the "RY" series of transformers. One of the principal measures of performance of a transformer was how high the temperature of the core of the transformer rose during operation. A temperature rise in the core to 60°C, under full load (referred to in the trade as 60°C or simply 60.0) was considered acceptable performance by the industry. The lower the full load temperature, the better the product. With this in mind Keller had set a much tighter factory specification of 50°C.

Production in the Cleveland plant for the last three weeks had been consistently above the 50°C cut-off point for acceptable output for shipping. Anything below 50°C was automatically shipped, but anything above was put on "hold" or "reject" depending on its test results. In the last three weeks production batches had been running in the 50.5 to 55.5 range. Keller was not unduly concerned about shipping transformers in this range, because the customer specification (the performance rating promised the customer) was 60°C. What concerned Keller more was whether or not to officially alter the factory standard.

Consolidated Transformer, with three U.S. manufacturing sites, was one of the largest domestic manufacturers of electrical transformers. The Cleveland plant produced 40 percent of total company output. Although

This case was prepared by Professor Ramchandran Jaikumar as the basis for class discussion rather than to illustrate either effective or ineffective handling of an administrative situation.

the plant produced 37 different product lines, almost 60 percent of Cleveland's sales dollars came from the RY series transformers. The Cleveland plant, a two-story facility, spanned almost 250,000 square feet. Total employment included 600 hourly production workers and 200 managerial and staff employees. The RY series was being produced on two semiautomated assembly lines, each with an output of four batches (of 2,500 pieces each) per shift.

Dan Keller, 53 years old, had been the plant manager of the Cleveland facility for just under six years. He had spent 17 years at Consolidated and had amassed a wealth of experience in almost every type of manufacturing job within the company—from supervisor to manager of manufacturing engineering. During his tenure as plant manager, he had installed automated assembly equipment and material handling systems, reduced plant overhead by almost 15 percent and in his opinion, succeeded in raising the status of manufacturing within the company. In addition, Keller had helped design and implement a rigid Quality Control System that he believed was the best in the industry. Keller's consistent record of quality had forced a number of suppliers to meet standards set by Consolidated and helped it meet the threat of foreign competition by providing performance guarantees.

QUALITY CONTROL SYSTEM

The Quality Control System at Consolidated Transformer was in Keller's words "my most important contribution to the company." To bring everyone's attention to the importance of quality he had personally pushed for and supported the institution of a strict system of inspections and corrective actions. For instance, on the RY series, he had originally worked with marketing and engineering to establish the 60°C performance level as the customer specification for that particular transformer. He had then deliberately set the factory specification level at 50°C, 10 points better than the marketing department requirement and customer expectations. He explained his reasoning:

> This process, like any manufacturing process, is made up of a number of discrete operations which individually have random fluctuations. My setting the factory spec higher than the customer spec was an attempt to capitalize on the information I had about the normal distribution of the process variations and to set the limit so that two standard deviations below the expected value of the processes fell right at the actual customer specification. I want to guarantee quality.
>
> I was aware that with bigger batch sizes, the standard deviation of the population increased. So in setting a performance specification I had to be concerned with the batch size as well. There were those natural breaks during a shift—a large break of 40 minutes and two shorter breaks of 10 minutes each.

The Distribution of Core Temperature Rise for Individual Transformers*

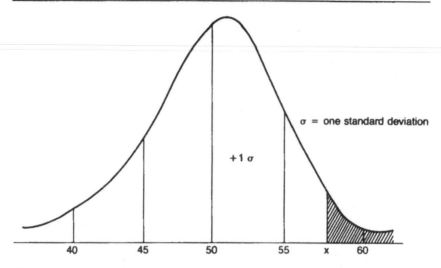

σ = one standard deviation

$+1\sigma$

40   45   50   55   x   60

*The ratio of the shaded area to the total area under the curve, is the probability that the temperature of the core will exceed temperature X.

During these breaks, adjustments could be made to the machines to maintain consistent quality. These adjustments took about 10 minutes, so having four production batches per shift on each of the two assembly lines worked out just fine. We could produce 2,590 units in the hour, and 45 minutes allotted for each batch. The standard deviation with this batch size was 5°C. In order that 95.5 percent of the units shipped met customer specifications of 60°C we needed to keep the factory specification 2 standard deviations (i.e., 10°C) below the customer specification.

I also wanted to instill the pride of workmanship in the work force with the knowledge that they consistently produced a product that was *better* than the customer needed. Finally, it is my opinion that the quality that we deliver is an intangible plus for our product in the market. It is worth something to us, but we can never measure the exact amount. It is a highly competitive market and anything that I can do to keep an edge is helpful.

Besides the establishment of tight product specifications for all product lines, Keller knew that he had to establish a system for detecting and then correcting any product defects that occurred. Keller had his manager of quality control set up a system that identified product defects by testing them right after final assembly. Keller explained his sampling procedure:

Of the units coming out of the assembly line, one in every 100 transformers is inspected. This comes to about one every five minutes. Once the production run is completed the entire sample lot of transformers to be inspected is sent to the office of Quality Assurance, where the temperature measurements under full load are made. This way I can keep my costs of inspection to 1 percent of the cost of the product. An average for the lot is computed and plotted on an X chart (Exhibit 1); we maintain an X chart for each assembly line. If the mean of the lot sample is below 53°C the lot is accepted, otherwise, the lot is logged and segregated in a different part of the finished goods warehouse. A "Quality Notice" is then written and circulated to all involved—production, marketing, and plant management.

If the lot sample was above 53°C but below 60°C, the lot was tagged "hold" and a corrective action team was set into action. This team, nick-named the "SWAT team," consisted of a product supervisor, a quality inspector, a maintenance engineer and a purchasing agent. This team met, performed an initial investigation, and completed a "Corrective Action Report" (CAR) within one working day after the lot was tagged and management was notified. (This team was staffed on a rotating basis with a six-month tour of duty.) The CAR identified potential causes of the problem and listed specific action steps to be taken to prevent the defect from happening again. In addition, within two weeks, a follow-up CAR was completed, summarizing the steps taken, results achieved and any other further insights of information that had been discovered since the first detection. Every one of the seven cases of "hold" lots that had occurred during the last 15 months had been eventually shipped to the customer and accepted. In six of the seven cases, there had been a "cause" found that had been corrected; the seventh was just a "random event" (as the report read.)

If the lot sample average was above 60°C specification, the lot was tagged "reject," placed not in the finished goods warehouse, but in a special area of the work-in-process storage area. Like the "hold" category, the "reject" notification (colored red and different than the "hold" notification) was distributed to all managers at the direct level (one below the plant manager), all involved production supervisors and quality inspectors and managers, and the plant manager. A different team than the "hold" corrective action team was initiated and they had two working days to do the initial investigation and complete a CAR. The CAR for a reject lot differed slightly from that for a "hold" lot in that it had to include the recommendation for rework or scrap and it had to include an estimate of the total cost of the rejected lot. The "reject SWAT team" also rotated like the "hold SWAT team," but operated under less pressure than the hold team because the shipment date wasn't in question—no rejects could be shipped. During the 15 months, only two lots of the standard product lines had been rejected. In both cases, the cause was reported as a "machine error."

Keller had further established that only he could sign off on the "hold" lots that were to be shipped. Each one was to come under his own scrutiny and, in his words, "each decision is a separate management decision which I make independently of any other consideration." To release a

EXHIBIT 1

	Shift 1			Shift 2	
Date	*Lot Number*	*Sample Average*	*Date*	*Lot Number*	*Sample Average*
Nov. 13	1	48.72	Nov. 13	1	48.82
	2	49.66		2	50.41
	3	48.60		3	49.50
	4	52.33		4	52.17
Day's average:		49.83	Day's average:		50.225
Nov. 16	1	53.34	Nov. 16	1	53.95
	2	53.61		2	54.18
	3	55.05		3	55.33
	4	51.77		4	53.38
Day's average:		53.44	Day's average:		54.21
Nov. 17	1	53.84	Nov. 17	1	54.34
	2	53.64		2	53.84
	3	53.50		3	52.85
	4	51.66		4	51.83
Day's average:		53.16	Day's average:		53.215
Nov. 18	1	53.34	Nov. 18	1	52.80
	2	53.28		2	52.16
	3	51.96		3	51.25
	4	54.56		4	50.00
Day's average:		53.285	Day's average:		51.55
Nov. 19	1	52.61	Nov. 19	1	53.50
	2	53.23		2	52.12
	3	52.90		3	52.77
	4	52.53		4	54.55
Day's average:		52.82	Day's average:		53.235
Nov. 20	1	55.04	Nov. 20	1	52.08
	2	52.56		2	53.64
	3	53.84		3	53.10
	4	53.81		4	51.77
Day's average:		53.81	Day's average:		52.65
Nov. 23	1	52.95	Nov. 23	1	53.84
	2	53.50		2	51.44
	3	51.26		3	51.44
	4	53.10		4	52.36
	5	52.80		5	53.18
	6	50.67		6	53.81
Day's average:		52.38	Day's average:		52.68

EXHIBIT 1 (continued)

	Shift 1			Shift 2	
Date	Lot Number	Sample Average	Date	Lot Number	Sample Average
Nov. 24	1	53.05	Nov. 24	1	51.82
	2	53.84		2	52.61
	3	52.87		3	54.34
	4	52.87		4	54.23
	5	53.23		5	52.98
	6	52.95		6	52.56
Day's average:		53.14	Day's average:		53.09
Nov. 25	1	51.72	Nov. 25	1	52.95
	2	53.13		2	53.15
	3	53.25		3	52.98
	4	52.12		4	53.10
	5	52.92		5	54.75
	6	53.13		6	53.84
Day's average:		52.71	Day's average:		53.46
Nov. 26	1	51.36	Nov. 26	1	51.59
	2	52.05		2	52.08
	3	52.26		3	54.65
	4	53.15		4	53.67
	5	51.96		5	53.36
	6	54.08		6	50.95
Day's average:		52.48	Day's average:		52.72
Nov. 27	1	52.05	Nov. 27	1	52.36
	2	53.41		2	53.18
	3	52.53		3	52.19
	4	51.52		4	53.05
	5	53.61		5	53.84
	6	52.16		6	52.87
Day's average:		52.55	Day's average:		52.915

"hold" lot for shipment, according to the procedure, the quality shift supervisor along with each member of the SWAT team had to sign a form indicating that they approved the shipment of the lot. Then the director of sales and the director of quality had to both initial the form and place the entire process in Keller's hands, who then made a decision whether or not to release it. If Keller approved of the release, he would sign a separate "Release Letter" which was forwarded to marketing, sales, quality, and shipping. After 48 hours had elapsed from Keller's approval

EXHIBIT 2   Performance Chart of Sample Lots

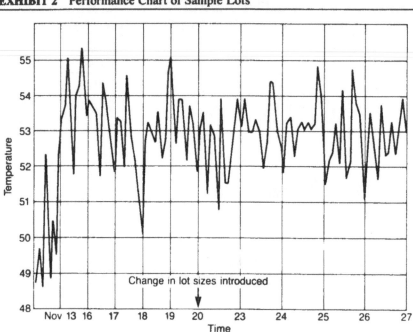

the shipping department, along with a quality inspector, had to record the serial numbers of the lot in a separate QC log and Shipping Release log and file the copy of Keller's release along with it. Only when this had happened could a forklift driver, accompanied by a quality control inspector, remove the "hold" tags and move the transformers into the "pre-ship" area where they were routed into the normal shipping procedure. This entire procedure, although painstaking and often "bureaucratic," had the necessary checks and balances (along with the widespread distribution of reports) that Keller wanted. It was designed to prevent either accidental or intentional shipment of a "bad lot." Keller was firm that the procedure be followed, and he personally followed the movement of the lot the first six times it happened, including going to the shipping dock and watching them load the lot.

THE NEW WIRE

Recently, however, due to the continuing rise of copper prices and energy costs, every major wire producer had shifted the alloy mix on most of the standard electrical wire sizes. An industrywide review board approved the specification for the wire. This shift in composition and processing caused a minor change in some electrical characteristics when the wire was used in a few applications. One of these was in the transformer process. Due to the way that wrapping was done for some models ("transverse wrap"), the rated output of the transformer dipped slightly under some higher temperature applications. Consolidated's RF, RY, and RP series transformers were affected by the shift in alloy mix.

The recent automation of the wire wrap process for the RP and RY series line meant that a change in wrapping procedure was going to be a six-month process including some new tooling. Some manual operations were changed for the RF series to resolve the wrapping problem. Once the new wire was used, the rated output performance of the RY series dropped. The recent dip in output rating during the last three weeks was acknowledged by all to be the result of this change in raw material. Since the wire change had been industrywide, it was not possible now to get wire of the old composition without going to a more expensive grade. Keller had his plant controller calculate the impact of the increase in raw material costs on the RY series, and it raised the manufacturing cost by 3 percent.

KELLER'S RESPONSE

At the end of the first week after the introduction of the new wire it was found that of the 80 production batches that week, 38 were on hold. At the Monday morning manufacturing management meeting a number of options were discussed, including:

1. Raising the cutoff point for acceptance from 53°C to 55°C.
2. Using a high grade wire.
3. Reducing batch size.

Of the three options, Keller was adamant in not wanting to change the acceptance cutoff. This left one or both of the other two options to be followed. The impact on quality of reducing batch size was not clear. By having more setups where machine settings could be inspected and adjusted it was felt that it would have an impact on quality but the magnitude of the change was not known. With more setups the downtime increased and production was lost. After much debate Keller felt that with a cost breakdown of the RY transformer—60 percent of material, 20 percent direct labor, and 20 percent overhead, he would like to experiment with increasing the number of batches from four on each line shift to six. This unusual step meant introducing two 10-minute breaks each half shift. It was felt that the workers would not object to this as long as they had a 40-minute lunch break. Ken Holz, the manufacturing director, felt that as a temporary measure such a step might be alright, but if they continued it for too long it would be difficult to go back to the original schedule of one break for each half shift. At the conclusion of the meeting the new schedule was announced and enthusiastically accepted by the workers.

TWO WEEKS LATER

At the Monday morning review meeting the quality problem was again intensively debated. After two weeks of the new schedule with 240 batches produced, 110 were on hold. Neither the mean nor the range of the sample seemed to have shifted. Ken Holz wanted to raise the acceptance level to 55°C and go back to the old schedule of one break each half shift. He felt that if they used the new schedule too long, group norms would be set and it would be difficult to go back to the old schedule.

The large number of "hold" lots the last three weeks had created a logistic and bureaucratic nightmare for the manufacturing and quality departments. Every other lot now coming off the RY line was on "hold." Holz was beginning to hear some grumbling from the workers and supervisors (including some of the quality inspectors) that the rigid system was becoming more trouble than it was worth. It seemed particularly annoying to the manufacturing director because "everyone knows that all the hold lots are OK for the customers and all the hold lots get shipped anyway." Keller even admitted to himself that he was getting tired of constantly seeing the release forms on his desk and had had his secretary sign the last four because he was out of town. His system was becoming a "rubber stamp" and Keller knew it.

Keller, on the other hand, despite the additional paperwork, was reluctant to lower the factory specification to 55°C for a number of reasons. He had too often in his career seen the specifications lowered for "good reasons" and never raised again. Keller was also not willing to change his whole system just with three weeks of production history with the new wire. He was hopeful that his engineers might be able to make some minor adjustments in the "wire lay angles" that would correct the problem. The corrective action team had been working with the maintenance department on this for three weeks. They had promised his manufacturing director a report in four more days. A conversation with an engineer in the corridor a day earlier was not very encouraging. Indications were that with some adjustments they could improve the output performance by 1°C. However, for any further improvement they needed some major changes in equipment and product design. Further, he didn't want to lower the specification because he had been told by both the marketing and engineering departments that the higher specification on the transformers was a waste of time when he had first established it. Both of those directors had told him that the market didn't "value the additional quality since they didn't need it." Finally, after working hard to instill a respect for his system and to have his people follow the procedures, he didn't want to start making changes in it. He believed that stability was the one production attribute which helped promote both higher quality and better work force attitudes. Of course, he could "temporarily" lower the factory specifications to 55°C until his process design people could have new tooling made—this was estimated to be approximately six months. Keller promised his managers a decision soon.

Marve Robertson, a recent hire in the Cleveland plant reporting to the director of quality, was the new manager of product planning and quality. This position had been created by the director of quality to ensure that quality was a consideration in all the decisions that had to be made as a new product moved from engineering through manufacturing to the markets. Marve Robertson, a recent MBA, had made a study of quality and product planning for the next five years. Keller's opinion of him was still forming, but Keller admitted that he was a little put off by Robertson's "executive style." Keller had mentioned to the director of quality that Robertson's recent 13-page report on the five-year plan for quality had been "complete, but a little impractical." What disturbed Keller was that Robertson, when he had heard of the wire problem, had written a memo to the director of quality (copying Keller) stating that he thought that in the long-term interests of the plant, the factory specification for the RY series ought to be increased to 60°C to coincide with the customer specification. He even implied that Keller's system was a travesty on the integrity of the manufacturing engineering-marketing relationship—the exact phrase he used was "an aura of mistrust in interfunctional matters." Needless to say, Keller had more than a mild reaction to Robertson's report, but he also knew that he was a bit defensive about his own system and was trying to be fair about the whole matter. But Keller had to admit that his system had been working flawlessly for over two years.

What Keller found interesting in one of the exhibits of Robertson's report was the vendor inspection policy of a major customer. The customer had classified vendors into grades A, B, and C, grade A being one where no inspection of incoming quality was necessary, grades B and C being customers whose products needed to be inspected under inspection plans with different degrees of intensity. A typical Grade B inspection plan was one where a sample of ten items from a lot was taken. The lot was rejected if more than two items from the lot failed to meet the test condition. Consolidated was the only vendor given a Grade A classification. The vendor also classified the RY series into three grades: RY-1, RY-2, and RY-3, depending on whether the test condition was set at 55°C, 60°C, or 65°C for the rise in temperature. The price premium for each class was about 8 percent over the next best class. Keller wanted to use this information in his own planning and convince everybody that quality does pay.

Later that afternoon, he had scheduled a meeting between his directors of manufacturing and quality. He knew that he could make his own decision stick no matter what it was, but he also wanted to do what was best for the plant. This whole thing seemed to be running deeper than it should and he thought to himself, "Is it more important than I realize?"

General Electric— Thermocouple Manufacturing (A)

At exactly two o'clock on the afternoon of January 15, 1982, William P. (Bill) Draper, manufacturing manager of General Electric's Aerospace Instruments and Electrical Systems Department, Wilmington, Massachusetts, stepped into the conference room, motioned for quiet, and began to speak.

> As you know, we're here to discuss the possibility of introducing a new system of production control, Toyota's "Just-In-Time" system, in our thermocouple product area. All of you, I hope, have reviewed the materials that I sent around last week, outlining the project and providing a brief description of how the system works in Japan. Since this is our first meeting, I'd like to spend the next few hours reviewing the major elements of "Just-In-Time" and considering how it might be applied here in Wilmington.
>
> To help answer your questions, I've invited Roger Hyatt to join us at today's meeting. Roger is the president and founder of Production Systems, Inc., a consulting firm specializing in operations planning and control, as well as a leading expert on inventory management. He was one of the instructors leading the GE program I attended in Japan last fall. Feel free to speak openly in front of Roger; after two weeks in Japan as his bridge partner, I can vouch for his ability to keep confidential matters to himself.

This case was prepared by Professor David Garvin as the basis for class discussion rather than to illustrate either effective or ineffective handling of an administrative situation. All data and names have been disguised.

I suppose some introductions are in order now. Working clockwise around the table, we have Sam Phillips, manager of production planning; David Hartwell, manager of planning for the thermocouple area; Bob Stone, production control supervisor for the third floor, which includes thermocouples; Eric Nelson, first-line supervisor in the thermocouple area; Steve Hansen, manager of advanced manufacturing systems; and Henry Malone, manager of shop operations in the thermocouple area.

One final comment before I give Roger a chance to speak. You all know that inventory has been a perennial problem for us, and that something must be done to bring inventory levels down. That's the whole purpose of this project. My personal view, after visiting a number of Japanese factories, is that our goals are relatively modest. We're shooting for a reduction in the thermocouple area's inventory from the present level of $6.2 million to $2.1 million after three years, with a further reduction to $1.4 million or less after five years. Some of the Japanese companies we visited carried only six to eight days worth of inventory; many didn't even have stockrooms. I don't see why we can't do as well.

That's about it from my end. You all know that I believe in this project and think it can succeed. Roger, is there anything that you would like to add?

Roger Hyatt leaned back in his chair, looked slowly around the room, and addressed the group:

I'm sure that you've all read the materials that Bill sent around and have formed some opinion of the project. I'd like to hear what you think. Exactly what will it take to make the "Just-In-Time" system work here at Wilmington?

THE GE IMPACT PROGRAM

Bill Draper and Roger Hyatt had first met at General Electric's Impact Program—the Manager of Manufacturing Course held in the fall of 1981. The program, staffed by a variety of consultants and educators specializing in manufacturing management, had been designed by General Electric to help the company improve the quality and productivity of its operations. Approximately 25 manufacturing executives attended the program each year. Their first four days were devoted to classroom work, including an overview of manufacturing strategy, a review of the major elements of operating systems (primarily production planning and material control, quality, processes and equipment, and work force), and a contrast of U.S. and Japanese management systems. The remaining 10 days of the program were spent in Japan, visiting three Japanese companies, analyzing their manufacturing and management practices, and discussing these findings as a group.

The participants then returned to the United States to begin applying what they had learned. After completing trip reports, each executive was expected to identify areas of opportunity within his or her own operation, and to propose a project embodying the principles and practices he or she had observed in Japan. All participants met again for a week to compare project proposals and strategies, and to receive updates on projects already initiated under the Impact Program. The group then disbanded, and each executive proceeded with his own project. If necessary, further assistance could be obtained from any of the program faculty.

It was after attending the Impact Program that Bill Draper had proposed a "Just-In-Time" system for the thermocouple area of General Electric's Aerospace Instruments and Electrical Systems Department. (See Exhibit 1 for organization chart.)

THE THERMOCOUPLE PRODUCT AREA

Thermocouples, like thermometers, were used to measure temperature. They operated on a simple physical principle: When two dissimilar metals were joined at the ends and heat was applied, an electrical current was generated. If a gauge was then attached to the thermocouple, an extremely accurate temperature reading could be obtained.

Although they had a variety of applications, thermocouples were especially useful in aircraft engines, where accurate temperature readings were critical to safe and efficient operation. An aircraft engine thermo-

EXHIBIT 1   Organization Chart

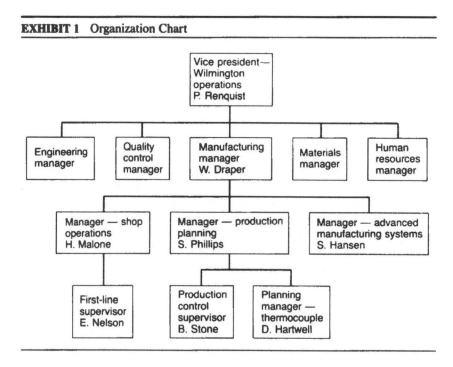

couple was generally roughly circular in shape, to conform to the interior dimensions of an engine, and consisted of a number of probes joined to a main harness. (See Exhibit 2.) These probes pointed toward the engine's exhaust gases; they, like all other segments of the harness, were constructed from swaged leads, which consisted of rods containing two wires made of dissimilar metals, separated by magnesium oxide or some other insulator. Current was generated when heat from the engine's exhaust gases passed by the probes. Since the probes were attached, via a junction box, to a gauge in the plane's cockpit, the engine's temperature could be monitored continuously.

The thermocouple product area was one of five at the Aerospace Instruments and Electrical Systems Department in Wilmington. With 1981 sales of $21.5 million, it ranked second in total dollar volume. (See Exhibit 3 for a more complete profile of the thermocouple business.) Nearly

EXHIBIT 2   A Typical Aircraft Engine Thermocouple

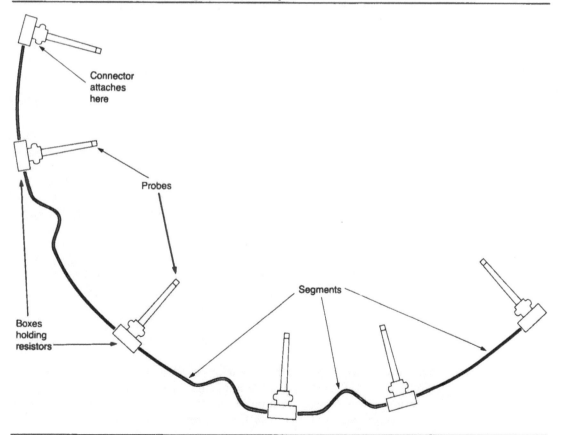

EXHIBIT 3   Thermocouple Business Profile, 1977–81
(in millions of dollars)

	1977	1978	1979	1980	1981
Sales	$9.9	$12.3	$13.1	$20.8	$21.5
Cost of sales					
Materials	2.5	3.0	3.1	5.1	5.6
Labor	1.0	1.3	1.8	2.8	2.2
Gross profit	.2	.6	(1.2)	(.5)	.8
Inventories (12/31)	2.7	4.1	5.4	5.3	6.2

30 major types of thermocouples were produced by the area, and 90 different models. The simplest of these, the J-79 product line, was regarded as a likely candidate for any experiment with Just-In-Time techniques because it was relatively easy to manufacture, employed swaged leads that were less fragile than those used by other product lines, and enjoyed a comparatively stable business outlook.

THE PRODUCTION PROCESS FOR THE J-79 THERMOCOUPLE

The J-79 thermocouple, like others manufactured at Wilmington, was composed of four subassemblies—probes, segments, resistors, and a connector. The connector was attached at one end, to plug into the engine; segments and probes then alternated around the main body of the thermocouple. Each probe contained two resistors. In total, the J-79 consisted of five segments, six probes, twelve resistors, and one connector.

The segments were built up from swaged leads. These were first cut to the proper length, and then passed twice through centerless grinding: once to reduce them to the proper diameter, and a second time to narrow their ends. The ends were then turned back to reveal the wires inside. After cleaning and annealing, the segments were "formed" (shaped), each in a slightly different pattern (i.e., there were five segments, and therefore, five shapes, per thermocouple). Finally, each segment was sandblasted and inspected. Many of these same steps were also followed in the manufacturing of connectors, another of the basic subassemblies. (For a more detailed description of the production process, see Exhibit 4.)

Probes also began with swaged leads. These leads, however, were purchased precut to the proper length. They then passed twice through centerless grinding, with both passes on the same machine. The initial setup of the grinder took between four and six hours; with a run time of 22 seconds per piece, this encouraged lot sizes of 1,000 or more. The second

EXHIBIT 4   The Production Process for the J-79 Thermocouple*

Subassembly 1: finished segments

Swaged lead → cut to length (12″) → centerless grind (12″) → turn back ends (3″) → sandblast (19″) → anneal (1.5′) → form (26″) → sandblast (7″) → inspect → finished segment

Subassembly 2: thermocouple box (probe)

Swaged lead → (setup 4 hours) → centerless grind (22″) → setup (4 hours) → centerless grind (14″) → inspect → clean (4 hours) → add flange → heat shrink flange (40.5″) → turn back ends (60″) → turn back ends (67″) → deburr and remove powder (38″) → sandblast (8″) → shape wires (42″) → form (39″) → weld (60″) → check (7′) → thermocouple box

Subassembly 3: resistors

Wire → (setup 4′38″) → cut and wind (60″) → inspect ⟶ assemble → back and check (7′35″) → resistors
                                              add components

Tubing → cut (8″)

Subassembly 4: connectors

Swaged lead → cut (13″) → centerless grind (9″) → turn back ends (52″) → sandblast (21″) → assemble components (2′13″) → check and form (27″) → assembly (3′35″) → weld (1′55″) → form (33″) → anneal (30″) → sandblast (11″) → inspect (27″) → connector

Bottom Assembly:

Sort parts onto building board (5″) → assemble 6 thermocouple boxes and 5 finished segments (4′55″) → weld (18′30″) → weld 2 resistors to each box (7′20″) → form wires (33′37″) → weld wires (13′9″) → weld connector (1′45″) → finish weld (1′34″) → form wires (connector) (7′9″) → weld wires (connector) (7′9″) → space wires (connector) (2′38″) → space wires (5′25″) → check welds (4′) → space probe loops (2′30″) → check resistance (1′18″) → other check (2′7″) → add stamp → pack and weld (15′45″) → check voltage (2′30″) → sandblast (7′33″) → check resistance (1′6″) → check (2′40″) → fix stamp (35″) → prepare shipping tag (2′11″) → inspection (3′) → prepare for shipping (1′15″)

Total Standard Run Time

Subassembly 1:	2.8 minutes
Subassembly 2:	13.5 minutes (excludes cleaning)
Subassembly 3:	8.6 minutes
Subassembly 4:	11.4 minutes
Bottom Assembly:	2.4 hours

*(x′y″) is x minutes and y seconds. Refers to standard hours.

pass through the grinder required another lengthy setup, and also involved a relatively brief run time. After cleaning and inspecting, a flange was welded to the end of the leads. The wires within the leads were then sandblasted, straightened, and separated. From this point on, it was necessary to keep the wires properly identified; each worker carried a small magnet to help determine which wires were positively and which were negatively charged. The probes were then formed and inspected, a protective casing (called the probe shield) was welded on and the weld inspected, and a box for holding the resistors was added. Additional cleaning and inspection followed, after which the probe assembly (now called a thermocouple box) was complete.

The simplest part of the production process was the making of resistors. Wire and cement-like substance were combined in a mold, then baked for several hours. After inspection, the resistors were ready to be used.

Once these four subassemblies were complete, they were combined on a building board, which helped to insure that all pieces were welded in the correct order. In addition to welding, final assembly involved a number of forming operations, in which wires were combined and twisted together. These activities often alternated with welding. Welders were frequently idle while the formers were at work, and vice versa. Various tests of resistance, voltage, and weld strength were performed next, a small piece of metal (called the stamp) was placed over the top of the probe box to protect the resistors, and the completed thermocouple was inspected a final time. It was then ready for shipment.

WORK FORCE ISSUES

Most hourly workers were paid on an incentive wage system, based on piece rates that had not been reviewed for several years. Standards were tighter for some parts than for others. Because workers normally had considerable autonomy in the parts they chose to make, they often focused on those with the loosest standards, rather than on those that were in the greatest demand. The result was excess supplies of some parts and shortages of others. Probe assemblies, for example, which an experienced worker could make at 150 percent of standard, were at an inventory level estimated at six weeks usage under normal demand conditions. One plant employee, commenting on the problem, observed: "How can anyone keep them [the workers] from making whatever they want? All of the materials they need are right there on the floor. The supervisors would have to watch every move those guys made to keep them making the right parts. And the union certainly won't stand for such intervention all of a sudden." In fact, several managers had noted a strong correlation between the tightness of standards and the percentage of time that parts appeared on the critical list (indicating a parts shortage).

Welders, who were the most highly paid and most senior group of hourly workers, presented an additional problem. Several managers suspected that they were using the incentive system to their own advantage. Because the rate for reworking defective parts was higher than the rate for welding them in the first place, welders benefited from building in their own defects. As one manager noted: "Although the welding equipment that the welders have to work with isn't the greatest, I don't see how there could be a 22 percent defect rate without some help from them. But what can we do? No one has supervised the welders for years."

ORDER ENTRY, PURCHASING, AND PRODUCTION CONTROL

Orders were received from three groups: the military, GE's Aircraft Engine Group (AEG), which dealt with both military and commercial customers, and a distribution warehousing operation (DWO) that served as an interface between Wilmington and its customers. DWO received orders from the airlines, grouped them, and then requested the necessary thermocouples from Wilmington. It was concerned solely with commercial customers.

Commercial customers accounted for approximately 50 percent of thermocouple sales. Because they expected goods to be delivered exactly on the contract date, neither early deliveries nor delinquencies were acceptable. To increase the probability that contract dates would be met, the thermocouple manufacturing plan called for commercial orders to be completed approximately one month before the actual contract date. Delinquencies had declined in recent years, but continued to be a problem. (See Exhibit 5 for a summary of 1979–1981 delinquency statistics.)

Military customers were willing to accept shipments at any time up to the due date. They also remitted advance payments based on Wilmington's investment in materials and labor. This gave the plant a bit more negotiating power than it had with commercial customers, who paid only after their goods were delivered. Military customers therefore allowed some flexibility in scheduling, which led to a smoothing of the manufacturing plan over time.

Manufacturing was not, however, permitted to build until a firm order was received. This kept inventories down because raw materials orders were placed only after thermocouple orders were confirmed. Until recently, DWO had been an exception to this rule. It had sometimes ordered from Wilmington on the basis of forecast, rather than confirmed, orders. DWO was now required to have an order in hand before contacting Wilmington, although cancellations remained an occasional problem.

EXHIBIT 5   Thermocouple Sales Delinquencies*
(in thousands of dollars)

	1979	1980	1981
January	$ 386	$845	$466
February	606	800	284
March	377	972	326
April	570	932	320
May	694	944	183
June	1,043	909	195
July	1,008	593	162
August	1,382	558	212
September	1,851	494	296
October	1,640	396	287
November	1,536	479	0
December	1,142	341	341

*Delinquencies that have not been shipped by month-end appear in the following month's total as well.

Wilmington dealt primarily with sole-source suppliers. Many were located outside the Northeast; several considered Wilmington to be a very minor customer. However, 12 of the 21 items purchased from outside suppliers by the thermocouple manufacturing area were provided by GE's West Lynn plant, located in a nearby suburb.

West Lynn's quality and reliability, which had historically been high, were currently in decline. Rumors were circulating that the plant was soon to be closed. Wilmington management, however, did not feel that replacement suppliers should be approached until there was some official announcement about West Lynn's future.

Because production was planned on a monthly basis, workers could theoretically be building product up to four weeks ahead, adding that much more work-in-process inventory. If production were planned on a weekly basis, the problem would be greatly reduced. However, the production planning department was not optimistic about this approach. One manager commented: "Those guys can't even level a monthly schedule! How do they expect to level a weekly one?"

Wilmington's sister division, AEG, had been responsible for some of the volatility in scheduling. In the past, AEG had been unwilling to commit in advance to a present schedule. It wanted to preserve its ability to respond to last-minute orders. It was primarily these orders that caused the thermocouple schedule to be "nervous" and difficult to predict.

INVENTORY MANAGEMENT

Wilmington used a system called the Inventory Control Package (ICP) to manage inventories. ICP was a variant of MRP, in that it exploded backwards to determine raw materials needs. The system had been ordered from an outside consulting firm, which had not fully integrated it into Wilmington's operations. In-house systems people had therefore found it necessary to introduce modifications. Many of these changes were undocumented, and the people who had introduced them were no longer at Wilmington.

Because materials control was relatively loose once materials left the stockroom—Wilmington staff had nicknamed the shop floor "no man's land" because goods seemed to disappear there without a trace—it was difficult to know the accuracy of the inventory balances reported by the ICP. While the system was reinitialized every two months to clear old information from the computer, this process did not reflect goods already on the shop floor. Once these goods left the stockroom, there was no further accounting of them until the yearly physical inventory. For the J-79 thermocouple, inventory on the production floor was estimated at between \$300,000 and \$315,000, approximately 80 to 85 percent of the total inventory on hand for that particular product line.

Loss factors, representing the percentage of raw materials expected to end up as scrap, had not been reviewed for several years. These normally reflected either yield losses on the production floor or rejects from incoming inspection. In 1981, these losses totaled 6.3 percent of sales. However, certain items were being purchased in as much as double the actual quantity needed because the yield on their processes had been only 50 percent when loss factors were initially established. Over the years, many of these processes had been streamlined; yields had often improved dramatically, although loss factors had not been similarly adjusted.

THE MEETING

Roger Hyatt glanced around the room as he waited for a response to his question. In the brief silence that followed, he added, "I'd like to walk out of this room with two things: a first cut at the problems that we're going to encounter when this program is implemented, and a reasonable action plan with responsibilities assigned to specific people. I'd also appreciate comments about what you perceive as being *good* about the plan."

Bill Draper spoke up immediately. "This plan is perfect for the thermocouple area. First, it is heavily process-oriented. That should make gains much easier to obtain. Second, thermocouple manufacturing is physically cut off from other areas, so it should be possible to isolate the working of the plan, as well as the results."

Nelson introduced the "people" aspect:

I don't see why it should be so natural to expect that our people will be willing to bust themselves for this plan. After all, what's in it for them? They get the same wage whether they do D— or A+ work. That would only change if we implemented an incentive system of some kind, and I think that would open a whole can of worms that we don't want to deal with right now. You all know that we promote on a seniority basis around here. Are we willing to jeopardize that system for potential improvements in one area?

Another thing that we need to deal with is whether workers will think that we'll begin laying them off as soon as productivity goes up. If the workers in the thermocouple area like you and believe that they'll keep their jobs even if productivity has improved, they *may* work with you. But, if they think that they are going to work themselves out of a job by improving the process, we're going to have some serious problems.

Draper replied:

We saw in Japan that financial incentives were not the primary motivation for working at a high level of quality. What could we do to make our workers want to do a better job? After all, most of what we've been hearing about Japan suggests that Japanese companies have been very successful in motivating American workers when they purchased factories here. Don't get me wrong. If an incentive system or profit-sharing system would help, I'd certainly consider it. On the other hand, we should remember that our workers know that there are problems in the thermocouple area, and that if things don't improve, we'd be forced to consider moving the area to another plant. We may have some unusual leverage as a result.

Hartwell joined the discussion at this point:

I agree with Bill that the people situation is manageable. I feel that the biggest problem will be with our vendors. It just doesn't make sense to talk about improving quality when our vendors are so unreliable. They neither produce good pieces nor deliver them on time.

We have four major vendors: GE-West Lynn, Shelbyville, Owensboro, and ITT-Harper, as well as a number of small ones. Our best vendor is supposedly GE-West Lynn. Today they're having problems with the J-box, one of the most important parts we purchase. Just imagine what would happen if we were on a Just-In-Time system, and the J-box or some other vendor quality problem came up. With the inventory reductions that you're describing, we would have to shut down the line. Are we really willing to take such risks? AEG wouldn't be very happy with us if your delinquency rate went any higher.

Talking about the Japanese and their great quality with so little inventory . . . do they have as much pressure to meet output goals as we do? Working in the shop, production comes first, with quality way behind. We would like quality to be better, but right now, output is the driver. Even Bill, as excited as he is about inventory levels and as much as he wants to improve quality, knows that when it comes to the bottom line, you've got to meet the shipment schedule. In other words, our next promotion has a lot more to do with meeting shipments than with improving quality or reducing inventory.

Phillips interrupted:

Don't we share some of the blame? Don't we keep changing the schedule? We should establish and then stick to a six- to eight-month schedule, but we're willing to change at the drop of a hat.

Draper nodded, then added:

Let me give an example. AEG is in here now with an inquiry that would make us delinquent for the month of January. Originally we were supposed to build 30 pieces. On January 7, they came in and told us to build 65 pieces in January and 50 a month in February, March, and April, rather than the 30 originally planned. They don't want anything in August, September, or October, but a lot more in November, then none in December because they don't want a high year-end inventory. That's the kind of schedule changes we deal with regularly. What's important is not that we get these occasionally, but that we get them on the average of one per week.

Hartwell observed:

We don't work with our vendors the way the Japanese do. Of course, we have a very different situation. In Japan, vendors tend to be much smaller and more easily controlled by customers. Can you just see us trying to control West Lynn? They'd laugh in our faces! We can't be 5 percent of their business.

Stone added:

Most of our parts are sole-sourced. We need to develop second sources, but until then we can't force the vendors that we have to meet our needs. Nor should we forget how difficult and time-consuming it is to develop new sources. Besides, what assurance do we have that new vendors would be any more responsive to our needs? If we're a very small part of their business, why should we have any more leverage than we currently have? We might even make our first-source suppliers less concerned about our problems. With that little inventory, we're going to be very exposed.

Malone agreed:

Right now we have a thermocouple with an oddball part. The schedule for this thermocouple has been firm for the past eight months. We're supposed to build 128 this month, but the vendor called last week and said that the entire batch of parts had to be scrapped. We won't get any for a month and a half. They simply don't feel any pressure to stand on their head to give us 128 oddball parts. They say, "So what? We're late. Who is thermocouple anyway? They buy 128 pieces every two years."

Roger Hyatt rejoined the discussion at this point: "We should probably spend some time discussing your production process. How much will the current process have to change to adapt to Just-In-Time?"

"In the future, people will have to be more versatile," said Malone. "Believe it or not, we've already started in that direction. We're now doing a lot more cross-training."

Draper observed: "In Japan, the most we saw were six labor grades, and they didn't even use the first one. We have 250–300 grades in this plant alone. In the thermocouple area, we have 12–18 grades, with hourly wages ranging from $8.60 to $10.62."

"We need to reduce setup costs," said Malone. "That's the only way that we will be able to reduce our lot sizes. For example, one of our grinders requires four hours of setup time for a typical run of eight hours. We'll have to completely change our thinking in this area. Until now, we've always assumed the large lot sizes were the key to profitability."

"What kind of gains could be achieved if set-up times were reduced from three hours to ten minutes?" asked Hyatt.

"It's impossible."

"Let's take an example to see if that's true. How long does it take you to change a flat tire on your car?"

"Fifteen minutes," replied Nelson.

"What if you were a racing car driver competing in the Indianapolis 500?"

"That's different. He has help. There it takes only about 15 seconds."

"Then," continued Hyatt, "why couldn't you make similar improvements in manufacturing? Couldn't you set up a special team to do set-ups? Or break down the tasks and have some performed in advance?"

"It could be done in stamping," replied Nelson. "We would have to set up a special rate for the setup team, but I guess that's not impossible."

Phillips added, "I hear that the workers in Japan do a lot of the thinking about reducing setup times. Perhaps there's some way for us to get more input from our workers."

Draper noted: "In one plant we visited in Japan, there were 88 suggestions per employee per year, even though workers received only token payments for their contributions. Management tried to respond to all suggestions within a month. Those inputs from the workers can make a big difference."

"Let me introduce a new subject," suggested Hyatt. "Tell me about your reject rates."

Malone replied: "As you might expect, they vary by stage in the production process. In-process rejections, between subassembly and final assembly, run about 15 percent. At final inspection, the reject rate is 5 percent. In the field, the rate is less than 1 percent. That's not strictly accurate because we rework a lot before parts get to the in-process stage.

At some early steps, our yield is only 60 percent. The major cause of this low yield is poor welds due to moisture."

Nelson disagreed: "No, the problem is in the cap. If you take ten caps, no two are the same. That's why there is an inconsistency in the way they weld. It's a vendor problem."

"What if," asked Hyatt, "we could get the yield losses at that operation down to 25 percent, at final inspection to 2 percent, and in the field to less than .5 percent? What would that mean?"

"Great savings! Productivity increases, better product, lower inventories," enthused Nelson. "I think that we should set some goals that will get us moving in that direction."

Malone interrupted: "I think that goals in those areas should be initiated by the quality control department. Quality hasn't yet given us any real targets to strive for."

"What kind of support are you going to need to make this project work?" asked Draper.

Malone replied: "We'll need engineering and quality control involvement. We'll also need a team dedicated solely to this project. That may be expensive, but I don't think that we can do this if we try to tackle it on a part-time basis. Most of us have had to cancel important meetings or else spend less time on the factory floor in order to be here today. Eric, for instance, was only able to spend two hours on the floor today, and he has 35 people. That will hurt his productivity by 2–3 percent."

Hyatt surveyed the group, then asked: "Are you all convinced that the project should move forward?"

Malone quickly replied: "I'm not."

"What will it take to sell you? You're the manager of the thermocouple department. It's not going to work if you're not sold. What about you, Eric? Are you on board?"

Nelson answered: "I'm sold. I want to give it a try. Of course, I have less to lose than Henry does. He's the manager. If this thing fails, he's going to be left holding the whole ball of wax. What's more, even if we do succeed in reducing inventory, the materials and inventory people will get all the credit. Henry and I will still be evaluated on whether the product gets out the door."

Malone added: "Don't get me wrong. I'm willing to try anything once. But I'm the one that has the most to lose here. I'm very open, I'm just not committed. I won't be a roadblock. But I don't have a KanBan stamped on my head either."

Hyatt stood up to terminate the meeting:

I've asked some very tough questions today, and you've been honest and open in your responses. If this project is going to work, that honesty will have to continue. Part of your job will be to lower the walls that now exist between the various functions represented here. You'll need to encourage a lot more cooperation for the common good. Somehow, Bill will have to figure out a way to reward you for your performance. He'll do that. But if you continue to develop cooperation and trust, you have a great opportunity. You're trying to discover a new way to manage the business. It will no longer be an inventory problem, a quality problem, a shipment problem, but *our* problem. I'm very excited about the possibilities.

The group was silent for a moment. Stone then observed: "One of the readings that Bill sent us contained a cartoon showing rocks that are exposed as the sea of inventory is lowered. The problem for us is that we don't want to know about the rocks."

"But," responded Hyatt, "you've told me about the rocks already. That's what we've been discussing for the last two hours."

Nelson chuckled. "We haven't told you about half of them! Come out on the floor with me and I'll show you what you're really in for!"

What's Your Excuse for Not Using JIT?

Richard C. Walleigh

Just-in-time production, or JIT, has probably received more attention in a short time than any other new manufacturing technique. The main reason is that JIT gets the credit for much of Japan's manufacturing success.

Despite the extensive publicity and interest, few companies have implemented JIT in their manufacturing operations. If JIT provides all the benefits claimed for it, why have so few factories adopted it?

JIT's widespread publicity has been a mixed blessing. The popular press, and even some technical articles, focus on the easily observable differences from batch production systems but ignore some of the more important but subtle features of JIT. Writers rarely get very far past the lower inventory costs attributable to JIT and seldom describe how the technique can improve the entire manufacturing process. Managers who have read only a little on JIT rarely understand how it can help their operations. Usually they focus on the fact that, in the end, JIT increases a company's ROI.

More important than the reduction of inventory and greater ROI are the improvements in manufacturing that result from operating with low inventories. JIT removes the security blanket of high inventory and thus exposes related operating problems. These are problems that need not be faced and solved—and therein JIT can be seen to create hurdles of its own.

Converting to JIT means a big change—in the culture of a company as well as in its manufacturing operations. Established routines and rules become obsolete. Where backup inventories were once considered to be insurance against unexpected shortages or delays, they are now viewed as evidence of lackluster planning or controls, even of laziness. Large production batches can no longer be viewed as beneficial because they help amortize setup costs. JIT forces the elimination of the waste inherent in long setups.

Few manufacturing organizations are very flexible, either in their operations or in the minds of their creators. A typical operation is like a huge steamship, for which a rapid change in course is difficult. Most factories have been making similar products using similar processes for many years; their managers are comfortable with what they know. In this environment, change comes slowly. This inflexibility combined with misperceptions of JIT keep a lot of executives from using JIT. They excuse themselves by saying: "I know JIT has done a lot for others, but our plant, and our processes, even our people, are different. In our situation, JIT won't work."

Since misperceptions create a roadblock to implementation of this valuable management technique, let's look at them first.

Problems with Suppliers

Excuse number 1: "Our suppliers won't support JIT by delivering our raw material in small batches on a daily basis."

Asking suppliers to make daily deliveries is a common mistake of managers who focus on

Reprinted with permission from *Harvard Business Review*, March–April 1986, pp. 38–54.

the inventory-reduction benefits of JIT. Ultimately, this is the right thing to do, but it's the wrong place to start. If manufacturing executives recognize JIT as a problem-solving technique rather than an inventory-reduction plan, the proper starting point will be clearer. JIT should be adopted and practiced inside the factory, where the company can control any problems, rather than outside, where close cooperation with another organization is necessary. Once a company begins to master JIT, it should then begin to work with suppliers to help them understand the benefits it holds for both parties.

Furthermore, JIT is a demand-pull system. Each operation produces only what is necessary to satisfy the demand of the succeeding operation—in contrast with the traditional batch-push system, in which parts are made in large batches and pushed to the next operation on a fixed schedule. Ultimately, every activity in the factory and every demand on vendors is driven by the final assembly operation. Final assembly is the control point for the entire manufacturing process, and it is the place to start implementing JIT.

JIT demands that the production process be rationalized and simplified. For uninterrupted flows in a demand-pull environment, the schedule for final assembly must be smoothed out. As I said earlier, this schedule will drive every activity in the factory and thus will level the operating rate of each. Operating rates of different processes can then be matched, and buffer inventories that separate processes eliminated. All processes will operate better without the wide swings in demand characteristic of a traditional batch-push production system.

As companies implement JIT in this way, suppliers will find a predictable demand for their products as a result of the smooth final assembly schedule. Ultimately, they will also enjoy a consistent long-term demand. The prospect of being able to plan their operations will influence them to sign up as JIT suppliers.

In the long run, the factory will want only one supplier for each purchased part. This long-term supplier must always deliver a quality product, on time, in small batches. In return, the factory will give the supplier a long-term purchase forecast with a guarantee to buy at a percentage of it. In addition, when the JIT operation is in place, suppliers will know that the company is serious about the concept and can share the experience of implementing the technique as they become JIT suppliers.

JIT's success depends on the high quality of incoming materials. If a supplier delivers a bad batch, the whole production line will stop! Once suppliers understand the consequences of failure, they will be sure to make on-time deliveries of high-quality materials. Although the relationship between manufacturer and supplier in a JIT setting entails risks, the rewards of perfect parts always delivered on time are tremendous.

At Hewlett-Packard's computer systems division, just-in-time production was off to a good start before the procurement department made any attempt to convert suppliers to JIT delivery. Although most of our suppliers are still not making JIT deliveries, we are helping them to consistently deliver quality parts on time. Once they have met all the prerequisites for a JIT relationship, we will convert them to JIT suppliers. In the meantime, our efforts have resulted in improved quality, more frequent and smaller deliveries, and a reduction in raw-material inventory.

In our initial efforts to convert suppliers to JIT deliveries, we concentrated on the manufacturers of large components that require a lot of storage space. Although most suppliers of these parts were willing and able to cooperate, in one case we changed to a supplier who was closer and more responsive to our needs. We introduced the vendors to JIT concepts and then showed them their implementation on our own production line. This demonstration convinced them to supply our needs with JIT deliveries.

Making JIT deliveries did not mean that in every case our suppliers immediately adopted JIT in their own factories. One vendor of large metal frames continued to produce them in monthly batches in his own shop while making daily deliveries to us. After several months, when we told him that some of the frames were out of square, he realized his tooling had worn out without his knowledge. He also discovered a month's worth of inventory of defective frames in his shop. Had he been producing frames just-in-time, he realized, he would have had to repair only one day's worth. Besides improving his inspection procedures, the vendor quickly adopted JIT and asked us to help him sell the concept to his other customers.

After attacking the problem of excessive large-part inventory, the computer systems division began working with suppliers of high-value parts. Most of them are distant vendors of unique parts. They were harder to convert to JIT. Most are still not operating on a strict just-in-time basis, but we have reduced our raw-material inventories, and multiple deliveries within a week are common. In addition, these vendors now ship more than half of their parts directly to stock without the need for incoming inspection.

In winning vendors over to our viewpoint, humor sometimes helps. Our materials manager explains that we understand how difficult it is to produce perfect parts and that we will accept a few rejects from our suppliers. We insist only that our suppliers first separate their bad parts from their good ones and ship them to us separately. Furthermore, if a delivery is going to be late, our materials manager requests only that the supplier's vice president of marketing give our production workers a presentation on satisfying customers while they wait for the material to arrive. Suppliers get the point.

By working with our suppliers and showing them what we expect, we have reduced the average number of parts back ordered (past delivery date) from more than 200 to 2. Despite this success, we still have a long way to go in converting many of our suppliers.

Late Production

Excuse number 2: "We will always have back orders in our factory. We are constantly expediting production to make up for these shortages and to complete products for shipment within the scheduled cycle time. If we go to JIT, the line will always be shut down, and our production will always be late."

This is a common lament of production managers who feel that they are always making up for the poor performance of the materials department. In their view, although late deliveries of incoming materials or subassemblies are undesirable, late shipments of finished products are unacceptable. On the other hand, while the materials department tries to make deliveries on time, it knows that products can be made in much less than their scheduled time and it often helps the production manager expedite late orders so that shipments will not be missed.

With long production cycles and large in-process inventories, this kind of adaptation will continue indefinitely. With large in-process inventories, the actual time spent working on a product represents a small percentage (often less than 5 percent) of the production cycle's length. During most of the cycle, a product sits in inventory waiting to be worked on. Although expediting enables products to bypass these inventory queues, a large factory requires an army of expediters.

With JIT, large inventory queues don't exist, so the production cycle can often be reduced by 90 percent or more. Then production schedules really mean something, and expediting is no longer possible. When production schedules represent the minimum time possible for a product to move through manufacturing, the materials department will understand that late deliveries mean late product shipments. Unwilling to be responsible for late product shipments, the materials department will strive to eliminate late deliveries. And when buyers or schedulers understand that a late delivery will stop production and generate a crowd of anxious managers around their desks, they will devote more effort to ensuring that materials arrive on time.

A safe method for introducing JIT into this type of environment is to leave the cycle times of the materials planning system intact while shortening the actual production cycle. The effect will be to move the inventory from the factory floor back to the stockrooms. While this strategy will not reduce inventory, it will improve the flow of materials through production. Indeed, that has been the objective of traditional queue-reduction programs. Parts that are late from suppliers will usually be delivered before production requests them from the storeroom.

With inventory pushed back into the storeroom, late deliveries will cause fewer problems and less confusion on the factory floor. Parts delivered after they are needed by production will cause late shipments. The results, however, will be no worse than they would have been under the old system. If there are no late product shipments under a batch system, there will be none under JIT. As deliveries improve, cycle times for materials planning can be matched to the production cycle, and inventory will be reduced.

Need for Software

Excuse number 3: "Our batch-oriented materials planning and control system won't allow us to operate in a just-in-time mode. We need to install a just-in-time software package before we can convert our production operation."

Experienced managers know that forcing a production process to fit a software system is a prescription for disaster. Designing processes to conform to the requirements of a particular software package often makes operations less effective. The process needs to be converted first.

Before installing a software system, a company naturally should understand the process that the system will be supporting, how the system should be designed, what features it should have, and how it should operate. This can be done only if the process is designed first and has been debugged through operating experience. This is especially true when implementing JIT because it is so different from a traditional batch-push operation. If manufacturers' basic philosophy and operations do not change, workers will figure out informal ways to get around any JIT computer system that may be installed; for example, they can accumulate materials at their individual workstations to serve as buffers.

What then can be done in the period between the start of JIT production and the implementation of a JIT computer system? How can managers be sure the batch materials system accurately reflects the status of materials in production?

There are many ways to synchronize a batch-oriented system to JIT production. The techniques depend on the most flexible system resource ever conceived—manual clerical effort. Obviously, better long-term solutions exist than reliance on pencil and paper, but when both information needs and production processes are changing rapidly, nothing is more flexible.

Simple forms can be designed and used to monitor the flow of materials on a piece-by-piece basis. Inputs can be made to the system in batches when a form is completed. Materials allocations automatically made in batches by a traditional MRP system can be manually adjusted for individual pieces. Japanese-style kanbans (or American chits) can be used to requisition parts on an individual basis, and systems transactions made only when a certain number of kanbans are accumulated.

These are just a few examples of how a manual system can temporarily serve in place of a formal materials planning and control system. The technique will depend, of course, on the operation it supports and will probably change as the process evolves. Once the operation has stabilized, the company will have a good understanding of the requirements for a materials system and will know how to customize a software system to fit its operation.

A hard-line adherent of JIT might say that a company doesn't need a computer system for materials planning and control because materials are pulled through the factory by demand. This demand is transmitted back to vendors, and materials are pulled into the factory's first operation. Although some Japanese companies do operate within guidelines that approach this theoretical limit, U.S. manufacturers will probably never feel comfortable with this type of operation. Although they may be able to replace complex computer systems with simpler ones, few will care to give up the planning capability they have today for a system driven by the often erratic demands of the marketplace.

Control of Inventory

Excuse number 4: "If we adopt just-in-time production, we won't be able to track materials through the factory with work orders. So we'll lose control of our inventory."

Accountants, production control schedulers, or expediters usually raise this objection. They are the ones who operate (and may have developed) an elaborate inventory control system. With hundreds or even thousands of batches of material in process, a system for tracking work orders is necessary to prevent chaos. Eliminating such a system in a traditional environment would mean a total loss of control.

If production is simplified, however, the tools required to monitor and control it can also be simplified. In a JIT setting, little inventory is on the floor, the flow of materials is clear, and the production cycle is short. Thus even without tracing work orders, accountants and schedulers actually have more control than in a traditional system. Inventory consists primarily of either raw materials or finished products, both of which are easy to count and value.

At Hewlett-Packard's computer systems division, the finance department was worried that control over inventory would be lost under JIT, and schedulers responsible for planning subassembly production feared they would lose the production tracking capability they felt was necessary for their work. To satisfy the finance department, we conducted an inventory of work in process at the beginning of each month. Because work in process was expected to be low, the inventory could be taken quickly. In the first month of JIT operation, the inventory took two hours. We have since reduced it to a half-hour exercise.

The subassembly production schedulers gave up their traditional scheduling method as no longer necessary. Now they do not need to maintain particular inventory levels within in-process line stocks as buffers between production operations. Calculations of safety stock and safety days no longer determine the right inventory level.

Low-Volume Operations

Excuse number 5: "We are a low-volume operation, so we couldn't benefit from JIT."

Most production operations involve low volumes. The principles of JIT are just as applicable here as in high-volume operations. Emphasis on reducing setup times, building products in smaller batches, and making things only on demand will improve a small operation as much as a large one.

In fact, a small-volume operation may find it easier to convert to JIT because it may already be making products in small batches using simple equipment with short setup times. Hewlett-Packard's computer systems division, which builds fewer than six computers a day, has realized tremendous benefits from JIT. These are, however, complex machines with many subassemblies and thousands of individual parts.

Batch Orientation

Excuse number 6: "We're a job shop, so our business is naturally oriented to batch production. We can't use JIT."

If a job shop got only unique orders whose patterns were unpredictable, just-in-time production would meet with little success (although reduced setup times could still be beneficial). Most job shops, however, have much more repetitive business than they realize. By not taking advantage of this fact, they are losing the chance to improve their operations.

As a first step to such improvements, job shops should separate their repetitive business from their unique orders and develop special production methods for the repetitive work. Equipment can often be dedicated to particular tasks as a way to eliminate the unproductive setups required to change machinery from one batch job to the next. A JIT production line set up to do this work will be much more efficient than all-purpose equipment in a batch environment. If demand for a particular product is insufficient to dedicate equipment to its fabrication, groups of dissimilar equipment in manufacturing cells can often be dedicated to making families of parts that require the same manufacturing sequence.

Use of "group technology" allows parts to be made in small batches (or even one at a time) with short setup times. The cycle time for producing a part in a small cell of closely spaced machines will be much shorter than it would be for a large batch that has to travel around the factory. The quality feedback for this cell will be much quicker, and the production operations can be more finely tuned.

Management Complacency

Excuse number 7: "Our factory is operating okay already. We don't need to put in the effort to convert our operations to JIT."

This is the most dangerous excuse of all but probably the most prevalent. It demonstrates the complacency that has already been the downfall of manufacturing in many U.S. industries and that threatens many more. Our international competitors are improving rapidly, and if we want to stay in business, we must keep pace with them. Waste and confusion in our factories harm both productivity and product quality. Converting to JIT manufacturing is an excellent way to expose problems and improve operations, but it requires top management commitment and effort.

Although commitment is always mentioned as a prerequisite for any organizational change or new program, it is especially needed when converting to JIT for two reasons. First, JIT exposes manufacturing problems. To develop an excellent factory, managers with the authority must be willing to commit resources to solving these problems. The best candidate for conversion to JIT is the organization that has a quality-control program and is already documenting its processes, measuring their performance, and eliminating problems. The computer systems division adopted statistical quality-control techniques two years before conversion to JIT began.

Second, JIT brings fundamental change to the organization's way of doing things. When a production line stops because of a product defect, the emphasis in JIT is on determining the cause of the defect and then fixing the process accordingly. No longer do production workers attempt a quick fix; instead, the focus is on continuous improvement. Workers must think not just about doing their jobs, but about doing them better. Thus training is needed to reduce fear and to enlist the cooperation of the entire organization.

Involving the Work Force

Management commitment by itself will not ensure an efficient or rapid transformation to JIT. First, the staff must have the time to implement JIT. In any organization, everybody is usually occupied by their existing jobs, and no one has time to work on improving things. Making the extra effort will be low on everyone's priority list, and progress toward JIT will be slow. The solution is to appoint a project leader who will champion JIT's implementation.

In some organizations, the person who initiated or has been most favorable toward JIT becomes the champion responsible for setting up training, making a conversion schedule, and forming a conversion team with representatives from every department. The JIT champion will have to deal with the fact that no paradigm exists so that everyone can pursue the same objectives. The best way to overcome any problems is through a common learning experience.

Such training can be tailored to the different levels in the organization. For example, a group may attend professional seminars outside the company or internal classes taught by an expert in the field. Other useful activities include jointly reading and discussing a book on JIT, jointly visiting another factory that has implemented JIT, and creating models or simulations of JIT production lines that all can observe. At least one division of Hewlett-Packard has used all these techniques either alone or in combination. The Greeley Division has

attracted some attention for the simulated production line it developed to show the differences between "push" and "pull" materials movement and the effect of moving materials in different-sized batches. The company has videotaped its demonstration of these effects, and the tape is available as the basis for a group learning experience.

Even after this conceptual training is complete and the organization has a common vision of the future, no company is prepared instantly to transform its entire factory into a JIT operation. To give everyone a better understanding of how JIT works in practice, the company should establish a pilot JIT project before converting the whole factory. In a plant with multiple products, this project can involve an entire product line (preferably not the largest). If it is impossible to separate out one product line or if the factory produces only one product, the pilot implementation can be done between two operations in the plant (preferably the last two).

The computer systems division, which builds only one major product, the 3000 Series 68 superminicomputer, adopted JIT between the last two processes in the factory. When employees felt comfortable with this arrangement, the division quickly implemented JIT back through the factory to the first production operation. Besides allowing an organization to experiment with various concepts and techniques before making commitments for the entire factory, the pilot allows mistakes to be made that are small enough to be easily fixed. Most important, it enables people to experience JIT in a real situation.

Visiting a factory using JIT is still another way to see the concept in action. If pictures are worth a thousand words, then observing another factory using JIT is worth several chapters in a book. Ideally, a company would visit a factory making the same product as it does, but if its competitors have already reaped the benefits of JIT, they will be unlikely to pass on their experience.

As the time for implementing JIT approaches, another problem will emerge: a small group of managers cannot anticipate all the many detailed changes that have to take place in production and overhead procedures. The obvious solution is to enlist more people in the implementation effort, and the right way to do this is to enlist everyone.

Nonsupervisory employees on the production line and in the office are the ones who know the most about how the system actually operates, and they probably have a lot of ideas on how to improve it. Their involvement will also make them more receptive to the change and will give everyone a greater sense of ownership of the new system.

Many organizations have recently begun to tap the tremendous potential in their first-line people by involving them in quality circles or quality-of-work-life teams. If such programs already exist in an organization, they can be built on in implementing JIT. If such involvement has not been a tradition, implementation of JIT presents a perfect opportunity to get started.

At the computer systems division, a team of managers outlined the new flow of materials through the factory. Then each supervisor and his or her people enumerated the details involved in building, testing, and moving one assembly at a time through each operation. Employees met often in teams with their supervisors and worked out the necessary operational changes.

When the changeover to JIT occurred, plans had already been made to accommodate the expected production problems. Actually, very few problems surfaced right away because the production workers had done an excellent job of planning. No group of managers, engineers, or planners could have done as well.

Staying flexible may mean having extra process engineering resources available to solve problems as they arise. It may also mean scheduling the production operation at less than its maximum capacity for a while so that changes can occur without compromising the factory's scheduled output.

The last important step in implementing JIT should be to position the organization for continuous improvement. One method of doing this is to continue to remove inventory between operations in the factory. This reduction will not only expose operational problems but also allow production departments to move closer together, reduce the effort devoted to materials handling between them, and improve communication.

Eventually, all production employees could work next to each other in a linear (or U-shaped) production flow. One person would hand material to the person performing the next production activity. This would minimize inventory and materials handling, maximize communication, and result in better feedback on quality and ideas for improvement. In addition, the teams set up to plan operating procedures should be maintained and encouraged to focus on continuous process improvement. After all, people on the production line will have the best information on how the process is operating and what can be done to improve it.

JIT cannot be quickly put in place and forgotten. Implementing it is a commitment to operate in a new way, a better way—a way that demands things be done right. This method will not forgive inattention to solving problems. Above all, JIT is not just a way to reduce inventory in order to get a better return on assets; rather, it is a means of solving the problems that block the building of an excellent manufacturing organization. Some enlightened manufacturing managers have already made the commitment to JIT. If you haven't started implementing JIT, what's your excuse?

EG&G Sealol (A)

For the past week I've been racking my brains over a problem most firms would welcome—two big orders. We just can't handle them now, but we can't turn them down.

In September 1981 Joe Robinson, manufacturing manager for EG&G Sealol, and Jim White, manufacturing engineering manager, were discussing the dilemma Robinson had posed above. Production growth at Sealol's Engineered Products Division in Warwick, Rhode Island, had exceeded 20 percent in each of the previous five years. The plant, the firm's principal facility for manufacturing mechanical seals, had been short capacity for several years and had begun routinely to subcontract excess work. Recently, two of Sealol's most valued customers, National Engine and United Jet Engine, had each received a sudden increase in orders for aircraft engines, triggered both by the demands of a new generation of medium-range passenger jets and by increased Pentagon spending on aircraft. These companies in turn placed large orders for aircraft engine gear box seals with Sealol. Such seals were high precision and made to rigid specifications. The company was recognized for its dedication to quality and precision. It was doubtful, though, that there was capacity in the Warwick plant to fulfill the orders on site, and it would be difficult for any of its subcontractors to duplicate the quality or make delivery date. Sealol could meet the requirements of either National Engine or United, but not both. Nonetheless, both companies' orders had been accepted. Moreover, the two were competitors in their industry and each demanded

This case was prepared as the basis for class discussion rather than to illustrate either effective or ineffective handling of an administrative situation. Names and some of the data have been disguised.

priority. Inspectors from both companies were permanently assigned to the Warwick facility to check the finished product's quality; they functioned, informally, as expeditors of their own orders as well. At Sealol, however, every order was given a number and flowed through the plant in sequence.

EG&G SEALOL

Over half a century ago, a small group of engineers and technicians started a company called Sealol to develop and manufacture the rotary face seal, a revolutionary improvement over the inefficient shaft packings of the time. The firm prospered during World War II, inventing the first pressure balanced seal, which quickly became standard in high-performance aircraft engines of the time. In 1968, Sealol joined EG&G, a highly diversified firm specializing in high-growth, high-technology businesses, which in 1980 was listed 428th in *Fortune*'s list of the 500 largest manufacturers in the United States and 100th in return on sales.

Sealol's worldwide operations were headquartered in Warwick, Rhode Island. The site also housed a manufacturing operation that employed approximately 750 people. The hourly employees were represented by the Sealol Shop Union (an independent union). See Exhibit 1 for an organization chart.

Sealol had the world's most complete line of mechanical seals, with products ranging from water pump seals costing pennies to sophisticated, high-temperature and corrosion-fighting welded bellows seals. It was the second largest manufacturer of such products in the world. Its Industrial Division and the Engineered Products Division served a variety of original equipment manufacturers and users in the automobile, aircraft, petrochemical, energy, electronic, nuclear, and marine industries. The Engineered Products Division had developed many original devices, supplied mainshaft seals for the U.S. Navy's nuclear submarines, and was a major supplier of seals to all jet engine manufacturers. The division had also designed 26 different bellows devices used in the first human landing on the moon.

MECHANICAL FACE SEALS

Every rotating shaft passing through a housing containing fluid under pressure must be sealed to prevent leakage. Often, a simple rubber packing or close tolerance bushing would suffice, but under severe conditions caused by speed, pressure, or temperature, these could not match the performance of a face seal. A properly designed face seal would often outlast the equipment in which it was installed. Standard Sealol seals

EXHIBIT 1   Engineered Products Division Organization Chart
January 1981

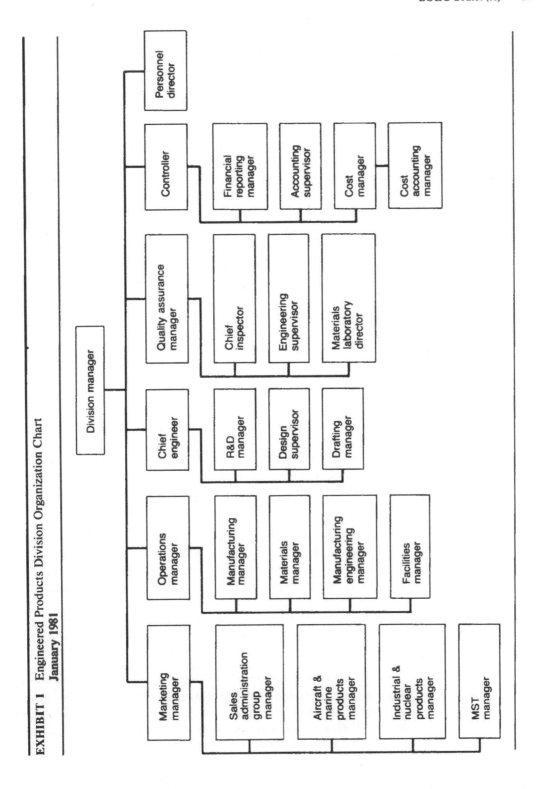

could withstand temperatures up to 1290°F, pressures to 1000 psi, and shaft speeds to 125,000 rpm. More demanding conditions could be handled by special design.

In most mechanical face seals, the seal remained stationary; it was secured in the housing by an interference fit. A rotating mating ring was mounted on the shaft. (See Exhibits 2 and 3.) Both components had an extremely flat, smooth sealing face. When the seal was installed these faces were pressed together and held in contact by compressed springs or bellows in the seal. Because of their closely controlled surface finish, virtually no fluid could pass between them.

To be sure the faces remained in intimate contact at all times, the seal had to accommodate mating ring wobble caused by misalignment or shaft runout. The springs or bellows provided flexibility to allow axial movement of the seal face, compensating for eccentricities in the rotation of the mating ring face.

There were three steps in the manufacture of an ordinary face seal:

1. Cup assembly.
2. Seal ring composite.
3. Seal assembly.

CUP PRODUCTION

The first stage of the process was to manufacture the metal casing in which the seal ring composite was to be housed. There were two methods of producing the cups.

1. For the larger seals (greater than 3″), raw material in the form of steel bar stock was first cut to length on a band saw machine, and then transferred to a lathe. There, excess material was removed, transforming the workpiece from a solid block of metal into a hollow casing of precise dimensional specification.

The degree of operator attention required in the machining operation was a function of the sophistication of the lathe. In the case of CNC tools, the operator positioned the workpiece in the chuck (holding device), called up the appropriate part program on the computer terminal attached to the machine, and watched as the part was machined. On occasion, he might be required to clear a blockage, replace a broken cutting tool, or deal with contingencies of this nature. On the manually operated lathe, however, the operator made all the adjustments to the cutting tool, workpiece, and machine based on the blueprint which was provided.

The machined cup was then transferred to the welding bay where two lugs were welded to the cup's inner surface, 180° apart. These metal lugs were necessary to hold the seal assembly in place and prevent rotation in the cup. Following welding, the cup was transferred to a surface grinding machine for removal of excess welding material.

EXHIBIT 2   Description of Rotary Face Seals

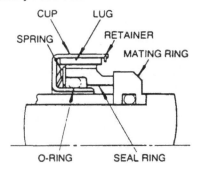

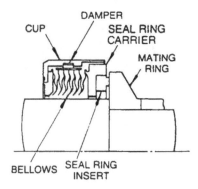

Standard Nomenclature

Cup—This is the casing of the seal itself. It may be either stamped from sheet metal or machined. The cup is statically mounted in the housing and is sealed by either an interference fit or "O" ring packing.

Lug—Lugs inside the stationary cup fit into matching slots in the seal ring. This arrangement prevents rotation of the seal ring, yet allows the axial movement necessary to accommodate shaft runout. The lugs may be welded in place or take the form of a simple clip or indentation in the cup.

Seal ring—This is the most critical member. Normally made from a suitable grade of carbon graphite, it forms the stationary sealing face that must remain in intimate contact with the rotating mating ring.

Spring—The springs provide the initial closing force that keeps the faces in contact. In bellows seals, the metal bellows acts as the spring.

Packing—In elastomeric types, an "O" ring forms a "secondary seal" to prevent leakage between the seal ring and cup. (Flexible bellows seals do not require this packing.)

Mating ring—This is the rotating sealing element.

Bellows—In welded metal bellows seals, the bellows is composed of metal discs welded together at their inner and outer diameters. The bellows eliminates the need for springs and secondary elastomers.

Damper—The damper is used in bellows seals to damp axial, radial, and torsional vibration.

Composite seal ring—In both bellows and some elastomeric seals, a carbon insert in a metal carrier is used instead of a solid carbon seal ring.

EXHIBIT 3   Advertisement Showing Build-Up of Mechanical Seals

Low Cost-High Performance
Mechanical Seals *for liquids and gases*

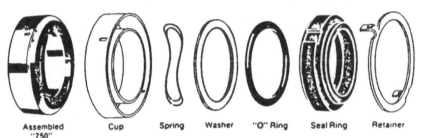

| Assembled "750" | Cup | Spring | Washer | "O" Ring | Seal Ring | Retainer |

The Sealol "750" is a low cost, high performance facetype mechanical seal, for use in sealing rotating shafts under normal operating conditions; pressures to 350 psi, temperatures from -65°F to +450°F and rubbing speeds to 10,000 fpm. Conditions in excess of these limits can be accommodated; contact your Sealol sales engineer.

This seal, within its range, has all the features of many of the most costly seals, in addition to being more compact and lighter in weight. Its weight is approximately one-half that of seals of traditional design and equivalent capabilities.

The "750" is available as a balanced or unbalanced seal, with low friction, low torque, minimum heat generation and long life. A special feature of this seal is the retaining ring which has anti-rotation, locking lugs which engage the full depth of the carbon seal ring — preventing "bayoneting" and malfunction.

The Sealol "750" is engineered to meet more than 70% of all shaft-sealing needs. Wherever pumps, compressors, motors, etc., or rotating shafts of any description — require a dependable, long-life, low-cost seal, the Sealol "750" should be considered first.

NO BAYONETING
Seal ring and retainer, showing anti-rotation locking lugs with full axial contact through the depth of the carbon ring.

2. The smaller cups (less than 3″) were stamped (cold formed) from sheet metal. Whenever a cup of a given design was produced, the hydraulic punch presses used in the stamping operation had to be setup.

This was a time-consuming operation. Consequently, these stamped cups were made in fairly large batches. They were then transferred to a second hydraulic press where lugs were formed by indenting the wall of the cup. Typically no further machining was required to the cup at this stage.

SEAL RING COMPOSITE

The seal ring composite comprised a carbon-steel spring, a steel washer, a rubber O-ring, the carbon seal ring, and a metal retainer. The spring and washer arrangement was required to compensate for eccentricities in the rotation of the mating ring face. The O-rings provided sufficient friction both to prevent seal ring rotation and to block leak paths between the seal ring and cup inner face.

Some special springs were manufactured in-house, while others were purchased from outside suppliers, as were all O-rings. The washers were stamped in a lathe to the specified level of tolerance. Usually springs and washers were made in fairly large batches.

In some seals, the springs and attachments were replaced by a bellows arrangement which also allowed for some shaft movement. More important, the bellows were designed to accommodate extreme temperature fluctuations which might have caused expansion or contraction. Essentially, the bellows were made from circular rings of metal foil welded to each other along first the inside and then the outside diameter in such a way that they acquired a spring-like quality. The bellows manufacturing process was performed in a separate section of the plant, and the process was highly proprietary and closely guarded by Sealol.

The seal ring was made by machining a block of carbon in a lathe. It was then transferred to a milling machine[1] for preparation of the faces and cutting of two slots for the lugs.

SEAL RING ASSEMBLY

In the seal ring assembly process, the cup and seal ring composite were brought together for final assembly. Once the spring assembly, O-ring, and seal had been properly inserted, a snap ring or retainer was placed over the shoulder of the seal ring to hold the seal ring composite and springs (or bellows) in place. In the case of stamped cups, the lip (or edge)

---

[1]A milling machine is similar to a lathe in many ways, except that in the latter, the workpiece rotates while the cutting tool is held stationary; in the former, the part is stationary while the tool rotates (or reciprocates in a straight line for the cutting slots, etc.).

of the cup was crimped (bent) on a hydraulic press, eliminating the need for a retaining ring.

The assembled seal was then transferred to a lapping machine where the carbon face of the seal ring was lapped (smoothed) to a high level of precision. On completion, the seal was cleaned to remove carbon dust and sent to storage.

PRODUCTION AND CAPACITY

With the large variety and the highly technical nature of many of its products, the lot sizes at Sealol were relatively small. A large run was about 200 parts at one time; the highest volume of a single product was about 5,000 per year. "We have many runs of 50 parts and many that are much smaller," said Robinson. Such a small-batch, job-shop manufacturing environment often resulted in difficult scheduling and routing problems with long lead times.

To make matters more complicated, actual orders had outstripped forecasts for the past few years. As a result, the Warwick facility constantly operated near or beyond the limits of its manufacturing capacity. To prevent layoffs, should demand decrease in an economic downturn, management policy had been to subcontract rather than allow physical expansion. Robinson summarized the recent subcontracting history:

> At the beginning we let purchasing decide what to subcontract out. Naturally, they did so without much knowledge of the production problems associated with different jobs. Often the job contracted out would be the one with the biggest lot size. We were basically at the mercy of any expeditor without regard to our production facilities. Clearly, this was wrong. We now let the supervisors decide which jobs to keep and which to subcontract. The supervisors keep those jobs which have a straightforward production line and those that do not require high setup/unit time. The rest are given out.[2]

Problems associated with such a strained operation had a snowballing effect: When one manufacturing operation had trouble, it complicated other aspects of the operation—planning, inventory, and machine tool capacity were all mutually affected. On top of all of this had come National Engine and United Jet Engine orders.

It was clear that capacity had to be expanded. However, Robinson was unsure about adding any new physical equipment. In the past, when new

---

[2]Sealol had to subcontract some of its work because it was in an Environmentally Protected Zone. The Pawtuxet River flowed by the plant and processes that might entail dumping of wastes (e.g., heat treating, hard surfacing) were always subcontracted. Also, processes which were not dependent on high technology and/or did not require great precision were often contracted out.

equipment had been acquired it had been located haphazardly wherever open space was available because relocation of existing machinery in the shop would have meant major disruptions, and the plant could ill afford a long shutdown. The result was a nightmare of parts flow across the plant (see Exhibit 4). Robinson observed:

> We have to bite the bullet this time. Any new capacity addition would require a more streamlined production flow. I need to know production flows as they exist and hope to rationalize process flows. Joe Clark (manager of manufacturing systems) keeps telling me that we need to introduce group technology cells to better control production. As a concept it sounds appealing, however, I don't know how I can pull it off.

PRODUCTION MEETING

Shortly after the two jet engine companies' orders had been accepted, Robinson called a meeting to address the capacity problem. Joe Clark, the manager of manufacturing systems, Fred Bluestone, the materials manager, and Jim White, the manufacturing engineering manager, were present. After much discussion three alternatives emerged:

1. Make standard items to inventory;
2. Expand subcontracting;
3. Introduce group technology.

The Inventory Option. Fred Bluestone observed that of the 6,300 active parts (representing 522,000 actual hours of work), 4,800 parts (representing 380,000 hours of work) were custom orders, made to the specifications of customer blueprints. Representing 130,000 work hours, 1,370 parts were standard machined parts made to Sealol design specifications, and the remainder were stamped (forged) parts. When the standard parts were ranked by volume and cumulated, a small percentage of the items accounted for the majority of the shop volume. A detailed breakdown is given below.

No. of Items	Percent of Standard Parts
150	50
300	60
600	80
1,200	98
1,370	100

Bluestone guessed that he could substantially reduce the time spent on setups if the company stopped making parts to order. A spot check for

EXHIBIT 4   Process Flow

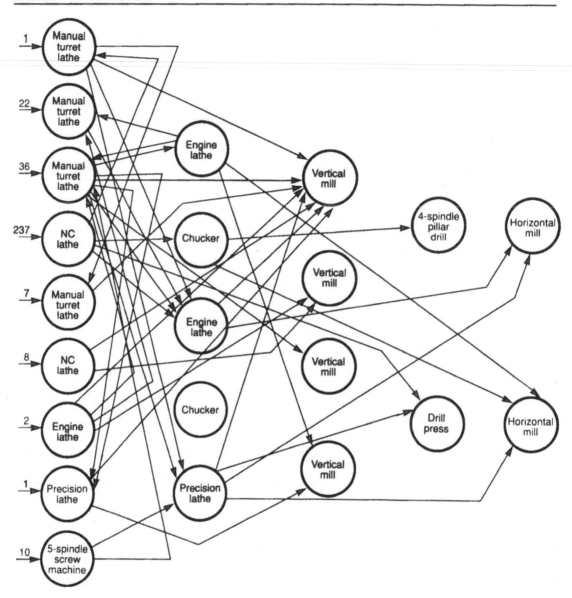

324 parts using 22 machine tools before GT

the previous month had shown that about 50 percent of the time was spent on machine setups. The order sizes for standard items ranged anywhere from 50 to 500 items with an average of about 80 items per lot. A customer ordered the same part from once to four times a year. All Sealol designs were customer-specific, but about 30 percent of the parts were common to more than one design. The average order frequency for a part by a customer was twice a year. Bluestone figured that with an ABC inventory control system and economic order quantities (EOQ), he might be able to reduce setup times by as much as 50 percent. This would provide a 25 percent increase in capacity of the plant, enough to accommodate the increase in sales. White, however, doubted that such an increase could be obtained.

The Subcontracting Option. Robinson was partial to the subcontracting option. Two of his best operators, along with an industrial engineer, had left the company to form a job shop. They were the best people he had had. He knew that if he subcontracted some high-volume items to them, they would perform to high-quality specifications and meet deadlines. He even suggested that Sealol lease some expensive NC machines to them to ensure that quality was maintained. Since Sealol was already subcontracting over 25 percent of its volume, he felt further increases in subcontracting would not lead to any control problems. White pointed out that the cost of subcontracting was $50 per hour, while direct labor costs in-house were $15 per hour (indirect labor was 50 percent of direct labor whether work was done in-house or not). For an average part costing $200 and priced at $300, the direct labor input was $50. Further subcontracting would cut into margins quite substantially.[3]

The Group Technology Option. The underlying principle of group technology (GT) was quite simple: group similar parts with appropriately similar manufacturing processes for a smooth manufacturing flow. The machines required to manufacture these parts would be located together and called a "manufacturing cell"; geographical proximity and a smooth linear flow would help achieve some of the advantages of mass production in a batch-oriented job shop. In a typical job shop, similar machines were grouped; for example, all lathes were together in one area, milling machines in another, and grinding equipment in yet another. A part to be fabricated and requiring all three operations would move from one area to another. In a typical job shop, the part spent 95 percent of the time waiting in a queue and was worked on only 5 percent of the time. It could wait in line for three weeks to be worked on for only one day. Only if all the machines for a group were in one geographical area, and if the capacity of the machines in the group were well balanced, could one hope to

---

[3]All cost data have been disguised.

reduce work in process substantially. Robinson was concerned about what group technology would mean to his operation:

> The problem I have with GT is that I cannot afford to dedicate machines to a group of products when not all machines will be fully utilized. Especially now when we are at capacity. Beside that, what really bothers me, though, is the shift from functional layout to a product focus. Our supervisors are trained to manage specific functions—turning, milling, grinding, etc. With a product focus I need a supervisor for the manufacturing cell who does all three. No matter who I choose to work in the cell, there is bound to be some elitism and with a union shop we have to tread carefully.

Joe Clark had been pushing the group technology concept. He had studied 336 different parts, some of which belonged to the United and National orders. He found that in the month of March, 40 parts, with an average lot size of 50 units, were made in-house; 12 other parts, with an average lot size of 100 units, were made outside by subcontractors. Of the units accounted for in making the 40 different parts, 30 percent of time

---

EXHIBIT 5   Process Flow for Sample of 12 Products

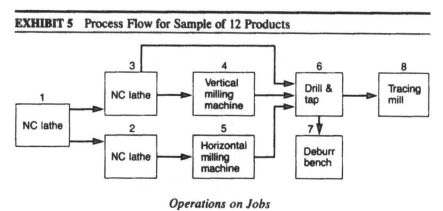

Operations on Jobs

Job # (Specific Parts)	Machine Sequence
1	1 - 3 - 6 - 8
2	1 - 3 - 4 - 6 - 8
3	1 - 3 - 4 - 6 - 8
4	1 - 2 - 5 - 6 - 8
5	1 - 3 - 4 - 6 - 8
6	1 - 3 - 4 - 6 - 8
7	1 - 2 - 5 - 6 - 8
8	1 - 3 - 6 - 8
9	1 - 3 - 4 - 6 - 6
10	1 - 3 - 6 - 8
11	1 - 3 - 4 - 6 - 8
12	1 - 2 - 5 - 6 - 7

was spent in setups, 30 percent in making the part, 10 percent in inspection, 10 percent in inspection, 10 percent in machine downtime, 20 percent in idle time. Of the total throughput, 20 percent was rework.

For a more comprehensive understanding of the process, Clark had studied one week's throughput of 12 different products. The process flows of the 12 products are given in Exhibit 5; the setup times for the key machines are given in Exhibits 6 and 7. He felt that with better control in the group technology cell, he could reduce the rework from 20 to 5 percent and cut the idle time from 20 to 10 percent. Furthermore, he would be able to sequence the jobs through the cell better.

EXHIBIT 6 Machine Setup Times on NC Lathe #1 and NC Lathe #3

Setup Times Between Jobs for a Sample of 12 Jobs on the NC Lathe #1

Job Number	1	2	3	4	5	6	7	8	9	10	11	12
1	0	1.8	1.8	2.5	.8	.8	2.5	.2	1.8	.1	1.8	2.5
2	1.8	0	.3	2.0	1.2	1.2	2.0	1.8	.5	1.8	.2	2.0
3	1.8	.4	0	2.0	1.2	1.2	2.0	1.8	.7	1.8	.1	2.0
4	2.5	2.0	2.0	0	1.0	1.0	.1	2.5	2.0	2.5	2.0	.2
5	.8	1.2	1.2	1.0	0	.05	1.0	.8	1.2	.8	1.2	1.0
6	.8	1.2	1.2	1.0	.05	0	1.0	.8	1.2	.8	1.2	1.0
7	2.5	2.0	2.0	.1	1.0	1.0	0	2.5	2.0	2.5	2.0	.15
8	.2	1.8	1.8	2.5	.8	.8	2.5	0	1.8	.05	1.8	2.5
9	1.8	.6	.8	2.0	1.2	1.2	2.0	1.8	0	1.8	.5	2.0
10	.1	1.8	1.8	2.5	.8	.8	2.5	.05	1.8	0	1.8	2.5
11	1.8	.3	.2	2.0	1.2	1.2	2.0	1.8	.6	1.8	0	2.0
12	2.5	2.0	2.0	.2	1.0	1.0	.15	2.5	2.0	2.5	2.0	0

Setup Times Between Jobs for a Sample of 12 Jobs on the NC Lathe #3

Job Number	1	2	3	4	5	6	7	8	9	10	11	12
1	0	1.0	1.0	—	1.2	1.2	—	.2	1.0	.1	1.0	—
2	1.0	0	.3	—	1.4	1.4	—	1.0	.5	1.0	.2	—
3	1.0	.3	0	—	1.4	1.4	—	1.0	.7	1.0	.1	—
4	—	—	—	0	—	—	—	—	—	—	—	—
5	1.2	1.4	1.4	—	0	.2	—	1.2	1.4	1.2	1.4	—
6	1.2	1.4	1.4	—	.2	0	—	1.2	1.4	1.2	1.4	—
7	—	—	—	—	—	—	0	—	—	—	—	—
8	.2	1.0	1.0	—	1.2	1.2	—	0	1.0	.05	1.0	—
9	1.0	.5	.7	—	1.4	1.4	—	1.0	0	1.0	.5	—
10	.1	1.0	1.0	—	1.2	1.2	—	.05	1.0	0	1.0	—
11	1.0	.2	.1	—	1.4	1.4	—	1.0	.5	1.0	0	—
12	—	—	—	—	—	—	—	—	—	—	—	0

EXHIBIT 7 Specification Sheet

PAGE 1 OF 1 DATE: 8/26/81

EG&G SEALOL ENGINEERED PRODUCTS DIVISION

SUMMARY OF OPERATIONS

EXECUTIVE ORDER NO. 11246

PAGE 1 OF 1

PART NUMBER		RV	DESCRIPTION			M.O.# PC M S.O.# GOVT.
3-1132-2-SS		01 SEAL RING SHELL				

RAW MATL NUMBER	RV	DESCRIPTION	QTY/PC	U	HEAT NO.	VENDOR & PO#
0110255	00	AMS-5640	.80	H		

SUB RAW MATERIAL DESCRIPTION	HEAT NO.	VENDOR & PO#	GOVT SOURCE INSP. MERCURY CLAUSE PHY & CHEM. HYDRO CERT. MERCURY FREE

OP NO	WORK CNTR	OPERATION DESCRIPTION TOOLS, SPECS. & DOCUMENTATION	S. TIME R. TIME	EMP F/N	QA PO#
01	0191	ISSUE FROM STOCK.	.00 .00		
05	0317	150 PCS. AND UP. PROCESS PER METHOD PRINT. TURN COMPLETE AND CUT OFF.	5.00 .014		
10	0221	FACE TO LENGTH, TURN SKIRT DIA. AND FACE FLANGE.	1.50 .01		
15	0221	BREAK TWO CORNERS AT NOSE END.	.75 .012		
17	0273	CLEAN.	.25 .005		
20	0146	STAMP TWO SLOTS. USE TOOL #902.	.50 .003		
25	0186	DEBURR SLOTS.	.50 .00		
30	0221	FINISH C'BORE. FORM UNDERCUT.	1.00 .01		
35	0070	INSPECT.	.50 .05		
40	0190	STOCK.	.00 .00		

MICLASS CODE NUMBER   BY   DATE

1833-2721-3112-   -   -   MJR 04/00/81 RT:00 CELL#:   FLOW#:

THE DECISION

Robinson asked White to evaluate the options and put together an action plan to address the capacity problem. Prior to joining EG&G Sealol, White had been involved in the initial coding of products to set up a group technology manufacturing cell for his former employer, and he believed that a similar approach was the best solution to EG&G Sealol's current problem. Nonetheless, he realized that to sell the idea to the manufacturing and materials operations, he would have to develop an extensive plan that addressed their concerns and showed sufficient economic justification. White knew he had several other issues to address beyond identifying which parts should be grouped together and which machines to dedicate to each part's grouping.

1. White's initial review of some of the part numbers that might be included in the first group technology cell indicated that there were several different machine routings for similar parts and that there was a large variance in setup and standard times (1½ to 4 hours for similar operations). He was also suspicious of some of the data coming out of the factory data collection system. The data were based on operators' keying on and off the system at the beginning and ending of a task. The system assumed that the elapsed time was completely spent on the task and, thus, did not account for indirect time. This was adjusted for by the supervisors.

2. The quality control organization would have to accept group technology totally. After each operation, a part had to be moved into a central bullpen to be inspected. To keep the work balanced within the GT cells, White wanted to keep all inspection within the cell. He felt with people working as a group in a cell he could eliminate inspection at the end of each operation and replace in-process inspection with only one inspection at the end when the part was completed. With the inspector physically present in the cell, identifying problems as they arose, rework could be reduced. This presented a potential problem with the operators, above and beyond the concerns that the quality organization were bound to raise. Due to delays in the inspection area it was often difficult to identify who made specific errors on some operations. If inspection were done within the cell, however, the cause of errors could be immediately identified and feedback given. Even though this would be a strong selling point to management, he doubted that the work force would be very positive.

3. In putting together a comprehensive plan, White also had to consider the ramifications that group technology might have on the work force and job classifications. Since the manufacturing operation was represented by the Sealol Shop Union, he had to assure that his plan did not negatively affect any labor agreements. See Exhibits 8 and 9 for a listing of the job classifications and excerpts from the labor contract.

EXHIBIT 8   Wage Rate Schedule

Labor Grade	Position	Labor Grade	Position
1	Maintenance electrician	6	Lapper bonder—large seals group leader
2	Tool and die maker		
3	Toolmaker	6	Inspector—AA1
3	Machine specialist	6	Lathe operator—engine (EPD)
3	Numerical control machine specialist—group leader	6	Lathe operator—engine (Ind/D)
		6	Assembler specialist (Ind/D)
3	BPD turning area—group leader	6	Inspector—tool and gage calibrator (Ind/D)
3	BPD machine shop—group leader		
4	Automatic chucking machine— setup and operate	6	Bellows and spot welding—group leader
4	Repair person	7	Assembler—hydraulics
4	Repair person—maintenance	7	Press operator—press room
4	Lathe operator—large turret	7	Grinder operator—blanchard
4	Lathe operator—AA turret	7	Deburring—group leader
4	Lathe operator—vertical turret	7	Bellows and spot welder
4	Tester—lab	7	Shipper/receiver/cutoff/driver (EPD)
4	Machinist—lab	7	Shipper/receiver/cutoff/driver (Ind/D)
4	Hydraulics—group leader	7	Inspector AA (EPD)
4	Numerical control lathe operator	7	Inspector AA (Ind/D)
4	Milling machinist—group leader (Ind/D)	7	Inspection coordinator
		7	Tool crib clerk
4	Grinding—group leader	7	Assembler and lapper
4	Universal grinder specialist	7	Tool control—bellows
4	Numerical control machine center (Ind/D)	7	Assembler/miscellaneous operator (PWA)
4	Vacuum oven grinder	7	Deburring—group leader
4	Bar machine operator—automatic 5 spindle	8	Utility machine operator
		8	Machine lapper—carbon and steel
5	Press operator—group leader	8	Utility person (Ind/D)
5	Universal grinder	8	Miscellaneous operator—BPD
5	Lathe operator—turret (EPD)	8	Pickler—group leader
5	Lathe operator—turret (Ind/D)	9	Pickler
5	Tool and cutter grinder	9	Machine deburring operator
5	Tool and gage inspector (EPD)	9	Stock clerk (EPD)
5	BPD—group leader	9	Stock clerk (Ind/D)
5	Repair and assembler specialist (Ind/D)	9	General floor person/tool crib attendant
5	Inspection—group leader (EPD)	10	Miscellaneous operator—K department
5	Inspection—group leader (Ind/D)		
5	K department—group leader	10	Tool crib attendant—SVD
5	Engine lathe specialist—bellows	10	General floor person
5	Mill and drill operator	10	Janitor—watchperson
6	Grinder carbon and internal	10	Helper—maintenance
6	Assembler specialist (EPD)	11	Cleaner/marker/packager

EXHIBIT 9   Excerpts from Labor Contract

ARTICLE 8.   Seniority

8.1

(a) The parties recognize and accept the principle of seniority as relating to layoffs, promotions, shift preference, and overtime on a rotation basis, and transfer of the work forces and the filling of vacancies, providing the employee shall possess the skill and ability to qualify for the basic requirements of the job.

(c) An employee may exercise seniority to bid at the same labor grade or any lower grade when there is an opening on a job other than his or her own job provided that the employee has not been awarded such a bid within five (5) years preceding the current bid.

8.10

(a) Job openings will be posted for three (3) days and filled in accordance with Section 8.1(a).

ARTICLE 10.   New or Changed Jobs

10.1   When the Company establishes a new or changed job, the following procedure will be followed:

(a) The Company will prepare a draft Job Description and present it to the Union.

(b) Union and Company will review the draft Job Description for the purpose of agreeing on a description of the new or changed job.

(c) After the Company and the Union have agreed on the Job Description, the Company will then evaluate the new or changed job using the AAIM/NMTA Manual and submit the evaluation to the Union. In the event that the Company and the Union disagree on the evaluation, the Job Description will be submitted to AAIM/NMTA for evaluation of the new or changed job.

(d) If the Union does not agree with the description as

8.13

(a) Temporary assignment is defined as job assignment whose expected duration is not more than sixty (60) days and such assignment shall not be for more than sixty (60) days. Any temporary assignments, which can reasonably be expected to extend beyond thirty (30) days, will be mutually agreed upon in advance by both parties—the Union and the Company.

(b) In the application of this section, the following understandings will apply:

1. If the transfer is to a higher rated job or one offering a greater earnings opportunity, the senior qualified employee (in the area from which the Company makes the transfer) who desires the transfer will be transferred. If no senior qualified employee desires such transfer, the Company can require the junior qualified employee to accept the transfer.

2. If the transfer is to a lower rated job, the Company will transfer the junior qualified employee in the area from which the transfer is to be made and this employee is required to accept the transfer.

6. The Company will not be required to observe seniority in making temporary transfers of an expected duration of not more than five (5) working days.

submitted to AAIM/NMTA or the evaluation, the Company may put the Job Description and Evaluation into effect. The Union shall then have the right to submit the matter in dispute to the grievance procedure. In such event, the grievance must be submitted within thirty (30) days after the operation of the new or changed job commences.

10.2 If a job changes and is grieved by the Union, and the Company and the Union disagree on the evaluation, the job will be submitted by the Company to the AAIM or NMTA for evaluation of the new or changed job.

4. Finally, White was concerned that, to make group technology successful, cooperation would be needed from the support functions. Physical relocation of equipment would mean a commitment of resources by maintenance, and rerouting parts would most likely require a major retooling effort. To keep the cells balanced, there would also have to be tight coordination with materials to decide which operations should be subcontracted.

Group Technology and Productivity

Nancy L. Hyer
Urban Wemmerlöv

Introducing a new part into manufacturing can cost from \$1,300 to \$12,000, including expenses for design, planning, and control, and tools and fixtures.[1] Clearly, if a company can reduce the number of new parts it needs, it would realize large cost savings. Consider a company that typically releases 2,000 new parts per year. If the company could substitute existing parts for only 10 percent of these new parts, it could reap an annual saving ranging from a relatively modest \$260,000 to a quite substantial \$2.4 million. But this saving hinges on one critical factor: the identification of parts that can be used with or without modification to meet the designer's need. A "group technology" manufacturing data base offers great assistance in this identification process.

Group technology (GT) is a concept that currently is attracting a lot of attention from the manufacturing community. The essence of GT is to capitalize on similarities in recurring tasks in three ways:

- By performing similar activities together, thereby avoiding wasteful time in changing from one unrelated activity to the next.
- By standardizing closely related activities, thereby focusing only on distinct differences and avoiding unnecessary duplication of effort.

- By efficiently storing and retrieving information related to recurring problems, thereby reducing the search time for the information and eliminating the need to solve the problem again.

GT offers a number of ways to improve productivity, according to studies of companies in batch manufacturing. One senior executive in the agriculture machinery business told us, "The fundamental reason for our adoption of GT was to improve cost and quality through reduction of design proliferation, response time, and work-in-process inventories through standardization and simplification of manufacturing planning and through creation of more efficient plant layouts."

Over the last two years, this company has saved more than \$9 million through GT. When 20 U.S. manufacturers using GT were surveyed, 17 of them indicated that the benefits of implementing GT equaled or exceeded their expectations.[2]

The management of manufacturing technologies represents a vital component in the com-

Reprinted with permission from *Harvard Business Review*, July–August 1984, pp. 140–149.

---

[1]Alexander Houtzeel, "Integrating CAD/CAM Through Group Technology," paper presented at the American Production and Inventory Control Society operations management workshop, Michigan State University, July 26–28, 1982; and Inyong Ham, "Introduction to Group Technology," Society of Manufacturing Engineers, technical report MMR76–03, 1976.

[2]Nancy L. Hyer, "The Potential of Group Technology for U.S. Manufacturing," *Journal of Operations Management*, forthcoming.

petitiveness of U.S. industry, one that should play a more important role in the formulation of strategic plans.[3] For this to happen, general management must become more familiar with emerging and promising technologies. In this article we discuss several such technologies, all tied together by group technology, and identify their wide-reaching applications to all areas of business operations. We describe the potential benefits that can be achieved as well as common implementation programs.

The Meaning of Group Technology

GT is, very simply, a philosophy holding that managers should exploit similarities and achieve efficiencies by grouping like problems. In most cases, a prerequisite for the recognition of similarities is a system by which the objects of interest can be classified and coded (that is, assigned symbols representing relevant information). As an analogy, books in a library catalog are classified and coded in such a way that one can easily find all books written by a particular author, covering a certain topic, or sharing the same title.

In design engineering, parts can be classified by geometric similarities using codes that contain design attributes. The purpose could be to retrieve all parts with certain features, such as rotational parts with a length-to-diameter ratio of less than 2. If one of these fits the need at hand, the engineers can thereby avoid having to design a new part.

Similarities between parts, captured in the GT code, can in like manner be used by manufacturing engineering, manufacturing, purchasing, and sales. For example, a manufacturer can drastically reduce the time and effort spent deciding how a part should be produced

if this information is available for a similar part.

A GT data base is a computerized filing system that speeds up the retrieval of parts information, facilitates the design process, improves the accuracy of process planning, aids in the creation and operation of manufacturing cells, and enhances the communication between functional areas.

An early use of GT was documented in the Soviet Union in the 1940s. It has since been implemented in many European and Asian countries, mainly in the manufacturing area. Interest among U.S. manufacturers took root in the mid–1970s, and by now many large corporations (John Deere, Caterpillar, Lockheed, General Electric, Black & Decker, and Cincinnati Milacron are a few) have taken advantage of GT or are planning GT programs.

The expansion of computer capabilities and the availability of software obviously have abetted the growth of GT applications. Storing and retrieving codes with 20 or 30 characters are unthinkable without the aid of a computer. But with advanced employment of the computer in any area of production operations also comes the need for coding and classification as a way to integrate tasks and even organizational units. This is the reason why many experts see GT as the missing link between CAD and CAM (computer-aided design and computer-aided manufacturing) and thereby as an important building block for CIM (computer-integrated manufacturing).

Classification and Coding

Among the range of materials that manufacturers handle—raw materials, purchased components, fabricated parts, subassemblies, and complete items—GT is predominantly applied to purchased items and fabricated parts. We will concentrate our discussion on these groups.

When engineers are classifying parts and as-

---

[3]See Alan M. Kantrow, "The Strategy-Technology Connection," *Harvard Business Review,* July–August 1980, p. 6.

signing those with closely related attributes to a particular family, they can determine similarities between items in several ways. From a design standpoint, for example, similarity can mean closely related geometric shapes and dimensions. From a manufacturing point of view, similarity between two parts means that they are processed through the factory in the same or almost the same way. Of course, parts that look alike are not always produced in the same way (it depends on variations in raw materials, tolerances, dimensions, and so on), while parts that are routed through the same machines can be quite dissimilar in geometric form.

Simple, informal parts classification techniques have been employed in companies where the sole intent was to identify families with similar manufacturing requirements to create dedicated lines or cells of machines. The Langston Division of Harris-Intertype Corporation in Camden, New Jersey, one of the early users of GT in the United States, took Polaroid snapshots of every seventh of some 21,000 fabricated parts. When inspected from a production processing point of view, about 93 percent of the sample could be allocated to five part families.[4]

While informal ways of grouping parts are not uncommon, the greatest potential of GT comes via a formal coding system in which each part gets a numeric or alphanumeric code describing the attributes of interest. For the widest use, the code should be able to describe the part from both a design and a manufacturing point of view. Such characteristics as the external and internal shapes, dimensions, and any threads, grooves, and splines describe the geometric form. The shape and chemistry of the raw material, the surface finish and tolerance requirements, the need for special processes like heat treatment, and parts demand—all these are manufacturing attributes. Exhibit 1 offers a simple example of a coded part. Clearly, to capture all significant attributes, a large number of characters is needed.

Numerous coding systems have been developed all over the world by university researchers and consulting firms and also by corporations for their own use. A handful of commercial systems is available on the U.S. market. Of all these, many have a short range of applications, such as coding sheet metals or forgings only. Modern systems, however, are often computerized, and coding takes place by having the planner work in a conversational mode with the computer, responding to a series of questions on the CRT screen.

Once the parts have been coded, classification, which simply means the grouping of parts with similar characteristics, follows. For example, one family could consist of all parts with a 2 in the third position of the code. Another could be of those with a 3 in the third position and a 2 in the fourth. It is easy to see that with a code length of 30 digits, it is possible to create a very large number of families.

Applying GT. . .

Although many areas of business operations can benefit from GT, manufacturing, the original application area, continues to be the place where GT is most widely practiced. Two important tasks in manufacturing planning and manufacturing engineering are scheduling and process planning. Job scheduling sets the order in which parts should be processed and can determine expected completion times for operations and orders. Process planning, on the other hand, decides the sequence of machines to which a part should be routed when it is manufactured and the operations that should be performed at each machine. Process

---

[4]Ken M. Gettelman, "Organize Production for Parts—Not Processes," *Modern Machine Shop*, November 1971, p. 50.

EXHIBIT 1   Simple Example of a Coded Part

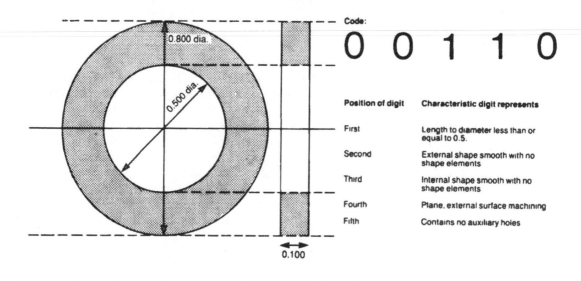

Code:

0 0 1 1 0

Position of digit	Characteristic digit represents
First	Length to diameter less than or equal to 0.5.
Second	External shape smooth with no shape elements
Third	Internal shape smooth with no shape elements
Fourth	Plane, external surface machining
Fifth	Contains no auxiliary holes

Note: Based on the first five digits of the Opitz coding system.

planning also encompasses tool, jig, and fixture selection as well as documentation of the time standards (run and setup time) associated with each operation. Process planning can directly affect scheduling efficiency and, thus, many of the performance measures normally associated with manufacturing planning and control.

. . . In Production Planning and Control

Grouping parts with similar manufacturing characteristics into families will reduce the time spent on setups of parts and tools. In small- to medium-batch manufacturing, striving for setup reduction is most important. This type of parts production usually is carried on in a job shop environment where general purpose machines are grouped according to function, such as lathes in one cluster and grinders in another. Job shops also usually have high

work-in-process inventories, long lead times, and an extremely low productive use of the time a part spends on the shop floor (normally no more than 5 percent of the total shop time). The following are among the ways GT can be carried out in production planning.

Sequencing of Parts Families. The simplest—and a highly informal—application of GT in a job shop setting is to sequence similar parts on a machine. This procedure, followed daily by foremen in most machine shops, often means overriding formal dispatch lists, which are made up with no consideration of efficiency. The saved setup time from running two or more related parts in a row can be converted to productive time. A more sophisticated application is the creation of families of parts (using the GT code) and the dedication of machines for exclusive processing of families. This approach has several advantages. First, there are fewer interfering flows of material at each machine. Second, setup time is reduced since common tooling and fixtures can be developed to handle all members of each family processed at the work station. Third, the quality of parts can be improved, since the variety of parts flowing through the work station has been reduced.

Cellular Production. The most advanced GT application is through the creation of manufacturing cells. A cell is a collection of machine tools and materials-handling equipment grouped to process one or several part families. Preferably, parts are completed within one cell. (The Japanese make much use of such cells, but apparently without formal classification and coding systems.) The advantages of cellular manufacturing are many, especially when the cells are designed with one dominant materials flow and with a fixed conveyor system connecting the work stations. A cell represents a hybrid production system, a mixture of a job shop producing a large variety of parts and a flow shop dedicated to mass production of one product. Exhibit 2 illustrates the difference between a job shop, based on a functional layout, and a cell shop.

The allocation of equipment to a subset of parts will reduce interference, improve quality, make materials handling more efficient, cut setup and run times, and therefore trim inventories and shorten lead times. Shortening parts manufacturing lead times can reduce the response time to customer orders and thus lead to smaller finished-goods inventories as well. These benefits are likely to be greater with a physical rearrangement of machinery into cells.

One U.S. manufacturer, EG&G Sealol in Warwick, Rhode Island, found that after producing 900 parts (representing about 30 percent of all standard hours-in-the-factory) in manufacturing cells, work in process dropped by 20 to 30 percent and the need for floor space declined by 15 percent. For one example, Sealol turned out 324 parts in one cell with seven machines, whereas before the parts had been routed to 22 machines. All of these improvements contributed to a 150 percent rise in total output.[5]

Management of Otis Engineering in Carrollton, Texas, estimated that at one time it had spent $5 million a year on setups. The magnitude of the potential saving is shown by the result of Otis's first cell installation—reduction in setup time of 35 percent.[6]

Cellular manufacturing offers other advantages too. The factory layout change has organizational and behavioral implications. Otis Engineering, for example, achieved a more efficient use of supervisory personnel, and equipment operators gained flexibility and thereby job enrichment. Otis also established centralized tool and gauge storage for the cell to permit easier access to the tools and better tool scheduling.

Managers can simplify production planning and control by considering the cell as one planning point for which capacity planning can be performed and to which jobs can be released. Cells commonly have more machines than operators, which means that the operators must balance the load in the cell. This, of course, represents a decentralization of tasks, requiring operators to handle several machine tools and processes. Further, with cellular manufacturing, tracing a part to its origin is easy, which facilitates accountability for quality. Management can exploit this by assigning the responsibility for quality inspection to the cell operators.

---

[5]James A. Nolen, "Cellular Manufacturing at EG&G Sealol," paper presented at a Society of Manufacturing Engineers seminar, Dallas, Texas, May 10–12, 1983.

[6]William H. Oliver, "Actual Implementation of Group Technology in a Machine Shop Environment," Society of Manufacturing Engineers, technical paper MF78-952, 1978.

EXHIBIT 2   Movement of Parts through a Job Shop and a Cell Shop

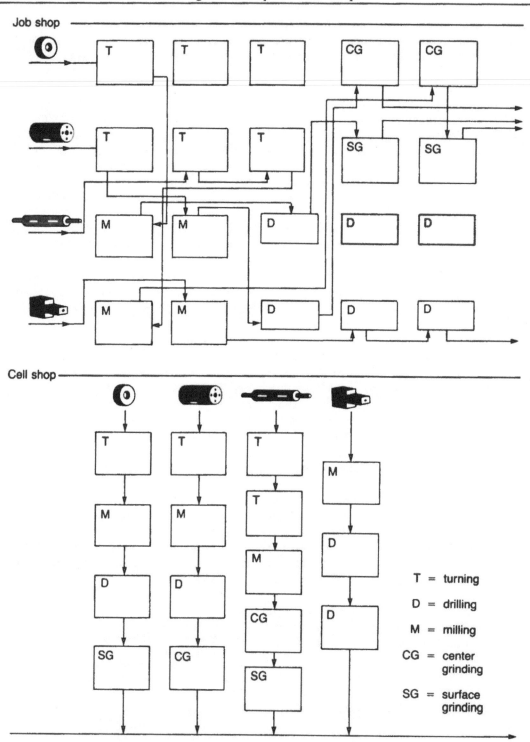

The capacity for wider task variety and the need for higher skill levels are features of the cellular approach, together with the opportunity for teamwork and the focusing of the production process from raw material to finished part. These advantages can add to operators' job satisfaction, which can lead to higher productivity and better quality.

Manufacturing cells also change the tasks production planners, schedulers, and manufacturing engineers perform and, most dramatically, alter the role of the foreman. Where before he or she was responsible for only one process, the foreman now supervises production of an entire part. This, too, affects accountability. In a job shop environment it is always possible to pass the buck by blaming other foremen for not having parts ready on time. With cells, timely completion becomes a responsibility solely of the cell foreman.

By mechanizing and automating the materials handling and the manufacturing process, engineers can create unmanned cells based on GT principles. A robotic work cell designed to process a small set of part families, for example, consists of computer-controlled machine tools located around one or more materials-handling robots. Since there is no fixed machine sequence, this type of cell can be quite flexible. A somewhat less flexible and often much larger cell, designed for higher volumes and more specialized parts, is called a flexible manufacturing system.

. . . In Process Planning

Some of the largest productivity gains have been reported in the creation of process plans that determine how a part should be produced. With computer-aided process planning (CAPP) and GT it is possible to standardize such plans, reduce the number of new ones, and store, retrieve, edit, and print them out very efficiently.

Process planning normally is not a formal procedure. Each time a new part is designed, a process planner will look at the drawing and decide which machine tools should process the parts, which operations should be performed, and in what sequence.

There are two reasons why companies often generate excess process plans. First, most companies have several planners, and each may come up with a different process plan for the very same part. Second, process planning is developed with the existing configuration of machine tools in mind. Over time, the addition of new equipment will change the suitability of existing plans. Rarely are alterations to old process plans made. One company reportedly had 477 process plans developed for 523 different gears. A close look revealed that more than 400 of the plans could be eliminated. Another company used 51 machine tools and 87 different process plans to produce 150 parts. An investigation determined that these parts could be produced on only 8 machines via 31 process plans. Process planning using CAPP can avoid these problems.

Process planning with CAPP takes two different forms:

With *variant-based* planning, one standardized plan (and possibly one or more alternate plans) is created and stored for each part family. When the planner enters the GT code for a part, the computer will retrieve the best process plan. If none exists, the computer will search for routings and operations sequences for similar parts. The planner can edit the scheme on the CRT screen before printout.

With *generative* planning, which can but does not necessarily rely on coded and classified parts, the computer forms the process plan through a series of questions the computer poses on the screen. The end product is also a standardized process plan, which is the best plan for a particular part.

The variant-based approach relies on established plans entered into the computer memory, while the generative technique creates the process plans interactively, relying on the same logic and knowledge that a planner has. Generative process planning is much more complex than variant-based planning; in fact, it approaches the art of artificial intelligence. It is also much more flexible: by simply changing the planning logic, for instance, engineers can consider the acquisition of a new machine tool. With the variant-based method, the engineers must look over and possibly correct all plans that the new tool might affect.

CAPP permits creation and documentation of process plans in a fraction of the time it would take a planner to do the work manually and vastly reduces the number of errors and the number of new plans that must be stored. When you consider that plans normally are handwritten and that process planners spend as much as 30 percent of their time preparing them, CAPP's contribution of standardized formats for plans and more readable documents is important. CAPP, in effect, functions as an advanced text editor. Furthermore, it can be linked with an automated standard data system that will calculate and record the run times and the setup times for each operation.

CAPP can lead to lower unit costs through production of parts in an optimal way. That is, cost savings come not only via more efficient process planning but also through reduced labor, material, tooling, and inventory costs. One manufacturer of lampmaking machine tools has developed an eight-digit coding scheme allowing selection of standard processes to produce certain components. This application has led to a 76 percent improvement in manufacturing productivity. At the same time, the use of CAPP boosted the process planners' productivity by 30 percent.

GT can help in the creation of programs that operate numerically controlled (NC) machinery, an area related to process planning. For example, after the engineers at Otis Engineering had formed part families and cells, the time to produce a new NC tape dropped from between 4 and 8 hours to 30 minutes. The company thereby improved the potential for use of NC equipment on batches with small manufacturing quantities.[7]

. . . In Parts Design

GT coding of parts is useful for the efficient retrieval of previous designs as well as for design standardization. These features help speed up the design process and curb design proliferation.

It is not unusual for a company to find several versions of basically the same part during a preliminary investigation of the part population. The parts can serve the same function but differ in terms of tolerances, radii, and so on. General Dynamics' Pomona Division, for example, came across a case where a virtually identical nut and coupling unit had been designed on five different occasions by five design engineers and then drawn by five draftsmen.[8] These parts were purchased from five suppliers at prices ranging from $.22 to $7.50 each. The company also investigated 2,891 parts with different part numbers and discovered that the number of distinct shapes leveled out fairly quickly to comprise a population of only 541 shapes.

Design proliferation of this kind occurs because of difficulties with design retrieval. While a part similar to the one that is needed may already exist, the designer has neither a system nor the patience to find it. It is easier to create a new part, which then means that a new part number must be assigned, a new process plan made up, new tools designed, and so on.

The aim of design standardization is to re-

[7]Ibid.

[8]Raymond J. Levulis, "Group Technology," K. W. Tunnell consulting company report, (Chicago: K. W. Tunnell, 1978).

duce variations, to make the parts efficiently, and to require justification for deviations from norms. Standardization does not mean that all parts with the same function must be identical. It does mean, however, that norms are established for tolerances, dimensions, angles, and other specifications. Setting these norms should be done with both manufacturing and design considerations in mind, bridging the gap between these two areas and making design engineers more aware of manufacturing costs and restrictions.

A GT coded parts population simplifies the cumbersome job of sifting through old drawings to find an already designed part. The designer can enter on a CRT a partial code describing the main characteristics of the needed part. The computer will then search the GT data base for all items with the same code and list them on the screen. The designer can go through the specifications of each part and select one that fits or can be modified. With modern computer graphics, each designed part can be displayed on the CRT so that the designer can inspect it. Once a part design has been selected or "edited," the designer can make the actual drawing manually or by computer.

In one company, the retrieved drawing of an already existing spur gear required only slight modification for a new design.[9] The overall saving in design time, manufacturing planning time, and tooling requirements was estimated to exceed $10,000. In the same company, an analysis of a shaft family revealed that all eight shafts used three different undercuts. By adopting a design standard to allow only one undercut design for parts in the family, the company saved about $12,000.

... *In Other Areas*

GT can also be applied in purchasing. Relying on the GT coding of purchased components and raw materials and on information from the production planning system, a purchasing manager can obtain statistics not directly available with a traditional parts numbering system. GT can help reduce proliferation of purchases of different kinds of parts, for example, by identifying components that serve the same function. It can also list identical parts for which designers have specified different brands. Companies that reduce the number of different parts they order and the suppliers they do business with can use the increased volume as leverage to negotiate better deals.

The Aerospace Group of VSI Corporation, which produces engine nuts, purchased blank slugs based on part number demand.[10] By using a coding system, the company found that fewer different parts could be purchased at higher volumes. This resulted in an average reduction of the purchase price per unit from 22 cents to 20 cents. This small reduction per unit, multiplied by the 4.8 million pieces purchased in a year, resulted in an annual saving of $96,000.

Another interesting application is in sales. The same company received a request for immediate delivery of an engine bolt that was not a stock item. A search of the GT data base, however, turned up a substitute part that fit the customer's need and could be delivered right away.

GT can also be used for cost estimation. A company that needs product cost estimates for bidding purposes, for example, can tentatively code the required parts and then search the

[9]W. Stephen Bucher, "An Integrated Group Technology Program," Society of Manufacturing Engineers, technical paper MS79–977, 1979.

[10]Harry G. Smart, "Group Technology and the Least Cost Method," Society of Manufacturing Engineers, technical paper MSR80-05, 1980.

GT data base. For parts falling into established families, standard cost data might already exist. If not, the CAPP system can help to determine the processes needed to manufacture the part, thereby arriving at cost data. Several companies have found that GT-generated cost estimates can be constructed more quickly and with greater accuracy than those made by traditional methods. The approach is also helpful during the design process to help select components that will lower the total cost of the proposed product.

GT can also assist in determining the economic consequences of anticipated changes in materials cost. Assume, for example, that the price of an already expensive alloy is expected to rise. With GT coding, a list of all parts that use this alloy can be produced within minutes, permitting a swift assessment of how the increased purchasing cost will affect the manufacturing cost of products made with the parts.

Implementing GT

GT is a philosophy calling for simplicity and standardization. Any serious attempt to take full advantage of it begins with the selection of one or more coding systems (each type of material can conceivably have its own coding system) and the subsequent coding of the material. As with any formal information system that requires changing ingrained methods and old procedures, GT cannot be casually decided on or instantaneously implemented. A GT program could require two or three years to install and will have far-reaching ramifications inside the organization, particularly if cellular manufacturing is instituted.

We recently surveyed 20 U.S. manufacturers using GT to discover the problems they had encountered during the implementation process.[11] The most common problems fell

---

[11]Hyer, "The Potential of Group Technology."

into three categories: organizational change and associated human resistance, classification and coding of parts, and planning and execution of the manufacturing cell concept. The discussion that follows draws to a large extent from our findings.

Resistance to change, of course, is a universal problem in any organization. Resistance can take different forms, depending on the employee's perception of job status and security, understanding of the new situation, and ability and willingness to adapt. In the case of GT, some examples illustrate the range of potential problems:

• The mandate for designers to reduce the number of new parts can conflict with a company's long-standing evaluation and reward system. In one company, designers had been evaluated on the number of new drawings they created. Instituting the variety-reduction concept, therefore, necessarily meant changing the incentive system. (Installing CAD by itself can lead to a surge in new parts, simply because of the speed with which new designs can be produced.)
• In manufacturing, problems commonly stem from the changing roles of operators and the new areas of responsibility for supervisors. Working as a team and participating in decision making puts employees in a new sociological setting. Operators should be able to move from work station to work station and to perform quality inspection. This requires additional skills and constitutes a new job design. Both workers and labor union representatives often resist these changes. The foreman's role also expands to cover many functions. The foreman must, therefore, know several manufacturing processes instead of only one and be responsible for the completion of the whole part and not for only a single operation.

Extensive education about GT concepts, hands-on training, and the early involvement of the affected individuals are the best ways to implement new work roles. Selling the idea of GT cells to labor unions can require great efforts, including restructuring payment systems. Personnel policies and training systems must change as well. Because of the different requirements of working in a cell, companies commonly rely on volunteers when forming teams. The Alfa-Laval company in Sweden lets the workers form their own teams and then trains them by simulating the movements and coordination inside the manufacturing cell on a scale model.

- Production planners and schedulers are also directly affected by cellular manufacturing. Once a cell has been established, parts not belonging to the appropriate family must be routed elsewhere. Some companies have found that planners have a tendency to break the family rule: They schedule a part to the cell containing machines they consider to be the most efficient for its production or to cells that have become so efficient that they seem underused. It is difficult to keep the new system up, and lapses like these can destroy the system's integrity and lead to backsliding to old ways. Regular monitoring should accompany the period immediately following a changeover, when these lapses are most likely to occur.

- The most common problem companies face in the area of coding and classification stems from the inability of codes to describe the material adequately. Some companies, for example, found that their coding systems were suitable for design but not for manufacturing purposes. A manager of one company commented that its coding system could not handle electrical parts. In fact, about two-thirds of the companies with externally developed systems had made modifications to suit their own needs. This finding reflects two facts. First, codes to handle both design and manufacturing are relatively new, and second, companies often need to develop procedures that reflect their own ways of doing things.

- One direct implication of cellular manufacturing is that the more rigid, flow-oriented system with drastically reduced work in process requires a stronger emphasis on machine maintenance. Cell formation, however, also creates a visibility that does not exist in job shops. This visibility makes it easier for managers and supervisors to identify load-balancing problems in the cells. When a manufacturing cell is designed, one obvious goal is to achieve a high utilization of all machines in the cell. The result is usually that one or two machines end up being bottlenecks, while the others are underused.

The load-balancing problem, also affected by the mix of parts entering the cell, can be alleviated somewhat by the way parts and part families are released to and sequenced through the cell. At least one company tried to attack this problem through its production planning and control system. Interestingly, however, the integration of GT cells and a scheduling system like MRP (materials requirements planning) can cause a whole new set of problems. These derive from the fact that an MRP system focuses on the completion dates of individual parts and assemblies, while cellular manufacturing focuses on the efficient production of part families. Bringing these two systems together will require new procedures for planning and scheduling.[12]

Another problem concerns buffer inventories, which usually stack up between work stations, and the operators' ability to eliminate these inventories in a balanced fashion. The unbalanced work load can become an advantage, however, by focusing capacity planning on these few bottlenecks. Even if the addi-

---

[12]See our article, "MRP/GT: A Framework for Production Planning and Control of Cellular Manufacturing," *Decision Sciences*, vol. 13, no. 4, p. 681.

tional, dedicated machines cause a lower overall machine usage, the increased throughput times, lower inventory levels, increased productivity, and higher quality associated with cells represent a net gain. One company actually reduced its machine population by 25 percent after switching to cellular manufacturing because of a significant increase in efficiency.

Two things, finally, must be pointed out regarding cellular production. First, there are, at present, no established and widely accepted formal procedures for creating cells. It is clear, however, that in the future computer simulation will increasingly be used for this purpose. Second, if cell manufacturing is implemented in an existing plant, the expectation that all parts can be allocated to part families and manufactured in production cells is unrealistic. Instead, the converted plant will be a mixture of a job shop and a cell shop, where the job shop area retains the flexibility to handle odd parts.

The Cost-Benefit Picture

Companies that have implemented GT have reported that it can produce impressive benefits. In the 20-company survey previously mentioned GT was given credit for reduced tooling and fixture expenses, reduced materials handling costs, reduced production planning and control efforts, reduced need for floor space, reduced lead times, reduced work-in-process inventories, improved quality, increased worker satisfaction, reduced design effort, easier design retrieval, and easier and more accurate cost estimates. One company stated that the rationale for using GT was "to simplify operations so they come within the bounds of human comprehension." Another company suggested that GT's most significant contribution was to provide "a tool for understanding what we manufacture."

Although manufacturing cells can be established without previous coding of the parts, coding systems are definitely necessary for a wider use of GT. Selecting coding systems and coding the parts is a long and costly exercise. When this milestone has been passed, however, the code opens a wide range of possible uses. Otis Engineering, for example, reportedly spent 18 months training people and analyzing parts populations with no apparent benefits. The company, however, quickly recaptured the cost of the preparation during the first nine months of manufacturing operations based on GT principles.[13]

Once coding has been done, it can be used in connection with design retrieval, CAPP, cell formation, purchasing, and tooling development. The more experience a company gathers, the more it can take advantage of the GT concept, and the more satisfied it will be with the result. According to the survey, an absolute majority of the GT users were reaping benefits that met or exceeded their initial expectations. And users were generally more satisfied with GT the longer they had been involved with it. Almost all of the companies had plans for expanding their use of GT, for example, by increasing the number of manufacturing cells, installing CAPP, or coding all material in their inventory files.

To make any type of investment worthwhile, the expected benefits must exceed the expected costs. According to one senior-ranking executive, few U.S. businesses have adopted GT because of a misperception of its cost-benefit picture. "The real causes of excessive manufacturing costs are either not understood at all or only poorly understood by

[13]Robert Alton, "Group Technology," paper presented at the American Production and Inventory Control Society operations management workshop, Michigan State University, July 26–28, 1982.

most manufacturing managements," he says. "Therefore, the benefits of the group technology approach to design and manufacturing are not appreciated. On the other hand, the time and money needed to adopt GT are readily apparent and discourage managers who cannot visualize the benefits."

In a recent Harvard Business Review article, Bela Gold warned against the uncritical use of traditional capital budgeting techniques in connection with CAM projects.[14] An investment in modern manufacturing technology should be a piece of a larger manufacturing system, in which each new part creates tangible and intangible synergisms. The system-wide approach also requires a broad level of analysis and a longer planning horizon. GT and associated technologies should be analyzed the same way. If the rewards are long-term, short-term financial projections are simply inappropriate—particularly if the ultimate objective is to create an integrated manufacturing system.

Implications for Management

The world of manufacturing is changing rapidly, due both to new applications for the computer developed largely in the United States and to unorthodox views of management principles and systems often taken from abroad. These changes put pressure on management to acquire new skills. Emerging technologies require executives to have at least a working knowledge of their applicability and the role they will play in corporate strategy. Top management cannot abdicate its responsibility and delegate these important decisions.

Launching a GT program is a major undertaking with long-term implications for the organizational structure and the people in it. It will be costly, but it will also make contributions that grow over time. GT is another example of the need for management to include technology in strategic decision making. As was the case in the past with installations of management information systems, so today in the case of GT, top management support and absolute commitment are critical for successful implementations, particularly because better communication and coordination between departments necessarily evolve as a result of the GT concept. Computerized information technologies represent an opportunity for many manufacturers to establish a competitive edge. The learning process is long, so there is no time to waste.

---

[14]Bela Gold. "CAM Sets New Rules for Production," *Harvard Business Review*, November–December 1982, p. 88.

Rio Bravo Electricos, General Motors Corporation

Despite the car's air conditioner, Alfonso Vazquez perspired impatiently in the line of cars waiting to pass through customs back into El Paso, Texas from Rio Bravo, Mexico. He was in a hurry to buy transparencies in El Paso as they were not readily available on the Mexican side. The phones had not been working again at the plant, so he had not been able to find out when the office supplies store would close. Tomorrow he was to make a presentation on his study of the centralization of the capital-intensive, lead preparation area of the plant. He would make his recommendations to the Latin American operations staff of Packard Electric Division of General Motors Corporation.

This was August 1981, the summer after Vazquez's first year of the MBA program at a well-known eastern business school. Previously, he had worked for three years in engineering at General Motors. In the summer of 1981, he was assigned to operations for the first time, studying the manufacturing process and organization at the one-year-old plant in Rio Bravo. His fluency in Spanish and his desire to work in operations had been key in this assignment. He was anxious to do well.

This case was prepared by Professor Roy Shapiro as the basis for class discussion rather than to illustrate either effective or ineffective handling of an administrative situation.

GENERAL MOTORS' STRATEGY FOR THE 1980s

Despite lagging auto sales in the early 1980s, GM had strengths that other automakers did not: a larger volume of sales over which to spread research and new product development, and practically no long-term debt. Based on these, GM's strategy for the 1980s was to continue to produce a full line, to improve quality, and to fund and develop new products and manufacturing techniques. Quoting from the *1981 General Motors Public Interest Report:*

> . . . General Motors has embarked upon an aggressive program to redesign nearly all its cars and most of the plants that produce them in the first half of the decade.

PACKARD ELECTRIC DIVISION, GENERAL MOTORS

Packard Electric Division was one of the major component divisions within General Motors, supplying automotive electrical systems including wiring harnesses (bundles of electrical cables with connectors designed to be readily inserted into the automobile on the assembly line), connectors, electronic modules, and wiring expertise to the numerous domestic car divisions and foreign GM subsidiaries as well as a small number of nonallied customers. Packard's main function was to supply harnesses in high volumes at just the right times to supply the largest manufacturing process in the world, General Motors automobile assembly system. It was highly integrated vertically, making complete wiring systems from raw copper cathode and train carloads of raw plastic. For example, Packard processed 100,000 pounds of plastic resin, 133,000 pounds of copper and 25 million feet of cable *daily*, producing a total of 29 million parts *daily* which were shipped to some 3,630 different destinations each month.

Based in Warren, Ohio, Packard had been expanding geographically during the 1970s to support a growing product line.

THE MEXICAN MAQUILAS

In Mexico, the word *maquila* (mah-KEE-lah) referred to the toll charged by small local mills for grinding corn into meal. In the late 1960s and 1970s, "maquila" began to be used to refer to the practice, sanctioned by the Mexican government, by which American firms were allowed to ship components into Mexico, add labor value, then ship the finished products back across the border. A small U.S. duty was charged on the value added in Mexico. American firms gained due to the lower cost of Mexican labor; Mexico gained the inflow of foreign exchange wages from the United States. Squeezed by rising labor costs, Packard saw the maquila as a way of saving on labor-intensive aspects of their business.

The first maquila experiments ended badly when a few unscrupulous textile companies cheated the locals by not paying the promised wages, but, by 1981, the maquila concept was well regarded by parties on both sides of the border. There were several major American corporations with new and modern plants in the El Paso/Rio Bravo area. Because of shipping expenses and the cost of training, usually only the labor-intensive operations were moved to Mexico. This implied that firms use whatever machinery they did install as efficiently as possible.

PLANT ORGANIZATION: CENTRALIZED OR MODULAR?

There were two Packard plants in the area, Rio Bravo Electricos (RBE), which had been built in 1980, and Conductores Y Componentes Electricos, a three-year-old plant. Both plants were considered extremely successful by all measures: cost, delivery (despite the shipping distances), and quality. RBE had the highest quality index of any North American or Mexican Packard plant. Both plants had a decentralized "module" organization which was also reflected physically in the plant layout.

The module system was a way to break up a plant into hopefully more manageable smaller plants within the same building. Rio Bravo currently had three modules, with a fourth about to be added in vacant plant space.

RBE's organizational hierarchy consisted of a plant manager (Gary Richardson, the only American at RBE other than Vazquez), a module manager for each module, a general supervisor on day shift for each module, and three first-level supervisors per module on each shift. Within each module, one supervisor ran an area called lead preparation (lead prep) and the other two each ran a final assembly area. Exhibit 1 shows the decentralized RBE organization. Exhibit 2 is an example of the proposed centralized organization. Exhibit 3 shows how the modular organization was reflected in plant layout.

Vazquez had been assigned to RBE to work for Richardson during the summer. He was to act as an outside consultant in the sense of providing an objective analysis. Richardson was one of Packard's most successful young managers. He had worked with both centralized and decentralized plants in Ohio and now felt that lead prep should be centralized to save on expensive machinery. He asked Vazquez to make his own study and include an analysis of the relevant cost differences.

During his first week at RBE, Vazquez was invited to a staff meeting where the centralization question was to be discussed. He found that several managers had strong opinions one way or the other, but John Wilson,

EXHIBIT 1 Present Modular Organization

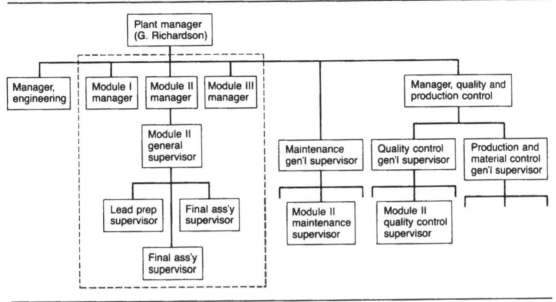

EXHIBIT 2 Proposed Centralized Organization*

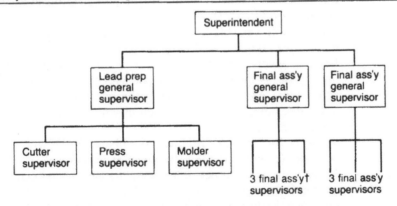

*Replaces that section of the organization chart of Exhibit 1 enclosed by dotted lines
†Three separate supervisors were needed due to the policy which limited the number of workers who could report to one supervisor.

EXHIBIT 3   Schematic of Modular Plant Layout*

Module I final assembly	Module II final assembly	Module III final assembly
↑	↑	↑
Module I lead prep	Module II lead prep	Module III lead prep
↓	↓	↓
Module I final assembly	Module II final assembly	Module III final assembly

*Arrows represent directions of product flow.

the director of Latin American operations, said he was still unconvinced and wanted more facts before he would commit himself. Gary Richardson made the best arguments for centralizing lead prep, emphasizing that the main reason for coming to Mexico in the first place was to reduce costs. Leonardo Ortiz, the manager of the Conductores plant, argued well for the module system. He had developed within that system and emphasized that it fostered teamwork. Furthermore, he pointed to the very successful record of both plants.

Ortiz, and others who supported retaining modularity at RBE, argued that, because problems within the module had to travel only one organizational level (from supervisor to general supervisor) for integration, the organization was able to respond quickly in a production crisis. Furthermore, they argued that although each module produced some slightly different products, there would be less risk of upsetting the entire plant during a process breakdown since the breakdown would be confined to a single area. As an added benefit, the proponents of modularity pointed out, three independent units, each of which incorporated the *entire* process, made it possible to train three managers quickly for the job of plant manager. Richardson's goal was to replace himself with a local as soon as possible. (Conductores Y Componentes was already run by a Mexican.)

Richardson, and others who thought a centralized plant would be more appropriate, argued that despite these advantages, there were several inefficiencies in the modular organization, especially in lead prep, which was the capital-intensive part of the process. Dividing the plant into three units would clearly increase in-process inventories. The actual cost of this increase was not known. Furthermore, certain leads common to all three final assembly areas could be made on the same machine with the same setup. Therefore, the module system required more machine capacity, since time was taken for duplicate setups. Finally, every time the number of machines was calculated for another module, one had to round up to the nearest whole machine.

Near the end of the meeting, Wilson noted that Vazquez had remained silent throughout the discussion. He asked him what he thought. Quickly remembering what he had learned in Organizational Behavior, Vazquez jokingly suggested that since he had now spent almost a week there and was therefore an expert, he would be glad to comment. Luckily for Vazquez, everyone laughed, realizing that they would have to wait for the MBA to commit himself as well.

PRODUCTION AT THE RIO BRAVO PLANT

Packard plants were designed to run two shifts, holding the third shift as reserve capacity. As a component supplier of parts that had low relative value in the automobile (as compared to the body, engine, drive train, etc.), RBE always had to be able to deliver to meet the needs of the car divisions. Similarly, high quality was considered to be essential, since once the wiring harness was installed in the car and covered with interior finishing, it was hard to service. A defective harness would hold up the entire car at the assembly plant until it could be repaired or replaced.

The Lead Preparation Process

The components received as inputs by RBE consisted of large spools of cable, plastic connectors, spools of terminals on headers designed for automatic or semiautomatic application, molding compounds for special connectors that had to be molded onto cables, and miscellaneous components such as splice clips, electrical tape, solder, plastic conduits, etc. The lead prep area prepared, cut, and partially or fully terminated cables which were sent to final assembly where they were inserted into connectors and bundled into complete wiring harnesses. Lead prep was also where large molding machines molded special connectors onto certain cables. Certain special operations such as splicing or soldering were shared by both lead prep and final assembly. Each of the three operations (cutting, pressing, and molding) is elaborated on below:

Large spools of cable first had to be cut to various specified lengths on high-volume, automatic cutters. Some not only cut the cable, but automatically applied (pressed on) terminals at one or both ends of the cable after stripping the ends. These cutters, while automatic, required careful adjustment during setup according to the cable gauge (conductor diameter), specified length, and terminal types to be applied.

Cutters were unique in the process since *all* cables had to pass through them first on their way to any other operation. As a result, "lots" were first created at cutters, since cut cables went into large plastic bins, each bin containing the quantity cut during one setup of the cutter. This quantity remained intact until it was delivered to final assembly. A cutter processed 4,000 leads during one setup and if those leads had to have a connector molded onto them, then the molder would also run 4,000 leads on a setup. This current cutter lot size was based on yearly inventory carrying cost/lead of $0.01, cutter running capacity of 2,000 cut leads per hour (as opposed to *effective* capacity which would be less due to the one half-hour setup incurred when changing from one lead type to another) and average consumption of these leads by downstream processes within each module (presses, molders, or paced final assembly lines) of 100 leads of *each* type per hour. Cutters were also special in that they were the only fully automatic machines in the process. One operator could run three machines at once. The total cost (including tooling, maintenance, labor, and depreciation) for a cutter was $5,000 annually.

Presses were machines that pressed terminals onto the stripped ends of cut cables. The total direct cost for a press was $4,000 annually. Presses were operated by one person who positioned the cable end in the machine and then stepped on a foot switch to apply the terminal. This operation required good manual dexterity and eye/hand coordination as well as a knowledge of the right method. Operators were also responsible for detecting when the machine was damaging the terminal or wire as this affected the reliability of the termination. Some of these malfunctions were very subtle so experience and a good eye were essential.

Molders were the other major area of lead prep. These were similar to presses in that each machine had one operator and was semiautomatic. Here, however, a cable with a terminal already applied was inserted into the machine and a connector was molded around it. Many of the molders at RBE were old and required frequent service. These machines, because of temperature controls and electronics, were the most expensive and the slowest. Only a few types of cables required molding. The total direct cost for a molder was $7,000 annually.

Certain cables had to pass through all three operations before assembly, while others went straight from the cutters to final assembly. Exhibit 4 illustrates examples of cutting, pressing, and molding in operation.

Component Shortages

Logistics were somewhat complicated by two factors: (1) the large distance between RBE's El Paso, Texas warehouse (shared by Conductores) and Warren, Ohio where components were made, and (2) poor communications across the border. The former was a fact that could not be

EXHIBIT 4   Rio Bravo Operations

Bundling cables after molding

A molder in operation

EXHIBIT 4   (continued)

Presses in operation

changed, and the latter resulted less from the language barrier (the Americans spoke Spanish rather well and the Mexicans were taking English classes at the plant) than from a very poor local telephone system. On some days it was impossible to get a call across the border, and on several occasions, the phones were shut down altogether.

When a supervisor recognized that supplies of a part were inadequate, he or she usually contacted either the general supervisor of the module or the materials general supervisor. If the part was not in receiving, the materials general supervisor would immediately order the part from the warehouse in El Paso, assuming the phones worked. If not, he or she might drive 30 minutes to ask for it personally. Even then, the part could be held up in Mexican customs for a couple of days before delivery.

An alternative was to check with the other modules to see if they had the part, which was often the case. Finally, they could also check with Conductores. Despite these alternatives, shortages sometimes occurred. There was some animosity between the materials general supervisor and the El Paso warehouse, each suspecting that the other was somehow at fault when a shortage occurred.

EXHIBIT 4 (continued)

An inspector

A worker on the assembly line

Cables ready for packing

Maintenance Problems

Vazquez talked to several lead prep supervisors and found that their major complaint was bad support from the maintenance people ("technicos"). It was the technicos' responsibility to do machine setups as well as repairs. The reporting relationship was under study at the time. Currently, they did not report to the lead prep supervisor but to a maintenance general supervisor. They considered themselves elite because they had received more training than operators. Part of their workday was still spent on continuing their training.

Supervisors complained that the technicos were sometimes nowhere to be found when they were needed. They claimed that some were somewhat insubordinate. Vazquez did observe technicos coming in late in the morning and reacting with disinterest to supervisors' urgent requests for service.

THE LABOR CONTEXT

There were about 700 unskilled hourly employees at RBE, all hired from the local population. The work force was very young, almost entirely between 17 and 22 years old with the average being about 19. Due to a combination of factors, both absenteeism and turnover were very high. Turnover exceeded 40 percent on an annualized basis despite excellent working conditions in the new, modern plant. Because of the number of new plants in the area, workers would often switch to a new plant closer to home (almost none of the hourly workers had cars). They typically came to work in vans run by one of the large Mexican unions, though neither RBE nor Conductores was unionized. Another cause of turnover, among the technicos in particular, was the hiring away of trained workers by new plants paying just above the prevailing wage to get a base crew of skilled workers, trained by the existing plants.

Absenteeism was high due to the fact that many of the workers came from very poor homes, where they sometimes filled the roles of breadwinner, mother, and daughter all in one. If other family members became ill, the worker might be the only one who could care for them. Working in an industrial environment was also a new experience altogether for many; they did not understand the impact that their absence might have on the work group. This was another factor which made the supervisor's job difficult.

A TYPICAL SUPERVISOR'S DAY

As his electronic watch awakened him at 4 A.M., Vazquez asked himself how anyone could do this every day. He had arranged to spend an entire day with the lead prep supervisor in Module 1, and he wanted to arrive before the shift started at 6 A.M.

When he walked into the plant at 5:30, Eugenio Batista, the supervisor, greeted him. Batista was surprised that Vazquez would come in so early and he enthusiastically began to show him the ropes.

Numerous things had to be done before the shift began. The molders had to be turned on immediately so that they would be warm when the operators arrived. Batista had to review the notebook used to communicate with the supervisor on the other shift. It contained information such as what setups certain machines had had, which machines were down, which components were low, or whether final assembly needed emergency runs of any particular cables.

At 5:40, one of several "utilities" arrived. These were skilled operators who helped the supervisor with material and information flows. They rarely operated machines, but spent their time seeing that operators did not run out of components. This utility was responsible for leaving "cut cards" on each cutter to tell the operators what cables to cut and in what quantities. The supervisor checked these cards before allowing the utility to distribute them. During the day, the utilities frequently came to the supervisor with problems which required his assistance, such as getting a technico to work on a machine or obtaining a component that could not be located.

By 6 A.M., the machines were running. Batista had to shift several operators around when he became aware of who was absent. He said Mondays were particularly bad for absences. At 6:15, the cutter technico appeared in the area and Batista already had work for him. Several times during the shift, final assembly utilities came to Batista requesting emergency runs of certain cables. These runs were needed occasionally because the cable inventory report (which was updated on a chart halfway through the turn by production control personnel) was not current enough, especially at the beginning of the shift, or because a batch of cables had been found to be defective by QC inspectors or by final assemblies, so that effective inventories were much lower than reported. Batista reacted quickly to these requests and new cables were started very soon.

By 3 P.M., the end of the shift, Vazquez was exhausted. The supervisor had spent the entire day on his feet, handling one crisis after another. Vazquez wondered how centralizing lead prep would affect Batista's job. Batista said he liked the module system and would not like a centralized area.

THE QUANTITATIVE ANALYSIS

Vazquez realized he had to calculate a dollar amount which would represent the savings possible by centralizing lead prep (as a result of eliminating current duplicate setups). He found that only cutters seemed to be running significant numbers of lead types that were common across modules. The results of this analysis are shown in Exhibit 5. Combining these

EXHIBIT 5   Commonality of Cutter Lead Types Across Modules

| | Number of Types per Module | | |
	Mod I	Mod II	Mod III
Class A	70	50	100
(lead types unique to each module)			
Class B*	50 ------------- 50		
(lead types common to all modules)			
Class C	30 ------------- 30 ---------------- 30		
(lead types common to all modules)			

*That is, 50 of the lead types cut in module I are also cut in module II, with duplicate setups.

types in a centralized area would allow longer runs between setups, increasing effective capacity.

He also wondered what effect, if any, centralization would have on lot size and therefore on total plant inventory. After some thought, he was convinced that the often-seen square root relationship applied here. Thus, he could easily determine the optimal inventory level or lot size for multiple locations as a function of the optimal level for a single site, or vice versa. For example, centralization would double the demand on lead prep for Class B lead types (see Exhibit 5), and thus the optimal lot size would increase by a factor equal to the square root of two. Similarly, since the demand, in a centralized environment, for Class C lead types would be triple the demand previously faced by any module's lead prep area, the optimal lot size would increase by 73 percent.[1]

[1]Since $\sqrt{3} = 1.73$.

Dulaney Sound Systems, Inc.

In one way, Rick Jerauld was refreshed by the present pause in Dulaney Sound Systems' previously steady and rapid growth. On the other hand, he was concerned about the changing emphasis in manufacturing which was filtering down through the organization, especially the emphasis on better management of assets.

For the past couple of years, Jerauld had been Manager of the Metals Department of Dulaney's Timonium, Maryland plant. In January 1978, he faced decisions affecting both the short- and long-term future of the Metals Department.

In the short term, the Metals Department had excess capacity. Jerauld had to decide which items, if any, he should pull back from vendors for fabrication in the Timonium plant. Figuring out what should be done in the longer term was more difficult. Surely inventory management would be important, but other procedures such as bidding for outside work could also aid the management of assets. It wasn't clear what the company's position would be on these matters, or whether the company would change any of its inventory and production control systems. Indeed, Jerauld was no longer sure how the company would now measure his performance. He wondered what, if any, influence he could have and what actions he should take.

This case was prepared by Assistant Professor Roger W. Schmenner as the basis for class discussion rather than to illustrate either effective or ineffective handling of an administrative situation.

A BRIEF HISTORY OF THE COMPANY

Dulaney Sound Systems had been founded in 1962 by Al Naney. He had sensed that the increasing affluence of youth and the success of rock and roll music would combine to make a large, sustainable market in stereos and high fidelity sound equipment. While Dulaney's initial products had been designed for the home market, the company had soon shifted most of its product development to the commercial and institutional market—sports complexes, auditoriums, shopping centers, discotheques, and the like. The company sold and serviced complete sound systems tailored to the needs of the customer. From about $1 million in sales in 1962, Dulaney Sound Systems had grown to about $100 million in sales in 1977, an average growth rate of over 35 percent per year.

For years, the company had been a high flyer on the American Stock Exchange and had experienced no difficulty in securing the cash for its growth by repeated stock issues. In 1977, however, money market conditions, the company's size, and a slowing in the company's growth due to a slackening of the discotheque craze restricted Dulaney to a bond offering only. This softening in Dulaney's ability to attract cash appeared to be compounded by a modest, if temporary, condition of excess capacity throughout many of the company's manufacturing operations. Both of these conditions were new to Dulaney, which had been used to scrambling for increased capacity in almost any form.

It became clear to Al Naney, Dulaney's President, that the company would need to generate more cash internally. In early 1978, this insight had spurred the concern for better asset management.

THE ORGANIZATION OF THE COMPANY

In 1978, Dulaney Sound Systems operated four plants, three in Maryland and a fourth, just starting up, in Santa Monica, California. The original plant was in Lutherville, Maryland, just north of Baltimore. In 1978, it served as the corporate headquarters, as the research and development center, and as a manufacturing plant for some components such as amplifiers. A second plant, in nearby Timonium, housed a metals shop; the speakers, receivers, and other components which formed the core of any sound system were also made in Timonium. The third plant, in Sticks, Maryland, integrated components from Lutherville and Timonium into complete sound systems and then tested them.

The Santa Monica plant had been intended for the manufacture of those components which Dulaney used in high volume, such as certain small speakers and cable and harness devices. These products were shipped to Sticks for integration into a sound system, although it was eventually planned that some small system sales might proceed directly from Santa Monica.

The company was organized into five market groups: Home Stereo Systems (the original product, which was now a small part of the business), which now concentrated on large, sophisticated custom systems; Outdoor Sports and Concert Systems (stadiums, outdoor music festivals); Indoor Sports and Concert Systems (gyms, auditoriums, domed stadiums); Shopping Malls; and Discotheques (the most recent product line and the fastest growing). These five market groups operated more or less autonomously at the levels of marketing and systems integration and test (the Sticks plant). However, all five could theoretically use any of Dulaney's speakers, amplifiers, etc., in any system. Thus, at the level of the Timonium, Lutherville, and Santa Monica plants, the market group division of the company was not a particularly useful delineation. For example, Jerauld's Metals Department served all five market groups.

Jerauld's dealings with his customers—product and market groups—were governed by a so-called ask/accept system. Requests for metal parts would come from these customers. Jerauld and his staff would then either accept the request or negotiate with the customer on quantity and delivery time. Jerauld had the option of making the part in-house or of securing it on the outside. In either case, he was responsible for delivery, price, quantity, and quality.

Al Naney had fostered a fairly free-wheeling style within the corporation under the corporate motto "Just do the right thing"; thus, negotiations in the ask/accept system were not marked by mutual trust. In fact, dealings throughout the corporation were characterized by second guessing—customer departments over-ordered or early-ordered because they did not trust supplier department capabilities; supplier departments under-produced because they did not believe customer department forecasts or inventory positions. Periodically, upper level managers cried for more "predictability," but, in the past, "doing the right thing" had been translated into making certain that deliveries were always met, even if this meant inventories were enlarged.

Inventories throughout the company were a puzzle anyway, since no one seemed to "own" them. There were few incentives to keep them low. Certain end-of-quarter targets were set, but this only seemed to stimulate such last-day behavior as refusing supplier deliveries at the plant gate, shipping orders early so they would be "in-transit," or in other ways trying to shift responsibilities. Some managers argued for much more direct responsibility for company inventories and average, as opposed to end-of-quarter, measurements.

THE METALS DEPARTMENT

Jerauld had been with Dulaney Sound Systems for two years, having been hired for his Metals Department job. For the previous 12 years, he had worked in one of the metals shops at nearby Black and Decker. As Jerauld saw it, the two companies were very different: "Dulaney is free and easy, at least compared with Black and Decker. There you could do it by the book. Demand was pretty steady, and it seemed you always had time to react. Here it is more of a challenge. I feel like I'm much more my own boss, though I don't supervise any more people. I know I'm working harder."

George Gernand had been Plant Manager at Timonium less than a year; he had been Assistant Plant Manager at Lutherville for the previous 5 years. His background was in electronics, however. He had spent 15 years with the Martin Marietta Company before it closed its doors in Baltimore. Gernand's experience at Dulaney had been restricted to individual products and their engineering. Only since his promotion to Plant Manager at Timonium had he supervised a supply function such as the Metals Department. He described his attitude towards managing the Metals Department as "seeing Metals is run as well as any metals shop around. That means comparing its performance against outside shops as much as possible."

Most of the staff were young managers "passing through" the Metals Department on their way to more lasting assignments. Only Cliff Osborn, the Engineering Manager, could be said to be an old hand. Of the younger staff, some had functional roles (Otis Milby, Controller, Mary Conrad, Purchasing Agent, and Eleanor Carson, Financial Manager) and were unlikely to branch out to different roles. Others like Fred Frey, Audrey Cheek, Bob Webster, and Frank Fucile had already led varied careers and expected to keep moving along in the organization.

Jerauld was well regarded by his staff. In Bob Webster's words:

> Rick is easy to work for. Doesn't get emotional. Keeps all of us informed. Polls us for views on all the big things going on. Isn't nosey. Yet we know we have to produce, because the Metals Department is getting judged more and more like it isn't even part of the corporation. Don't get me wrong, Rick isn't Mr. Charisma and he isn't gunning to be the next President of this company, but he's a good guy and if Metals screws up, chances are it won't be his fault.

The Metals Department occupied about 150,000 square feet in the Timonium plant and employed about 225 workers on three shifts. Jerauld's department was not the only metal-working facility of the company; another, much smaller shop at the Lutherville plant supplied the Engineering Department's needs for experimental and prototype products.

The metal work required by Dulaney's market groups was not highly sophisticated; it was mostly the forming of sheet metal. Nevertheless, the

EXHIBIT 1   Metals Department Organization Chart

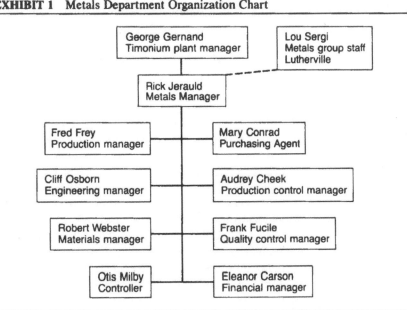

breadth of Dulaney's product line, coupled with the customization for which the company was famed, meant that over 4,000 parts could be made in the shop. Of these, about 1,500 were very active. In any week, about 400 different parts were scheduled through the shop.

The weekly loading of the shop was triggered by orders placed on the Metals Department by the materials managers of the market groups and of specific component products. Since the various market groups might use the same components and parts, some grouping of part orders was possible within each week. These groupings helped reduce the number of machine setups and the problems of handling materials through the shop. Groupings of more than one week's orders, however, were not possible under the present system. In a very real sense, then, Jerauld's metals shop was at the mercy of the materials managers of the product and market groups. As Audrey Cheek of Production Control put it, Metals was in the stock replenishment business.

There was little or no coordination of inventories between product or market groups and the Metals Department. Everyone suspected that there were excess buffers in the system, since even poor delivery performance by Metals in the past (only 75 percent of deliveries were on time) had not elicited much yelling and screaming within the company. Gernand, the Timonium Plant Manager, was trying to remedy this coordination. In the past, new equipment had been fairly easy to secure. New capital expenditures had also been easy to justify (e.g., by citing a need to increase capacity). Some managers even went as far as labelling previous requests as "wish lists." No one could remember an appropriation request which had actually been rejected; relatively few had even been sent back for further elaboration. To satisfy the new fiat, most managers felt, would require "sharper pencils" and justifications centering on cost reduction.

Some managers at all levels of the Metals Department also felt that the moment was ripe for some post-installation investigation of the returns generated by previous appropriations. Apparently, none of the tough capital decisions of the past had been investigated and these managers doubted the claims made for many of them. Post-installation investigations would also serve to scale a manager's credibility on capital appropriations and in that way might perhaps improve his handling of that management task.

Jerauld could not help but think about the recent start-up of the Santa Monica plant. That plant was dedicated to a very few high-volume products such as small speakers. Unfortunately, the small speaker components segment of the plant was not sized to match the assembly operations but was much larger than necessary. Jerauld had heard the manager of the components segment say that a more concerted attempt to match the capital appropriations of the products actually designated for the plant might have saved something like $1.5 million in plant and equipment.

THE EVALUATION SYSTEM

Rick Jerauld's performance evaluation was the immediate responsibility of George Gernand, the Plant Manager, with the advice of Lou Sergi. The evaluation had a number of dimensions, many of which were perceived as changing. As Jerauld put it: "Something's changed, but it's hard to tell exactly what. Everyone tries to tune his or her ear toward Lutherville, that's certain, but sometimes the messages from the top are a little confusing."

Because of Dulaney's history of rapid growth, meeting deliveries had been by far the most important aspect of any manager's job. Sergi felt that fully half of any evaluation before 1977 was determined by the record on deliveries. With Dulaney's changing fortunes, however, the weight attached to deliveries had been reduced. Estimates ran from 25 percent (Gernand) to 30 percent (Sergi) to 35 percent (Jerauld). Delivery performance was measured as the ability to meet the targets set in the ask/accept system.

The Dulaney groups which were customers of the Metals Department requested certain metal parts. Jerauld then accepted the request or negotiated a change. At some point, subject to negotiation, all requests for a given month or quarter were frozen; after that date, no changes in requests were honored. Naturally, this system encouraged Jerauld to be a hard bargainer with his customers and to be conservative in estimating what he could deliver. To counteract this incentive of the system, plant managers like Gernand would poll customer groups within Dulaney to ascertain how pliable and reasonable the managers under them had been. As Gernand had put it: "The plant managers see to it that the department managers all play the game fair."

The decline in importance of delivery performance was replaced in part by increased emphasis on product cost. Before, there had been some nominal concern for cost variance performance; Fred Frey, Metals Production Manager, observed, however, that variances could be manipulated readily by a choice of standards that could easily be beaten, thus generating positive variances. The present emphasis on product cost was for actual product costs. The Metals Department was almost unique within the corporation in that its costs for making certain standard parts could be compared rather easily with those of outside vendors. Such was not the case for most other departments, whose products were custom designed with no suitable comparisons outside the corporation. The cost comparisons for the Metals Department were being made on parts agreed to between Gernand and Jerauld.

Other matters on which managers were evaluated included quality, employee relations, general perceptions of the individual's ability to grow with the job and the company, and—new in 1978—asset management. This latter subject was ambiguous. What were assets anyway and which assets did a manager really control—plant? equipment? inventories? (whose? where?) people, especially considering Dulaney's no-layoff policy? For that matter, what wasn't an asset? And what sort of measurements could be unambiguously defined? Or would another fuzzy measurement be added to the increasing list by which Jerauld was measured?

Jerauld already held a healthy skepticism about the applicability of return-on-investment or return-on-assets measures, since ratio measures of that type often meant that (1) the ratio could be altered by adopting certain policies affecting either numerator or denominator in ways which, by other criteria, might not be considered in the best interests of the corporation, and (2) the ratio could be altered by forces totally beyond a manager's control.

Possible measures of asset management which Jerauld had heard offered included:

- Return on investment.
- Capacity utilization measures (but was capacity defined?).
- Worker utilization measures.

- Combination of machine and worker utilization, say the product of two ratios.
- Asset turns measured, say, as value of department output/value of plant, equipment, and inventories.
- Inventory turns or inventory targets, especially on internal factory inventories.
- Ratios of indirect to direct labor (as compared to industry standards, past history, and product costs).
- After-the-fact evaluations of capital expenditures.
- Yields.

From discussions with Gernand, Jerauld knew that Gernand, too, was not keen on ratio measures. Since the plant itself was measured on inventories, Jerauld also knew that Gernand valued a department's ability to meet its inventory targets. With asset management so new, more than this Jerauld did not know and Gernand was not yet tipping his hand. Jerauld knew Gernand as a stickler for costs. "He may have to pay lip service to asset management, but he's going to be taking a close look at costs. Hell, we've spent hours discussing exactly which of the parts we make can be accurately compared with vendors. I just hope the comparisons are fair. I don't mind pulling my share of the load, but I've got to carry corporate overhead which I'm sure our vendors don't have to carry."

COPING WITH EXCESS CAPACITY

Among other things, the tailing off of discotheque sales and the associated inventory adjustments meant that the Metals Department would soon have half its capacity unused, if nothing were done. All overtime and Saturday work had already been halted. Jerauld now confronted several options for coping with the situation:

1. *Pulling products back to the factory.* Jerauld could withdraw parts orders placed with vendors and make the parts at Timonium. There were several variations of this scenario:

   a. Pull back a larger portion of the volume of parts which were both purchased and made at Timonium (20 percent of vendor volume).

   b. Pull back parts which were on a long-term vendor contract but had previously been made at Timonium (30 percent of vendor volume).

   c. Pull back parts which were *not* on long-term contract and had previously been made at Timonium (15 percent of vendor volume).

   d. Pull back parts which were assembled into comparatively new products and which had initially bypassed the Metals Department because the department did not have the capacity for them (10 percent of vendor volume).

e. Pull back mature parts which had never been made at Timonium (25 percent of vendor volume).

Jerauld was contemplating a list of candidate parts that could be pulled back to the department (Exhibit 2). Several criteria seemed applicable to an evaluation of which products to pull back:

> The parts we pull back need to be important ones—lots of labor hours—and ones where the economics make sense, parts where we are at least competitive with vendors. [The corporation's present conventions on the matter called for comparing the vendor's price for the part, adjusted upward for the expected costs Dulaney would incur in purchasing the part, having it transported, and inspecting it, against the internal factory cost for the part which included material, labor, and overhead expenses.] We also need to pick parts that won't throw the shop into turmoil just trying to get them set up. But you have to be careful. You can't yo-yo a vendor for very long before he quits being your vendor, so you don't want to shut off your best ones completely.

2. *Transfer products from one plant to another.* The Timonium Metals Department made a number of parts which were also made by the metals shop at Lutherville for the engineering operation there. On many of these parts, Timonium could offer lower variable costs than Lutherville. Nevertheless, since the transfer of any of these parts would result in idle time in the Lutherville shop, Lutherville resisted such transfers. Given the organization of manufacturing, only the vice president of Manufacturing had the authority to transfer products in such a situation.

3. *Build up inventory.* Many managers at Dulaney believed the present situation of excess capacity was temporary. To the extent that it was, it was possible to contemplate using capacity to build up inventories of those parts which could reasonably be anticipated not to become obsolete. Given this condition, it remained to decide what kinds of parts would best comply with "good asset management" and to what extent this inventory build-up strategy could be relied on, relative to other options.

4. *Supply the outside world.* It was conceivable that the Metals Department could bid on jobs outside the corporation. The door had not been shut completely on this idea. However, some of the corporate staff were known to dislike the image such an action might generate, even though it could absorb some of the department's fixed costs.

5. *Transfer workers intrashop and intershop.* It would be fairly easy to transfer workers within the department, but somewhat more difficult to make transfers involving other departments.

6. *Unload assets.* If things looked slow for a year or more into the future, Jerauld could sell off some of the excess assets. However, he was not aware of this having ever been done, and wasn't sure he wanted to be the first manager to suggest it.

EXHIBIT 2 Candidate Parts for Pulling Back to Factory

Part	Volume	Current Status	Vendor Price Including Adjustment*	Internal Cost (Actual or Estimate)				Comments
				Mat'ls	Labor	Over-Head	Total	
A. Speaker cone	4,000/month	Long-term contract, made in shop before	$ 11.97	$ 3.20	$ 1.37	$ 5.20	$ 9.77	Vendor problems with quality and delivery
B. Amplifier and panel	2,500/month	Short-term contract, made in shop before	24.64	9.56	2.75	10.44	22.75	
C. Speaker cabinet	4,000/month	Long-term contract for 3,000/month, 1,000/month also currently made in-house	38.59	12.56	4.27	16.21	33.04	Vendor makes D part as well
D. Receiver chassis	420/month	Long-term contract, made in shop before	49.50	24.95	12.66	48.58	86.19	Same vendor as C part
E. Speaker stand	670/month	Long-term contract, new product, not made in house before	140.15	28.65	17.90	68.03	114.58	High expectations for vendor, some doubt about internal cost estimate; tooling costs of $22,000 for vendor though Dulaney owns tooling itself
F. Amplifier back plate	4,000/month	Long-term contract, made in shop before	1.87	1.28	.20	.65	2.13	—
G. Speaker box frame	1,600/month	Long-term contract, always made on outside	10.72	5.52	1.02	3.88	10.42	$2,600 tooling needed

*Vendor price adjustment is the mark-up applied to the vendor's quoted price to account for freight, order, and inspection costs.

LONGER-TERM CONCERNS

The recent start-up of the Santa Monica plant posed a longer-term concern for Jerauld. Its charter dictated that the Santa Monica plant would concentrate exclusively on the higher volume items which could be assembled into a Dulaney sound system. Many of the plant's candidate products, however, made significant use of both the sheet metal and machining shops of the Timonium Metals Department, accounting for perhaps a quarter of the output of those shops. Since it was unlikely that Timonium could economically supply Santa Monica with its metals needs, Jerauld faced a potential further decline in the capacity utilization of his department.

Another longer-term concern centered on changing priorities within the corporation. It seemed to be the view of Al Naney and other top level managers at Lutherville that the manufacturing priorities at Dulaney Sound Systems would be shifting from the historical emphasis on growth and flexibility to a concern for cost, even at the expense of continued high growth. For Jerauld, this shift in priorities meant more than increased attention to managing existing assets. The Timonium Metals Department had been organized to accommodate product and volume flexibility within the corporation—general-purpose machines had been purchased and laid out for multiple uses, parts contracts let to outside vendors were generally for standard, long-run parts, while quick deliveries and low volumes were produced in-house. If, in fact, Dulaney was serious about realigning its competitive priorities, Jerauld would have to contemplate a host of possible changes in his department and in the organization and management of his vendor network.

Andreas Stihl Maschinenfabrik

In late October of 1983, Erik von Woedtke, Vice President—Manufacturing of Andreas Stihl, got up from his desk and walked over to his chalkboard. On the chalkboard was a sketch of the floor plan of the new Stihl final assembly plant (WERK II) in Waiblingen, West Germany. (See Exhibit 1.) Von Woedtke was working on the final draft of a paper he would present at the upcoming conference on new work structures, sponsored by the Fraunhofer-Institut fur Arbeitswirtschaft und Organisation of Stuttgart. One thing was certain: The assembled group of economists, sociologists, government officials, and engineers at the conference would be very interested in the results of Stihl's experience with designing jobs according to workers' interests.

ANDREAS STIHL

Stihl, the world's leading producer of chain saws in 1982, was a privately controlled firm headquartered in Waiblingen, West Germany. Stihl produced a broad line of premium quality saws. Its products, universally acknowledged as the quality standard of the industry, had long been a leader in safety and comfort features. Worldwide sales were DM 743,000,000[1] in 1982, up from DM 556,000,000 in 1979. Of that amount, DM 422,000,000 were produced in West Germany (83 percent exported to over 100 coun-

---

[1] $1.00 = 2.6251DM in December 1982; $1.00 = 2.7821DM in October 1983.

This case was prepared by Professor W. Earl Sasser as a basis for class discussion rather than to illustrate either effective or ineffective handling of an administrative situation.

EXHIBIT 1 Plant Layout—Werk II (26 employees per line)

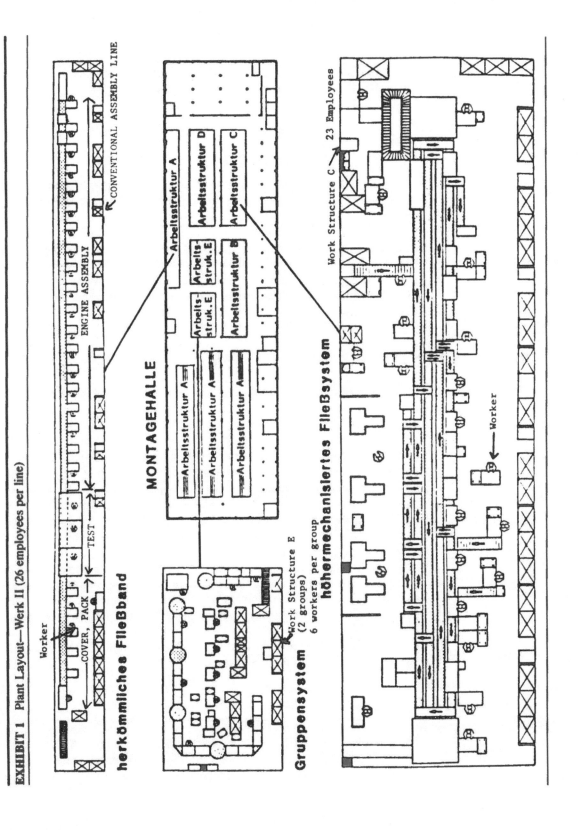

tries) and the remainder was produced in Stihl's four foreign plants in Australia, Brazil, Switzerland and the U.S.A.

Stihl sold its products through servicing dealers, a policy to which it strictly adhered. As a result, Stihl dealers were acknowledged by industry observers to be particularly knowledgeable and loyal to Stihl. Those same observers readily admitted that Stihl's dealer organization was also first in quality of servicing.

Stihl was the most fully integrated of the chain saw manufacturers, producing most engine parts (the cylinders were one of the few items purchased from an outside source), all its bars and sprockets and all of its saw chain. Stihl had its own tooling group which designed and manufactured most of the machinery used in Stihl production facilities. This group had developed proprietary machinery and processes for magnesium die casting of engine parts and for several machining operations. Stihl's production strategy was characterized by industry observers as being one of extremely high quality at relatively high cost.

THE PRODUCTION PROCESS FOR CHAIN SAWS

The production process for chain saws was a sequence of batch processes feeding into a moving assembly line. Manufacture of the major parts was done in separate departments or Werks (plants). For example, the major parts of the engine were machined in the connecting rod department, the crankshaft department, and magnesium machining department. These departments were quite capital intensive, often employing only a single worker per large machine. Chain broaches which machined flat spots on parts for purpose of proper fit and which required operators only to load and unload unmachined parts on its moving chains were used extensively in the connecting rod department; an automated machine tool turned crankshafts.

Once all parts had been machined in their respective departments or die cast in the magnesium die casting facility, they moved to the subassembly stage. After components were built into subassemblies, they moved to final assembly. After engines were assembled, they were tested. Finally, the cover was added and the unit was packed with the bar, guards, chain, other accessories, and operating manual. The assembly operations were all contained in the new plant.

THE NEW PLANT (WERK II)

The new plant (Werk II), construction on which started in 1981, was completed in 1982. The engineering and equipment expenses were partially funded by the West German government as part of an effort to encourage experimentation by West German firms with alternative work structures. All the products produced in Germany were assembled in this plant. Prior to the completion of Werk II, all assembly was done in Werk I on seven conventional moving belt assembly lines (A-lines) and another more automated line. (This B-line will be described below.) Seven of the A's and the B-line were transferred to the new plant. Twelve different models of chain saws were assembled at Werk II and each model had three to five possible modifications, depending on various end-use conditions and requirements. See Exhibit 2 for the distribution of models/types for 1982. According to Von Woedtke:

> In general, marketing is trying to increase the number of models and our technical and financial people are trying to reduce the number of models. We have held constant the number of models we manufacture in Germany for several years, although we have expanded the number of models of smaller saws we produce in the United States for the casual market. We export those small saws worldwide from the United States.

The new plant was designed as an effort to accommodate a wide variety of interests in job enlargement among workers. As well, Stihl executives hoped that the newly designed production "lines" would enhance its manufacturing flexibility, allowing the company to respond more quickly to market demand with the "right" chain saw models, with their various modifications, at the "right" time. Such production processes would have to be able to change models quickly, to maintain Stihl's record of high quality and reliability, to increase volume very quickly, while at the same time holding the line on unit costs. Von Woedtke noted:

> When we were designing our new plant, I could see that the investment in several of the new structures because of several automated stations and complicated workpiece carriers (to be discussed below) was going to be much higher than the investment in a conventional line. These systems would naturally produce a higher depreciation, higher indirect labor, and lower direct labor costs per unit compared to the unit costs of the conventional line. I would not accept one penny more cost per unit. What could we do? The only possibility was to increase output to two shifts, reduce the amount of breaks, or go to overtime. We decided that two shifts was the best option, but our demand per model for most models was not high enough to support two-shift operations. That led us to realize that we had to design our new capital intensive lines with the flexibility to produce several models. But, even with this flexibility, the investments are risky. You need to think long term; the payback is certainly not short term.

The new plant was located approximately two kilometers from the original Stihl assembly complex. Stihl staffed the new plant entirely with employees from the older Werk I in Waiblingen. There were 17 labor grades in the plants, of which the lowest 4 were not used at Werk II. The highest grade was held by the most experienced tool maker; the lowest by a floor

EXHIBIT 2   Distribution of 1982 Volume by Model and Type

Chainsaw Model	Type		Percent of Annual Unit Volume	Chainsaw Model	Type		Percent of Annual Unit Volume
A	1		.3	G	1		.3
	2		.4		2		.6
	3		.8		3		.5
	4		.7			Total G:	1.4
		Total A:	2.2				
				H	1		6.1
B	1		2.4		2		2.9
	2		2.0		3		3.6
	3		.7		4		2.4
		Total B:	5.1		5		2.0
						Total H:	17.0
C	1		3.6				
	2		2.2	I	1		1.1
	3		1.6		2		1.0
	4		2.5		3		.6
	5		.7		4		.5
		Total C:	10.6			Total I:	3.2
D	1		9.1	J	1		1.0
	2		3.7		2		.9
	3		1.3		3		.9
	4		1.7			Total J:	2.8
	5		1.0				
		Total D:	16.8	K	1		1.4
					2		.9
E	1		10.6		3		.2
	2		11.8		4		.3
	3		2.7			Total K:	2.8
		Total E:	25.1				
				L	1		.3
F	1		7.2		2		.6
	2		3.0		3		.5
	3		2.4			Total L:	1.4
		Total F:	12.6			Total:	100.0%

sweeper. In designing the new job structures, management tried to match the production setup to their employees' abilities and desires to enlarge their jobs. According to management's research, the distribution of the employees' desire for job content and job rotation was as shown in Exhibit 3. Von Woedtke commented: "Our younger, better educated em-

EXHIBIT 3   Worker Preference for Job Content and Job Rotation*

Desire For Job Rotation	*Job Content Preference (Work Cycle)*			
(number of jobs done per month)	*1.5 min.*	*3.0 min.*	*6.0 min.*	*Total*
1	26	6	2	34%
2	14	4	3	21
3	8	3	3	14
4	6	4	4	14
5	7	2	3	12
6	3	1	1	5
Total	64%	20%	16%	100%

*Based on questionnaires administered to assembly line workers in WERK I during early 1980.

ployees are asking for both job enrichment (more independence at the workplace) and job enlargement (longer cycle times). And, we expect that the demand for job enrichment and job enlargement will increase even more in the future." With this distribution in mind, Stihl designed three alternative work structures, "conventional," "rectangular," and "group."

THE CONVENTIONAL LINE

The job structure design began with the conventional line as a base. Stihl attributed an investment index, quality[2] index and cost per unit index of 100 each to the conventional setup.[3] The conventional line was a simple moving belt with 26 work stations (see Exhibit 1). Model specific tooling for this line was about 30 percent of the investment for nonmodel specific equipment. Each station had a cycle time of 1.5 minutes, the minimum required by the labor union contract in that region of West Germany. The characteristics of the various work structures are summarized in Exhibit 4. None of the work stations were automated,[4] and each worker performed the job task at his or her station only. The conventional line em-

---

[2]Quality, according to Von Woedtke, was measured "like most firms around the world. Our quality people, first, sample the output and, next, using a checklist, clarify the defects discovered and, finally, calculate a composite quality index by weighting the number of defects by the seriousness of the defects."

[3]The higher the investment, unit cost, and quality, the higher the index.

[4]Earlier attempts to automate individual work stations (a mechanized assembly operation) on the conventional line had proven very difficult because automation required the use of a carrier for the exact placement of the work piece.

EXHIBIT 4　Work Structure Characteristics

| | | | | | | Indices | |
| | | | | | | | Investment |
Work Structure	Cycle Time	Number Employees (per line shift)	Daily Output* (per line hour)	Cost Per Unit	Quality†	Tooling (per model)	Equipment
A	1.5 min.	26	28.5	100	100	100	100
B	1.5 min.	24	34.3	96–98	105–107	100	150
C	1.5 min.	23	40	94–96	106–108	80	400
D	1.5 min.	21	40	92–94	108–110	50	300
E	6.0 min.	6	10	110–115	85–90	60	10

*The daily output represented good output per line. While the expected output for each of the A, B, C, and D lines was 40 units per hour, the actual good output per hour differed among the lines. Two factors accounted for much of the differences. a) The A line was shut down each morning and afternoon for a 10-minute break. The B, C, and D lines had three permanently assigned rovers to provide breaks for personnel on those lines. b) Those units which did not pass inspection in the test area of the A line were pulled aside and members of the A line worked on those units at the beginning of the next day's shift. On the B, C, and D lines more in-line inspection was added and rework was done at different points along the process.

†The quality index was quality control's indicator of quality based upon a sample of the line's "good" output.

ployees were from labor grades 7 to 12. A grade 7 received an hourly wage of DM 11.20 plus the 80 percent fringe benefits; each higher grade earned an average DM.30 more than the grade just below, i.e. grade 8 = DM 11.50, grade 9 = DM 11.80, and so forth. The new plant had seven conventional lines, but only three of them were staffed in the fall of 1983.

THE RECTANGULAR LINE

The rectangular line was essentially a closed loop design. (See Exhibit 1.) There were three variations of this design, the B-line transferred from Werk I, the C-line which began operation when the new plant opened in 1982, and the D-line which began operation in September of 1983. Rather than having the 26 work stations in a row, this second design was arranged rectangularly, enabling workers to see and interact with each other. In addition, a number of the work stations were automated screwing stations so that with a cycle time of 1.5 minutes per station, the rectangular lines had a lower labor content than did the conventional lines. Von Woedtke explained the rationale for the rectangular system:

> I personally recommended that the system should be a closed loop, rectangular design. I wanted the workers at the beginning of the line to see what was going off the end of the line. I also thought a rectangular line would encourage more of a group identity than a long line stretching from one end of the plant to the other. And, remember the technical problem of getting the work piece carrier back to the start of the process. It's not very difficult if the last station is near the first station.

The line was staffed from labor grades 8 to 12 and these employees were cross-trained to perform four or five different job tasks. Workers rotated among stations during the day by weekly job assignment. As could be expected, the rectangular line required a much higher investment than the conventional moving belt, in fact four times as much for the C-line. Von Woedtke noted, "We would not have made this investment without government help. The government paid 50 percent of the engineering expenses and approximately 50 percent of the investment in tooling and equipment for the new lines. The investment for the new lines appeared to be paying off; the two new lines' quality indexes ranged from 106 to 110 and their total cost per unit figures were less than 96. Von Woedtke was quite pleased with these results. The new plant's C and D lines were each running on double shifts; the B line was on a single shift.

Both lines C and D were designed to be more flexible with regards to model changes than lines A and B. Model changes on C and D required only 2 hours, while those on A and B required 2 days. Line C was more costly because it used more hard automation and more expensive tooling. Von Woedtke explained the evolution from A to D:

You can see how we learned from A to D. On the A line we used a moving conveyor belt without a work piece carrier. On line B we added a simple carrier. The worker still had to take the work piece off the carrier to work on it and then put the work piece back on the line. The simple carrier did allow us to add several automated screwing stations.

On the C & D lines, we work right on the line because we can rotate the work piece carrier. The process design required much more capital to allow us to rotate the workpiece. The products running on line D require less assembly labor than on B and C because we redesigned the models run on this line to enable us to use more automation and to enable us to achieve substantial tooling savings.

We believe we can do even a little better. We have a need for highly automated, highly flexible systems. But, these new systems drastically change our workforce requirements. For example, on lines C and D, we have a much higher need both for mechanical maintenance skills because of the new work piece carriers and the more complex automated screw machines and for electronic technicians because of the numerical control equipment utilized on these two lines. (These technicians were in grades 15-17.)

In the rectangular system there was a small buffer stock between stations but a larger buffer stock between major sections (10 or so stations) of the process. In the B line, there was one section buffer of 60 minutes and about 15 minutes of buffer between individual stations. Because of their uncertainty about the new equipment in line C, Stihl management added two additional section buffers each with 60 minutes of inventory, while reducing the between-station buffers to 7.5 minutes. In the recently completed D line, they reverted back to a single section buffer of 1 hour and kept the between station buffers to 7.5 minutes of stock. The buffers were designed, according to Von Woedtke, "to allow us to react to various problems at the work stations without shutting down the entire line."

GROUP BUILD

The group build production setup was designed on the concept of grouped jobs. This third alternative had six work stations or stalls, where a worker performed four to five of the job tasks of the conventional line at a single work station. (See Work Structure E in Exhibit 1.) Consequently, the group alternative was staffed with more highly skilled workers from labor grades 10 and 11; these levels corresponded to hourly wages of DM 12.00 and DM 12.50, respectively, plus 80 percent fringes. [Labor grade 11 was the grade at which qualified apprentices entered the Stihl workforce.] Once a worker had mastered all six job stations, that is, all 30 job tasks, he or she attained level 11 rank. As with the rectangular line, the group build setup was arranged in a separate work area so that the six workers could see and interact among themselves. With no automation (the work place was a well-designed table) and only six work stations, the investment index for this alternative was only 10; it also quite flexible to model changes. Also, cycle time per station was six minutes, yielding a lower labor content, than the conventional line. The lower labor content was the result of less handling of the product such as picking the chain saw up from the line and then returning it to the line. The group concept seemed to von Woedtke as though it should be an ideal production setup: it employed highly skilled workers, it produced an opportunity for employee advancement, and it required a low investment. But he was puzzled why the quality index on the group structure was below 90 and the cost per unit index well over 110. Von Woedtke explained why he thought there were problems:

> The group structure requires the workers to learn too many tasks. Only one third of the workers assigned to the group have achieved our quality and output targets. The group structures require so much more training—at first, 4 to 5 times the training required on the conventional lines and, eventually, 30 times the training if the operator is to perform successfully all the tasks of assembling a chain saw. Remember, most of these workers were volunteers or transfers from the conventional A lines in the old plant.

Von Woedtke continued his discussions of the workers, especially their reaction to the second shift[5]:

> We have little trouble finding people to work on the second shift. Ninety-five percent of our workers are guest workers;[6] most desire to earn as much money as possible. They like shift work because they receive DM400 (tax free) for shift work and because shift work allows one parent to be home with the children at all times if the husband and wife can work opposite shifts.

He also commented on the works council involvement in the project:

> The works council has been very involved; they have participated from the very beginning when the new plant was just an idea. Although they have been critical at times, they have been very supportive of our efforts. We have shared with them all our results.

But, Von Woedtke wasn't as pleased with the union's position:

> The union has taken a more negative stance against the experiment. They expected more money for the workers and more discretion for the workers over their job environment than it was possible to give them. In contrast, the works council understands our constraints and more or less accepts the new structures as they are designed.

---

[5]Shiftwork for Stihl was two weeks on first shift; then two weeks on second shift.

[6]Guest workers were workers who did not have German passports; the workers in the Stihl plant were mainly from Turkey with some workers from Spain, Italy, and Yugoslavia.

Von Woedtke then addressed the concerns of the individual workers:

Before and during the project we have surveyed employees about every six months. We have five years of responses. The basic measure has been the degree of worker dissatisfaction with their jobs. The amount of dissatisfaction has been significantly reduced on the new B, C, D lines but not on the A and E lines. The E-line workers are under great stress. They know they aren't doing well. We share with them the figures on actual labor times and quality levels versus the targets. The workers who remained on the A-line have been steady in their level of dissatisfaction. Some workers on B, C, and D lines have requested to be moved back to A lines; some on E have requested to move back to B, C, or D. In most cases, this means a reduction in pay due to a job grade reduction and the giving up of the monthly shift bonus.

Von Woedtke sat back down at his desk. Yes, the professionals at the work structure conference would find the results from Stihl's alternative work structures most interesting indeed. The question was, would they be able to explain the results?

Century Paper Corporation

In August 1976, John Murphy, manufacturing manager for Century Paper's Michigan Pulp Mill operations, reflected on his past two years' experiences. It had been an exciting period of activity for the Pulp Mill; an $11-million capacity-increase project moved toward completion, and a joint labor-management committee had recently designed and instituted several substantial changes in work schedules, task assignments, and technical training requirements. These changes were directed at further improvement of what was already generally recognized as one of Century's best-managed operations.

While most managers and hourly employees were recognizing significantly improved outcomes from the new changes, the Pulp Mill supervisors were experiencing considerable difficulty in adapting themselves to their new work situation. The four supervisors felt that the organizational changes required them to adopt a whole new set of skills in order to "survive." Over the next month John Murphy, along with his two department managers, planned to devote a great deal of their attention to the supervisor issue. They hoped that well-planned and executed remedial action could improve the situation and bring the supervisors "up to speed."

This case was prepared by Assistant Professor Leonard Schlesinger under the direction of Professors John Kotter and Richard Walton as a basis for class discussion rather than to illustrate either effective or ineffective handling of an administrative situation.

BACKGROUND OF THE PULP MILL PROJECT

The Michigan operations of the Century Paper Corporation included one sulfite pulp mill, nine papermaking machines, and a number of end-product converting operations. Until 1974 the pulp mill had served as the sole supplier of the sulfite pulp used in the paper making process.

Rising sales of Century's products resulted in an expansion of papermaking and converting operations throughout 1974 and 1975. These expansions raised the total amount of sulfite pulp used to 300 tons per day. The pulp mill, however, was able to produce only 240 tons per day, even when operating a 7-day-a-week, 24-hour-a-day schedule. As a result, Century was presented with a major make-or-buy decision relative to future pulp needs.

Despite an oversupply of sulfite pulp on the world market in 1974, Century chose to authorize an expenditure of $11 million to increase the capacity of the Michigan pulp mill to 310 tons per day. Bob Jensen, the Michigan operation's general manager, reflected on the appropriation:

> The $11-million appropriation was a testament to our ability to manage well. However, given that this appropriation is the largest single upgrade expenditure in the company's history, there is even greater pressure on us to pull it off without any hitches.

THE PULP MILL PROCESS

Pulp Mill operations were housed in a series of 20 interconnected buildings of various shapes and sizes which encompassed a concentrated area of approximately four large city blocks. The complex represented a capital investment in excess of $35 million.

Exhibit 1 represents a flow diagram of the pulp mill manufacturing process which can be briefly described as follows:

1. Wood is purchased from area foresters in 100-inch lengths and is peeled prior to entering the woodroom where the logs are chipped and placed into storage.
2. The wood chips are removed from storage as needed and transferred to the digesters. Digesting is a high-temperature, high-pressure chemical cooking process designed to break down the tree material (lignin) which holds the wood's cellulose fiber together. The digester cooking solution (ammonia bisulfite) is manufactured in the acid plant located within the pulp mill complex. The digester cooking process is critical to the successful flow of pulp production. An error at this stage of the process affects all of the other processes downstream and limits production. However, as a continuous flow process, breakdowns at almost any stage of the process can slow down or shut down the mill.

EXHIBIT 1 Pulp Mill Process Flow Design

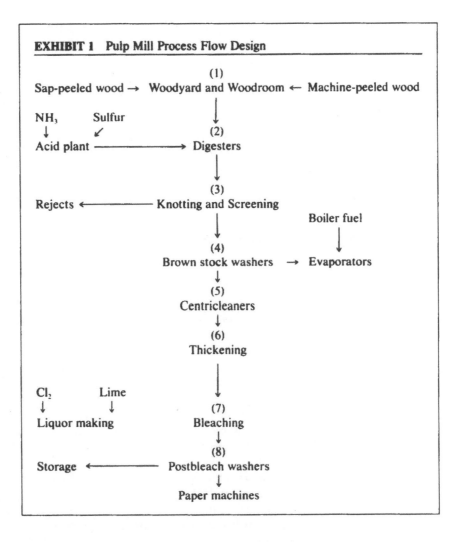

(1)

Sap-peeled wood → Woodyard and Woodroom ← Machine-peeled wood

NH₃   Sulfur

Acid plant ⟶ Digesters (2)

Knotting and Screening (3)

Rejects ←

Brown stock washers → Evaporators (4)

Boiler fuel

Centricleaners (5)

Thickening (6)

Cl₂   Lime

Liquor making   Bleaching (7)

Storage ← Postbleach washers (8)

Paper machines

---

3. At the conclusion of the digester cooking process, the now pasty wood solution is transferred to a knotting and screening operation where the tree knots and incompletely digested wood are removed as rejects.

4. Next the "brown stock" washing process begins. A series of high-pressure showers "wash" the wood "paste" to separate the cellulose fibers from the remaining digester cooking solution. The recovered digester cooking solution in turn is sent to the evaporators where it is converted to fuel to be burned in the mill's boilers.

5. From the washers the "paste" undergoes a purification process called centricleaning, which removes sand, grit, and foreign material present.

6. Next, the paste goes through a thickening process designed to increase the consistency of the pulp stock.

7. The pulp is oxidized in a bleaching process to achieve a desired white-ness. The bleach "liquor," as it is called, is a mixture of lime, chlorine, and water, which is produced on site.
8. The stock goes through another washing process and is either placed into storage for later use or sent on to the paper machines for use in the papermaking process.

The technology used in this manufacturing process continually changed in incremental ways through the installation of new equipment. The direction of this change was consistent; the Pulp Mill had been becoming more of an automatic process control facility.

THE PULP MILL ORGANIZATION

Structure

Exhibit 2 contains an organization chart for the Pulp Mill management group. The Pulp Mill *manufacturing manager* assumed overall responsibility for operations and liaison with the rest of the Michigan facility. Reporting directly to the Pulp Mill manufacturing manager were the chemical engineer, industrial engineer, and the two department managers.

The *chemical engineer* had various responsibilities, including process improvement, chemical usage and management, quality monitoring, and control. Any new chemical materials entering the Pulp Mill were tested and approved by him. In addition, the chemical engineer played a major role in cost reduction programs through adopting chemical substitutes or reductions in usage rates.

The *industrial engineer* served as the Pulp Mill's fiscal and budgetary officer. A good deal of his time was spent in the preparation of budgets and forecasts and in the preparation of cost information data. In addition the Pulp Mill Industrial Engineer coordinated the Mill's cost reduction programs.

The two *department managers* subdivided the task of managing the Pulp Mill's daily operations by dividing the mill into a pulp-making part (up through centricleaning) and a pulp-finishing part (up until the pulp is accepted by the paper machines). Each of the department managers had a maintenance manager, two area managers, and two supervisors reporting to them. A good deal of the department managers' attention was devoted to supervising and developing the managers who reported to them. Regular performance feedback discussions were routinely held.

The *maintenance managers* each supervised a crew of from six to eight mechanics and maintained responsibility for the upkeep and maintenance of the machinery and equipment within their area of the mill.

The *area managers* were technical experts in complex portions of the

EXHIBIT 2 Pulp Mill Management Organization

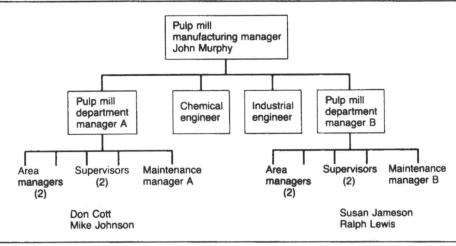

Pulp Mill process. There was one for the digesters, one for the brown stock washers and acid plant, one for bleaching, and one for all of the screening, centricleaning, and so forth. In the role of the technical experts the area managers worked with the maintenance managers on equipment upkeep and trouble-shooting, designed process improvements or equipment alterations, and now coordinated the capacity increase project work in their areas. The area manager's job was a fairly new position in the Pulp Mill (begun in 1973). As the technology of pulp production became increasingly complex, such a position was deemed necessary to "keep on top" of process changes and improvements. However, the area managers as a group lacked the official authority to secure maintenance and managerial assistance and relied heavily on interpersonal skills to accomplish these tasks.

The *supervisors* were responsible for shift coverage of pulp mill operations. Although they reported to a single department manager they were responsible for supervision of the whole pulp mill on their shift. A common shift's duties included several tours of the pulp mill to check on equipment, employees, etc. and responding to calls on equipment problems by inspection, contacting maintenance employees, supervising repair work, and documenting their actions. In addition supervisors were the primary disciplinarians and "monitors" of hourly employees. They checked attendance and lateness, followed up on employees absent from work, admonished employees for sub-par performance, etc.

The supervisor's job served as the entry-level position for those interested in careers in manufacturing management. All of the area managers and maintenance managers, along with one of the department managers, the Pulp Mill manufacturing manager, and the chemical engineer had been Pulp Mill supervisors earlier in their careers.

The Pulp Mill operated with three shift crews with at least 10 hourly employees on each shift. Exhibit 3 contains an hourly organization chart.

The employees were spread throughout the Pulp Mill and relied mainly on an internal telephone system for communication. However, many employees had the flexibility to visit with the employee(s) nearest to their work area.

The Work Schedule

Each Pulp Mill employee was assigned to work a six-day week with a rotating day off. Employees also had the opportunity to work each Sunday at double-time premium pay. This alignment of work schedules resulted in an average paid work week for pulp mill employees of 56 hours versus the 40 to 48 hours that were common in other parts of the Michigan facility. With each employee having differing days off Pulp Mill employees were distributed among three shift crews which changed virtually every day.

Pulp Mill managers (aside from the shift supervisors) generally worked a regular Monday to Friday work week. Shift supervisors worked under a "Southern Swing" rotation which was common for the rest of the Michigan facilities. The "Southern Swing" work schedule called for seven days on then four days off, seven days on then two days off, and seven days on then one day off.

The schedule resulted in the shift supervisor providing the only supervision in the Pulp Mill outside of regular daytime hours. Other Pulp Mill managers were "on call" 24 hours a day and could get into the plant on a half hour's notice.

Compensation

All managers were compensated on a salaried basis with a regular provision for salary reviews at least yearly. Salaries were generally regarded as being among the highest in the paper industry.

Pulp Mill employees were compensated on an hourly basis with rates negotiated with the local union. In 1975 the lowest-rated Pulp Mill job paid just under $5/hour with the top Pulp Mill hourly rate approaching $6.60/hour.

There was no bonus or incentive compensation program for either managers or hourly employees.

EXHIBIT 3   Pulp Mill Hourly Organization (pre-changeover)

1. Digester cook	Responsible for the wood-chip cooking process. Controls input, cooking cycle, and flow of pulp stock through the four digesters. Basically tasks are process-control-oriented. Mistakes at this level are the most costly to production.
2. Acid maker—Brown stock washer	Responsible for the manufacture of digester cooking acid and the brown stock washing process. Monitors process flow and controls mixing equipment.
3. Digester cook helper	Assists the digester cook in the function of his/her duties.
4. Bleacher	Responsible for the bleaching process. Conducts frequent quality and brightness tests on the pulp stock in addition to calibrating bleaching equipment.
5. Bleach liquor maker	Responsible for the manufacture of the bleach liquor used in the bleaching process. A process control activity.
6. Digester cook helper relief	As a result of the six-day-on-rotating-day-off schedule, the digester cook helper relief's role was to serve as a "fill in" operator on other employees' days off.
7. Evaporator operator	Responsible for controlling the process by which used-up digester cooking solution is converted to boiler fuel.
8. Rejects operator	Responsible for the accumulation and storage of reject stock eliminated in the knotting and screening process. Fairly routine tasks with heavy direct labor component.
9. Pulp storage operators (2)	For pulp which does not go directly to the paper machines there are two pulp storage operators responsible for preparing the pulp for storage. Tasks are highly routine and mechanical. Job involves continuous heavy lifting.

Selection and Development

Century Paper recruited nationally for college graduates interested in careers in paper manufacturing management. The bulk of their new managers were degreed engineers from Midwestern schools. Century practiced a strict promote-from-within policy, with virtually all new hirees beginning their careers as shift supervisors. This initial assignment generally lasted from one to three years. In addition to recent college hirees, several longer-term supervisors had been promoted from the ranks of the hourly work force. Century believed that these individuals served to provide a sense of continuity in the first level of supervisory management.

Most of the hourly employees were born and raised in the plant community with newer employees being attracted from local high schools. The community was rural in character with its citizens strongly influenced by a European Catholic background. This background was evident in the people's strong work ethic, in the pride they took in their work, and in the importance they put on relationships and on liking the people they worked with, who were often the same people with whom they socialized. It could also be seen in their strong family orientation and respect for those older and more experienced. A significant number of the older employees owned and operated local farms in addition to working at Century.

The pulp mill had a considerable age range among its employees. At one extreme were men with 20 to 25 years' experience, at the other, young men and women who had just started at Century. Many of the older employees could be found discussing "the way things used to be" and were often drawing comparisons with their past experiences at Century.

The Union

The Pulp Mill employees were represented by the International Paperworkers Union. Union-management relations were generally perceived as being positive at the Michigan facility. They had suffered no strikes or work stoppages in the past 30 years. However, recently, the union had decided to clog up the existing grievance machinery in many of the departments by reporting virtually all indications of contract infractions (the Pulp Mill was not affected). Much of the tension in the relationship could be attributed to the large amount of change occurring at the plant and the continuing pressure it placed on the union to "look after the interests of its members." However, communications with union officials were continuing; they were being notified by management well in advance of changes in order to jointly discuss them, and each operating department, as well as the plant as a whole, had developed formal mechanisms for ongoing union-management dialogue.

THE WORK RESTRUCTURING PROJECT

Prior to 1974 all changes in the Pulp Mill operations, both technical and environmental, were met with what could best be described as "patchwork" changes to the mill social system. A good example of this phenomenon was the schedule with stable crews working five-day weeks. The 1960s brought increasing demands for pulp and a six-day operating schedule. The employee work week was simply extended to six days, again with stable crews. In the late 1960s the Pulp Mill moved to a seven-day operating schedule, and Pulp Mill management simply introduced the rotating-day-off concept; a simple solution which led to the crew instability described earlier.

Now there was a great deal of pressure on the Pulp Mill management to make the capacity increase project an enormous success. With this in mind, managers expressed a desire to go beyond upgrading, replacing, or adding equipment and were beginning to look at making some inroads to changing the Pulp Mill work climate. Especially at the department and manufacturing manager levels, there were strong feelings that the capacity increase project provided a vehicle for making some fundamental work improvements. Highest on the management agenda was changing the work schedule for all employees to the Southern Swing schedule. Other items managers wished to address were:

1. Upgrading the employee's technical expertise.
2. Maximizing the employee's job flexibility.
3. Promoting teamwork.
4. Providing employees with jobs that are "meaningful."
5. Attracting and keeping talented employees.

After several months of management discussion on these issues, the union leadership was invited to participate in a joint union-management committee to design and implement "social system" changes in conjunction with the capacity increase project. The union, desirous of having input into any such process, readily agreed to participate in the joint venture.

From February to September 1975 a joint union-management committee, consisting of the Pulp Mill manufacturing manager, the two department managers, the Michigan Facility industrial relations manager, the local union president and vice president, the Pulp Mill committeeperson and the two Pulp Mill shop stewards, met regularly to discuss potential changes. In addition, area managers and employees participated in various task forces designed to provide input to the joint committee. The changes finally agreed upon included:

1. Schedule Change to Southern Swing. As was indicated, management entered the process determined to change the employee work schedule to the "Southern Swing" work schedule more common to the rest of the plant. Such a shift would allow for the formation of four stable work crews which rotated regularly along with a supervisor who worked the same schedule. As Bob Jensen stated:

> Our existing work schedule was designed in the late 50s for a five- or six-day operation. Times have changed and our existing schedule no longer made sense. Perhaps we should have changed earlier but now is clearly the time to update the way we schedule our people for work.

A common work schedule would also allow for the development of supervisor-operator teams, which was viewed as being extremely desirable.

2. New Job Designs. The Michigan plant had traditionally relied upon detailed, task-oriented job descriptions which always ended with the catchall phrase "and any other duties as assigned by management." The vagueness of statements such as this had resulted in several unnecessary disputes between line managers and employees over the years. In the new job descriptions, however, the committee agreed to drop such detailed task listings in favor of one which simply listed the processes the employee was accountable for, the outcomes which the company expected from the employee's efforts, and the skills necessary for performance of the job. In addition, to assist management in setting new wage rates, for each position a "Project Job Change Summary" was completed which detailed the equipment changes and their impact on the performance of each job.

It was determined that seven of the current jobs were to be replaced by eight new or revised ones:

Pre-project	Project changes
1. Digester Cook	1. Digester Cook
2. Acid Maker-Brown Stock Washer	2. Pulp Making Operator A
3. Digester Cook Helper	3. Pulp Making Operator B
4. Rejects Operator	4. Digester Cook Helper
5. Bleacher	5. Pulp Finishing Operator
6. Bleach Liquor Maker	6. Bleach Liquor Maker
7. Evaporator Operator	7. Evaporator Operator
	8. Pulp Relief Operator

There were three major changes in the job structure: the change to Pulp Making Operator A & B, the addition of a Pulp Relief Operator, and the combination of the Rejects Operator and Bleacher jobs into a Pulp Finishing Operator.

Equipment additions in the Acid Plant and Brown Stock Washers area necessitated breaking the job up into two separate jobs. The A and B operator concept was designed to gain maximum flexibility from and for the employees affected. The new job descriptions increased the flexibility an employee had in the performance of his/her duties and the A and B concept called for cross training, mutual assistance, and overlap of responsibilities. It was envisioned that a successful test of this concept would lead to further efforts of this nature.

Process improvements justified merging the Rejects and Bleacher jobs into a single all-encompassing position.

The addition of a Pulp Relief Operator was designed simply as an outgrowth of the desire to see what could productively be accomplished with an extra person on each shift. It was expected that this individual could be used to free up other employees for training, committee work, or special projects. In addition, the Union agreed that it was a viable option to utilize this operator to cover for absent employees so that the company would not have to pay overtime.

3. Compensation Changes.   Inherent in the schedule change to Southern Swing was a wage loss averaging eight hours pay per employee per week. Plant management was concerned about how this could possibly create resistance to a schedule change. At the same time, however, they did not want to give the impression that they were going to place themselves in the position of simply increasing salaries to "grease the wheel" for smooth implementation of the changes agreed upon. They finally decided that a management task force consisting of the plant industrial relations manager, the plant industrial engineer, and the Pulp Mill managers would review all plant jobs in light of recent equipment and process changes and adjust salaries accordingly. The adjustments were to be made by comparison with the original jobs, other comparable plant jobs and similar jobs in the paper industry in the area.

As a result of the evaluation, the committee recommended wage increases ranging from 5 percent to 8 percent (28¢ to 45¢) on most of the new jobs. The wage increases represented approximately 50 percent of the wages which employees would lose through the schedule change. The Union appeared pleased with the results, and (with the exception of a few minor concerns which were dealt with immediately by the management) accepted the changes in their entirety. The ease in gaining Union acceptance of the revised wage rates could be explained by the fact that the raises granted represented the largest (outside of collective bargaining) wage increases ever given at the Michigan facility. In addition, employees generally were very willing to "trade off" some loss of earnings for their new free time.

4. Shift Teams.   The schedule change and revised job descriptions provided for the establishment of shift teams under the guidance of the supervisors. It was jointly agreed that overtime, training, and education efforts in the area of team development and effectiveness would be directed at the Pulp Mill employees.

5. Training.   A $60,000 appropriation was approved for the development of new training materials to significantly upgrade the knowledge of Pulp Mill employees of the pulping process in general and their area in particular, beyond their present knowledge base of, "If A happens, push knob B."

6. Additional Items.   In addition, it was agreed that in many areas equipment would be provided to allow operators to begin performing their own quality tests as a means of gaining more control of the process.

There were few, if any, difficulties in the joint union-management resolution process. A significant proportion of the Pulp Mill employees had demonstrated their desire to participate in the project via their work on the committee or the various task forces, thereby resulting in a good deal of joint ownership in the outcomes.

Implementation and Development

All prior events appeared to set the stage for a smooth transition to the new system in the fall of 1975. No one was able to foresee, however, that technical problems would plague the first phase of the start-up. Managers worked day and night over a six-week period to attempt to stabilize the new processes, and employees were instrumental in assisting their troubleshooting efforts. Employees also exhibited much concern over the status of the process and the lost production. They appointed a group of employees to approach the plant management and volunteer the crews to work through the Thanksgiving holiday rather than shut the Pulp Mill down. They indicated that their "new" understanding of the business (generated via participation in task forces) convinced them of the desirability of not shutting the mill down when production levels had been so low.

Each week, Pulp Mill management-union committees continued to meet to work out some of the details and fine-tune the various changes. Union grievances dropped, too. Morale was high as was the energy level of employees.

Project Outcomes

The results of the project clearly matched the expectations of the joint committee; employee morale improved, employees became more active in problem troubleshooting and technical process work, attendance improved, and desired production improvements were steadily being recognized. Much of the positive outcome of the project was ascribed to the broad-based participation in the development of the changes. Exhibit 4 contains an assessment of the Pulp Mill social system in March of 1976. Managers and employees alike viewed the bulk of the changes in a fairly positive light. Representative comments included:

Bleacher operator:

> It's about time that managers around here realized that I don't have to turn off my mind when I come to work. I'm enjoying the opportunity to solve problems on my own and get recognized for it.

EXHIBIT 4 Pulp Mill System Assessment—March 1976 Data Generated by Pulp Mill Department Managers

---

I. *Job Designs*
   Positives
   1. There is a great deal of flexibility that is being utilized—people are no longer saying "that's not my job."
   2. Employees are housekeeping their own areas.
   3. Operators can be seen assisting each other regularly.
   Negatives
   1. Supervisors are still not comfortable with the system and are not utilizing the flexibility to its most productive ends.
   2. Operators are still not comfortable doing on-the-job problem solving to the extent desired.

II. *Teams*
   Positives
   1. People are always talking about *their* team.
   2. Some teams have already begun safety and clean-up programs.
   3. Supervisors are now taking more responsibility to communicate with employees—operators say they are getting more information and are informed of the status of follow-up items.
   4. The Union continues to give the effort its active support by continuing to work on task forces, solving problems before they become grievances, and making flattering comments about Pulp Mill management at Union meetings. The general membership does not perceive the project as a threat because of the lack of "buzz words," technically difficult behavioral science concepts, and outside interferences by corporate staff.
   Negatives
   1. Area teams haven't developed to work on some parts of the technical process.
   2. Supervisors are having a difficult time in coping with their new role.

III. *Supervisors*
   Positives
   1. They are enthusiastic and *want* to understand.

EXHIBIT 4 (continued)

2. They are not waiting for engineers to solve technical problems, but are working on them jointly with their crews.

Negatives

1. They still need to develop better skills in leading meetings, handling conflict and change.
2. They have difficulties in communicating from shift to shift.
3. Supervisors, unlike their teams, are not taking the responsibility of making timely decisions when they are needed.
4. They spend a great deal of time as "go-fers" for operators.

IV. *Union Management Interaction*

Positives

1. Successful schedule change.
2. Union and management members are above board and do what they say—they feel part of an important team and take ownership in their actions and decisions.

Negatives

1. Need to expand the membership base for joint activities (supervisors, shift team representatives).

V. *Management of Change*

Positives

1. Getting results (with more to come).
2. Communication channels are wide open.

Negatives

1. Often have problems reaching timely resolution on issues.
2. There is often no explicit direction on several items to be worked.

VI. *Union*

Positives

1. Officials take ownership in the changes and have received a large part of the credit for the positive outcomes.
2. The membership rates Union activity on this project as excellent—they are satisfied with the salary changes.
3. The union can point to this project in other operating areas when complaining about conditions.

VII. *Hourly Operators* (Quality of working life)

Positives

1. Information is now reaching the operator level more often.
2. Workloads and the distribution of dull tasks are better balanced.
3. Operators have been able to upgrade their technical competence.
4. A reasonable work schedule which provides for more leisure time exists.
5. They have a greater say in how the Pulp Mill is run.

Negatives

1. Operators need more training to develop these skills even further.

Union representative:

> What we've accomplished here in the past few months is really good. The employees are generally happy with the wage increases and really like the new schedule. We keep solving problems before they blow up into big things.

Chemical engineer:

> If someone told me we could have pulled off such a big change a year ago I would have told him he was crazy. But we did it and it seems to work.

However, one area of continuing concern surrounded the role of the supervisor in the new operating system. Supervisors were not an important part of the original design process and it appeared that employees were more aware of the flexibility inherent in the new operating system than were their supervisors. The simple change of having stable work crews day in and day out fostered a strong team spirit and placed the supervisor in a position where the traditional role of disciplinarian and technical problem solver in all areas no longer fit. One supervisor related a recent occurrence:

> On the night of the mill manager's picnic we ran the mill without a supervisor on the premises. A serious injury occurred to an employee on the shift. One employee coordinated coverage of the equipment while another escorted the employee to the hospital for treatment. It wouldn't have been any smoother had a supervisor been here. In fact, it might not have been as smooth.

It was obvious to managers and employees alike that the new systems required the supervisors to have "people development" skills which were not present within the existing supervisory ranks. One supervisor responded:

> Supervisors need to work on several skills that weren't crucial to being successful before:
> 1. Team building—both for our crews and ourselves.
> 2. Communication skills.
> 3. One-to-one feedback skills.

Unfortunately, the existing work schedule for Pulp Mill supervisors precluded their spending additional work hours on developing these skills.

SUPERVISOR MEETING—JULY 1976

Three of the four supervisors attended a meeting in July 1976 where they presented their views of the situation as it currently existed. (See Exhibit 5 for biographical data on the four supervisors and other Pulp Mill managers.)

EXHIBIT 5   Biographical Data on Selected Pulp Mill Managers

Supervisors

Donald Cott
—55 years old, 30 years with Century Paper, local high school graduate, promoted to management from hourly ranks in 1965, Pulp Mill supervisor since 1965.

Mike Johnson
—28 years old, 2 years with Century Paper, two years of college, promoted to management from clerical work force in 1975, Pulp Mill supervisor since May 1975.

Susan Jameson
—25 years old, 1 year with Century Paper, 1975 graduate of Purdue with Bachelor's in Chemical Engineering. In training September to December 1975, assumed crew responsibilities January 1976.

Ralph Lewis
—47 years old, 1 year with Century Paper, 20 year veteran of U.S. Air Force. In training June to September 1975, assumed crew responsibilities September 1975.

Pulp Mill Manufacturing Manager

John Murphy
—32 years old, 9 years with Century Paper, graduate of Indiana with Bachelor's in Chemical Engineering. Previous assignments as paper machine supervisor, paper machine chemical engineer, pulp mill department manager. Pulp mill manufacturing manager since December 1974.

Michigan Operations General Manager

Bob Jensen
—37 years old, 12 years with Century Paper, graduate of General Motors Institute, MBA from Michigan. Previous assignments as pulp mill supervisor, pulp mill maintenance manager, pulp mill department manager, paper machine department manager, converting department manager, pulp mill manufacturing manager, Michigan General Manager since 1972.

Susan Jameson:

There are closer relationships now between the supervisors and their crews. You get a chance to get acquainted with your people rather than working with different people almost every day. The supervisor now knows how the people react in different situations, how quickly they respond to changes, and how they "troubleshoot" problems. Thus the supervisor can *manage* more and *supervise* less. Knowing the crew better, along with their strengths and weaknesses allows the supervisor to develop his/her people and ultimately to abolish the supervisor's job completely!

Mike Johnson:

> The traditional image of the Pulp Mill supervisor is rapidly fading away. In months to come, if we develop our people's skills, we will be fighting far fewer "fires" and thus managing the business better. Our role is changing from a supervisor to a manager where we will be training our crews to do our job. So the shift is from meeting crises to teaching. Now we're caught in the transition and are doing a pretty lousy job of both. We don't want to continue doing all of the problem solving but our crews haven't really developed their skills in this area to the point that they can do it alone. And because of that situation we can't get off of shifts to get the necessary training to improve our "people skills."

Donald Cott:

> The supervisor is still necessary today as the crews aren't developed enough to take on full responsibility. But this is changing, perhaps more rapidly than we know. They are making more and more decisions on their own and doing their own troubleshooting and problem solving. There are many nights when the mill would have run quite nicely without me. But there have been other nights when my crew needed me there if for nothing more than to bolster their confidence. Eventually shift supervisors will be obsolete. A coordinator on shifts, especially days, however, might still be desirable. This could be a crew member though or perhaps a rotating job so everyone could do it.

One Pulp Mill department manager, while recognizing the positive tone of most of the supervisors' comments, stated the following:

> I think they're wrong when they talk about eliminating their job. The supervisor's job, no matter what level their crew is at, is essential as a learning experience for Century Paper management careers. This experience provides reality training that we all need regularly. It allows the manager the opportunity to develop in areas like group effectiveness, problem solving, priority setting, standards communication, motivation, and so forth. It allows the supervisor a first good look at everything that is production management.
>
> Now, I do see the supervisor's role changing in the coming months and years. I see them as taking the key role in lowering decision making to the hourly employee level. This must be a total Pulp Mill effort but the implementation of going from a supervisor to a resource must come from the supervisor. The move to Southern Swing is a first step. The second step is to make people development a goal for the Pulp Mill. The third step is to implement.

THE SITUATION IN AUGUST 1976

Supervisors had been meeting regularly as a group over the past three months in an attempt to come to grips with the new operating system as individuals and as a group. However, these meetings appeared to have had little or no impact on their performance as managers.

In an analysis of the situation as it presently existed John Murphy viewed a number of options to remedy the situation:

1. *Work toward a rapid elimination of the supervisor's job*—This appeared to be an attractive option. However, Century Paper utilized a "promote from within" policy and relied heavily on the supervisor's position as a training spot for promotions to higher level manufacturing positions. In addition, it was likely that the existing supervisors could not be integrated into the Pulp Mill organization and would be transferred to the papermaking or converting areas as supervisors.
2. *Emphasize additional training for supervisors over the next several months*—Murphy wondered about the benefits of a crash training effort and was concerned about the time investment it would take.
3. *Let supervisors work out a role for themselves*—Such a laissez-faire approach might be costly in terms of continuing ambiguity and dissatisfaction for supervisors but had its advantages.

Murphy planned to have several meetings with other Pulp Mill managers over the next several weeks before settling on an alternative.

Why Supervisors Resist Employee Involvement

Janice A. Klein

In 1941 many of the nation's factory supervisors, unhappy with the changing climate in the workplace, banded together to create a union, the Foreman's Association of America. Today, as the climate in the workplace is again very much in flux, resistance by supervisors is once more making itself felt. In general, when managers initiated employee involvement programs, they expected to face some resistance—but to face it primarily from workers and their unions. Surprisingly, the real foot-dragging is coming from a different quarter: first-line supervisors. Indeed, as one manager noted: "We were so worried that the employees would not accept employee involvement that we spent all our time trying to convince them of the benefits. We assumed that since supervisors are managers, they would just accept the program. We found that it was much easier to sell it to the employees than to the supervisors."

This article reports on a study of responses by first-line supervisors to employee involvement programs (details of the research are outlined in the accompanying insert). Most revealing, perhaps, is the finding that although nearly three-quarters (72 percent) of the supervisors view these programs as being good for their companies and more than half (60 percent) see them as good for employees, less than a third (31 percent) view them as beneficial to themselves (see Exhibit I).

Reprinted with permission from *Harvard Business Review*, September–October 1984, pp. 87–95.

When I asked supervisors to explain the responses summarized in Exhibit I, this was the refrain: because top management should know what is good for the company, employee involvement must be good for the company. Because the program is aimed at involving employees, it must be good for employees. But what about us? How is it going to help us?

As one supervisor pointed out: "For five years we have been beaten over the head about the need for more participation by workers. By this time we know we'd better believe, or at least say we believe, that it is good for the company and for employees. No one has really stressed that it would be good for us, just that we had better believe it or we don't have a job."

This division of perception is a genuine cause for concern. The support of first-line supervisors is essential if meaningful changes in the workplace are to take root. But if supervisors view programs that increase employee involvement as detrimental to themselves, they will withhold their support, potentially dooming the initiative. When asked how much say workers should have about things that affect them on the job (safety, for example, or wages or layoff practices), the supervisors who responded in favor of workers' participation are much more confident (75 percent to 24 percent) that such involvement would benefit them as supervisors than are those who said that workers should have little or no say.

Managers must, therefore, pay attention to what supervisors think about programs like these and try to understand why they are often reluctant to support them.

EXHIBIT 1 Supervisors' Attitudes toward Employee Involvement Programs

Percentage of supervisors responding that employee involvement programs are . . .	Plant							
	A	*B*	*C*	*D*	*E*	*F*	*G*	*Average*
Good for the company.	46%	75%	44%	67%	53%	96%	83%	94%	72%
Good for employees.	42	58	22	44	53	81	78	72	60
Good for supervisors.	15	42	33	44	32	23	39	44	31
Sample size	26	12	9	9	19	26	20	18	139
Type of program	Quality circles	Quality circles	Quality circles	QWL program	Quality circles	Quality circles	Cross-functional task teams	Semiautonomous work teams	

Sources of Resistance

Supervisors rarely show open resistance to programs top management initiates. After all, they are part of management, too. More to the point, few have access to formal mechanisms for voicing disenchantment, and most perceive that their job security depends in no small measure on following upper management's instructions. Nonetheless, the negative attitudes are not far below the surface—negative not only toward proposed changes in management style but also toward the process of change itself.[1] True, supervisors occasionally criticize a program in discussions with peers or subordinates. More often, however, they remain silent or demonstrate only mild enthusiasm, which workers quickly interpret as a questionable show of support for the program.

In one plant this nonsupport hastened the death of a quality-of-work-life (QWL) program that encouraged but did not require employees to form teams for discussing work-related issues. Because no manager or supervisor was necessary on each team, the foremen felt no need to volunteer for the program and refused to participate. With no knowledge of or influence over what occurred when the teams met, the foremen saw the activity as a threat to their control and authority, which they tried to regain by bad-mouthing the program. This bad-mouthing, in turn, discouraged many of their subordinates from participating. In the end, the whole effort just faded away for lack of interest.

In another plant, resistance took the form of supervisors keeping "hands off" as newly formed, semiautonomous work teams attempted to solve problems. When questions arose that the teams were unable to handle, supervisors replied, "That's not my job; it's the team's problem." In essence, the supervisors were undermining the teams so that they could resume their traditional position of authority.

Although managers regularly complain that it is the older, more senior foremen who cannot accept or adjust to the new participative programs, each of the plants in this study had such individuals who were very supportive of them. It may be, of course, that by virtue of their personal influence, older supervisors have a disproportionately great impact when their response is negative. Even so, the study found no consistent pattern of negative attitudes associated with age or seniority. The major causes of resistance lie elsewhere—in concerns shared by most supervisors, regardless of their age, background, or leadership style.

Understandably, the first concern has to do with job security. A question often raised by the popular press—will supervisors become redundant under a system of participative management?—is on the minds of supervisors themselves. When one plant began forming semiautonomous work teams, management guaranteed that the teams would pose no threat to the job security of hourly workers but offered no comparable guarantee to supervisors. Although none of these jobs was actually in danger, rumors to that effect began circulating.

A second area of concern is job definition. What are supervisors really expected to do, and how are they to be measured? In the plant just mentioned, management took more than three years to articulate clearly what it expected of first-line supervisors. Lack of a well-defined set of responsibilities also gave special trouble to supervisors who had to balance the egalitarian position required of them by participation in quality circles one hour per

[1]Their suspicion of change is documented in a survey of some 7,000 supervisors reported in Lester R. Bittel and Jackson E. Ramsey, "The Limited, Traditional World of Supervisors," *Harvard Business Review*, July–August 1982, p. 26.

week with the authority they exercised on a daily basis.

A third concern is the additional work generated by implementing these programs, work that ultimately falls on supervisors either for short periods of time (as with team development and training) or for extended periods (as with quality circles). One group of foremen, for example, who were not allowed to pay employees overtime for quality circle activities, found that all coordination and follow-up for the program fell on their own shoulders, especially on their time off and usually without extra pay. This was particularly true for second- and third-shift supervisors. In addition, they felt management expected them to do many things methods engineers had been responsible for.

Even after managers addressed the issues of job security, job definition, and extra work, some supervisors were reluctant to accept the concept of employee involvement programs. Evidence gathered at the plant level indicates that each of the five types or categories of supervisors outlined in Exhibit II has its own reasons for opposing employee involvement programs. Although these categories are not intended to be mutually exclusive and a given supervisor may fit into several groupings, they do serve a useful diagnostic purpose. Supervisory resisters, then, can be roughly divided into . . .

. . . Proponents of Theory X

Leadership styles, according to Douglas McGregor, grow out of the allegiance of managers to Theory X or Theory Y beliefs. Theory Y managers rely on participative techniques and value the opinions of their employees; proponents of Theory X believe workers need to be controlled closely and told exactly what to do. In each of the plants I studied, Theory X supervisors were usually among the resisters and, in many cases, were the first to be sorted out and replaced. A few

were able to modify their behavior sufficiently to "play the game." Their attitudes, however, remained unchanged. "Workers are really just children, and if you don't watch them, they will take advantage of the employee involvement program" is a fair sample of the general sentiment.

. . . Status Seekers

These resisters enjoy the prestige of their positions and do not want to relinquish any of their status or prerogative. "A supervisor will always end up the leader in a group," they typically insisted, "because you can't be an equal within a quality circle or quality-of-work-life team and a supervisor outside it." Believing that workers can be self-motivated, but personally enjoying the prestige of being the boss, status seekers regard as belittling the idea of being equal to workers or of acting in a support function. Thus, when one plant manager circulated an inverted organization chart that showed supervisors in their role as resource persons reporting to QWL teams, opposition came through loud and clear.

Further, because most organizations reward outstanding performance by promoting individuals to supervisory positions, status seekers view the sharing of their tasks with workers as a reward that has not been earned—and that diminishes their own standing in the eyes of fellow workers. Worse still, when ill-constructed programs lead employees to bypass supervisors and bring concerns or make recommendations directly to plant managers, the perceived threat to status encourages supervisors to dig in their heels.

In one plant culturally influenced by "respect for those above you in the hierarchy," supervisors viewed themselves as the leaders in choosing what should be discussed in quality circles, and the circles were designed to support that leadership role. Workers could voice their opinions and offer ideas but could not participate in decision making or in deter-

EXHIBIT II Supervisors as 'Resisters'

Type	*Why They Resist*	*Clues to Behavior*
Proponents of Theory X	The concept goes against their belief system.	Comments such as "Employees are children, not adults" and "Employees will just take advantage of the program to get out of work."
Status seekers	They fear losing prestige.	Unwillingness to let go of behavior associated with control. Fear of losing leadership role. Comments such as "Foremen can't be equal members of a team; they will always be the leader."
Skeptics	They doubt the sincerity and the support of upper management.	Comments such as "This program is no different from past ones. It will fade away in a few months"; "The problem is the next level up"; and "They don't really practice what they preach."
Equality seekers	They feel that they are being bypassed and left out of the program.	Comments such as "Why do we have to change before the employees do?" Nonsupport and "hands off" as problems arise.
Deal makers	The program interferes with one-on-one relationships with workers.	Comments such as "We've been stripped of our power" and "We have no control over the process."

mining which problems to discuss. This arrangement worked because it fit the organization's culture and structure and did not threaten the supervisors, but it failed to give employees any real say in their workplace.

. . . Skeptics

While proponents of Theory X and status seekers oppose the very concept of employee involvement, skeptics question the ability and desire of an entire organization to change. Not surprisingly, my research confirms that the sincerity of top managers determines whether supervisors view a given initiative as just another program to rally up the troops or as evidence of a real change in managerial philosophy. Indeed, one of the reasons supervisors most commonly gave for the limited success of quality circles at their plants was the lack of upper management support in setting priorities, attending team presentations, and allocating staff and other resources.

When, for example, one plant decided to embark on a broad QWL program, supervisors wondered, "Since quality circles didn't last, why should we believe management will support this program?" Similarly, at a plant where the training process for employee involvement stretched out over seven years, many supervisors (and their managers) came to believe that the talk of change was just that—all talk and no action. Only at a plant where the shop had been completely reorganized into semiautonomous teams was skepticism minimal, for the thorough alteration of how work was done had convinced all employees that the change was permanent. The team concept had ceased to be a "program"; it was now simply the way the work force was managed.

All too often middle managers are every bit as skeptical as supervisors, and their lack of commitment can stop a program in its tracks. As one manager noted: "One of the main problems early on was that we didn't have

the support of the middle managers. Often a manager would walk up to one of the employees and sarcastically remark, 'Oh, aren't the teams running well,' which would be viewed as the manager making a joke out of the teams. When the supervisors heard this, they questioned the wisdom of supporting the teams if their boss didn't believe in the idea."

At the outset, of course, such support is always fragile and can easily be destroyed by anything that rekindles the skepticism often rooted in supervisors' past experience. A history of short-lived programs, incongruent actions by upper management, or superiors not genuinely committed to change is more than enough to light the spark. When asked how concerned management was about the morale and working conditions at their plants, supervisors who rated their managers highly were significantly more positive (39 percent versus 25 percent) toward employee involvement programs than were those who gave their superiors low marks. This result should not be surprising. When managers seem not to care about the quality of work life, supervisors are reluctant to stick their necks out for programs designed to improve it because they cannot count on getting long-term support.

Where managers and workers are traditional adversaries, supervisors may legitimately wonder if workers or unions will want to participate in employee involvement activities. There is also the question of who should be the first to compromise. In one plant with a history of poor labor-management relations, supervisors openly questioned why they should have to change their behavior before the union formally agreed to the program under discussion.

. . . Equality Seekers

A fourth group of resisters, equality seekers, want more involvement for themselves, because employee-oriented programs should not be merely "the top telling the middle what to do for the bottom." They want the middle to have some say both in the decisions made above them and in the process of involving those below. Hence, although they object to most programs as currently designed, they are not opposed in principle to increased employee involvement—provided that supervisors are included as employees.

At base, these supervisors are asking, "Aren't we as important as workers?" As one manager recalled: "In the design process we decided we should take the reserved parking spaces away from the foremen as a show of good faith to the union. However, the foremen made an issue out of it and we had to reverse our decision. In essence, they were asking where the quality of work life was for them."

The concern of equality seekers is not that employees are finally getting recognition but that they are getting it at supervisors' expense. Concern can rapidly become frustration if complaints are ignored. "They ask for our opinion," runs an all too familiar criticism, "but then ignore it and do whatever they were planning in the first place."

Because employee involvement inevitably means that first-line supervisors must give up some of their traditional power, it makes sense that their attitudes toward the programs reflect the amount of power they perceive themselves as having. If they feel secure in their positions and in the authority they wield, they should be willing to share some of it; if, however, they feel relatively powerless, they will be less inclined to welcome changes that reduce what little authority they do have. The survey results confirm these expectations. Supervisors who see themselves as participating in decision making were much more positive toward employee involvement programs than were those who did not: 44 percent versus 25 percent. Supervisors who perceive themselves as having a say in their own activities were also more positive toward the programs: 39 percent versus 19 percent.

. . . Deal Makers

The power of foremen has long rested on their ability to reward or penalize workers. Before the rise of trade unions and personnel departments, foremen had almost total control over the workplace.[2] To stimulate production they could use the threat of dismissal or the favor of hiring a friend or relative. With the development of workplace rules and union contracts, supervisors would make such informal, one-on-one deals as: "Meet the production quota for the shift, and you're free to do as you like until the bell rings." In addition, they had to rely on more subtle exchanges, such as the promise of good job assignments or time off from the job.

Employee involvement programs are changing and challenging what supervisors can do informally. Instead of being able to deal one-on-one with employees, supervisors must increasingly manage through a team or group. Everyday tasks like job assignments are now often the responsibility of the teams themselves. Further, supervisors, who have long been the prime conduit for the flow of information between their superiors and workers, find that quality circles and QWL programs encourage a direct flow of communication from top management to the shop floor.

As a result, deal-making supervisors need to find new mechanisms of influence. Their resistance stems from their unwillingness to let go of the old ones. A team that reported to a supervisor who was notorious for making deals with the shop steward discovered that the supervisor was attempting to undermine its efforts to reform job assignments. If, as the team discovered, the supervisor handed out work requirements at the beginning of the

---

[2]For a historical review of supervisory power shifts, see Leonard A. Schlesinger and Janice A. Klein, "The First Line Supervisor: Past, Present, and Future," in *Handbook of Organizational Behavior*, Jay W. Lorsch, ed. (Englewood Cliffs, N.J.: Prentice-Hall, 1987).

week and allowed the team to do the assignments, it could sharply reduce manpower needs. After several unsuccessful attempts to get the supervisor to support the idea, management finally transferred the supervisor to an area in the plant where there were no teams. Shortly thereafter, the team regrouped and began generating new ideas to improve productivity.

When asked how they preferred to deal with employees, 85 percent of the supervisors responded that they prefer one-on-one (as opposed to group) interactions. Asked whether employee involvement programs improved their ability to have such interactions, less than half (45 percent) said yes. Not surprisingly, supervisors who thought the programs increased their ability to deal one-on-one with workers were quite well disposed (54 percent versus 18 percent) toward the programs.

The study also asked supervisors how useful they found formal policies and procedures in doing their jobs. Since employee involvement programs require management to be more flexible, one might expect supervisors who find formal rules least useful to favor the new programs most strongly. The exact opposite proved to be the case: supervisors who find rules useful like the programs better than those who do not.

This result lends support to the deal-making argument. Supervisors who prefer to handle their employees by means of informal arrangements have little use for formal rules. Deal makers resist employee involvement programs because they interfere with their own style of management. As one supervisor noted: "Under a quality-of-work-life program, everyone gets involved and knows the rules and, in some cases, is involved in making the rules. It is an extremely democratic process. But because everyone knows what is going on, things become more formalized. Therefore, those who do not like rules are negative."

Managerial Responses

Most managers respond to supervisors' resistance by giving the problem lip service and hoping it will fade away. After all, or so the reasoning goes, resistance to change is perfectly understandable—and temporary. Indeed, this is how most managers in this study felt, but with few exceptions, the resistance did not dissipate.

One exception was a case involving supervisors hired fresh out of college who were, as a team member explained, "really gung ho, thinking they were the boss. It took them about six months to learn that if they treated us with respect we'd perform and they'd look better." With most supervisors, however, the reasons for resistance short-circuited the development of a "this is the way things are now, so we just have to accept it" attitude, and managers had no idea what to do.

One plant repeatedly tried to sell the concept of semiautonomous work teams through formal presentations and informal discussions, but even this approach backfired. One supervisor recalled, "The main problem was that they [upper management] kept telling us how helpful it would be for us. They built up our expectations. Then we found out that it wasn't all that good for us. It would have been better if they hadn't said anything."

The most prevalent approach was some form of training. One plant sent some of its disbelieving supervisors to sister plants that had institutionalized employee involvement. Seeing it in action helped convert several hard-liners. By itself, however, the usual kinds of classroom training proved relatively ineffective.

Instead, successful responses focused primarily on providing employee involvement for foremen and giving them recognition similar to that given workers. One plant had supervisors make presentations about team activities to upper management. Another found that a QWL team for foremen was a good way to regain their support for its program. A third began annual off-site meetings for supervisors as one reward for their participation. In plants with semiautonomous work teams, the crucial action was to delineate clearly the duties and managerial expectations of supervisors—in part by asking them what they perceived their job to be and what they believed it should be (see Exhibit III).

As one executive noted: "Two years ago the people-development issue got out of hand among the supervisors. We left them out of the process and forgot to give them any special training. Production managers held meetings with all the operators, and the supervisors would just have to go along for the ride. The production managers held all the control. Today, we try to get the supervisors involved. This started about two years ago in an off-site meeting with the plant manager for new supervisors. This is now a one-day session done yearly."

What's Needed

This study uncovered at most plants a Band-Aid approach to the concerns of supervisors and not a coordinated strategy to make supervisors an integral part of the change process. What might such a strategy look like? Surely, it must include:

Support-Based Training

Training is an indispensable first step in defining the common language and tactics necessary for change, but classroom or seminar-based training is not enough. Many supervisors complained, "I basically agree with this new approach, but the boss won't back me up when problems arise." The plight of these skeptics is real: few supervisors have seen their support and reward systems modified. When they walk out of a training session, they walk right back into the lion's den.

The best training includes ongoing consultation with official "change agents" (often consultants) and managers, both of whom provide continual feedback and coaching. The quickest way to modify supervisors' behavior is for managers to become role models because supervisors need to be shown that their superiors are committed to a participative style of management.

Accordingly, managers should explain why employee involvement is different from other programs that may have come and gone. Another possibility, albeit risky, is to ask supervisors what it would take to convince them that management is truly committed and supportive, for if managers can meet this challenge, they can erase much lingering skepticism.

Supervisory Involvement

First-line supervisors seek two types of involvement. The first includes them in the design and implementation of employee-oriented programs; the second gives them a say in decisions that affect their own jobs, as with supervisory QWL programs, for example. Executives can help ease opposition from supervisors by getting them involved early in the definition of employee programs as well as in the definition of their own role. There is, of course, substantial risk for superiors, a potential loss of power, in sharing these decisions with supervisors, but it is much the same kind of risk that supervisors face with the introduction of employee involvement programs.

Responsibility with Authority

The perceived loss of power is quite real to those supervisors excluded from overall decision making when employees take over many day-to-day decisions. Thus, it is important

EXHIBIT III	Duties of First-Line Supervisors in a Plant with Semiautonomous Work Teams
Commitment	Supervisors must promote teamwork and convey a genuine interest in and support for the concept. They should encourage cooperation and work with employees in developing team competence and cooperation.
Communication	Supervisors are the major link in the communications channel between management and hourly employees. Supervisors must learn to relay daily instructions through the team to reach team members.
Training	Supervisors are in a good position to assess the training needs of their employees. They are responsible for identifying, coordinating, and where possible, conducting the necessary training.
Human relations	Since supervision is key to maintaining positive employee attitudes and high morale, supervisors must develop good human relations skills. This means becoming "employee centered" by showing an interest in employees' problems, emphasizing communications and team spirit, and demonstrating a sincere concern for workers' welfare.
Motivation	Supervisors must learn to be motivators, not just disciplinarians. They can motivate employees by giving them responsibility (with accountability) and a sense of contribution. Supervisors should learn to use such phrases as "What's your opinion?"; "What can I do to help?"; and "Thank you."

that managers delegate increased responsibility along with appropriate authority and not merely give supervisors additional administrative tasks, such as the paperwork associated with quality circles. Supervisors will need to regain their self-respect and the prestige they once enjoyed in the eyes of subordinates. Only through enhanced decision-making authority can they do so constructively. With such authority, they can act on employee requests and provide necessary support without having to function as a go-between among employees, support groups, and upper management.

Supervisory Networks

Supervisors must begin to recognize the potential benefits that employee involvement programs hold for them. Today, when managers attempt to convince supervisors of the merits of these programs, many supervisors perceive their efforts to be only a "sell job" or propaganda. A useful alternative is to encourage peer networking among supervisors through such support group mechanisms as periodic dinners with no high-level managers present. Peer assistance of this sort can help resisters see the value of employee involvement programs.

EXHIBIT III	(continued)
Delegation	This allows supervisors to take on additional duties by passing on to employees many of the routine details that they can handle efficiently. Delegation helps to bolster both individual and team morale by giving everyone an opportunity to share responsibilities and rewards.
Decision making	Supervisors' performance ultimately hinges on the outcomes of decisions. Teams can be allowed to make routine decisions, but supervisors must set priorities and explain why some decisions take precedence over others.
Discipline	Supervisors must always remember that employees are individuals as well as members of a group. Most employees will respond to techniques aimed at motivating them, but a few may not. In those cases, the supervisor may have to administer discipline.
Feedback	There are times when every team needs guidance, support, and reinforcement. Both oral and written feedback is necessary to make individuals and teams aware of their strengths and weaknesses.

From the plant G internal communication "The Role of the Supervisor," June 23, 1961.

Replacement

An organization can provide awareness or behavior-modification training, but it is questionable whether the values that individuals hold once they reach maturity can be altered by a short-term structural change in their work environment. Unfortunately, some supervisors will just not accept the concept of employee involvement. When all else fails, their bosses may have to replace them, but replacement need not mean demotion or termination if the only problem is resistance to employee involvement. Even supervisors who are genuinely supportive of the notions behind these programs may turn against them if, for example, they perceive a threat to their own job security or that of their peers. In such cases, managers can move supervisors laterally to what one company called "less damaging" positions.

Final Thoughts

Resistance by first-line supervisors is real, as this study verifies, but much of it is understandable and even justifiable. Supervisors do not as a rule undermine change because they are obstinate. Organizations have always placed them in the middle of a no-man's-land, and most employee involvement programs have made their position even more precarious. Designed to boost productivity by increasing the participation of workers, these programs have rarely had the interests and concerns of supervisors in mind. The outcome was predictable: seeing nothing in the programs for themselves, most supervisors resent the loss of power and control and, in one way or another, fall into a pattern of resistance.

That pattern can be broken, however. When their interests are taken seriously, supervisors will indeed give the programs a chance and often find them of real value. Listen to some of their comments:

- "Many foremen have found quality circles useful on two counts. First, they have found that they can get things done through quality circles that otherwise would never get a high enough priority to get attention. Second, they use the circles as a two-way communication vehicle. They can gather information on the work force through the circle members and they can use the members to pass on information to their peers."
- "It becomes expected that team members will provide inputs for decision making. This is extremely helpful for the supervisor because it gets employees to help things run smoothly."
- "Teams help you get to know everyone better, so you just naturally talk with them more often. This is especially true for some of the soft-spoken people who end up showing you how knowledgeable they are. In this way, supervisors start interacting with those individuals they may not have paid attention to in the past."
- "Teams break down the barriers between supervisors and hourly people. By giving us a common reference point, they make it easier to communicate."
- "Teams help most in upward communication. The program helps team members communicate with supervisors, whereas in the past they were often reluctant to talk with a supervisor."

Two additional problems came to light in the course of this study. First, many of the issues first-line supervisors raised are also on the minds of administrators and managers one and two levels up; and second, both supervisors and middle managers often resist the introduction of new technology for many of the same reasons they oppose employee involvement programs. Thus, this way of thinking about re-

sistance to particular organizational changes may have more general applicability.

If so, the findings reported here raise anew a critical—and open-ended—question for managers: Is resistance to change a function of the individuals involved or of the way the organization manages them?

Research Methods

This article is based on data collected during 1981 and 1982 from eight U.S. manufacturing plants belonging to four multinational corporations (see the Table for a tabulation of the plants' characteristics). Each of the plants had begun some form of employee involvement program at least three years earlier. These programs ranged from quality circles to semi-autonomous work teams. In seven of the plants, the programs were designed to increase the participation of an existing work force; in the eighth case, the plant had included work teams from the outset.

Plants A, B, C, and D had begun quality circles in 1976 as a way to improve labor-management relations. The president of the company had made a five-year commitment to support them, and the circles flourished for several years. When the top management team changed, however, priorities did too. By the time the research began, the quality circles had been almost completely phased out, although most managers viewed the program as a building block for future efforts. In 1981 Plant C introduced a QWL program that encouraged employees to form loosely structured teams to discuss any work-related issue (except for contractually negotiated subjects). The program, opposed by supervisors as well as the union, was discontinued after approximately one year.

Plant E began quality circles in 1979 during a corporate drive to improve productivity. Before that time, the company had introduced a number of less structured programs to in-

TABLE   Plant Characteristics

				Plant				
	A	B	C	D	E	F	G	X
Total employees	540	475	350	800	1,600	265	275	2,785
Hourly employees	400	300	200	350	1,300	215	230	2,440
Age of facility (years)	55	55	15	7	25	40	10	16
Union	yes	yes	yes	yes	no	yes	no	no
Type of program	Quality circles	Quality circles	Quality circles & QWL program	Quality circles	Quality circles	Teams	Teams	Teams
Type of manufacturing	Machining	Fabrication	Fabrication	Assembly	Fabrication & assembly	Processing	Processing	Processing
Number of supervisors:								
Level 4, plant managers	1	1	1	1	1	1	1	2
Level 3	5	3	1	2	1	2	1	16
Level 2	2	1	3	4	6	6	6	56
Level 1, supervisors	27	13	11	20	26	20	18	273

crease employee involvement in an effort to remain nonunion. By the time of the research (three years into the program), all supervisors were leading at least one circle, and management viewed the program as a success and was investigating other ways to boost employee involvement.

Plant F attempted to change supervisory behavior through training programs and cross-functional teams, an intermediate step between the narrow scope of quality circles and broader programs such as semiautonomous work teams. The process, which began in 1974, had been slow, and there were questions about its effectiveness.

Plant G redesigned its job structure in 1975 to use its human resources more effectively. The plant had begun operation three years earlier under a traditional job design, with each employee doing an assigned job at one station on the line. Management used the new design to gradually reorganize the entire plant into semiautonomous work teams. Although there had been much resistance, particularly from first-line supervisors, managers rated the team structure as highly successful.

Plant X was a mature organization that had long since institutionalized a highly integrated work system employing semiautonomous work teams.

I conducted a multiphased data-gathering process in each plant. First, I interviewed key line managers from the plant-manager level down about the evolution of the employee involvement programs, the history of employee-management relations at the plant, and managers' expectations and perceptions of foremen. Next I gave all first-line supervisors (except those at Plant X) an eight-page survey that explored attitudes toward and changes in their jobs that had occurred as a result of the employee involvement programs.

After tabulating the survey responses, I met with all the respondents (four or five supervisors at a time) in 90-minute interviews to discuss the results. These sessions allowed me to validate the survey and to probe for whatever underlay the responses. Since there had been no change in management philosophy or style within Plant X, my interviews attempted simply to understand the nature of the work system and its relationships. Where feasible (as it was in four of the eight plants), I interviewed a sample of hourly employees. In all, I interviewed more than 260 managers, supervisors, and hourly employees.

Traditional versus New Work Systems Supervision: Is There a Difference?

Janice A. Klein
Pamela A. Posey

Joe, a first-line supervisor at Plant A, starts his day 20 minutes before the beginning of his shift. His first task is to meet with supervisors going off duty to discuss problems and get status reports. He then walks through his area to check on people, parts supplies, production status, and quality reports. After making sure line workers know what the production requirements are for the day, he returns to his desk to work on some paperwork. As problems arise, he deals with them on an individual basis, calling in support when needed. At times he may speak to other supervisors about problems, but most of his attention is focused solely on his area of responsibility. Several times during the shift, he calls for status reports on production and quality checks. Except for the time he spends on paperwork (his most time-consuming activity), he spends most of his time on the line, observing production and people.

Jack, a team advisor in Plant B (a sister facility to Plant A, which produces the same product) has no prescribed arrival time at the plant. Upon arrival he checks in with his team to let them know where he can be found, asks if there are problems he can help with, and reviews status reports. Jack's first formal activity of the day is a production meeting attended by other team advisors and several team members. He also holds a periodic team meeting, usually once or twice a week, which provides a forum for sharing information related to team and individual development needs, team performance, production issues, and plant-wide concerns. Sometimes Jack pitches in to help on the line if the team needs extra help, other times he moves on to complete some paperwork. The team members, however, do much of the report writing and take on the tasks of checking parts, quality, and production needs. Jack can often be found with other team advisors talking about problems concerning their teams or the work. Involvement in plant-wide activities, which is an accepted and desirable activity, often takes him away from the line. Although he walks through his area during the day, unless there are problems he does not feel compelled to spend a great deal of time on the line.

Joe and Jack both work for the same company and both hold front-line management positions. But how they spend their days is vastly different due to the work structures of the plants to which they are assigned. Plant A is a typical manufacturing operation, whereas Plant B is structured around semiautonomous work teams. Joe's role is straightforward—he performs the traditional job of a first-line supervisor. Jack's role, however, is less clear. His work team performs many of the traditional supervisory tasks.

There is a general consensus that work systems such as Plant B, often referred to as high commitment work systems,[1] produce im-

---

[1] Richard E. Walton, "From Control to Commitment in the Workplace," *Harvard Business Review*, March–April, 1985.

proved efficiencies at the workplace, but there remain many questions as to what role first-line supervisors can or should play in these new systems. Initially when these systems were designed, the role of first-line supervisors was defined as transitory: Supervisors were to work themselves out of a job and transfer their duties to the work teams. Many a manager could be heard saying the following:

> Workers formed into teams can do the job of supervision as well as, if not better than, the supervisors. Therefore, let's eliminate the first level of management. In addition to giving employees more say in their jobs, we can also reduce our indirect headcount and get a cost savings, too.

Experimentation began. Employees were organized into teams and given a chance to run their own show. Problems, though, began to arise. Teams seemed to lack direction. They didn't have the skills to solve many of the technical problems that arose. Many team members balked at evaluating and disciplining their peers. Getting functional support groups to respond was often difficult. One food processing plant which set up work teams, but neglected to clearly define the supervisory role, experienced what they referred to as "supervisory depression" for over three years. Managers described the situation in this way:

- Supervisors started saying, "that's not my job, it's the team's problem." In essence, they became deserters and many used teams as an excuse not to be concerned with any problems.
- Supervisors took a back-seat and as problems arose would let the teams try to handle things. As it got to a point where teams couldn't handle it, supervisors would come in with a heavy hand. It got to a point where supervisors would hope that the teams would fail so that there would be a need shown for supervisors.

Many believed that the problem was only short term and that it reflected team learning or development. But as time went on and teams matured, the old adage that "every team needs a coach" continued to hold true. While the title varies from one organization to another—team leader, team manager, team advisor, and a host of other labels are used[2]—managers have found that there is still a need for first-line supervisors. One manager described the new role as follows:

> There's a completely new role in our new systems plants. The team manager is a facilitator, resource and trainer for the work teams. The new role incumbents are much more participative in their management style.

But, what exactly does this mean? There is still much confusion as to what this new position entails and what type of individual should be selected to fill the role. How do supervisors in these new plants carry out their job? How do they manage their teams on a day-to-day basis? How does this differ from styles of supervision in traditional systems? This article reports on a recent study which begins to sort out some of these questions. (See Appendix for details of the research method.)

What Are Supervisors Expected to Do? The Formal Job Description

As the descriptions of Joe and Jack illustrated, the jobs of first-line supervisors in traditional and new work systems differ in significant ways. Supervisors in traditional plants perform different tasks than those in the new systems. For example, traditional supervisors usually perform the record-keeping function

---

[2]Leonard Schlesinger and Janice Klein, "The First-Line Supervisor: Past, Present and Future," in *Handbook of Organizational Behavior*, ed. J. Lorsch (Englewood Cliffs, NJ: Prentice-Hall, 1987), pp. 370–384.

in their areas while supervisors in high commitment plants often delegate this to their work teams. Traditional supervisors assign workers to specific tasks while those in new systems often allow their teams to make decisions about how the tasks will be divided. In each plant, an internal language develops which reflects the philosophy of the organization and the way in which people there think about the people and the work. For example, supervisors in traditional systems describe their workers in terms of the tasks they perform individually (i.e., Bill is the lathe operator); supervisors in innovative systems refer to teams and group functions (i.e., this is the final parts mounting team).

A comparison of the formal job descriptions clearly highlight the differences between supervisors in the traditional and new work systems plants. Exhibit 1 lists the key elements of these job descriptions. The primary focus of the supervisor's role in Plant A is task oriented. Supervisors there talk about their work in terms of pieces produced and quality levels. In contrast, the role descriptions of supervisors in Plant B, where supervisors are referred to as "team advisors," are more relationship oriented. Team advisors speak about the people in their work teams, and describe their jobs in terms of managing people.

In the traditional plant, nearly all aspects of the written job description relate directly to the production efforts of the unit for which the supervisor has direct responsibility. Supervisors are held personally responsible for the cost, quality, and delivery of the product. They have clearly bounded areas of authority which they are assigned to manage. Implicit in the wording and focus of the job description is the expectation that supervisors supply the controlling influence over production activities and direct work effort internally to the work units to which they are assigned. One supervisor noted,

> I spend my time telling people what to do and how to do it. I have to make sure the line is staffed, that we can run with the people who are here. We don't switch jobs, and I don't fill in on the line. I need people who know what they are doing so I can get the pieces out without having

EXHIBIT 1   First-Line Supervisor Job Descriptions

Traditional	*New Work System*
1. Plan, organize, direct, and control line	1. Insure resources are available for team to produce on-time, quality product.
2. Meet cost, quality, and delivery objectives	2. Develop team maturity—coach and counsel
3. Manage daily variance	3. Represent team in plant-wide activities
4. Coordinate activities and resources	4. Train and lead team in problem solving
5. Plan/implement line improvements	5. Motivate team toward goal achievement
6. Administrative tasks • Safety • Housekeeping • Communications	6. Assume responsibility for indirect tasks

them rejected by someone else. If there's something wrong in the area or with the pieces, they let me know, and I correct the problem. It's my job to get things done right, and to get the pieces out.

In the new work system plant, written job descriptions are more general than in the traditional plant. They represent goal expectations rather than behavioral imperatives. Performance expectations imply a support function to the teams which are held accountable for production outcomes. Production control is not the explicit focus of the supervisory job. Team advisors are expected to train their teams to manage the production process and to serve as advisors and resources to aid that effort. Team advisors are also expected to manage the interface between their team and others—both other production teams as well as support teams such as maintenance and materials handling. Job descriptions are focused on the development of the team and team effort; a critical part of the role, for example, is to train teams in group problem solving.

> We have regular team meetings to talk about how we are doing, to share ideas for solving problems, and to make sure everyone has the information they need to do their work. I'm a facilitator in those meetings; I help the team understand how to work out their problems. I don't jump in and solve them alone.

Individual team advisors are expected to be involved in more plant-wide activities: Some production team advisors are assigned to special task forces to address plant-wide problems. As one team advisor explained,

> I have a lot of freedom here, freedom to get involved in activities I'm interested in. I can work on plant budgets or materials problems even though my job is as an assembly team advisor. Everyone here takes on some vertical tasks. We have our own assigned jobs, but we're expected to contribute to the plant as a whole, not just to our teams.

Another key difference between the roles of supervisors in the two types of plants is related to expectations about the manner in which people should be treated. The traditional job description is clearly task oriented rather than people centered. When supervisors are asked how they judge performance in their areas, they point out production numbers, quality levels, and how close they came to meeting their targets.

In the new system job description, five of six major responsibility categories focus on work teams and the relationship between the team advisor and the team. It is clear that team advisors are expected to work with and to support team efforts. This is related to the conceptual underpinnings of the plant design: a system designed to promote organizational effectiveness through participation and commitment of the people involved regardless of their level in the plant. One team advisor, who had come to the innovative plant after several years as a supervisor in the traditional plant, summed up these differences in this way,

> I really enjoy the participation here. People count, and we do a lot to support their needs. That was one of the hardest things to learn when I first came here . . . that I had the time to help the team learn from something that went wrong. I didn't have to take over; in fact, I was expected to let the team manage the problem solving process. Back in [the traditional plant] I would have solved it myself and told the workers what to do. Here, that's their job; I just help when I'm needed.

Beyond the Formal Job Description

The differences in job descriptions between traditional and new work systems settings define different task and performance requirements for supervisors in the two systems. But there are other forces which also help to define the differences between plants, forces which define the informal operating systems within which supervisors function. Hence, not only do performance expectations differ, but what the environments demand of the supervisors also differs. Supervisors are being pulled in different directions because of the expectations placed on them by the two systems. The contrast between operating environments can be seen in a number of ways.

First, the operating system in the traditional plant explicitly recognizes the supervisor's authority over the production process, including resources and people. As one supervisor explained,

> I have to get the work out on time. To make sure that's done, I check for parts, I check quality, and I make sure the people are doing what they're supposed to be doing. I'm responsible for what comes out of this area, and I've been given the authority to make sure it's done right. If I have to sit on people to do that, then I sit on them. They know and I know I'm the boss.

If questions are raised over performance or output in a particular area, the supervisors are approached to provide the necessary information because they are recognized as the authority figures in the area.

Team advisors in the new work system, however, do not always perceive that they have clear authority and control over their units. Team advisors, when asked about their authority over teams, responded with "I'm responsible for helping . . ." or "I'm a coach, not a director. . . ." Team members were told that the team advisor's role is one of support and that the actual production responsibility lies with the team members. As one team advisor explained,

> Sometimes it's hard to remember that I am really the leader, the one with the power in this team. We are so committed to the team concept and team decisions that it slows us down sometimes. I guess that's a trade-off you have to make to have a system where everyone has a say. It's frustrating sometimes, but the atmosphere here is one that promotes involvement. It's difficult to cross that line and act on your own when you are expected to pass responsibility and authority on to the team.

A second area in which clear differences emerge is related to the drive to achieve and improve the productive effort and output. In the traditional plant, supervisory attention is focused on meeting or exceeding production targets and managing those factors which detract from the ability to meet goals. While planning ongoing line improvement efforts is an explicit objective in the traditional job description, it is interpreted as an efficiency goal predefined by the system, not necessarily an improvement which stems from some other need.

> I guess I can do my job in any way I want, but when it comes right down to it, it's the numbers that count. So that's what I aim for . . .

In contrast, team advisors at the new work systems plant have a broader perspective of improvement. Here, striving for improvement is an implicit and accepted part of the culture of the organization; employees at all levels are aware that the work system was created with the guiding philosophy that the plant should be a place in which people would be motivated to perform at their best. The team advisors refer often to the atmosphere or environment that has been created within the plant—an atmosphere that is supportive of personal efforts to achieve and improve.

> There's a feeling here that we are expected to contribute to making the plant a better place than it is today. How we do that is really up to us; we can contribute to the team, to support teams, or to the plant as a whole. And we do it because it's the right thing to do, not because someone told us to do it. Maybe it comes from commitment to the concept here.

Finally, both plants expect supervisors to exhibit some degree of flexibility in the way they manage their tasks or people. There is a difference, however, in the type of flexibility between sites. Where supervisors in the traditional plant are expected to plan, organize, direct, and control the production process (all tasks which require some degree of flexibility to adapt to changing conditions), the range of behaviors is narrowly focused on accomplishing the production task. One supervisor described the concept of flexibility in this way:

> Problems come up all the time, that's part of the challenge in manufacturing; and I have to be ready to deal with it. It may be something simple like absenteeism where I have to find different people to fill in, or it may be a major machine failure that's going to hold up the whole line. I've got to be able to deal with it fast so we can keep working.

Team advisors in the high commitment plant operate in a more ambiguous environment than do those in the traditional plant. Supervisors have to be flexible enough in their methods to ensure that team learning occurs, to perform a wide range of tasks, and to interpret the job description so that it fits not only personal style but also the needs of the team and others in the plant. One team advisor explained:

> In this environment, the ability to deal with ambiguity is very important. We force people into unstructured situations, to grab the responsibility to do something with it, to put a structure around it. Those situations may differ on a weekly or a daily basis, and they have to be able to adapt to the difference.

As seen in the descriptions of Joe and Jack, supervisors in the traditional system not only work under different task assignments, they perform under a different set of environmental expectations about how they would do their jobs: They are expected to report at specific times and work complete shifts, their presence on the line is expected, and at all times they are concerned with quantity and quality of production. In the high commitment plant, supervisors set their own work schedules according to the tasks they define as important. Because the working environment explicitly recognizes and rewards participation from all

levels, much of the day-to-day responsibility for production is taken on by team members. This frees the team advisors to focus on managing relationships between teams and functions in the plant. To a great extent then, differences in the operating environments between the two plants reinforce and emphasize the differences found in the written job descriptions.

What Defines Success? Outstanding Versus Average Supervisors

Given these differences in expectations and demands between plants, the remaining question is how do first-line supervisors do their job? What characteristics or skills do they exhibit on the job? In particular, what leads to successful supervisory performance in each operation? How do outstanding performers manage and lead their employees? How do they solve problems? In general, do supervisors in new work systems approach their job differently than their peers in traditional plants?

Normative expectations tend to predict strong differences in the way supervisors in traditional and new work systems perform successfully. We can see major differences in the shop floor organizations, in the authority structures, in the job descriptions, and in the informal operating environments between the two systems. But, when we look closely at what it takes to do a first-rate job in human terms, these differences become blurred.

At the traditional site, outstanding supervisors are characterized as being strongly goal-oriented self-starters. They like their jobs and let others know it. They push for quality goals, provide clear direction to the workers, and give timely and accurate feedback to motivate them. They tend to coach their workers and share information with them. They enjoy challenges, and look on problems as such rather than as stumbling blocks. They are flexible people who have high tolerance for the stress inherent to the work they do. In addition, they take responsibility for the actions and outcomes of their units, know how to get the right people involved in the problem solving process, and take the initiative to do so. Finally, they are the supervisors who look beyond the immediate boundaries of their areas to understand the plant and company as a whole. They push to achieve company goals, not just the targets in their immediate areas.

The profile of outstanding supervisors in the high commitment plant is strikingly similar. They have developed reputations for delivering what they said they will, and this gives them credibility which allows them to shape and influence performance. They understand what developing the teams means in practical terms, and are able to share their skills and knowledge willingly with team members. They believe in and demonstrate power sharing and turn decision making into learning experiences for the teams. They are enthusiastic, committed, and flexible; they view ambiguity and the lack of structure as a challenge rather than as a frustration. They are not reluctant to take control in a crisis, and recognize that they have the responsibility and authority to do so. Yet, they are committed to the goals of teamwork and the participative spirit, and find ways to foster the development of this spirit within the team and plant.

In essence, outstanding supervisors in both plants are characterized as competent, caring, and committed to both the work and the people. They are highly respected within the plants, and are perceived as credible, honest, and trustworthy. Outstanding supervisors find ways to motivate their workers toward better performance; they all use some participative methods regardless of the type of system in which they work. They are described by workers, peers, and their bosses as goal oriented and people centered. All share information and responsibility with workers, al-

though the two systems place different expectations and restrictions on how they could do so. One manager in the traditional plant, in fact, explained that,

[A particular supervisor] is clearly the best one we have, and I don't know how he does it sometimes. This is not a participative system; it's a large and unwieldy hierarchy with a lot of direct control built into it. [This supervisor] has managed to work within the formal structures that we've built in here, and still give his people more responsibility and power than we thought possible. He is our best, and he's done it without a lot of structural or system support.

Discussions with supervisors and managers in both plants emphasized the fact that those qualities which contribute to outstanding first-line supervision regardless of type of plant are very similar, and that the real contrast is within plants between outstanding and average supervisors.

The average supervisors in the traditional plant are less enthusiastic and less committed to their work than their outstanding counterparts. They are viewed as directors rather than coaches, are less likely to provide regular feedback to their workers, and share less information with them. They tend to be less flexible and innovative in their problem solving approaches, and slower to change than outstanding supervisors. These individuals are less likely to initiate actions not clearly required by the situation, or to make decisions when complete information is not available. Their concern tends to focus on the unit for which they are responsible rather than the plant or company as a whole. The focus on the unit or business allows the average supervisors to adequately perform the job, but detracts from their ability to understand the plant or company as a whole.

The average supervisors in the new system plant also present a different profile than the outstanding ones. They have intuitive knowledge about developing teams, but have difficulties translating this into effective mechanisms for development. They tend to maintain tighter control over teams than their outstanding counterparts, and are not fully comfortable with the participative processes in the plant. The goals and performance standards set by the average supervisors here tend to be narrowly defined in terms of specific team tasks, and they are less concerned with or attuned to the overall goals and needs of the plant and company. In addition, these supervisors tend to focus on short-run issues, and are more reactive than proactive in performing their jobs. This certainly allows them to fulfill the requirements of their jobs, but hinders the development of a broad company perspective which would help them set broader, yet realistic, goals for the teams.

In comparing supervisory and managerial comments as to what differentiated outstanding from average supervisors, we found three important aspects which contributed to success in both plants. The first is a strong mutual respect between outstanding supervisors and their employees. Outstanding supervisors develop a rapport with their employees which is based, in part, on their respect for their employees' abilities and efforts. They are viewed as honest and trustworthy and are noted for treating their people fairly. This credibility, in turn, leads to their ability to motivate their employees. In the traditional plant, this is referred to as motivating their people; in the new work system, team advisors refer to this as motivating their team.

The second key aspect involves accepting responsibility for getting the job done. Outstanding supervisors are characterized as self-starters who are committed to getting the job done well. In the traditional plant they accept and follow through on their responsibilities. They believe that their job is important and they take pride in accomplishing their tasks. Although team advisor responsibilities are not

as well defined, outstanding team advisors willingly assume responsibility for getting tasks accomplished. They strive for improvement because they view it as the right thing to do, not because they are told to.

Lastly, outstanding supervisors are described as top-notch problem solvers. In the traditional plant, supervisors do much of the problem solving themselves, while outstanding team advisors are skillful in helping the teams do much of the problem solving. But in both plants, outstanding supervisors take the lead in process improvement and possess the knowledge and skills to pull together the needed resources to get problems solved.

As the comparison in Exhibit 2 suggests, those characteristics perceived to be most important to outstanding supervision in traditional and new work system plants are surprisingly similar despite major and clear differences between job descriptions and systems. Although the common belief among managers is that the supervisory role is distinctly different in new work systems, we have occasionally heard the opposite, such as the following comment by a senior manufacturing executive,

> We haven't seen any difference between the first layer of supervision in our new open systems plants versus our traditional shops. A good supervisor is a good supervisor regardless of the work structure.

Our findings suggest that this manager may be referring to the style of supervision versus the substance, or task structure, of the supervisory job.[3] Despite the structural differences between traditional and new systems jobs, outstanding supervisors perform their job in much the same way regardless of the system in which they work. In both plants, outstand-

---

[3]Saul W. Gellerman, "Supervision: Substance and Style," *Harvard Business Review*, March–April 1976.

EXHIBIT 2   Comparison of Supervisory Jobs

	Traditional	*New Work System*
Job Description:	Task-oriented	Relationship-oriented
Focus:	Production control	Team development
Skills:	Motivates people	Motivates team
	Follows through on responsibility	Assumes responsibility
	Individual problem solver	Team problem solver

ing supervisors are described in many of the same terms that have been used to describe outstanding managers throughout time. Our conclusion: A good supervisor is a good supervisor is a good supervisor . . .

Inquiries from Managers

These findings, when presented to manufacturing managers, from the executive level down to the first-line, tend to raise many questions. Below we address some of the more frequently raised questions.

Are the differences real or just semantics? Is the difference between referring to employees as individuals versus a team merely a reflection of the language of their environment?

Supervisors in new systems have been trained and socialized to refer to their employees as team members. Supervisors in traditional systems rarely hear the term and usually refer to their subordinates as individual employees. Outstanding supervisors, though, often think of their employees as members of their total work group or team. Although traditional systems do not formally recognize teams, many outstanding supervisors manage their employees as if they were a team.

While this may imply that the differences are merely semantics, we believe there is one subtle difference. The new systems force supervisors to interact with their work groups on

a team basis through regular formal team meetings. This requires supervisors to lead team meetings, make formal presentations to the team, handle group dynamics during problem solving exercises, and gain group consensus. Although outstanding supervisors in traditional systems may view their work groups as teams, they usually interact with each employee individually on a day-to-day basis. Managing group interaction is a very different skill.

Do new systems require a higher proportion of outstanding supervisors?

The evidence from this study says no, but the teams were quite mature having been in place for over eleven years. The teams, in many cases, ran themselves and when a supervisor was less than outstanding, the team compensated for it. Experiences in other mature team organizations validate this finding.[4] In some innovative plants, the teams take on the job of training new or suboptimal supervisors.

The answer, however, would probably be affirmative in plants that are in the start-up

---

[4]Leonard A. Schlesinger, *Quality of Work Life and the Supervisor,* New York: Praeger, 1982; Janice A. Klein, "First Line Supervisors and Shop Floor Involvement," in *Human Resource Management and Industrial Relations,* ed. T. Kochan and T. Barrocci (Boston: Little, Brown & Company), 1985, pp. 403–413.

phase. During team formation and maturation, stronger supervision or leadership is a must. Early on, team members lack technical and group problem solving skills[5] and require outstanding supervisory guidance. To be fair, however, the same may be true of start-ups in the traditional systems.

Can supervisors be transferred from traditional to new systems plants?

One of the most intriguing comments we have encountered came from a vice president of operations who reflected on his company's experience in transferring supervisors between new and old systems:

> We've found that as we move supervisors or team managers between plants, supervisors moving from traditional systems to our newer, more innovative ones find the transition quite easy and fit right in with their new environment. The reverse, however, isn't true. Supervisors from new systems stumble and fall for the first few months in our traditional plants.

At first, this appeared counter-intuitive. In line with the belief that there was a need for a "new breed" of supervisor, we expected that supervisors trained to survive in traditional autocratic work settings would find the new role of team facilitator to be a difficult transition. The conclusions from this study, however, help to explain this manager's experience.

Outstanding supervisors in traditional systems are often frustrated by the inflexibility of traditional structures (strict work rules, lines of demarcation between management and labor, etc.) and find new systems to be a breath of fresh air. They no longer have to work around the system to manage in a participative

manner. The structure of the new work system supports their way of managing.

Outstanding supervisors from new systems, on the other hand, find traditional systems too confining and are unaware of how to "play the system." It often takes them several months to learn the politics and numerous rules and policies with which they have to live. While the traditional system appears quite simple on the surface with its rigid behavioral guidelines, there is much skill in knowing how to navigate the waters. In addition, supervisors in traditional work structures must manage individuals who often do not understand the participative perspective these supervisors bring to the job.

It is, therefore, not surprising that outstanding supervisors find it easier to transfer from traditional to new systems. But, does the same hold true for average supervisors? Experience says no. The "looseness" of new systems is often overwhelming to average supervisors from traditional organizations. They need the rigid rules and guidelines to fall back on when problems arise. Here is where the subtle difference between "assumes responsibility" versus "follows through on responsibility" is key. Supervisors in new systems must be able to assume responsibility even if it is not spelled out for them.

What type of training is required for supervisors in organizations in transition to new work systems designs?

Part of the confusion to date has been that the terms "coach," "facilitator," and "trainer" are too abstract for many first-line supervisors and managers. They are nice organizational development jargon but managers and supervisors have had a difficult time translating them into action. In addition, many first-line supervisors are threatened when they are told they must now perform a new role.

Our study shows that the new role is not all that new and that there are role models even

[5]Richard Walton and Leonard Schlesinger, "Do Supervisors Thrive in Participative Work Systems?" *Organizational Dynamics*, Winter 1979.

within traditional plants. Pointing to these outstanding supervisors, identifying their performance as the new role, and structuring the organization to encourage such behavior is much less threatening than saying that the supervisory job has changed. This message will reinforce the fact that supervisors in new work systems must continue to maintain influence in the new setting. This is a signal which is often lacking as companies begin to introduce participative work structures.

What needs to change is the criteria for evaluation and the organizational infrastructure that supports supervisors. As for training, there are few new answers. The old standards of group dynamics and problem solving still apply. There is also a need to emphasize group presentation skills and how to run effective meetings.[6] Lastly, it may be useful to use some of the outstanding supervisors as trainers.

What should be the selection criteria for first-line supervisors in new work systems?

Our findings indicate that the "new breed" of supervisor may not be all that new. The new systems merely reinforce the behavior of outstanding supervisors of the past. Therefore, managers should reflect upon their prior hiring practices and identify those criteria which lead to the selection of their outstanding supervisors. Interpersonal skills need to be emphasized along with technical knowledge. Although outstanding supervisors in both systems tend to be good coaches, this trait in tra-

ditional plants is often more by accident than by design. Whereas being a good coach was not a prerequisite for supervisory selection in the past, it is today.

Lastly, it is important that managers do not overlook their current supervisory workforce when making the transition to more participative work systems. One company with a mature new systems plant specifically recruits its supervisory workforce from older traditional facilities because it wants to find supervisors who were dissatisfied with the old and want to find a better way to manage.

Postscript

Since completion of the study reported in this article, managers from the new work systems plant report one significant change in the system. Faced with increasing competitive cost pressures, managers within the new work systems plant felt a need to enhance the position of team advisors. With the emphasis on team accountability some team members failed to recognize that team advisors had formal authority over the production process. This affected the ability of the team advisors to truly manage the production process. In addition, in an effort to maximize team member participation, the organization, at times, tended to emphasize individual self-development over production requirements. To lessen this tendency, the title of the first-line supervisory position was changed from team advisor to team manager. Some of the incumbents noted that it made a difference in their ability to influence the work process, others saw no change. Although we did not do a formal study of the change, we suspect that those who saw little change were the more outstanding supervisors who had previously assumed authority.

---

[6]Audrey E. Bean, Carolyn Ordowich and William A. Westley, "Including the Supervisor in Employee Involvement Efforts," *National Productivity Review*, Winter 1985–86.

Appendix: Research Methods

The supervisory data from Plants A & B were collected during an eighteen month period from late 1983 to early 1985 in two manufacturing plants of the same company. One plant was a traditional, hierarchically designed and managed site; the other was a high commitment work system, organized and operated in a participative manner.

The two plants were closely matched in available technology and production processes; each had both a components machining and an assembly process. Plant A was the oldest facility in the corporation, and had approximately 4,000 employees. It was organized using a "business" concept in which distinct manufacturing functions operated semiautonomously within the larger plant structure. The five-layer management system (plant manager, directors, business managers, unit managers, and supervisors) was representative of many traditional manufacturing plant structures. Production workers belonged to an independent local union affiliated only with this corporation.

Plant B was a relatively new plant, having been acquired and brought on line approximately eleven years before the study; there were about 1,000 employees in this participatively designed system. Plant B also used the business concept, but was organized on a team format in which every employee was a team member. The management structure contained four levels; plant manager, directors, business managers, and team advisors (the team advisor role which focused on facilitation rather than direction combined the supervisor and unit manager roles of

Demographic Comparison of Research Sites

	Plant A *(traditional)*	Plant B *(new work system)*
Total employees	4,000	1,000
Age of facility	60 years	15 years
Union	Yes	No
Type of manufacturing	Machining and assembly	Machining and assembly
Number of supervisors		
plant managers	1	1
directors	2	2
business managers	8	7
unit managers	30	—
supervisors	150	35
Supervisor profile (averages based on survey responses)		
Age	44	34
Length of service at plant (years)	20	6
Years as supervisor	8	3

the traditional plant). Each business in this plant was comprised of both direct production teams and support teams which were responsible for functions such as setup, maintenance, and materials handling. Team members were also expected to take on responsibility for support functions (quality control, finance, and staffing, for example) in addition to standard production functions; these vertical tasks were an important part of the job at Plant B. This plant was not unionized although it was located in a heavily unionized area.

The research used multiple methods to generate the data base for analysis. First, group interviews with managers and supervisors in each plant (12 groups in Plant A and 6 groups in Plant B) were held to identify plant-specific perceptions of the supervisory job and to isolate important performance measurement criteria. Next, a survey (developed independently for each plant) was administered to all supervisors and production/operations managers which explored performance differences between outstanding and average supervisors (the response rate was 43 percent in Plant A and 54 percent in Plant B). Third, samples of outstanding and average supervisors (10 supervisors in Plant A and 9 team advisors in Plant B) were interviewed individually to generate detailed examples of excellent and poor performance. Finally, those supervisors interviewed individually were observed during their regular work shifts. In all, nearly 75 managers and supervisors were interviewed during the clinical phase of this study.

Sedalia Engine Plant (A)

The transition period for the Sedalia Engine Plant (SEP) would not be easy; of that, SEP's newly appointed plant manager, Danney Goble, was certain. The plant had been manufacturing and assembling diesel engines in Sedalia, Minnesota, since 1974, and SEP's parent company, American Diesel, had allowed the plant to chart an exciting and innovative course. A bold venture not only in the redesign of work, but also in the environment in which that work took place, SEP had emphasized a participatory style of management. Since its inception, SEP had been under the firm hand of the original plant manager, Donald St. Clair. Often described by coworkers as charismatic, St. Clair had served as a sort of father figure for the new plant and its employees.

In the fall of 1979, however, St. Clair had been promoted and moved to American Diesel's corporate headquarters in Beacon, Illinois. His successor, Goble, formerly SEP's director of assembly and test, offered a sharp—and to some, quite disturbing—contrast to St. Clair's managerial style. But the challenges facing SEP went beyond adjustments to a new plant manager. Not only St. Clair, but practically all of SEP's top management team had left, saddling Goble with a massive rebuilding program. That turnover, although a normal step in the career development of those involved, could not have come at a worse time. The declining national economy was forcing SEP to meet its greatest challenges yet. How would these challenges affect the level of commitment that had been from the beginning such a critical feature of life in the plant? Goble could not be sure.

This case was prepared by Bert A. Spector, research associate, under the direction of Lecturer Michael Beer as the basis for class discussion rather than to illustrate either effective or ineffective handling of an administrative situation.

GUIDING PHILOSOPHY OF SEDALIA ENGINE PLANT

The notion of experimenting with innovative work systems had taken root at American Diesel in the early 1970s. Following a 60-day strike at the Beacon plant in 1972, American Diesel undertook several innovations in its new plants. In 1974 American Diesel decided to move into Sedalia, Minnesota, and opened the Sedalia Engine Plant. The SEP philosophy, unique within American Diesel, grew out of a series of discussions between Don St. Clair and his operating team. Because of the oil embargo of 1974, American Diesel's business had been reduced sharply. As a result, the company allowed St. Clair and his team considerable slack time in which to work through and carefully develop an operating philosophy.

After visiting a number of innovative plants throughout the country and working with an outside consultant, St. Clair and his operating team selected four words to help identify the type of organization they hoped to create: excellence, trust, growth, and equity (see Exhibit 1 for a detailed definition of these goals as provided by the organizing team). Essentially, he hoped to create an organization where the self-interest and creative talents of all employees would be directed to the general well-being of the plant. Allowed a maximum amount of freedom, responsibility, and flexibility, employees would dedicate themselves to a high level of performance. "People want to work hard, perform well, learn new skills, and be involved in the decision-making processes that affect their jobs," said St. Clair in expressing the basic assumptions of his operating team. The work ethic was a powerful force that could be released if the proper atmosphere was created. Thus, the main point of the organizational structure at SEP would be to release that full potential inherent in most workers.

Although American Diesel's main plant at Beacon was unionized, there was to be no union at SEP when it opened. There would be ample protection for the rights of employees, St. Clair's team believed, in an elaborate governance system that would allow all employees to air grievances freely, speak their minds, and seek remedies to perceived inequities. Unions would not be needed to perform such a task, the team hoped. Besides, the operating group felt that at SEP, individuals would represent themselves through their teams and other plant organization. The adversarial relationship which they thought would result from a unionized plant would get in the way of the experimentation in job design. As a result, there would be no union.

The basic unit of organization at SEP would be the team. All employees would be grouped into small teams, and within each team emphasis would be on self-management, learning new skills, and performance. Teams would be trusted to regulate themselves, keep track of their own performance, and encourage growth and the acquisition of new skills on the part of individual members.

EXHIBIT 1   How We Do Business

Our business goals are basically the same as those of any successful business:

- Produce a quality product
- On-time shipment of our product to our customers
- Be efficient in our operations
- Be profitable

Each of us has ideas on how we can best reach these overall goals, and on how we can do business to reach more specific goals which will contribute to reaching these overall general goals.

A way to do business can mean many different things, such as a way to: run a machine, prepare a report, train an employee, solve a problem, evaluate plant performance, communicate an idea, arrange a work area, or practice safety. Each of these and all other ways to do business can be accomplished in many ways. Each different way to do business is assumed to have some advantages and some disadvantages and may be compared to other ways for accomplishing the same thing. The way with the strongest advantages and most acceptable disadvantages will be selected.

A good question to answer at this point is "How do we decide?" because we must do many things in unison as a single plant; we as a plant organization have come up with four basic guidelines to help us decide the *best* way of doing business or the way we think has the best chance of success. These guidelines or key questions which have served as the building blocks of our organization are:

Excellence:

Does the way we do business allow every employee to perform to the best of his/her ability? Does the way to do business allow the plant to function as effectively as possible? Our assumption is that all people who belong to our organization will want to do their best and will expect the same of others even in the performance of many repetitive, routine tasks which are part of our work. We assume nothing less than such effort will allow us to be an effective competitor in the diesel engine business.

Trust:

Does the way to do business reinforce the idea that employees are expected to behave as responsible adults, and therefore, information, equipment, and materials are made accessible to them? If employees are assumed to be responsible adults, then the risks of abuse of information are low and the advantages of accessibility and openness are great. For example, it is assumed that we can solve any problem once it is raised and that all employees will bring us problems, issues, and sensitive information because they are confident that this information will be treated effectively with no harm coming to them simply because they raised an issue. Trust does not mean that information will be handled carelessly or that everything will be available to everyone. However, it is

EXHIBIT 1   (continued)

assumed that ownership in and a commitment to accomplishing objectives is strongest when relevant information is available. Therefore, information will be made available as appropriate.

Growth:

Does the way to do business encourage both the *learning and performing* of many tasks by said employees? We assume that human resources are too valuable to waste. People have been educated more and are capable of learning and performing more at work. It is assumed that work can be more interesting when people are challenged to perform a series of related tasks that add up to a measurable end product or service, rather than routinely performing only the smallest parts of total jobs. This does not mean that uninteresting or repetitive work is eliminated; for a fact, many such tasks are required to make diesel engine parts, however we organize our work. It is assumed that as employees, we can recognize the need to continue to perform the skills that we have learned, even though without new learning, some tasks may eventually lose some of the initial excitement or interest they held. In our business, we must learn in order to perform. We do not learn for the sake of learning.

It is assumed that we can recognize the value of sharing the skills which we have learned with others even though our instincts and past experience may mislead us to think that protecting some unique skills which we have will make us more valuable employees. A valuable employee is capable of, and willing to train others.

Equity:

Does the way to do business treat all employees as adults and as fairly as possible? Our assumption is that we will tend to perform better as an organization to the extent that artificial differences in the way people are treated are eliminated. *Note that Equity does not mean Equality.* There are some differences in such areas as pay, benefits, and work areas, based on levels of responsibility and functional needs. However, equity does mean that *where there is no good business reason* to have differences, such as special parking privileges or less medical insurance, there *are* no differences.

Our business is growing rapidly and our ways of doing business are evolving as we learn from experience. Our experience in orienting new employees has taught us to be careful not to exaggerate the differences between our organization and other business organizations. It is our intention to do business in a few ways which we think will be effective and these ways may be different from what some of us have experienced in the past.

However, we are part of a plant startup situation. This means that many of the intentions that we have are to be found only partially in practice at this time. It is assumed that each of us is committed to devote the extra effort that it will take to make our organization work as it is intended. It is certain that this effort from each of us will succeed.

Groups of between three and four teams would then be clustered into what were called "businesses." One of the basic assumptions of St. Clair's organizing group was that the maximum number of people any business unit ought to employ should range between 200 and 300 people. That size would allow the business manager to know personally each and every employee. Since it was assumed that SEP would quickly grow to 1,000 employees and eventually to 2,000, the plant was divided into five "businesses," each operating as autonomously as possible under its own business manager. These business managers would oversee and direct the operations of various teams assigned to them.

As a reflection of their belief that, given the goal of creating an open and trusting environment, people would motivate themselves, St. Clair's organizing team replaced the traditional first-line supervisor with a "team advisor." While first-line supervisors usually act as spokesmen for company policy and policemen for the workers assigned to them, SEP's team advisors would work *with*, not above, their teams to facilitate performance and the acquisition of new skills. To fill these positions, St. Clair wanted to find individuals not just technically proficient but also able to work well with people in a nonauthoritarian, nonthreatening manner.

The compensation system at SEP would also reflect the organizing team's belief in growth and responsibility, and a shared commitment to the attainment of plant goals. To reduce the traditional gulf between workers and management, all SEP employees were to be salaried. Machine and assembly workers who traditionally are paid by the hour (as was the case in all other American Diesel plants) would be salaried along with management personnel. Everyone would be expected to work 40 hours a week and to make up any missed time. Further, the compensation system was based not on seniority, but on the acquisition of new skills and the willingness to perform various tasks. Wages at SEP for machine and assembly workers would be relatively high compared with the prevailing wages of the community—in the seventy-fifth percentile—but, because of the generally depressed condition of the Sedalia area, low in comparison with other American Diesel plants. St. Clair and his team viewed pay not as a motivator, but as a way of achieving equity among plant employees.

Participation at all levels—that was the key. "Participation," St. Clair said, "is a way of doing business effectively through achieving quality decisions or a commitment to carrying them out." St. Clair was adamant on the point that SEP was *not* a social experiment: "We are dealing with ways to develop and utilize the best skills and abilities of our human resources, while at the same time creating a more satisfying work environment."

INSIDE SEDALIA ENGINE PLANT

SEP's parent company, American Diesel, had been producing, manufacturing, and selling diesel engines, components, and parts since 1919. From corporate headquarters in Beacon, Illinois, American Diesel had constantly held onto a sizable share of the market—usually fluctuating between 40 and 50 percent—despite competition from such corporate giants as John Deere, Caterpillar, and even General Motors. Over the years, American Diesel had built its reputation on high-priced engines of excellent quality and exceptional service follow-up.

The town American Diesel selected in 1974 as the site for its new plant—Sedalia, Minnesota—was home to 55,000 people. The work force was both highly skilled and strongly unionized. Sedalia Engine Plant set up shop in a 930,000-square-foot plant vacated three years earlier. "Just to give you an idea of how enormous this place is," explained Connie Kelleher, one of SEP's directors, "they tell me you could put nine football fields on the roof. All you have to do is walk back and forth between the front office and the assembly lines once or twice and you'll have no trouble believing that." (See Exhibit 2 for a floor design of the plant.)

Once it was operational, the physical layout of SEP reflected its operating philosophy. The south end of the plant was made up of compact, self-contained manufacturing lines for various components of the diesel engine: pistons, piston liners, camshafts, camboxes, flywheels, and so

EXHIBIT 2 Plant Layout, May 1979

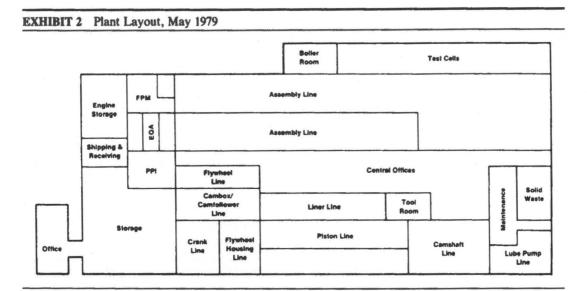

forth. The northern half of the facility housed the elongated assembly lines, added after the plant opened, where between 18 and 22 diesel engines were assembled daily on their way to the small test rooms in the back and, from there, directly to the loading docks.

The 24-foot-high building was open from top to bottom except for a core of offices dividing the plant down the middle. A few of those core offices were built when the plant first opened. Later, in an effort to link staff support personnel more closely to the manufacturing functions, SEP began to enlarge those core offices so that even more of the support people might be placed right in the middle of the plant. When construction was completed on the offices, they stretched from the east end to the west end of the plant, effectively shutting off the 10 manufacturing lines from the 2 assembly lines. Sitting at the front of the plant in a small cluster of offices were the plant manager, his director of organizational development and training, and the finance department.

Internal Plant Structure

There were two separate but overlapping structures set up within SEP: the operating organization and the governance structure.

Operating Organization.   The operating organization was divided into five levels: plant manager, directors, business managers, team advisors, and team members (see Exhibit 3). The plant manager, first St. Clair and then Goble, had overall responsibility for the operations at SEP. The seven directors immediately under him served as his operating team and were responsible for manufacturing and/or staff support functions. The combining of these two functions was done in order to create and maintain close ties between support and manufacturing operations, and to develop managerial talent. Thus, three of the directors—purchasing and materials, manufacturing services, and reliability—had both staff support and manufacturing functions under them. The director of purchasing and materials, for instance, had responsibility not only for staff support functions like transportation, shipping, and purchasing, but he also had one of the plant's manufacturing businesses reporting to him. Three directors—finance, personnel, and organizational development and training—had only staff support functions under them; and the director of assembly and test oversaw two manufacturing businesses.

Dropping down another level to business manager, the five managers in the plant reported on a direct line to their assigned director, and they served as heads of the five semiautonomous businesses in the plant. The assembly-related businesses were named engine business and operations business; the other three were merely assigned letters: *A, B,* and *C* (see Exhibit 3). The business manager for operations had responsibility for

EXHIBIT 3 Plant Organization Chart

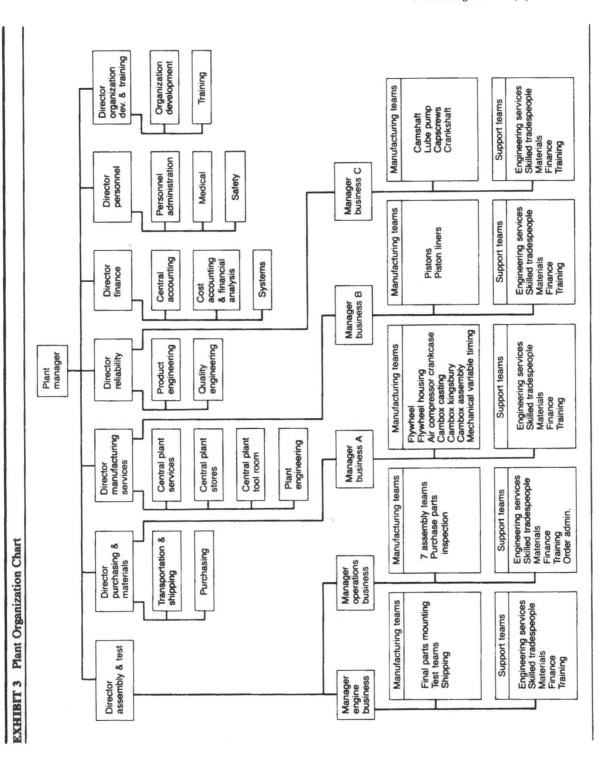

seven assembly teams as well as a manufacturing support group made up of engineering services, specialized skills trainers, materials, finance, training, and order administrations. In fact, every business within SEP had a number of staff support functions as well as a group of manufacturing teams (see Exhibit 4).

In point of fact, during St. Clair's tenure as plant manager, the business managers became an important part of the plant's leadership. Tom O'Donnell, business manager, said: "We were a real tight-knit group when Don was here. The business managers spent a lot of time together and a lot of time with Don. I know the organization charts didn't show it, but we really reported on a direct line to St. Clair." St. Clair relied heavily on his business managers for the day-to-day plant operations, while he and two or three of his directors—Goble among them—provided the overall leadership. That close relationship between St. Clair and his business managers tended to blur the precise functions of the director level.

Next, there were team advisors—SEP's first-line supervisors—and, finally, team members. (Their functions will be defined below in the section on teams.)

Governance Structure.   St. Clair was convinced that the operating organization of SEP would not be enough to insure a high level of commitment

EXHIBIT 4   Representative Mature Manufacturing Business

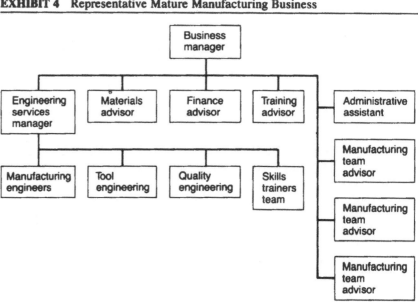

and participation. In addition, he felt the need for some sort of safety-value mechanism that would allow people to air grievances and work-related problems. There needed to be another structure that would allow people from all five levels to interact, and to work to resolve plantwide issues affecting the work environment. There needed to be some formal forum for workers to air grievances as well as participate in the shaping of management decisions. For that reason, his original team developed a complex and multilevel governance structure to parallel the operating organization.

Atop the entire governance structure sat the plant operating team (POT). Made up of the plant manager and his directors, POT was charged with giving a general sense of direction to the plant. It worked with American Diesel on plant objectives and commitments, and it was responsible for specific strategy, policy decisions, and other matters of plantwide concern.

The second major governance organization was the board of representatives (BOR), made up of 20 elected representatives from all the plant's business and functional areas. BOR acted as a forum or sounding board for SEP employees. New ideas, unique viewpoints, and lingering complaints were aired at regular BOR meetings. Employees used BOR to debate matters ranging from concerns over the compensation system and layoffs to complaints of petty thefts within the plant. BOR could then take suggestions, comments, or complaints to POT to seek some kind of resolution.

St. Clair had hoped that the atmosphere at BOR meetings would be such that all employees would feel free to speak their minds. But when Goble became plant manager, he was somewhat surprised to find a considerable amount of dissatisfaction with the way BOR operated. Staff support personnel in BOR complained that manufacturing representatives often clogged BOR meetings with complaints about petty problems like overflowing toilets and disciplinary matters. Manufacturing team representatives responded by insisting that they had nowhere else to go with such problems. BOR was the only structure in SEP that offered them the opportunity to get together and jointly air their grievances and concerns. Besides, they countered, most BOR discussions were monopolized by staff support people.

Numerous other groups and task forces within the plant focused on specific issues. One informal safety-valve mechanism begun by St. Clair and continued by Goble was the "fireside chat." Once or twice a week, the plant manager invited into his office a small group of employees representing a cross-section of the plant. St. Clair started the chats in order to keep "tuned in" to the plant. By speaking confidentially and keeping those confidences, he also hoped to create a feeling of trust that would allow any plant employee to come to him with their problems and concerns.

Plant Leadership

For five years, the personality of plant manager Don St. Clair galvanized SEP. The plant had been from its founding an intimate and important part of St. Clair's life, and he in turn provided a model of enthusiasm and complete commitment to SEP employees. "He was a father figure to us," explained one of the original business managers. "Whatever 'charisma' means, Don had it," added an advisor.

Everyone, from director to team members, leaned heavily on St. Clair during those start-up years, relying on the power of his personality and the force of his commitment to steer SEP through the hurdles of its early development. "This place looked and smelled like Don St. Clair," noted director Doug Pippy. "The whole thing was an extension of his personality."

St. Clair saw himself as a plantwide "agitator." Every day, he spent some time on the lines: talking to people, questioning them, encouraging, prodding them to take responsibility and to live up to the expectations of the plant. "He was one hell of a guy," said a team member who had been with the plant from the beginning. "And we worked harder for him."

St. Clair's managerial style was wide open; Goble observed of his predecessor:

> He was the kind of guy who liked to have his hand in everything. He liked to know everything that was going on. Because everybody in the plant—directors, managers, team members—trusted him, they would come to him with their problems. Anything they couldn't work out on the floor, they brought it right to Don. And he was glad to talk to them and work with them. His door was always open.

SEP's Teams

Manufacturing and staff support personnel at SEP were all organized into small teams, each with its own team advisor. Teams were usually made up of 20 people, although they occasionally grew to 50 or 80 people over three shifts. If a particular line or machine operated over three shifts, workers on all three shifts would belong to the same team with only one team advisor. At least once a month, team meetings would have to be arranged so that members from all three shifts could gather at the same time. Since all decisions at SEP were meant to be made at the lowest possible level, each team was designed to be as functionally autonomous as possible.

Manufacturing Teams.   Each of the 40 manufacturing teams was clustered around a business manager, reporting to that manager through its own team advisor. Team members undertook the usual task of operating on assigned machines or assembly tasks. At SEP, however, a premium was placed on the ability of each team member to perform not just one but a variety of tasks. Thus, each team was expected to handle setup and maintenance as well as operation of the machines or assembly tasks assigned to it. Furthermore, each individual team member was expected, over the course of five years, to learn how to perform these functions on each of the machines. The members of a piston line team, for instance, should eventually be able to perform not just one but each job and support function assigned to the team as a whole. (See Exhibit 5 for a description of team responsibilities.) One of the signs of team maturity was the degree to which the team members did, in fact, encourage each other to learn and perform new tasks.

Out of necessity, the assembly teams operated somewhat differently than the machining teams. Teams like the piston team were involved in the manufacturing of a whole product. Each of the seven assembly line teams, however, was assigned just one stage in the assembly of a diesel engine. They completed that stage and then passed the engine along a conveyor to the next team. The line was constructed in such a way that each team could build up a surplus of from 8 to 10 engines. The plant manager, together with the director of assembly and the business managers from assembly and test and shipping, set monthly rates of completion based on corporate demand. The entire assembly line was expected to schedule and pace their work in order to meet that monthly goal. Another less formal motivator existed for assembly teams. If one team worked so much more slowly than the next that the faster team depleted its surplus, that second, quicker team would find itself with nothing to do. "If that happens," explained Dave Palmer, director of organizational development and training, "members from the faster team will walk over

---

EXHIBIT 5   Team Responsibilities

1. Participation in the selection of team members:
   a. Departing team members
   b. Additional manpower requirements

2. Participation in the setting and administering of rules governing behavior affecting the accomplishment of the team's and organization's objectives:
   a. Attendance
   b. Housekeeping
   c. Safety practices
   d. Quality and quantity
   e. Training and team participation

EXHIBIT 5   (continued)

---

3. Training of its members:
   a. Evaluation of competency levels of each member
   b. Planning needs of team and individuals
   c. Assume active role in training programs

4. Distribution and assignment of tasks among team members:
   a. To cover for nonparticipation
   b. To provide training opportunities
   c. To assign individuals to committees

5. Coping with production problems that occur within or between the teams' areas of responsibilities:
   a. Quality
   b. Delivery
   c. Service

6. Regulations and control of process functions that cross crew boundaries, including planning and scheduling material requirements.

7. Participation in setting of organization's goals and objectives:
   a. Cost reduction
   b. Quality levels
   c. Budget control
   d. Product output

8. Achievement of team's objectives and insuring that members contribute toward the objectives.

9. Document and communicate the achievements and needs of the team to the necessary organizations in the systems.

Following Will Be Provided to Team

1. All relevant resources and information needed to carry out team's responsibility.
2. Rewards according to the individual level of competence attained and contribution to the team's achievement of its goals. These levels of competence will be based on objective measures of the ability to perform more than one combination of tasks.

Guarantees

1. No differences in privileges between team members and other members of the organization except as required externally.
2. No barriers to attainment of higher competence levels.

to the other group and not so politely ask, 'What the hell are you people doing?' "

The functions of SEP's manufacturing teams went far beyond the operation and maintenance of machines, however. Each team was also expected to do its own administrative housekeeping. Team members kept track of their own team's inventory; ordered and inspected necessary materials; documented their team's production, costs, attendance, and performance; oversaw safety practices; and participated in budgeting and forecasting. In addition, each team was responsible for hiring its own members. Teams were to work with their advisor in setting their own performance appraisal, disciplinary actions, and even determining raises to a higher salary level.

These administrative, housekeeping, and personnel functions were known in SEP as vertical tasks. One or more members of each team was expected to devote some working time to a vertical task such as material ordering, budgeting, inventory control, or keeping cost records. Who does what vertical task? What percentage of the work day should be spent on a vertical task? These were decisions that each team was to make for itself. Such assignments usually lasted a year before being rotated to another team member.

The degree to which individual teams performed up to the ideal established by St. Clair's organizing team varied considerably. Mature teams encouraged individual members to learn new tasks and participate fully in devising ways to reduce costs, evaluating performance of individual members, and helping to correct the behavior of individuals that might be harmful to the team. The most mature teams—about one-quarter of the teams, Palmer estimated in 1979—were virtually self-managing and autonomous. The less mature teams depended almost entirely on the intervention of the team advisor to handle matters relating to training, housekeeping, and evaluation.

Perhaps the most dramatic example of a team handling its own affairs occurred in the cambox team in 1979. Cambox production had begun three years earlier and grown rapidly to 1,500 camboxes a day. The size of the team grew with the production rate: over 80 people spread over three shifts with only one advisor. While the advisor felt his team was operating well, many team members thought otherwise. Said team member Sally Moore: "Everything was chaos. There was absolutely no communication between shifts. You would leave a message for the next shift, telling them what needed to be done. They either didn't get the message, or would ignore it. Nobody really gave a damn."

Signs of poor performance seemed to spread. Low morale led to low energy, high turnovers, and constant overtime. Quality suffered, and there was no way of knowing who or what was at fault. "There were so many people and so little communication," said Moore, "that there was no accountability." Finally, in one month American Diesel had to recall 35,000 camboxes, all made at SEP.

Even then, the team advisor refused to admit that he had a problem. But three months after the recall, Moore and fellow team member Dick Smith decided to take some action. With 80 people crowded into a meeting room for the regular team meeting, and with the advisor not present, the two stood up and expressed their feelings. In the middle of the meeting, Smith phoned St. Clair and asked him to join them. St. Clair did, and he encouraged the team to develop, on their own, some plan for reorganization. Moore and Smith then called for an informal committee to consider the possibilities. Two representatives from each shift met every night after work, sometimes for five hours at a sitting, for five weeks.

Their first conclusion was that the team was too big and should be divided into four smaller teams. Recalled Dick Smith: "Not everyone favored dividing us up like that. They said you'd lose product identification that way. But the committee decided that the number of people on a team is more important than product identity. It's not looking at a cambox and saying 'I made that' that gives me satisfaction. It's teamwork and cooperation that makes this place special."

In dividing up the tasks among the four teams—casting and drilling (kingsbury), assembly, variable machine timing, and inventory control— the committee made sure that each team still retained some variety of tasks. And to insure better communication between shifts, a coordinator was appointed for the second and third shifts of each team.

The impact of the cambox reorganization was both immediate and dramatic. Averaging 920 camboxes per day for the year prior to reorganization, the reorganized cambox line upped its daily average to between 1,100 and 1,200 camboxes per day. Overtime dropped by 50 percent, scrap was reduced from 17 percent to 5 percent, turnover from 30 percent to 4 percent, while machine utilization increased from 42 percent to 56 percent (a measurement of the total time available over three shifts a machine is in use). Subjective assessments—team morale, communications, and problem solving—were all positive as well. "It seems that whenever we have a chaotic situation, and a team is able to deal with that chaos, then productivity rises," said St. Clair. "I'm not sure why that is, but it always happens."

While nobody at SEP could say for sure what made one team mature and another not, there seemed to be general agreement on the importance of three factors: the size of the team, the length of time the team had been together, and the willingness of the advisor to encourage team growth.

Staff Support Teams.   Like the manufacturing people, staff support personnel in the plant also worked as part of a team. Each team had its own advisor. Staff support teams provided specific services—quality and product engineering; financial analysis; materials planning, purchasing, and handling; performance and/or training of maintenance duties; engineering, procurement, and installation of machine tools—in support of the manufacturing teams. If a manufacturing team member assigned to the vertical task of forecasting and budgeting, for instance, felt the need for expert consultation, he or she could call directly (or indirectly, through the team advisor) on staff support personnel in order to seek their advice.

Orientation.   St. Clair felt that in order to acclimate new employees to the drastically different work environment of SEP, all new employees should go through an intensive orientation program. Therefore, once hired, each new team member went through an orientation program consisting of 13 sessions to be attended during the initial six months of employment. They discussed plant history, philosophy, goals, and practices. St. Clair candidly admitted, "As hard as we tried to keep their orientations realistic, we really ended up building unrealistic expectations of the amount of freedom they were going to find, of the excitement and the challenge of working here. When they went back out onto the floor, we hoped they'd turn whatever disappointment they might have felt into a determination to work twice as hard to change their expectations into reality."

The reaction of newly hired SEP personnel to the orientation program was decidedly mixed. Some complained about precisely what St. Clair talked about—building unrealistic expectations. Team member Dick Kirkendall said: "We would be told all about teams, about how independent they are. But then I went to work and found out there were a lot of decisions we just weren't allowed to make as a team." Others found the courses to be over their heads. Said team member Bob Reed: "They kept talking about 'work systems.' I know what that means now. But when I first got here, I was fresh off the streets. I had just barely finished high school, you know? They'd say 'work systems,' and I didn't know what the hell they were talking about."

How Did Teams Work?   Most of the workers at SEP came there after some experience in traditional, often unionized, plants. The idea of maximum flexibility, of responsibilities ranging far beyond the operation of a single machine, and of a commitment to learning new skills often struck them as rather alien when they first arrived at SEP. "I used to work in a plant," said Mike Cassity, whose experience was typical of most SEP workers. "There, I had a job to do and I did the same thing every day." Still, the vast majority of the team members responded enthusiastically to the challenge of increased expectations.

Frank McCarthy, team member

Honest to God, I'm excited to get here every morning. I always hated to work, but not now. I like living in Sedalia and raising my family here. But there weren't many good jobs around, and I thought I'd have to move. Then this came along. I'm as happy as can be about it.

Ed Purcell, team member

Before I came here, I was a clothing salesman. I made good money, but I was bored crazy. When I came here I took almost a ten-thousand-dollar cut in pay! Can you believe that? And I've never regretted that decision for a minute.

Dave Thelen, team member

This is the best place I've ever worked, no doubt about it. A lot of the guys around here say they'd like to move up eventually, maybe become an advisor. But not me. As long as I can make enough money on this assembly line, this is where I'd like to stay.

One of the team problems that attracted Goble's immediate attention had to do with discipline. St. Clair had created a mechanism, called the correction action process, for correcting "unacceptable" behavior—substandard performance, abuse of paid time off and property, disrespect for others. This involved informal counseling with the team's advisor, but some business managers complained that the process was not working. They identified absenteeism as a key concern and heard reports of other disciplinary problems. Some team members on third shifts were sleeping on the job, and there were occasional outbreaks of fighting. In one case, the piston team decided to fire an unruly member—a dismissal upheld by the advisor, business manager, and director. The discharged team member, however, appealed directly to St. Clair who ordered him reinstated. (He lasted another month before being fired again, this time with no appeal.)

"One of the problems," said director Connie Kelleher, "was that we just didn't have any common definitions. What *is* excessive absenteeism? What *is* disrespect for others? Nobody could say for sure, and many teams were asking for better definitions."

Stress: A Side-Effect of High Commitment.   "I'm going to tell you something about this place you probably won't want to hear," said a team member. "There's a lot of stress here, a hell of a lot of stress."

By pushing down levels of responsibility and increasing expectations of commitment on the part of team members, SEP had introduced a new element into the lives of many line workers. Most team members had previously worked in traditional factory settings. Once they came to SEP, they were told to help set production and cost goals, maintain machines as well as run them, learn budgeting, and make up missed time on weekends. That high level of commitment had its rewards—enriched, interesting work—but also its cost—personal stress. There were no facts or figures on the level of stress among SEP's team members, or the impact that stress had on their families. But most SEP team members recognized the problem after a few years of working there, and they spoke candidly about it.

Mike Cassity, team member

> My old job was pretty dull, but at least I knew what I was supposed to do and did it, day in and day out. When I got interviewed for my job here, they told me all the things I would be expected to do: keep books, order materials, go to meetings, things like that. They asked me whether I could do that kind of work, whether I'd enjoy it? I said sure! I wanted the job, right? But when I got here and found out they really meant it. . . . My God!

Ed Purcell, team member

> Sometimes, to make sure we get everything done, our team will decide to work some extra hours or come in on the weekend. But my wife just can't understand why we do that. She gets really angry.

Henry Wallace, team member

> We're always being taught and encouraged to talk things out and share our feelings. That was all new to me at first, but eventually it become part of the way I did things. But then I'd go home and want to talk to my wife about things that were happening at the plant. I never did that at my other job. I'd just come home and forget about work. But there's something about this place that makes you want to share everything. And my wife would just look at me like I was nuts.

Harry Holmes, team member

> We all work so hard together and spend so much time together that we become sort of a family. Maybe that's why there are so many romances and divorces here. People on my team call this "little Peyton Place." I know. That's happened to me. I got divorced last month.

Team Advisors.   SEP's first-line supervisors, the team advisors, were given responsibility for the training of individual team members and the development of a mature, well-functioning, self-managing team. The SEP advisor was to act as a team builder, communicator, trainer, and occasional fill-in; but not as an autocratic overseer and decision maker. The plant sought advisors with a firm grounding in technical know-how, combined with interpersonal skills.

The team advisors sat in on the daily meetings of their teams, usually held in the first half hour of the work day, but their function there was to lead discussions rather than issue commands. They held personnel files for their teams, arranged for technical and professional advice from staff support personnel within the plant, and coordinated training for their teams.

When Goble became plant manager, he worried that several of his team advisors did not seem to be effective. The position had always been a difficult one to fill and maintain. While the original advisors came almost exclusively from supervisory positions in traditionally organized plants, their replacements came, for the most part, from within the plant. St.

Clair felt tremendous pressure from team members to provide a place within the organization for them to rise to, and that place was the advisor position. But the practice of promotion from within was not entirely satisfactory. Between 25 percent and 50 percent of all the advisors who had been promoted from team members quit their positions within a year or two to return to their team. Why did this happen?

Bob Kerr, team advisor

> What I'm finding is that the more mature my team gets, the duller my job is. I used to help out on all the machines and teach people how to use them. Now, my team doesn't need my help, and they do their own training. So I spend most of my time settling arguments and talking to people about why they missed a day or two of work. That gets old, fast.

Jim Gilbert, team advisor

> When I think of all the extra hours I put in, coming around on second and third shift, attending meetings—I often get here at 5:30 in the morning and leave at 5:30 at night—I figure that I'm getting paid less on an hourly basis than when I was a team member.

Goble worried that while team advisors from the inside seemed to be somewhat dissatisfied, he was not getting new people to come in from the outside to fill new openings. Since advisors rarely rose to a higher rank at SEP, it seemed like a dead-end job. And the old problem of finding someone with both technical and people skills made it difficult to select the right person. One of the constant complaints Goble heard from team members had to do with the competing pressures placed on the team member by the demands of their advisor for maximum effort on the machine or assembly lines versus the expectation that all team members would participate in vertical tasks. Time spent by a team member on budgeting and forecasting, ordering materials, or attending BOR meetings was time away from the line. Some advisors understood, even encouraged, this as a necessary and significant part of the work day; others did not.

Mike Cassity, team member

> Most advisors here put too much pressure on you. You go off to do a vertical task, and the advisor says to you, "Why the hell are you gone so often?" And then, when you're being evaluated, they say, "Why the hell haven't you done this or that vertical task?"

The pressure from some advisors for members to stay on the line was so great that a few members openly questioned whether the plant was really committed to its professed ideals. "I'll tell you this," said one disillusioned team member, "around here, production is number one. Quality of work life? That's two, three, or even four."

Performance Measurement.   The most significant continuing measurement of team performance came from cost-per-piece figures. St. Clair and his organizing team rejected a cost system based on standard costs established by industrial engineers. Instead, SEP had established a cost system based on improvements from previous performance. There would be a base cost for each piece that would be the real cost from the previous year. Each team in the plant would report monthly on its own cost-per-piece, and that figure would be measured against the base cost as well as the previous month's performance. Each manufacturing team would be responsible for computing its own cost-per-piece and for working to reduce that cost.

The monthly reports compiled by each manufacturing team were simple, yet complete. Costs were given for variable manufacturing expenses (broken down to such items as rework, maintenance, scrap, operating supplies, tools, and freight), semivariable expenses (salary, gas, travel, taxes, insurance), and total team costs. (See Exhibit 6 for a two-month cost-per-piece report of one SEP team). Those figures were then compared to the previously supplied base costs. At the same time, the plant manager and his directors set plantwide goals for cost reductions on specific line items (direct materials, team expenses, etc.). In 1980 the primary goal was to reduce costs by the rate of inflation.

Dave Palmer explained the reasoning behind the continuing stress on this type of measurement: "The cost-per-piece computation is important because it gives each team a feeling of autonomy, a belief that they are key to our productivity." The cost-per-piece figures allowed St. Clair to assess the performance of teams at SEP in terms both of actual costs and cost reductions, and allowed teams access to the information necessary to make good economic decisions.

The Compensation System at SEP

The main distinction made in the plant was not the usual one between hourly and salaried employees, but the legal one between those who must be paid extra for overtime work—nonexempt—and those who are not paid for working more than 40 hours—exempt. (For overtime work—more than 40 hours—SEP's nonexempt workers were paid on an hourly basis.) Team members constituted the nonexempt employees, while advisors, business managers, directors, and the plant manager made up the exempt group.

In keeping with this attempt to minimize distinctions among workers, many of the status symbols typically associated with a plant hierarchy were nowhere in sight at SEP. Dress codes for exempt employees along with special parking spaces and dining facilities were never introduced. St. Clair was convinced that the removal of such artificial distinctions enhanced communications within the plant. At least some team members, however, found this attempt to downplay the gulf between management and workers to be somewhat superficial. "It's as simple as this," com-

EXHIBIT 6 Cost Per Piece Cambox/Camfollower

	Base Cost	March Total	March Per Piece	February Total	February Per Piece	Cost Reduction	Total
Quantity, production		35,556		26,038			
Direct material @ base cost	$25.193	$ 895,762	$25.193	$655,975	$25.193		$33,849
Direct labor @ base rate	4.624	130,542	3.672	103,257	3.966	$ .952	
Team manufacturing expense							
Variable							
Rework	.009	155	.004	53	.002		
Premium	.633	23,699	.667	18,313	.703		
Maintenance	1.138	45,409	1.277	39,920	1.533		
Manufacturing, tools, gages	.086	6,430	.181	965	.037		
Operating supplies	1.090	41,805	1.176	29,983	1.152		
Scrap, manufacturing	1.090	27,357	.769	27,465	1.055		
Scrap, supplier	.522	42,585	1.198	173,716	6.672		
Scrap recovery	(.455)*	(46,117)*	(1.297)*	(209,754)*	(8.056)*		
Others	.047	3,519	.099	14,314	.550		
Subtotal variable	4.160	144,842	4.074	94,975	3.648		
Freight	.716	28,150	.791	14,608	.561		
Total variable	4.876	172,992	4.865	109,583	4.209		
Semivariable							
Salaries, wages, fringes	1.287	42,071	1.183	38,162	1.465		
Power	.268	21,535	.606	6,995	.269		
Gas	.011	535	.015	233	.009		
Travel	.011	305	.008	—	—		
Depreciation	1.210	29,892	.841	29,643	1.138		
Taxes, insurance	.076	2,407	.068	1,980	.076		
Total semivariable	2.863	96,745	2.721	77,013	2.957		
Total team manufacturing expense	7.739	269,737	7.586	186,596	7.166	.153	5,440
Total team cost	$37.556	$1,296,041	$36.451	$945,828	$36.325	$1.105	$39,289

*Figures in parentheses represent increases.

mented one team member. "When we have plant athletic teams, they [managers] sign up for the golf team, and we sign up for the bowling team."

Exempt Compensation System.   Initially, SEP placed all exempt employees into three broad pay categories (as opposed to the 13 narrowly defined categories at American Diesel's Beacon plant). This, it was thought, would allow people to progress through a series of pay increases without quickly coming to the top rate for their category. "We hoped this would encourage stability and development," explained Dave Palmer. "People could get their raises without seeking other jobs and moving to different plants."

"That just didn't work," admitted Palmer. "People here were getting the same money as other executives within American Diesel, but they wanted the promotions as well. They complained that their careers were moving more slowly here than they would elsewhere in the American Diesel system." Just before St. Clair left in 1979, SEP moved to a seven-level exempt compensation system. Team advisors would be placed on one of the first three levels depending on a combination of education and job experience. Their annual salaries would range from $15,120 to $30,420. Staff support managers were assigned to level four ($21,840 to $34,920); business managers to level five ($26,220 to $41,940); plant directors to level six ($32,280 to $51,600), and the plant manager to level seven ($40,680 to $65,040). A progression matrix was then constructed by POT, clarifying the combination of skills, experience, and performance that would allow individuals to progress from the minimum through the midpoint to the maximum of their salary levels.

Nonexempt Compensation System.   The compensation of nonexempt employees was supposed to be based entirely on the acquisition of skills by individual team members, and the willingness of the individual to perform those skills. "It's a mistake to base pay solely on the acquisition of new skills," explained Palmer. "People will learn something just to get more money, but then never put what they learned to use." There were five skill levels that each team member was expected to reach, one year at a time (there was also a six-month increment for the first year only, contingent on attending the 13 orientation sessions). All nonexempts worked out a yearly performance plan with their advisor, stating in writing what skills they should acquire and tasks they should perform during the year. At the end of each of the first five years, team members would be evaluated—by the team advisor, except in the case of an extremely mature team, in which case the entire team would participate—on how well they had met those expectations. Promotion to a higher level depended on meeting these expectations for performance and growth.

On some teams, members were reluctant to oppose openly the awarding of an increment to a fellow member. "Nobody wants to stand up at a meeting and say so-and-so shouldn't get a raise this year," said one team member. "If you do that, what's going to happen to you when it's time for your raise to be considered?" On those teams, the yearly increments designed to recognize the acquisition of skills became strictly seniority advances. Some team members, advisors, even business managers, insisted that they knew of no instances when a team member was denied a yearly raise. St. Clair acknowledged that while there were several instances when team members were denied a raise, compensation had indeed become a seniority-based system.

Another shortcoming of the five-level plan was the question of what happened after the fifth year. By the time Goble became plant manager, there were a number of five-year employees who wondered about that point. They could still receive general raises along with Beacon; however, they could not receive increments based on the acquisition of new skills.

Each of the five levels was given a flat rate. At first, POT pegged that rate almost entirely to the prevailing wages in the Sedalia area. SEP's wages were competitive in comparison with similar workers in the community, but because it was a depressed area, they tended to be rather low when compared to American Diesel's Beacon employees. In order to achieve greater equity within the corporation, POT sought to upgrade nonexempt wages in 1977. (See Exhibit 7 for nonexempt wage scale, prior to and immediately following this upgrading.) According to plant policy, POT reviewed the entire compensation system twice a year. In the summer of 1977 POT decided to tie Sedalia's wages to the union-negotiated wage agreements in Beacon.

A four-year employee at Sedalia was to be given a salary derived loosely from the average hourly rate for all four-year employees at Beacon. Five-year Sedalia employees, in recognition of their broader skills and responsibilities, would receive more than the average five-year Beacon employee. And because of that tie-in, any negotiated increase in the hourly wages at Beacon would result in an increase at Sedalia. Thus, SEP employees received raises in two ways: an annual advance in salary level for their first five years and a negotiated increase in the Beacon contract. Between 1977 and 1979 those increases often reached 5 percent and 6 percent every six months. That shift moved SEP into the ninetieth percentile of wages in the Sedalia community.

St. Clair hastened to add that the tie-in to the Beacon wage was not absolute. He and POT members felt no hesitation about adjusting wage rates up or down. "Compensation is one of those areas that I don't think should be too participatory," said St. Clair. "I tried to keep people informed about what we were doing, but me and my directors made final decisions ourselves. Seeking too much participation on compensation issues can get you in a lot of trouble."

One special category among SEP's nonexempt personnel was created

EXHIBIT 7   Nonexempt Compensation System

	1976	
Level	*Weekly Salary*	*Hourly Equivalent*
Entry	$148	$3.70
6 months	154	3.85
1 year	160	4.00
2 years	170	4.25
3 years	180	4.50
4 years	190	4.75
5 years	200	5.00

	1977	
Level	*Weekly Salary*	*Hourly Equivalent*
Entry	$186	$4.65
6 months	198	4.95
1 year	208	5.20
2 years	224	5.60
3 years	236	5.90
4 years	246	6.15
5 years	256	6.40

for skilled tradespeople like electricians, machine repairmen, or draftsmen, who were hired to train team members in their skills. Because SEP was having a problem attracting skilled tradespeople, POT created a special wage scale for them in 1976. They were placed on a wage scale considerably higher than the scale used for other nonexempts. POT also tied Sedalia's 90 tradespeople to the hourly wages for the skilled trades at Beacon. Thus, in 1977, the skilled trades entered at $246 a week and topped at $290. From the beginning, that distinction caused resentment within the plant.

Concept and Reality at SEP

A minor but revealing example of tension between concept and reality occurred over the question of precisely how to translate into the reality of plant life one of the plant's key philosophical commitments—that of trusting all employees. Team advisor Ed Fremder explained the flap that occurred in the plant over the question of locks:

> We say we trust people around here to act like adults. Because of that, we give people access to whatever tools they need to do their jobs. Somehow, that got translated to mean no locks anywhere in the plant. Doors, equipment, files,

everything was kept open. If Don saw a lock anywhere in the plant on *anything,* he'd rip it off. But that's not the real world, is it? People do steal things out there and in here.

Now, at one point, the plant bought three-wheeled bicycles, one for each team, to be used by their members in getting around. Right away, those bikes started disappearing. One team would "borrow" a bike from another without asking, and then "forget" to return it. Teams started hiding their bikes so that others couldn't find them. It got a little ridiculous. Oh, we spent hours debating that one! Meetings all the time. You should have seen it. And we never really decided anything. The bikes just drifted away, and we never bothered replacing them.

That tension between the concept and the realities, and the danger of allowing one to blur the other, had always been a matter of concern at SEP.

Performance at Sedalia Engine Plant

"It may be too early to tell, but there are encouraging signs that our style of management is starting to pay off." That evaluation was offered in the fall of 1979 by St. Clair as he prepared to move on to corporate headquarters in Beacon and pass on managership of SEP to Goble. St. Clair's hopeful appraisal of the plant's performance included the following specific points:

1. Absenteeism, including both excused and nonexcused, was down to about 3 percent, as opposed to about 6 percent at American Diesel's Beacon plant, and 8 percent in the Sedalia community.
2. SEP's safety record, while poor at first, was improving steadily.
3. Initial warranty data on SEP's engines were extremely favorable.
4. Plantwide machine utilization usually ran between 60 percent and 70 percent, and sometimes as high as 75 percent, compared with 50 percent at Beacon.
5. While technological differences existed between the two plants, indirect labor costs were significantly less at Sedalia than the Beacon plant. As the plant reached maturity, that savings could reach 20 percent. Because of the advanced skills of some team members, the need for skilled tradespeople at Sedalia was considerably less than for Beacon. Sedalia also operated with one-half the first-line supervisors at Beacon.
6. Team members were continually performing major machine overhauls and minor maintenance.
7. Except for some start-up problems, quality seemed to be running high at Sedalia. For example, the number of engines rejected by testing at Sedalia was 25 percent of the Beacon number.
8. The general climate was positive and focused on plant excellence.

9. The work system and the governance system seemed to be working to the satisfaction of SEP employees. Job satisfaction seemed to be higher than at other American Diesel plants. To support that conclusion, St. Clair pointed to the fact that no serious union drive had been launched at SEP.

10. SEP enjoyed support from American Diesel's CEO, although there was still some skepticism and lack of understanding about SEP from some key people in upper management.

"By far," St. Clair concluded, "this has been the best plant start-up American Diesel has ever had."

Danney Goble Takes Over

"He really walked into a mess," said one of the business managers about Goble's first months as plant manager. "You have to feel bad for the guy." Goble himself spoke of the challenge not just to him but to the plant. "We now face our most serious test ever of our strength and moral fiber," he observed.

Leadership Turnover. The contrasts in the personalities and styles of the old and new plant managers were dramatic and obvious to everyone in the plant.

Tom O'Donnell, business manager

> I would characterize the difference this way. Don was people-oriented, while Danney is process-oriented. He seems to have less tolerance for ambiguity and more need for structure. Don looked at results, like most manufacturing people. Danney's background is engineering, and he seems more concerned with details than Don was. Don was a visionary; Danney is a tactical leader.

Some worried that Goble could not lead the plant in the same way St. Clair had been able to—among them, Doug Pippy, director, who said: "Danney seems to have problems relating to other people. Like the other new management people he brought in with him, like me in fact, he's a little uncomfortable with other people."

Complicating matters even further was the fact that SEP was undergoing a large turnover among its top management team. Directors, business managers, even some advisors were leaving SEP in large numbers. Out of its top 24 slots, SEP lost 15 people. Only 3 of those left American Diesel; the other 12 went to plants within the American Diesel system. "I can understand that," said Dave Palmer. "People left here because they wanted to get more attention from corporate headquarters. So they went to Beacon, or some of the new plants that American Diesel was opening." Moreover, it was customary for managers to move every several years.

Also, American Diesel actively sought experienced SEP managers to help them start up new plants.

"The extent of the turnover and the short time in which it happened was completely unanticipated," said Goble. "I feel like I have to reinvent the wheel all over again, to teach these people from scratch what our operating philosophy is all about. This certainly won't make things any easier."

Economic Downturn.   "Our company business analysts still are projecting a major downturn 'soon.' Our company president stated two weeks ago we can expect it to be the second worst decline in the last 25 years for American Diesel (second only to 1974–1975). All our planning is based on this assumption." (See Exhibit 8 for summary of American Diesel's economic picture.) That was Goble's gloomy assessment of economic conditions, communicated to SEP employees just after he became plant manager. Already, the indicators were unmistakable. In the six months prior to Goble's becoming plant manager, engine orders had declined 18 percent, and projections indicated only a worsening of conditions.

"What we're faced with," said Goble, "is the possibility of our first layoffs at SEP. The whole corporation is suffering, and we're going to have to shoulder our fair share of that suffering. I'm almost certain that there are layoffs coming, and that those layoffs will be sizable."

In considering the possibility of layoffs, Goble felt that he had an option of two broad policies. SEP could join other American Diesel plants in laying off workers. Although estimates of the extent of that reduction varied from week to week, Goble figured he would have to reduce SEP's work force by about 4 percent, or 20 people. On the other hand, he could commit SEP to attempt at least to maintain current levels of employment. What layoffs would do to the high level of commitment that employees had built up over the years, he could not be sure. What layoffs would do to the fragile team member-management relationship of trust, already severely tested by the turnovers, he was even less sure.

EXHIBIT 8   Economic Performance of American Diesel ($ thousands except earnings per share)

	1973	1974	1975	1976	1977	1978	1979
Net sales	$637,330	$801,566	$761,504	$1,030,532	$1,268,814	$1,520,742	$1,770,851
Profit on sales	48,739	63,510	36,763	127,726	136,468	130,396	106,991
Net earnings	26,592	23,775	491	58,622	67,022	64,399	57,938
Earnings per share	$3.87	$3.31	$0.21	$7.66	$8.22	$7.62	$6.84

That second course—protecting all jobs—would still require sacrifice on the part of the plant, and a good deal of creativity in deciding how to absorb losses without cutting employment. Goble continued:

> If I go that route, I'm going to ask for even more commitment on the part of our employees. They're going to have to devote their energies to thinking of ways to cut costs. We might ask, for instance, that some people take temporary, voluntary layoffs. Perhaps we could all go on shorter work weeks. We might have to move people around from one team to another, maybe ask them to do work that we've previously contracted out for. We're getting ready to paint the plant, for instance. I wonder if we could get some of our teams to do that rather than hiring outside painters. But what will that do to our team structure?
>
> Another question I've got to decide, and decide right away, is how much to tell people in the plant? Right now, all of this is speculation. I don't know for sure that we're going to have layoffs. Should I tell people now that it's a distinct possibility? I might be getting them upset over nothing. And with all the concern now with me and the other new people, that kind of bad news might be too unsettling. On the other hand, I'd like to get people involved in thinking about the alternatives. How do I do that unless I tell them everything?

Nonexempt Compensation. "Danney keeps telling us that we're being paid fairly compared to the people at Beacon," said team member Bob Reed in the fall of 1979. "Now, I'm getting paid well, but not for the kind of work that I do. I do a hell of a lot more than the people at Beacon. Do they have the vertical tasks that we do? No! Most of the guys here don't even look at Beacon for a comparison. Instead, we look at the autoworkers over in Calhoun. They're getting nine and ten dollars an hour, while we get seven or eight."

With the enriched work and higher expectations of commitment, perhaps it was inevitable that nonexempt workers would begin to reconsider their compensation system. But the economic downturn brought the issue to a head just as Goble became plant manager.

Starting in 1977 nonexempt pay levels at SEP had been pegged loosely to the average wage of all four-year employees at Beacon. The economic downturn of 1979 had a dramatic impact on salaries, both at Beacon and Sedalia. Any large-scale layoff at Beacon could hurt Sedalia wages. If employees at Beacon were laid off in large numbers, quite a few employees there would find themselves bumped down to lower paying assignments; that would significantly reduce the average four-year wage at Beacon. If Goble and POT elected to adhere tightly to that average, it would negatively affect wages at Sedalia. Compounding the problem was the fact that the wages of the skilled trades at Beacon would not be affected by the downturn. Skilled tradespeople at Sedalia were tied independently to those at Beacon, which meant that they were still in line for significant raises.

Goble had an immediate question to consider in the fall of 1979. If he followed the formula of basing four-year Sedalia wages on the precise average of Beacon without making any adjustments, team members would receive only about two thirds of the increase they had received over the previous several years, while the skilled trades would receive a significantly higher raise. Team members already concerned about the fact that overtime had been virtually eliminated in 1979 might be even more upset by the enhanced inequity between themselves and the skilled tradespeople. Could Goble's leadership withstand the disruption that might be caused by such unhappiness among team members?

Goble was clearly leaning toward the idea of sticking literally to the old formula for 1979 and then rethinking it the following year. He anticipated that such a course would raise some concern among team members, so he needed to devise some sort of a process for bringing them into the discussions. On the issue of equity between team members and the skilled tradespeople, Goble figured it was an old problem that would cause no special concern now.

In fact, he worried more about the costs of suddenly abandoning the formula. Some POT directors were suggesting just that. "Let's make an exception in this case," one director argued, "and give the same raises to everyone." But Goble was skeptical: "If POT went into a meeting and suddenly changed the formula, we would be setting a horrible precedent for the future. And I think that would really upset people. Sure, they would be happy if they got a little more money. But in the long run, they'd be suspicious of us. If we could change this long-held policy behind their backs, so to speak, what else would we change without asking them?"

Besides, as one director told Goble, "Our people supported the old logic when it led to good raises. They should be willing to support it now that it means some sacrifice. Besides, people know what the economic situation is. They're not expecting much this time."

The possibility of opening up the process of compensation review to team members immediately after springing on them the news that raises would be reduced promised some disturbing times. Employees might come to view such participation as a kind of formal collective bargaining process over wages. There would be ticklish questions of deciding how employees should be brought into the process. Team members were dissatisfied with their representation in BOR, yet no other in-plant organization included nonexempts.

Then, too, Goble wondered just how high the expectations of Sedalia's nonexempts had risen. They already insisted that five-year Sedalia employees should earn more than their Beacon counterparts because Sedalia expected more of its employees. But just how high had his employees' evaluation of themselves risen, Goble wondered, and how would that affect their salary expectations?

A Possible Organizing Campaign. Goble was aware of one final development. During the same week that he was considering what direction to take with the compensation system, the following article appeared in the *Sedalia Free Press:*

> Production and maintenance personnel of the Sedalia Engine Plant have been invited to attend an informal meeting at 7 P.M. Thursday at the Holiday Inn, being held by the Machine Workers of America union.
>
> Robert Reinhold, a union representative, said his union represents about 6,800 workers in American Diesel's Beacon plant, and that the purpose of Thursday's meeting is not to organize the workers in the nonunion plant here, but simply to provide information on wages and benefits being given union workers in Beacon.
>
> "Of course, if the workers here wanted to organize a union, we would be interested," Reinhold said, "because we feel everyone would be better off if all the American Diesel workers were represented."

Sedalia Revisited

Ivan Gargarin, third plant manager of American Diesel's Sedalia Engine Plant (SEP), was facing one of his biggest challenges. Changes were happening on several major fronts as a result of a major restructuring of the diesel engine industry; worldwide competition had forced American Diesel to become more cost and delivery conscious. The corporation had embarked on a new strategy of just-in-time (JIT) manufacturing and of total quality systems (TQS), and there were numerous new cost reduction projects underway. In addition, American Diesel had an estimated 1,800 excess employees for 1986, 250 to 300 of whom were in SEP. And SEP was currently operating at about 40 percent of capacity. Gargarin was pulling in components that had been subcontracted to outside vendors and attempting to sell some of SEP's services and/or capacities (such as tool sharpening and microfilming engineering drawings) to other local manufacturers in an effort to reduce the number of excess people. But he knew he would still have to make a decision on work force reductions in early 1986.

During its 11-year history, SEP had grown to over 900 employees; it had introduced a new engine line and weathered economic downturns in 1979, 1981, 1982, and mid–year 1985. The plant had avoided layoffs through a freeze on hiring, voluntary one-month leaves without pay, work hour reductions (down to 35 hours/week for several months), and by the creation of a "swing team" comprising excess people who were assigned as needed to miscellaneous tasks throughout the plant. (See Exhibit 1 for

---

EXHIBIT 1   SEP Employment Statistics*

	Year-End Employment			Yearly Attrition (Number of employees)		OSHA Incident Rate
	Exempt	Nonexempt	Total	Exempt	Nonexempt	
1975	unavailable			1	—	unavailable
1976	75	68	143	3	2	7.4
1977	140	259	399	7	8	18.5
1978	138	413	551	12	25	13.9
1979	152	518	670	31	20	11.8
1980	155	527	682	16	15	9.9
1981	179	661	840	24	11	7.4
1982	180	631	811	15	17	9.1
1983	181	627	808	18	9	9.4
1984	175	741	916	23	21	9.1
1985	154	729	883	34	22	9.2

*Attendance as of year-end 1985 (on a rolling 12-month basis) equaled 97.6 percent. Although historical figures were not available, the personnel team manager who had tracked the statistics from the inception of the plant noted that there had been very little variation in the number.

historical employment statistics.) There had been no serious threat either to SEP's nonunion status or to its "team concept."[1]

But would the innovative plant design survive the problems and changes currently facing the company and SEP? Donald St. Clair, SEP's first plant manager and now vice president of operations with overall responsibility for SEP, had been quoted as saying that there was a need to reexamine the application of the precepts of the original organization. The plant had been originally designed with the intention of continual midcourse improvements to keep pace with changes in the market, and he believed that it was time to redefine how the principles were applied to the new business environment. Many team members and team managers[2] also expressed concerns about the current application of the concept:

- "We've lost sight of the concept."
- "Management is reverting to the traditional control mentality."
- "We're losing our team identity and individual freedoms with pull manufacturing [JIT]."

[1] See Sedalia Engine Plant (A) for details of SEP design and start-up.
[2] As of March 1985, team advisors became team managers. Further details appear later in this case.

- "There's less emphasis on vertical tasks and training because they're not viewed as value added."
- "The shift in the plant is from a human focus towards more business basics for survival."
- "The team concept is changing. This used to be a fun place to work. No rules. The changes are tightening things up and bringing the plant back in line with its original business focus. Now it's getting back to what the concept was intended to be."

Many of the negative comments were made by the more senior team members, as one machining team manager described:

> The veterans, or those nonexempts who have been at SEP since we started operations, came into a plant that told them that they would have a lot of freedom in the way the plant was managed and that they would have a lot of say. This raised their expectations about involvement and their freedom in how they spend their time. The newer people came into a plant that was much more structured, so their expectations are lower. As a result, they are much happier with the current state of affairs.

CHANGES IN BUSINESS ENVIRONMENT

Over the years, American Diesel had established a reputation for high-quality engines with superior reliability and durability records. This reputation and the company's ability to tailor engines to customers' specific needs enabled it to maintain a market niche and a price premium. Like most other American manufacturing concerns, especially those in the transportation industry, American Diesel's market changed in the early 1980s. Before then, American Diesel had held approximately 55 percent of the market share and competed primarily against other domestic manufacturers. In the late 1970s, European engine makers began aggressively to pursue the truck market and were followed closely thereafter by the Japanese. American Diesel chose to license one of its engines to a Japanese firm which quickly learned to manufacture similar engines at a significantly lower cost. It was estimated that the Japanese firm could produce an engine of roughly comparable quality for as much as one-third less than the cost of an American Diesel engine.

In addition, with a flattening of overall market growth, American Diesel had to reshape its strategy, from attempting to gain market share to holding onto the replacement market. As a result, the company had to find ways to cut costs significantly to remain competitive.

In 1979 a new engine, the A200, was introduced. The A200 was designed for the same applications as American Diesel's standard over-the-road diesel truck engine but offered the customer a smaller, lighter weight model with improved fuel efficiency. It was also more suitable for the bus market, a major new opportunity. Although many of the A200's major components were manufactured at the Beacon plant, the final assembly of the new engine was located in SEP. As a result, Gargarin was assigned the engine's overall cost responsibility. (A facility in the United Kingdom was assigned final assembly for engines sold exclusively in the UK and Europe, but the costs of these engines also fell under Gargarin's purview.) This brought SEP into the corporate limelight: it would be responsible for a critical part of the organization's business. Gargarin and his staff would now be dealing with the coordination of interplant and interfunctional relations, and this meant Gargarin would spend a significant part of his time on matters beyond daily SEP activities.

In July 1984, the president of American Diesel made a commitment to sell the A200 engine at a reduced price to gain market penetration and keep out foreign competition, primarily Japanese. The manufacturing organization was informed it had 30 months to cut the cost of each engine by $9,500 from approximately $16,200 to $6,700 per engine. (Similar goals were placed on American Diesel's other engine lines.) This "30-month sprint," as it was called, placed significant pressures on SEP. In response, Gargarin established cross-functional teams, led by the business managers, to concentrate on the cost, quality, and delivery of the new engine. He set up monthly (and sometimes weekly) review sessions with representatives from various parts of the corporation. They had to present the steps (including Gantt charts[3]) they were taking to meet cost reduction targets, focusing on each component and each cost category (labor, materials, and overhead). By October 1985, SEP had increased production to 75–80 engines per day from an initial start-up of 15–20 engines per day and had reduced the costs to $10,300 per engine. A year-end market slackening, which reduced daily production to 60 engines, further complicated the task. The traditional cost-per-piece measurements evaluated against an historical base were modified to reflect market prices.

JIT AT SEP

Faced with foreign competition, the "30-month sprint" challenge, and a corporate push to instill the Japanese philosophy of just-in-time manufacturing and total quality control, Gargarin established a "JIT improvement group" in the assembly and test (A&T) area. The group, formed in 1984, consisted of a team advisor, an engineer, a materials specialist, plus a team leader who came from material control. They began by scanning the literature, attending seminars, and visiting other companies who had im-

---

[3]A Gantt chart is a way of displaying a schedule of interrelated tasks with each activity having a designated start and end date.

plemented JIT. They found that other plants had focused on three aspects of the operation: quality, material flow, and work flow. SEP's group chose to focus its efforts on work flow.

Within a month, an implementation group, made up of A&T exempt and nonexempt team members, was established. This group's task was to work its way through all assembly teams by doing further analysis, recommending solutions, and trying to engage team participation—in essence, being a resource to all A&T teams. Their first step was to improve fixtures and tooling to reduce setups on individual work stations. They videotaped every operation in every team, beginning at the front of the assembly line; all the team members then viewed the tapes and brainstormed to find ways of reducing "nonvalue-added" work. After all the ideas were documented, the team estimated savings from each and identified it as short, medium, or long term.

In July 1985, a consulting firm specializing in JIT was hired to provide training and guidance for the JIT implementation project. They suggested that in order to achieve synchronous flow or JIT manufacturing, the plant should eliminate buffers and provide a visual means of communication for people on the line. One month later, a design group was formed to condense the base assembly line. (The assembly line consisted of three portions: base engine assembly, engine test, and final parts mounting which mounted customer specific parts or decals and painted the engines.) The group comprised four members from the implementation group, plus a manufacturing engineer, a quality engineer, a safety representative, and a material coordinator representative. Their task was to plan the physical rearrangement of the line and to lay out an implementation schedule. They began by eliminating work-in-process buffers between teams 6 and 7. Gradually, they then compressed teams 4 and 5 and teams 1, 2, and 3, leaving buffer inventories in only two locations—between teams 3 and 4 and between teams 5 and 6. The final step, taken during the Christmas shutdown, eliminated all in-process buffers and physically reduced the base assembly line by 60 percent of its original size. In addition, 27 work stations were eliminated, leaving 52 as of January 2, 1986. What remained was an open area of about 67,000 square feet which management hoped to fill with new business it was trying to entice corporate management to transfer to SEP. Throughout the plant other areas were opened up by inventory reductions; these were roped off and also designated for new business. See Exhibit 2 for a layout of the facility as of January 1986.

CHANGES IN A&T TEAM AUTONOMY

According to one team manager:

> The plant was designed based on a concept of "total team flexibility." Each team was considered a "mini-plant" with its own set of cost and delivery goals.

EXHIBIT 2 Layout of SEP as of January 1986

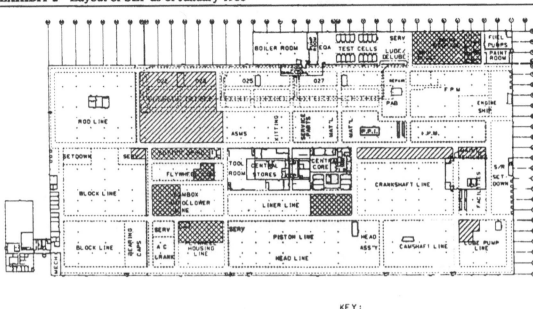

KEY:
▨ ALREADY IDENTIFIED NEW BUSINESS
▨ AVAILABLE FOR NEW BUSINESS

Each team could decide when it would work; that is, it could pretty much set its own starting time, when it wanted to take breaks and when it would stop for the day. JIT has eliminated much of the team autonomy relative to individual team members' freedom.

To allow for semiautonomous teams, SEP designers had inserted a two- to four-hour inventory buffer between assembly teams and up to a five-day buffer between A&T and the machining teams. Prior to JIT, there also had been 20–30 days of raw material or purchased parts inventory, plus significant levels of work-in-process inventory within each team area. And, to improve American Diesel's reputation for poor delivery (traditionally, meeting 20–25 percent of its daily delivery requirements to a broad range of customers while maintaining excellent delivery to a few key customers), SEP had held a finished goods inventory of approximately 600 engines in 1985. Delivery performance thereby improved to 70 percent of daily requirements, and some new sales opportunities were explored.

The implementation of JIT, or "pull manufacturing," as it was often called at SEP, and the rearrangement of the assembly line in December of 1985, had a significant impact on the amount of inventory held in the A&T operation. Inventory between teams was reduced essentially to zero, and inventory between A&T and machining was reduced to a maximum level of one day (with cams, blocks, and heads holding only four hours of inventory). Reductions in finished goods inventory were planned for later in the process and thus remained unchanged, but raw materials or purchased parts inventory was reduced to 10–15 days. The throughput of a complete engine from the first station on the assembly line through test and final parts mounting was reduced from nine days to four, and the goal was to reduce that to one day.

All managers assigned to A&T (team managers, business managers, and the director of A&T) were issued a beeper so they could be paged whenever problems arose. During January it often seemed there were more managers than assembly operators on the line. Although most team members viewed this as positive because resources were available whenever needed, others felt they were under constant surveillance.

A system of red and amber lights and buzzers was set up all along the line to signal any work station with more than two boxes (each box contained one engine as it progressed down the assembly line) accumulated ahead of it. The line was programmed automatically to stop whenever this occurred. Then all managers would swarm to the troubled work station to try to solve the problem, while assembly workers, other than those at the source of the problem, took the opportunity to clean up their work areas and rearrange materials to prepare for the resumed operation. Although a formal 30-minute break had been added to the morning schedule, some operators used the line downtime to run to the restroom or get a cup of coffee.

Initially, team members did not believe the JIT changes would be physically possible, and some reacted negatively when they returned from the December shutdown (when the rearrangement had taken place). The seven assembly teams would maintain their identities but now also had to function as one large entity; many team members felt their individual freedom had been taken away. Not only did they all have to start at the same time and take breaks together, they had to work at a more even pace and meet a six-minute cycle time. Many perceived this as working faster. One team member began wearing a T-shirt saying "I'm in cell block #3." Another nonexempt team member, whom a colleague described as one of the best workers on his team, was particularly resistant to the ideas of the JIT improvement group. As a result, he was assigned to the performance improvement process (PIP) counseling program to deal with his obstruction to the JIT program.

IMPACT OF JIT ON MACHINING

Although the JIT improvement group focused its initial efforts on the assembly and test operation, the plan was to extend JIT throughout the plant and to outside suppliers. The goal was to implement JIT in the head line by the end of the first quarter of 1986 and then expand into other areas at the rate of one per quarter. In the meantime, all machining teams continued to build to inventory as they had always done. In most areas, however, the overall inventory level had diminished significantly (ranging from 4 to 24 hours), and the teams had begun analyzing "nonvalue-added" operations in anticipation of JIT. The changes in A&T, however, had an immediate impact on the machining teams. The camshaft's material coordinator commented:

> JIT has had both positive and negative aspects. On the positive side it has made people more aware of our assembly line. They really do build engines here! It has forced more changeover of machinery, therefore forcing people to find more efficient ways to make changeovers. It has also made more team members involved in team production and scheduling. Lastly, it has made us more independent of upper management. Therefore, I think it has eased some of the load on management.
>
> But JIT has meant frustration among all team members. Since assembly stopped holding any extra inventory, the lines must hold a small amount. There is never a comfort zone. Just when they think they've got good numbers, they change them. Because the corporation allows customers to dictate when they want or don't want their engine, orders change daily. Since we can't bank any inventory, it is more difficult to respond to a schedule change. Communication must go on continuously between the assembly line, the coordinators, and the managers and be given to all team members. Just when you think you have communicated the right information, it changes.

There was a general feeling throughout the machining teams that JIT had been much easier to implement in A&T than it would be in machining, primarily because A&T was less machine-intensive. One team manager stated:

> A&T is basically just a people issue. Sure, they have a few torque wrenches to worry about, and they have to be concerned with their suppliers and their customers. But we have to worry about suppliers and our customers—A&T, Beacon and overseas—plus 45 to 50 machines that can break down at any time. They tell us we have to have "planned downtime," but I don't see how that's possible. You can do all the preventive maintenance you want but sometimes a machine still breaks down, and you can't tell it to wait until the end of the shift.

In anticipation of JIT, the block line had reduced its end of the line inventory to a four-hour level, a significant improvement from the rate the line started production with in 1981, when there had been a minimum of one to two weeks' supply at the front-end along with one to two weeks of finished goods inventory. In addition, the conveyors between machines (plus the spurs) were full of in-process inventory—in total around 400 blocks (daily production averaged 20 blocks). By year-end 1985, the total block line inventory had been reduced to four and a half days, a result attributable more to volume increases than to inventory reduction. Members of the block line team, however, felt that further reductions were not feasible because of machine unreliability (frequent downtime ranged from 10 minutes to 10 hours), procurement of many raw materials, such as castings from overseas, and lack of communication between the front and back of the line.

A major bottleneck came from an unreliable finishing machine. Operating at 100 percent efficiency it could produce 27 blocks per hour; realistically, a good day was 10 blocks per hour. When the machine was not operating for any length of time, blocks coming down the line would be taken off the conveyor and stacked near the machine so that the rest of the line could continue production. With a daily production requirement of 90 blocks, the team manager had introduced a split shift to staff the finishing operation 12 hours a day.

As blocks came off the end of the line they were loaded onto skids, to be transferred to A&T or the UK facility. (The UK plant required approximately 30 blocks per day. There was an agreement that the UK operation would provide a firm three-week order, but it often changed weekly.) The A&T materials coordinator used a Kan-Ban type poker chip arrangement to signal production requirements. Every time the coordinator would take a skid of blocks from the block line inventory, he or she would leave the poker chip on an inventory board. This in turn would signal the need for further production. The block line, however, continued to produce to a monthly schedule when A&T had to shut down, unless A&T was down for more than a day. The reasoning was that A&T could catch up by working overtime or working a little faster, but that the block line was paced by its machine times. The block line and A&T communicated in three ways: informal talks between the team managers, the material coordinator's poker chip system, and daily production reports.

TOTAL QUALITY SYSTEMS

Concurrent with implementing JIT, SEP was translating and implementing a three-year corporate program on total quality systems (TQS) which had been launched in October 1984. A corporate task force, including one member from SEP, had established 47 procedures (a large percentage of which concerned engineering) to be implemented throughout all the functions and facilities. A primary manufacturing issue was total process control (TPC), which included 10 elements such as statistical process control (SPC), training, maintenance, procedures, and equipment.

SPC was not new to SEP; it had been used in the block line from the first day the line began operating in 1981. The line was 90 percent of the way to a goal of having all durability and safety characteristics on SPC by the end of 1986. Other areas of the plant, however, had a long way to go. The manager in charge of SEP's TQS efforts commented:

> Over the years, the plant got a little loosey-goosey on how things are done. A team might choose to change a procedure because it found an easier way to do

an operation but never document it. They might have thought they were buying a cheaper tool but were unaware of the total impact it might have on the product. The teams need to think differently about the process. Change must be controlled and evaluated. The operators may not know all—they aren't trained in engineering or don't know all about metallurgical properties. We also tend to reward fire fighters and never measure fire fighting control. We need to think in a preventive way but don't know how to reward it. I expect we will have some problems as we implement SPC across all of SEP. We are taking away some flexibility the teams used to have.

CHANGES IN PLANT LEADERSHIP AND STRUCTURE

During the early 1980s, the plant went through several major realignments. Danney Goble, Gargarin's predecessor as SEP plant manager, had determined that the machinery businesses needed more common direction and established the role of director of machining with responsibility over Businesses A, B, and C. This position also inherited responsibility for other new machining businesses that were added to SEP, such as the block line in 1981–82 and the rod line in 1983.

In late May of 1983, Goble stunned the plant by suddenly announcing he was leaving American Diesel to head up a small manufacturing firm located on the outskirts of Sedalia. One month later, he showed up at Gargarin's house with a bottle of champagne and informed Gargarin of his selection to be the third plant manager of SEP.

Gargarin, who had worked for American Diesel for 19 years, joined SEP in 1979 as the director of assembly and test. At that time, according to Goble, Gargarin was overly sensitive that his actions would "screw things up." Gargarin recalled, for example, that when he needed to hire a new business manager, he had tried to involve a cross-section of that business in selecting a candidate. But he also had had some strong ideas about the kind of person he wanted and, as a consequence, had rejected the consensus candidate. Much gut-wrenching followed. Although he restarted the selection process and eventually got what he wanted, he felt he had mishandled the situation by not stating more clearly at the outset either what he wanted or what role the "participative" process would play.

As plant manager, Gargarin was considered less charismatic than his predecessors. When Goble returned for a visit several months after his departure, he was treated like a hero, with employees noting that the plant had not been the same since he had left. Seeing that Gargarin was a bit taken aback by these remarks, Goble took him aside and said, "Don't worry about it. I went through the same thing when I took over from St. Clair."

Gargarin's style, however, was quite different from his predecessors'. Staff members described him as more soft-spoken, conservative, and procedure-oriented. On the other hand, he made a special effort to attend weddings, confirmations, and funerals of employees' family members. When addressing personnel problems he tried to understand issues in other than a "cold business approach." Gargarin was concerned about his image and personally frustrated with people's perceptions that the current more formal approach (increasing emphasis on process controls and cost savings) was a "betrayal" of the plant's original concepts. He knew he had to find ways to bring the whole system into a better "fit."

When Gargarin took over the plant's operation, he had five directors reporting to him—machining, assembly and test, finance, human resources, and reliability. In August 1984, he eliminated the role of director of machining because of head count restrictions and a desire to link the business managers closer to the plant manager now that the plant start-up had been completed. This, he felt, would also give recognition to the business manager level, something that had been a problem in the 1979–80 period. Although this move was viewed positively by most, especially those within the machining businesses, a few managers (and an external JIT consultant) voiced concern that the machining businesses were wandering off in their own directions and lacked a common vision. Exhibit 3 shows the organization charts as of December 1985.

To strengthen internal communications, Gargarin continued to hold the "fireside chats" St. Clair had initiated and added monthly informational meetings for team managers and nonexempt representatives from each team. Gargarin, however, believed that plantwide information meetings were inappropriate because of the plant's size and its pressing business constraints. As an alternative, he chose to restructure internal communications, making the business and team managers the key information link in an effort to strengthen their positions. This process was frustrating at times, for information could lose its meaning or comprehensiveness as it was passed down the hierarchy. And although some team members seemed to miss the meetings, he felt they also questioned the business need for plantwide gatherings.

FORMAL GOVERNANCE STRUCTURES

Although Gargarin had made few changes in the formal structures of the governance systems, there was a general consensus throughout the plant that there was more top-down decision making. One team member commented:

> Most people feel that there is less involvement in decision making than there used to be. By the time team members get involved the decisions have already been made, and they are trying to convince you that it's the right decision.

EXHIBIT 3 SEP Organization, December 1985

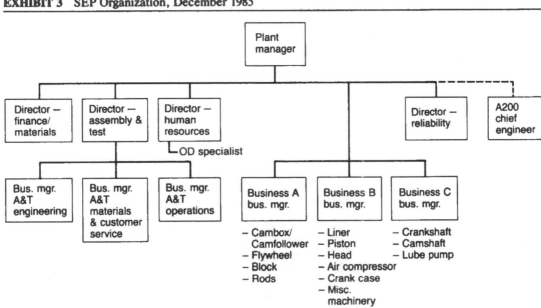

That's really more up-front than it used to be when they got you involved but led you, through the participative process, to come up with the decisions they wanted in the first place.

The Organizational Review Group (ORG), comprising the top 20 managers in the plant (business manager level and above), met regularly to work through plantwide guidelines. Gargarin noted:

I use ORG to get all the managers in the plant involved in decisions whenever there is something major which will affect the entire plant. I do believe, though, that many key business decisions should be raised and worked out in smaller groups like the plant operating team (POT) before being presented to the plant as a whole. ORG is a good vehicle for data gathering, but using it to set plant guidelines usually means six months of pain while the group wrestles with reaching a consensus.

The Board of Representatives (BOR) was also still active, but the group chose to meet less frequently than it had during the start-up years because it had fewer issues to confront. Exempt and nonexempt issues task groups, established during the early years, had atrophied. There was a proposal underway to combine the three structures into a new plantwide group entitled "Board of Employees" (BOE). See Exhibit 4 for details of the proposal.

EXHIBIT 4   Memo Concerning the Creation of BOE

Date: January 27, 1986

Subject: Logic/History BOE

In May of 1985, Joan Little of Organizational Development contacted the chairpersons of BOR, NEIG (non-exempt issues task group), & EITG (exempt issues task group) with an idea she had for streamlining the current plantwide groups. The logic developed for plantwide group consolidation came about based on the following:

- Too many people in too many groups and limited issues.
- A general need to review vertical task efficiency as we go to JIT/PULL.
- Fragmented/competitive efforts are occurring, i. e. issues handled by NEIG would be just as appropriate for BOR, and vice versa.

Current membership for BOR, NEIG & EITG consists of 50 employees and 3 POT representatives. This represents 2,067 hours per year if each group meets for 1½ hours biweekly. The proposed group would consist of 13 employees and 1 POT representative. This represents a significant reduction of hours, 2,067 to 507, and a savings of approximately $22K per year.

In November, information on this plantwide group consolidation was presented to BOR, NEIG & EITG for input. Concerns and questions were brought back to the study group and discussed.

Recently the group consolidation proposal was presented to ORG. The presentation included a charter and guideline which ORG reviewed and supported.

A new name, BOE (Board of Employees), was chosen because it felt that "wiping the slate clean" and starting fresh would renew membership vigor and would lend to group effectiveness and efficient use of time.

In conclusion, it should be noted that this entire process has been a SEP employee driven process. As stated earlier this has been an ongoing process since May of 1985 when Joan Little conceived the idea and contacted the chairpeople of BOR, NEIG & EITG. The idea has been carried on by representatives of those groups. It is our hope that the employees of SEP will view this as objectively as possible and support the process.

TEAM ADVISORS BECOME TEAM MANAGERS

Over the history of the plant, the team advisor role had presented a continual problem, and ORG still wrestled with the selection criteria and the training and development of team advisors. Although many team advisors felt they lacked authority to influence or manage their teams, Gargarin was convinced that they were a key element in SEP's success. Based on his encouragement, ORG decided in March 1985 to change the title to team manager. In addition, he rebuilt the Team Manager Forum (created by Goble) to focus on training and development issues. In a communication to all team managers Gargarin stated:

- The Team Manager role has been the critical role in the success of this plant and will take on growing importance.
- This change does not imply a concept change. It *should not be construed as a shift toward more controlling management.*
- To the contrary, we believe that the most effective Team Managers will *continue to delegate* as much decision-making responsibility as the team is able to handle. The critical decision point is making the determination *what decisions teams are capable of and are prepared to make.* The fundamental premise of SEP, which is individual responsibilities within a team setting and directed toward goal attainment, remains our operational norm.
- The Team Manager *retains prime responsibility for his/her team's performance,* but this too is a responsibility delegated to the team. All team members hold responsibility for improvement.
- The Team Manager title more accurately describes what is expected. We expect *team members to manage their work and Team Managers to manage the work of their team.* "Advisor" has at times been misunderstood. Team Advisors have always been an integral part of the management of this plant and the new title reflects this.

JIT also had a major impact on the A&T team manager role. Before the physical rearrangement, in early 1985, team managers had been reassigned so that one team manager had responsibility for teams 1, 2, and 3, another for teams 4 and 5, and a third for teams 6 and 7. (Remaining team managers were assigned to JIT improvement projects.) In addition, JIT implementation drastically changed the team managers' activities. Prior to JIT, they had spent only one hour a day on the assembly floor; the rest was devoted to meetings or special projects. As of January 1, 1986 they were required to spend 100 percent of their time on the line—problem solving, acting as a filler, leading repair work, etc. Gargarin was concerned about this shift. While he recognized a need for constant team manager attention to the line during the start-up phase, he felt a better balance would be needed. In commenting about the change he noted, "But clearly the team manager needs more involvement and driving on the hour-to-hour issues."

VERTICAL TASKS

The continual tension between maximum team output and team member participation in vertical tasks intensified with the pressure to reduce costs. Initially all team members were to rotate through all vertical tasks. Although no one could pinpoint a time or decision that signaled a change, most members of the SEP community believed that vertical tasks were deemphasized. In some cases, they were compressed and restructured to promote efficiency.

Piston Team Manager

There is less emphasis in the plant on vertical tasks because we no longer require all team members to hold all tasks. Also, with the downsizing that is occurring, there are fewer full-time vertical task assignments. This has reduced some of the pressures on the team members and pushed many of these tasks up to the team managers.

Support Team Manager

The concept used to be that everyone got into everything. With production pressures, that's changed a little at a time. Now if you want to be involved, you can be. Some people don't want to take on vertical tasks—I'd guess that only about 30 percent of the team members really want to take them on—and team managers don't feel any support to take those people off the job. An example of the production pressures was last week when each team manager was required to hold a team meeting to review the annual operating plan. The block line held their meeting from 10 to 11 on Thursday morning, but at least 10 people couldn't attend because they had to keep the machines running. The team manager had to hold another meeting to give the same package to them.

Camshaft Team Manager

We're now down to 22 people [from 43] and need everyone on the machines. Many vertical tasks are now being done during lunch. Degree of team member involvement in vertical tasks varies a great deal by team. Generally, though, the veteran team members don't want to be assigned to vertical tasks. They say, "I did it, now it's your [newer team members] turn."

Piston Team Member

We used to talk directly with suppliers and customers and directly with people at the Beacon plant. No more. Now we just get information from our team manager. We used to have control over what we ordered, how we ordered it, whom we talked to, and how much we could produce. That's all gone now.

These ideas were echoed throughout the plant. One machining business manager noted that his area was not as heavily staffed on vertical tasks as it had been during the early days. Vertical task assignments were now limited to a full-time administrative team member who was responsible for receiving and shipping, a part-time safety representative, a full-time materials and incoming product inspection representative, and a few team members dedicated to quality checks.

JIT also reduced the time that team members could devote to vertical tasks. On the assembly line (with the exception of quality and materials, which were full-time tasks), vertical tasks became almost nonexistent. The finance representative assignment had dwindled, and training was done by the quality representative. The equipment representative, responsible primarily for preventive maintenance, spent all but two hours a day on other tasks.

The assembly line was balanced to allow for an hour and a half each day for vertical tasks and improvement efforts (six-minute cycle with a daily production requirement of 60 engines). But by the fourth week of January, the line had finished early only one day, and that was with only 30 minutes to spare. Much of the time designed for vertical tasks was being eaten up during line start-ups at the beginning of the day and after breaks and lunch. Because team members were not accustomed to having to be at their work station at the very beginning of their shift, the line usually took at least a half-hour to get started; 6–12 minutes (and sometimes 18 minutes) were lost at breaks and lunch start-ups. The equipment on the line had a capacity to manufacture up to 100 engines per day, and the design team planned to rebalance the line in increments of 10 as demand fluctuated. But if corporate pushed the plant to balance the line to maximize labor productivity, vertical tasks would have to be assigned to full-time support personnel. One manager noted that the trend toward more job specialization had already started.

JOB ROTATION

The emphasis on job rotation within teams had also lessened and shifted toward "controlled rotation" to build skills and control product integrity. Although there was job rotation among all team members in a few mature teams, on average, a team member in the machining teams performed about 30–40 percent of the operations; in A&T, team members did about 60–70 percent of the assembly operations within their team. With the growing pressure to eliminate all non-value-added tasks, several team managers were worried that training was considered too long a payoff item, that it had no immediate value to add to the product. According to one team member:

> They seem to be questioning the value of training, but then they go and say that some of us have to attend hydraulics training. I'm on a vertical task assignment right now and won't use the training for at least a year. It just doesn't make good business sense.

Another problem stemmed from a practice of using group rather than team seniority for determining shift preference. The camshaft team manager explained:

> In the camshaft team we have a number of different work groups such as bearing and contour grinding. Group seniority (bearing versus contour), not team seniority, determines who ends up on second shift. If you take a first shift contour grinder and train him for the bearing area, he may resist being reassigned to the bearing area because he may have less seniority than the first shift

bearing grinders and end up on the second shift. This is pretty prevalent across the plant except in one-shift operations or where there is a small off-shift group. It's much easier to rotate people on the second shift.

Finally, there were tensions between functional organizations arising from the amount of rotation. The quality group felt that job rotation had a negative impact on the quality level of the process and argued for minimal or no rotation. The medical department preferred continual rotation (several times a day) to alleviate back and wrist injuries caused by heavy lifting and constant use of torque wrenches. In between were team members who typically preferred rotating positions (particularly in A&T) every few days.

EMPLOYMENT SECURITY AND COMPENSATION

In June 1985 the corporate offices mandated a 10 percent across-the-board reduction in the exempt work force. SEP was forced to make its first-ever forced terminations; 18 white-collar professionals, including three team managers, were terminated. Although these were marginal performers or people with limited capacity to grow in their jobs or in the organization, their terminations had implications for any future decisions on nonexempt reductions. First, the layoffs exacerbated an already high level of exempt frustration and alienation—the feeling that the system was built more for nonexempts than for exempts. Most exempts felt less secure in their jobs than nonexempts and compensation compression reinforced feelings of being treated differently. Second, the layoffs turned job security into the number one issue for nonexempt as well as for exempt employees. An announcement in December 1985 that American Diesel planned to close another nonunion, participative team concept satellite plant that had been opened in the South shortly before SEP further heightened concerns over job security. It was rumored that over 100 nonexempts at SEP would be laid off in the first quarter of 1986. And, there was a mistrust of the process by which the layoff decisions would be made. Many nonexempts were convinced that a decision had already been made and would be "sneakily" delivered.

Seeing the handwriting on the wall, Gargarin had asked POT to review SEP's reduction-in-force procedures and make recommendations on any needed revisions (see Exhibit 5 for excerpts from the procedure). In response, the human resources director noted that he was reluctant to make any changes from the established policy for three reasons:

First there is a matter of ethics and the fact that we have set people's expectations one way and should not change that at the last minute. Second, we are faced with a legal issue of implicit contracts. SEP has made an implicit contract over the years that we would go by seniority. If we as management choose to

EXHIBIT 5 Excerpts from Employment Stability and Work Reduction Guidelines (dated 6/22/82)

I. *Employment stability*

Since beginning operations, the Sedalia Engine Plant has made and will continue to make every effort to provide long-term employment to all employees. This emphasis on employment stability is built on the premise that the Plant must attract and retain a skilled and motivated work force, and that the employee who is constantly worried about his/her job security can not give full attention to the job and perform the highest quality work.

The company considers many things in its day-to-day operations to maintain employment stability. Some specific examples are:

- Develop flexibility in the work force through job design.
- Maintain a "lean" work force by avoiding overstaffing.
- Use temporary rather than permanent employees where appropriate.
- Buffer against a downturn through the use of overtime.
- Guard against future demand shifts by attempting to have a balanced product mix.
- Contract out certain of the more routine functions (i.e., security, office cleaning, etc.).
- Keep finished goods inventories and stocks at low level so that they can be built up in the event of a downturn.
- Build as much planning and lead time as possible into production schedules.
- Continue to develop and introduce new and improved production methods in order to operate the business successfully so that opportunities for employment can be maximized.

Such planning, however, cannot prevent a work reduction when business conditions become so unfavorable as to force a significant reduction in the demand for our product.

During a period of work reduction, the basic objective will be to retain as much of the skill base in the Plant for as long as possible.

A work reduction represents an exceptional hardship for the Plant and will require the support of everyone. We will need to work smarter, harder, and more cooperatively for as long as a work reduction lasts. It should be *remembered* that a work reduction will be an emotional time and a time of worry. Trust in each other that actions taken are intended to be as fair and equitable as possible will be necessary.

All employees will be kept informed of business conditions and Plant circumstances requiring a reduction in employment. Communications will include Corporate conditions, plantwide impact, the duration of a work reduction (if known), a review of major policy steps involved and a date when the work reduction will go into effect.

EXHIBIT 5   (continued)

II. *Work Reduction Sequence*

In the event that business conditions require a reduction in work force, every attempt will be made to make this process as fair and equitable as possible to all employees within the constraints of Plant production requirements. It is our intent that as many methods of providing employment as possible will be utilized. These will include the following actions which will generally be implemented in sequence unless severe business conditions require bypassing some steps:

1. Eliminate overtime where possible.
2. Encourage people to take voluntary temporary layoffs (VTL). Less than one month VTL's may be taken in conjunction with remaining unused vacation to total one month.
3. Eliminate temporary hires.
4. Redeploy people by using a plantwide swing team to offset overtime and to perform Plant facilities upkeep, maintenance, and support projects.
5. Redeploy people by using a plantwide swing team to take over a portion of the contract services (Janitorial, Security, Reception, etc.).
6. Implement a shared work reduction (fewer hours/reduced salary).

Only after taking the above steps and if a further reduction in hours to be worked becomes necessary, will layoffs be considered. Plantwide meetings (such as BOR, etc.), training, and orientation will be continued. These are important functions and may possibly become more important during a period of work reduction.

If a work reduction cannot be managed without a layoff, persons from the Office and Factory category will be designated for layoff first by lowest skill level and secondly by length of service within a skill level. If a layoff decision must be made among two or more employees who are at the same skill level and with the same length of service within that skill level, the decision will be made by lottery. It is intended to give at least one week's notice to those designated for layoff.

---

lay off by performance, we need formally to change and communicate the reduction guidelines. Third, I'm concerned about implications for the organization. If we change the system now, how do we get the teams to remain motivated to make further cost reductions in the face of pending layoffs?

Gargarin was also convinced that the link to Beacon's wage increases had to be cut, but he was extremely worried about the timing, particularly with the pending layoffs. Due to the tie to the Beacon compensation system, nonexempt compensation at SEP had risen to 30–40 percent above the community. In addition, 60–70 percent of the nonexempts were at the top of their pay scale, which meant they could receive a raise only through a cost-of-living adjustment.

Team members also voiced frustration that virtually no attention was being paid to individual performance. A senior team member noted:

> Pay raises occur automatically. Team managers seem either unwilling or unable to do much about the number of "loafers" in the plant and that number is growing. Punishment for nonperformance is virtually nonexistent.

A business manager also expressed his concern about the way the most talented people in the plant were being treated:

> The best performers have been hurt the most with the cutbacks. In the past, whenever someone was a good performer, he would be moved into an office or technical role in one of the support teams. That's an area that is being cut back by 50 percent. As these people are cut, there's no place for them to go and we end up losing the best talent in the plant.

JANUARY 1986

Gargarin was preoccupied with maintaining productivity and trust while so many changes were underway. He saw his biggest challenges to be compensation, employee security, confusion around the team manager role, and the implementation of JIT. He also realized he must face the inevitable: He could no longer avoid the issue of employment reductions. He believed he now had to make decisions on the method by which the compensation system would be changed, the number of people who needed to be reduced, the process by which they would be laid off, and the timing. He also had to deal with communicating these decisions to the work force. A common belief throughout the plant was that people were literally working themselves out of a job. One business manager felt "widespread paranoia" was about to break out. A dozen employees had made inquiries to local unions, although they were unwilling to become active in a solicitation campaign. Their main issue appeared to be employment security, fueled by the corporate decision to close the Southern plant. Gargarin also believed that other concerns about plant changes in the level of nonexempt participation and compensation and benefits were adding to the unrest.

In early January, SEP's human resources director accepted the job of plant manager at another satellite plant. To replace him, Gargarin hired a former colleague who had been both a team manager and an internal organizational development consultant at SEP, prior to having left the company two years earlier. In describing his reasons for returning to SEP, the new human relations director stated:

> There are basically four reasons why I chose to return to SEP. First, I have a lot of respect for the leadership and ability of the people at the top of American Diesel. They are committed and take definitive actions. The second reason is

the caliber of the people at SEP. They have strong values and a high level of capability and take an open, experimental approach to problems. Thirdly, there is an extremely strong work force at SEP—the people are great. Finally, SEP is on the cutting edge of becoming a worldwide competitor, while still maintaining the values and commitment to the initial concept. I view it as a valuable learning experience and I want to be part of it.

Looking ahead to 1986, Gargarin concluded that stronger leadership within the plant would be necessary to implement and manage the challenges facing SEP. He believed that one of the problems was that people did not know the limits within which they could participate.

The plant needs to be more focused. I believe that you need to give employees the boxes, tell them their involvement in the box and be clear on the parameters of the box. Some key areas which are outside team member control are compensation, employment security, goal setting, and manufacturing strategy. But, there are no limits on team involvement in communications and the improvement process. Currently the teams view this negatively and feel they are being restrained. We need to demonstrate that we are not taking away the most important elements of their involvement. I believe, though, that there is a need for team managers to move more towards a command role within the plant with more focus on cost, quality, and delivery and more streamlined decision making. At the same time, we need to keep the culture of SEP alive. The problem is how to do everything at once.

The Council Bluffs Plant (A)

In October 1985, Maxwell Industry's[1] CEO Jake Macover attended a presentation on Japanese investment in human resources. Impressed by the concepts he learned, he decided to send Steve Manning, the corporation's director of productivity and quality, to Japan for further investigation. Manning was in charge of an internal group of six consultants, located at corporate headquarters in Denver, Colorado, who worked as project specialists in introducing improvements in such areas as productivity, quality, and just-in-time (JIT). In the spring of 1986, Manning traveled to Japan and concluded that the Japanese were superior in establishing "focused leadership" (focusing all employees' attention on a common set of tactics) within their companies. After additional investigation, Manning recommended that Maxwell select a plant within its Transcorp subsidiary and attempt to develop it to the world-class performance level. The strategy for improving performance to a world-class level involved integrating JIT, total quality control (TQC), technology, and employee involvement (EI) tactics. The goal was to validate an assumption that successful implementation of these tactics would lead to extraordinary performance. If achieved, the plant could then serve as a model for other plants.

Transcorp, a wholly owned Maxwell subsidiary, manufactured off-road vehicles. Because the majority of Transcorp's plants were unionized, corporate management initially envisioned a unionized facility as the model

---

[1] In 1987, Maxwell was a highly diversified company.

This case was prepared as the basis for class discussion rather than to illustrate either effective or ineffective handling of an administrative situation. Names and certain data have been disguised.

plant site; they felt that would be a more valid test. However, they then decided that the integrated tactics (i.e., JIT, TQC, technology, and EI) needed experimentation, and experiments could be conducted more easily in a nonunion environment where flexible work rules existed. Council Bluffs, Iowa, one of Transcorp's two nonunion operations, was ultimately suggested as a potential site. The planned introduction of a new product line, the passenger transport cart (PTC), a vehicle used for transporting elderly or handicapped people through airport terminals, would provide a ready opportunity for changes to be made at the plant. Macover discussed the possibility of turning Council Bluffs into a model facility with Transcorp's CEO Michael Benchimol, who agreed to the idea in the summer of 1986. (See Exhibit 1 for the corporate organization chart.)

Rumors of the world-class model plant concept filtered down to the Council Bluffs plant in early summer 1986. The first official validation of them came in late summer when Manning visited the facility to conduct a "readiness survey" in which he asked the plant staff to describe improvements made in the past and where they perceived further opportu-

EXHIBIT 1   Corporate Organization

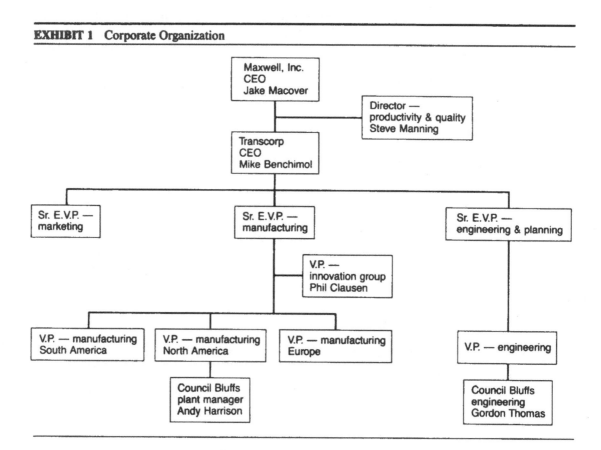

nities to lie. Shortly thereafter, on August 1, Andy Harrison, former controller and plant manager from a Transcorp Cleveland foundry, became Council Bluffs' new plant manager.

Harrison's arrival came as a surprise to plant employees, many of whom were upset that Nick Longobardi, the current plant manager, was being replaced.[2] Longobardi, a longtime employee, had won the respect and admiration of many people while managing the facility, and several key staff members had been particularly loyal to him.

The environment into which Harrison would be introducing both world-class manufacturing and the PTC line and its new technology consisted of many longtime employees who took pride in their past accomplishments and their current skills and abilities. Describing what the plant had been like since he had arrived in August, Harrison commented:

> It hasn't been easy trying to balance everything that needs to be done right now. With the transferring in of the PTC, setting up the equipment, and producing pilots, I haven't had enough time to focus on the world-class manufacturing model. Our target date for having the plant operating as world-class is September 1988, and it's already the end of February (1987). We sure have a lot left to do. The question is, how are we going to get everything accomplished in time and still meet our production requirements?

THE COUNCIL BLUFFS PLANT

The Council Bluffs plant was the smallest Transcorp facility and located the farthest from Transcorp's headquarters in Cleveland, Ohio; plant personnel liked it that way. Council Bluffs' origins dated to 1944, when an Iowa entrepreneur, Frank Charles, began a small manufacturing firm producing a line of off-road vehicles. In 1968, Maxwell bought Charles Manufacturing and assigned it to Transcorp as a separate division.

When Maxwell acquired Charles, the Council Bluffs plant exclusively produced golf carts. The plant's sales volume was approximately $30–35 MM, and it employed 330 people. By 1973, income had tripled, and plans for a new plant in Council Bluffs were announced. The facility was opened two years later, and assembly and shipping operations were moved, while the Cedar Avenue plant, six miles away, continued to do fabrication and engineering. Together, the two plants employed 542 people. Sales continued to increase, and by 1978 the employment level had climbed to 789. Transcorp was producing 5,000 units per day for a total of $60–$70 MM sales, of which approximately 17 percent represented international business.

---

[2]Longobardi was to remain at Council Bluffs as a transition manager for the PTC, reporting directly to the VP-North American manufacturing.

THE DECLINE

Around 1980, there was a downturn in the golf cart market, and the plant was hit with the first in a series of layoffs (see Exhibit 2). In the summer of 1981, Transcorp implemented several cost-cutting organizational changes. Council Bluffs thereby lost its divisional status, and its marketing, customer service, and technical publications departments were moved to Transcorp headquarters in Cleveland and other locations. Engineering support for golf carts remained but reported directly to Cleveland rather than to Council Bluffs management. Longobardi, who had been plant manager since 1969, became the top-ranking manager at the facility (reporting into Cleveland); and since the plant was no longer a P & L center, productivity (measured as labor efficiency × machine utilization) became the number one measurement, followed by inventory control, product cost, and product quality (warranty costs).

The consequences of Council Bluffs' losing its marketing department were explained by a supervisor:

> In the late 1970s, we had our own marketing system separate from Transcorp. They got rid of that system and started going through the Transcorp dealerships. Ever since, our golf cart sales have been going downhill, and our competition's market share has been going up. We used to make eight or nine models of machines, and now we're down to five.

Indeed, around 1981 Transcorp and its chief competitor, Illinois-based Goler Corporation, each had approximately 40 percent of the golf cart market while the remaining 20 percent was shared by Rectrans, Movewell, and others. By 1983, the changes in the distribution system resulted in a 50 percent drop in Transcorp's market share and a corresponding increase in Goler's share to approximately 55 percent of the total market. Other competitors also picked up some of Transcorp's lost share. Asset utilization at the Council Bluffs operation fell to 25 percent, and plant personnel began expecting the facilities to be shut down and the product reassigned.

Meanwhile, in the late 1970s and early 1980s, the market for other Transcorp products was also dramatically reduced. In 1982, in an effort to reduce operating costs, several domestic and foreign operations were closed and consolidated. In Council Bluffs, in 1984, the operations were consolidated to reduce duplicate overhead, and thereafter, in December, the Cedar Avenue plant was put up for sale.

In February 1985, in order to increase its market position, Maxwell acquired selected assets (including two domestic manufacturing plants) of ORV Corporation's Off-Road Equipment Division and integrated them with Transcorp. In 1986, as a result of the ORV acquisition, Transcorp announced an extensive manufacturing cost-reduction and facility-ration-

EXHIBIT 2   Employment Levels

Average Employment			
Year	*Total*	*Hrly*	*Comments*
1960	65	34	
1961	82	46	
1962	141	95	
1963	156	103	
1964	202	127	
1965	251	162	
1966	268	172	
1967	265	159	
1968	330	221	Charles acquired by Transcorp
1969	471	385	
1970	464	360	Less 215 employees from 1/70 to 12/70
1971	411	306	
1972	482	370	
1973	643	441	
1974	786	517	Less 180 employees from 9/74 to 12/74
1975	542	297	
1976	477	250	
1977	618	341	Plus 223 employees from 1/77 to 12/77
1978	789	430	Additionally, in the years 1978 through mid-1981 an average of 125 retail employees throughout the country reported to Council Bluffs
1979	823	444	
1980	669	311	Less 261 employees from 1/80 to 7/80
1981	488	247	All retail, marketing, and technical publications employees transferred to Denver or laid off. Less 146 employees excluding retail, marketing, and technical publications
1982	322	208	Less 119 employees from 1/82 to 7/82
1983	244	147	
1984	199	110	
1985	170	89	
1986	192	106	
1987	242	135	PTC production begins

alization strategy to improve asset utilization; overall, Transcorp was operating at 20 percent capacity. North American manufacturing space was to be reduced by 25 percent; hence plans were made to close three facilities. The product lines at the affected plants were to be shifted to other Transcorp operations. To free up space at the Warren, Indiana, facility which was to receive some of the displaced work, the PTC line would be transferred to Council Bluffs; this would raise plant capacity utilization from 25 percent to 85 percent and would lead to a tripling of Council Bluffs' work force.

Golf carts and PTCs had several similar features including size, manufacturing process, and the percentage of labor in the product. The major differences between the two products involved their parts, volumes, and the variety of models offered. (See Exhibit 3 for forecasted production volumes.) Two models of PTCs would be built at Council Bluffs: the 729, the largest model, which was targeted to begin production at Council Bluffs in March 1987 (Warren was building up a bank of 65 sets of welded parts and 270 sets of piece parts prior to transferring its tooling to Council Bluffs); and a new version of the 719, replacing a 719 presently being built at Warren and scheduled to start in Council Bluffs in July 1987.

GOLF CART MANUFACTURING

The golf cart manufacturing operation comprised three main areas: machining/shearing, welding, and assembly. Due to space and process flow considerations, the receiving area for the sheet stock was separate from the receiving area for all other incoming material (raw material and purchased parts). More than 75 percent of Council Bluffs' product cost was in purchased components, many of which were produced at other Transcorp "feeder" plants.[3] Ten percent of all incoming material, except sheet stock, was sampled for conformance to specifications. Although the plant performed Rockwell hardness tests, other types of metallurgical analysis had to be done at Cleveland.

Approximately 85 percent of the fabricated component parts (e.g., frames, hoods, covers, consoles, and mountings) for the golf cart were made in-house. The make-buy decision criteria, established by the plant's manufacturing engineering group, factored in the number of open hours available internally, the ability of the plant to compete on cost, and a concern for warranty requirements or the frequency of engineering change orders.

Shearing/Machining.   For sheet stock, the lead person in the shear area received the material, counted it, and performed a visual inspection. The sheet stock was then sheared and/or punched prior to being moved into the machine shop. When the machining area had been transferred from the Cedar Avenue facility in 1984, consideration was given to future in-

---

[3]The feeder plants were unionized; the union contract was due to expire in February 1987 but had been extended until May 1.

EXHIBIT 3   Forecasted Production Volumes

Golf Carts	*1987 Requirements*
Model 200	680
Model 300	440
Model 400	255
Model 500	320
Model 600	295

PTCs	
Model 719	2,575
Model 729	1,745

house expansion by leaving small areas, referred to as "oases," that were temporarily used for work-in-process inventory and tooling storage. Because of the oases, new equipment brought in for the PTC could be located within the area without having to totally rearrange existing equipment. Prior to the addition of the PTC line, 90 percent of the tooling was made in-house by six toolmakers; with the increased tooling load, a significant amount of the tooling had to be outsourced to local tool and die shops, often at a 30–40 percent cost premium.

Welding.   Prior to leaving the machine shop and before welding, a 2–5 percent sample of each batch was checked for critical dimensions. Welding was normally the final process for individual finished parts; golf cart welding comprised spot, spud, stud, and tack welding. When the operation moved from the Cedar Avenue facility, curtains were removed from the weld booths because it was hard for the supervisor to see what was going on inside the booths. In addition, this saved space, was less messy, and made it easier to move tooling in and out of the booths.

Upon completion of the subassemblies, the welds were randomly visually checked. Following inspection, more than 95 percent of the subassemblies went through the paint shop prior to assembly.

Assembly.   Due to the diversity of components in the various golf cart models, three subassembly areas fed the main assembly line. This allowed for staffing needs to be leveled out across the models; i.e., operators could be assigned to build subassemblies prior to the final assembly run. Golf cart models ranged from 4 to 14 assembly hours per unit, with a daily volume of from 3 to 11 units. Also, golf cart sales were relatively cyclical.

For the final assembly, golf carts were built on dollies to facilitate the movement of the golf cart from one station to another and to raise it to the proper working height. The process began with connecting the frame, axles, and driveshaft. Next, units were moved down the line for the installation of the motor. Fourplates and handbrakes were then put in, followed by batteries and other components. Operators controlled the speed of the assembly line, moving units to the next station when needed. At the final station, tires and the canopy were installed.

At the end of the line, assemblers checked each golf cart, filled out a "squawk" (or problem) sheet, and if need be, sent the sheet back up the line for feedback to the assemblers while the golf cart was repaired off-line. The completed "basic" golf cart would then move to an off-line finishing area where paint was touched up, decals were added, and the unit was steam cleaned. A final inspection assured that all problems listed on the "squawk" sheet had been corrected; in addition, critical torques and lubricants were checked. Finally, each unit was functionally operated for five minutes to check its performance.

Engineering change orders were the responsibility of the product engineering organization, which consisted of a design department, experimental shop, and a small group in test; ideas flowed from one group to another. Before designs were released to the shop, manufacturing engineers met with design engineers to evaluate them, which meant that changes did not often come back from the floor. However, the engineering manager predicted that the manufacturing engineers would not be as familiar with the PTC line which would mean that numerous changes would be coming back to the engineering group even after the designs had been signed off.

DESIGN AND START-UP OF THE PTC LINE

The Warren plant was considered by many levels of Transcorp management to be the company's best facility; it sold the most profitable products. After the transfer of the PTC line to Council Bluffs was announced, Harrison noted:

> Product performance personnel said, "You folks better build them as good as Warren." My response was, "The people working here have more to lose than people in Warren so you'd better believe we'll do a good job. In fact, we'll do even better."

The introduction of a new product line into the Council Bluffs plant provided an opportunity to design a new assembly line for PTCs. As a result, the supervisor was selected for the area in September 1986 and assigned, along with two manufacturing engineers, the task of designing the layout of the PTC assembly line. Their only instructions were to eliminate the use of dollies (a continual material handling headache on the golf cart line). The supervisor and the engineers spent two months building five PTCs (model 729) by hand and designing the new line. The supervisor noted:

> This was the first time that I had ever been given the chance to get away full-time from supervising and allowed to be involved in laying out a work area. It really helped me to better understand manufacturing engineering's problems.

The resultant design used a raised conveyor to move units from station to station; it was also sloped at the end to allow the unit to roll off automatically when complete. There would be 20 stations, each staffed by two operators per shift; every 10 feet a walk-through was provided so the operators could maneuver around both sides of the unit.

It was expected that a model 719 would require eight hours to assemble, whereas a model 729 would take approximately 13 hours. The first hourly assembler was hired in the second week of February 1987, and it was projected that by July, 40 assemblers would be producing 25 PTCs per day. Since the PTC would be produced in much larger volumes than golf carts, the PTC supervisor was concerned that the assemblers might find the work repetitive and boring in contrast to golf carts, where there was continual shifting because of model changeovers. As a result, he planned to rotate assemblers between work stations.

Rather than a final inspection at the end of the line, there would be "sequential inspection," i.e., each operator would check the work of the prior work station. Consequently, there would be no "squawk" sheets. Also, there would be no off-line finishing due to less model diversity; this meant that there would also be no touch-up painting or rework.

A semiautomated chassis line, costing $750,000, was on order to weld PTC frames. Aside from the assembly and chassis lines, all other in-house manufacturing of the PTC would be integrated into the existing operation (e.g., shearing, machining, welding, and painting). (See Exhibit 4 for an estimated machine loading.) The new product, though, required the purchase of additional equipment. A committee consisting of the manufacturing manager, manufacturing engineering manager, supervisor, maintenance manager, and purchasing personnel was formed to evaluate and decide on such investments. Although Transcorp's Innovation Group (IG)[4] was available for advice, the choice of equipment was left entirely up to the plant.

On the new chassis weld line, each station would have swing-up fix-

---

[4] The Innovation Group was headed by vice president Phil Clausen who had previously worked with advanced technology at another leading off-road vehicle firm and who had been hired by Transcorp in 1984 primarily to promote new technology within the company. He had later been assigned to lead the consolidation study of all Transcorp facilities, and in that role, he was responsible for tracking the progress of the plants and assuring that the overall consolidation plan was being achieved.

EXHIBIT 4   Machine Load Summary

Cost Center	Machine Description	Machines	Total Hours	PTC (percent)	Golf Cart (percent)	Equiv. Shifts Needed
611	Miscellaneous hand	—	1439	99	1	0.8
611	Band saw—peerless	1	1970	63	37	1.1
611	Cold saw—trennagher	2	1606	7	93	0.9
611	Shear—finish	3	6279	67	33	3.4
611	Shear—rough	2	3397	37	63	1.8
611	Shear—bar	1	1516	49	51	0.8
611	Flame cut—oxygraph	1	3307	22	78	1.8
611	Flame cut—plasma	1	1439	58	42	0.8
612	Engine lathe	1	513	27	73	0.3
612	Turret lathe	1	1785	71	29	1.0
612	N/C lathe	2	6004	9	91	3.2
612	Spline mill/hob	1	574	0	100	0.3
612	Pipe threader	1	135	0	100	0.1
612	Endomatic	1	126	0	100	0.1
612	Drill—single speed	3	5167	35	65	2.8
612	Drill—multi spindle	2	1255	44	56	0.7
612	Drill—turret	1	447	64	36	0.2
612	Drill—radial 3 ft.	2	3028	63	37	1.6
612	Drill—radial 5 ft.	2	1648	39	61	0.9
612	Mill—horizontal	4	1312	19	81	0.7
612	Mill—bridgeport	1	65	0	100	.0
612	Broach	1	273	0	100	0.1
612	Borematic	1	329	0	100	0.2
612	N/C HMC	2	9473	37	63	5.1
612	N/C VMC—small	1	2235	11	89	1.2
612	N/C VMC—large	1	4857	19	81	2.6
612	Misc hand—burr	—	421	99	1	0.2
612	N/C grinder	1	474	11	89	0.3
612	Shaft straightener	1	36	0	100	.0
612	Arbor press	1	1214	47	53	0.7
613	Misc hand operations	—	1206	68	32	0.6
613	Saw	1	20	0	100	.0
613	Deburr	—	2185	7	93	1.2
613	250T form press	3	5500	48	52	3.0
613	350T form press	1	3965	53	47	2.1
613	400/450T blank press	2	3911	38	62	2.1
613	65/115T punch press	2	3924	55	45	2.1
613	Plasma/punch	1	13670	89	11	7.4
613	200T blank press	1	2379	60	40	1.3
613	Flattening press	1	951	48	52	0.5
613	Forge press—hot form	1	140	0	100	0.1
613	8 ft × ½ roll	1	1263	93	7	0.7
613	4 ft × ¼ roll	1	30	0	100	.0

EXHIBIT 4   (continued)

Cost Center	Machine Description	Machines	Total Hours	PTC (percent)	Golf Cart (percent)	Equiv. Shifts Needed
613	Deburr	—	1169	4	96	0.6
613	Deburr hand grind	—	3888	45	55	2.1
614	Hand grind	—	5000	76	24	2.7
614	Size press	1	86	0	100	.0
614	Heat furnace	1	14	0	100	.0
614	Sub arc weld	1	560	63	37	0.3
614	Spot weld	1	182	0	100	0.1
614	Spud weld	1	1085	30	70	0.6
614	Stud weld	1	283	69	31	0.2
614	Robot weld	1	1519	0	100	0.8
614	Tack weld	—	1428	76	24	0.8
614	Golf cart chassis weld	—	9199	0	100	5.0
614	PTC chassis weld	—	12487	100	0	6.7
614	Weld—attachments	—	25035	33	67	13.5
614	Weld—small parts	—	7997	29	71	4.3
614	Weld—small parts	—	58343	95	5	31.4
614	Weld—small parts	—	1001	0	100	0.5
614	Hardface	—	2378	0	100	1.3
614	Weld—track	—	14	0	100	.0
614	Weld—drill stem	1	217	0	100	0.1
614	Weld—hoods	—	488	0	100	0.3
614	Weld—miscellaneous	—	35	0	100	.0
621	Subassembly	—	3711	0	100	2.0
621	Kit packing	—	1504	0	100	0.8
621	Engine build-up	—	1412	0	100	0.8
621	Assembly—golf cart	—	732	0	100	0.4
621	Assembly—accessories	—	462	0	100	0.2
621	Assembly—service parts	—	402	33	67	0.2
621	Assembly—accessories	—	4180	4	96	2.3
621	Assembly—accessories	—	80	0	100	.0
621	Saw—tubes	1	709	77	23	0.4
621	Deburr/flare tubes	2	2481	77	23	1.3
621	Assembly press + transmission	1	3144	37	63	1.7
621	Tube bend	2	2643	83	17	1.4
622	Assembly line north	—	10145	49	51	5.5
622	Assembly line south	—	81488	78	22	43.9
622	Assembly—attachment area	—	97457	71	29	52.5
623	Paint/blast	—	13065	72	28	7.0
623	Paint/no blast	—	4711	35	65	2.5
623	Line blast	—	144	85	15	0.1
623	Tumble blast	—	10	0	100	.0
623	Paint final touch-up	—	18389	61	39	9.9

EXHIBIT 4   (continued)

Cost Center	Machine Description	Machines	Total Hours	PTC (percent)	Golf Cart (percent)	Equiv. Shifts Needed
611	Shear/saw	Cost center total:	21120	51	49	
612	Machine	Cost center total:	42661	33	67	
613	Punch/press	Cost center total:	44578	60	40	
614	Weld	Cost center total:	127393	66	34	
		Fabrication total:	235752	58	42	
621	Sub-assy	Cost center total:	21517	29	71	
622	Line assy	Cost center total:	97457	71	29	
623	Paint	Cost center total:	18389	61	39	
		Assembly total:	137363	63	37	
		Plant total:	373115	59	41	

tures; all power controls, heat exchanges, and the weld wire would be housed in a mezzanine above the line to save floor space. Heat exchangers were necessary because the line would utilize water-cooled, in contrast to air-cooled, weld guns. This would allow weld wire to be laid at a rate of 550 inches/minute (typically air-cooled guns welded at a rate of 350 inches/minute). The gun being used, however, was experimental and was an improvement over prior designs which had been too bulky and inflexible. There were several other risks: 1) the weld wire, which was electronically charged and could therefore not touch anything metal, would have to be pulled over 100 feet via plastic pulleys; 2) if the line did not work, the Council Bluffs plant had no other way to weld PTC frames other than pulling old tooling (used at the Warren facility) out of storage; and 3) the plant had rejected a suggestion by the IG to totally automate the process. IG's suggestion was to use two robots, which would significantly reduce staffing requirements. The plant disagreed because the chassis had not been designed for robotics, there was not enough time for such a development project, it was believed that more than two robots would be needed, and the use of robotics would require a consistent fit of parts, something that had been a major problem at Warren. Therefore, the plant opted to go with the semiautomated line which would require 11 people on two shifts. Two engineers and two welders had spent five weeks at the vendor site working to ensure that the line would function properly before being sent to Council Bluffs; the agreement required all tool proofing to be done at the vendor site. The vendor had promised to have the equipment on site by March 2, 1987, but it appeared, in early February, that that date would slip by several weeks.

With PTC production soon to begin, the manufacturing group had begun inspecting the parts and tooling that had been sent to Council Bluffs from the Warren plant. They discovered that many of the parts needed rework. Warren had not been making parts according to print, but Council Bluffs personnel did not know whether the prints or the tooling was wrong.

EQUIPMENT MAINTENANCE

Machine downtime had been a problem for the plant in the past and threatened to become a bigger problem in the future because of increased capacity loading. Although the facilities manager noted that he would like to keep track of downtime, no system currently existed; he felt it would be too time consuming to have the maintenance workers document their every move. The maintenance organization comprised four workers and one lead person who were responsible for doing the assigned inspections; but with more numerically controlled (NC) machines at the plant, the facilities manager felt that he needed a bigger staff. The plant had only one highly skilled electronics worker (the lead person), and finding another electronics expert was proving difficult. The facilities manager had been looking for one since he first learned about the PTC line, but competition from local unionized manufacturers who offered higher wages decreased the pool of available applicants.[5]

Around 1975, when NC machines were introduced in the machine shop, a preventive maintenance program had been started at Council Bluffs; in late 1985, the system began to be automated. At present, all preventive maintenance activities were performed by the maintenance group, but the facilities manager noted, "By the end of the year, my goal is to have a checklist of maintenance tasks on each machine and to have the operators attend to them." One manager feared that implementing such a program might be problematic:

> There may be an attitude problem. The supervisors and others say the hourly weren't hired to do that job, yet the hourly people here do more than at other

---

[5]Though Transcorp's wages were lower than local manufacturers', they were still significantly higher than their competitors'. In an effort to bring wages more into line with the competition, no general wage increase had been granted since September 1985, and on January 1, 1986, the plant had instituted a new salary scale whereby new employees would be hired at a substantially lower rate than incumbents. Also, their progression up to the maximum pay within their job classification would be stretched from 24 months to four years. With the projected increase in employment, it was estimated that by year-end, 75 percent of the hourly work force would be on the lower pay scale.

Transcorp plants since we're nonunion. The hourly people will jump at the chance to do some maintenance work, but there may be a problem with getting some of the managers to go along with it.

PRODUCTION PLANNING AND SCHEDULING

The Council Bluffs plant built its products to dealer or dealer stock orders, independent of customer orders; profits were based upon shipments to dealers. In order to leverage overhead costs, the standard cost system required the plant to maintain a sufficient direct labor base. As a result, the plant would build to the marketing forecast, regardless of the customer orders on hand. The marketing forecast was put into a materials requirements plan (MRP) used by production planning for material procurement. Each cost center was allocated five days regardless of the actual run time required on a unit; for example, even if the shearing operation required only 15 minutes or the machining time on a particular part totalled one day, five days were allocated for the part to be processed by each cost center. An additional five days were added for any outsourced operations such as heat-treat or plating. Time spent in painting, de-burring, or inspection, however, was ignored. In addition, approximately 15 days (maximum up to 25 days depending on the model) were added for golf cart assembly; 10 days would be allocated for PTC assembly.

Work flow through the shop was scheduled by the manufacturing manager via an informal manual system. The objective was to keep the machine and weld shops one to two days ahead of the assembly line; assembly throughput time for golf carts ranged from 7 to 10 days (PTCs were planned to be assembled in 3 to 5 days). Typically, one and a half days of work-in-process for golf cart frames was maintained: a half day supply in front of the line, a half day in the paint shop, and a half day in welding. Due to space restrictions, similar inventory levels were planned for PTC frames and other major weldments.

This level of inventory was much reduced from 1973–74 when raw material and work-in-process inventory levels were at 2+ turns; prior to the consolidation of the operations in 1984, raw material and work-in-process inventory filled over three to four acres of outside racks at the Cedar Avenue facility. By mid–1987, inventory levels were expected to be at 9.5 turns with an ultimate goal of 12–15 turns. With the exception of some stockpiling of material due to the transfer of the PTC, all raw material or work-in-process inventory was stored indoors. Outside storage was primarily limited to finished goods. On average, 100 carts were held as finished goods inventory, though inventory typically fluctuated between 12 and 250 carts.

Council Bluffs had been selected as the pilot site for the introduction of the new data processing system comprising several modules. In March 1983, management had installed the first modules which included a bill of materials and shop floor controls such as manufacturing engineering routing sheets. In April 1986, modules on materials requirements planning (MRP), cost systems, purchasing, inventory control, and shop floor work orders were added. Additional modules on master production scheduling and capacity planning were planned. One staff member described the data processing system by saying:

> We consider it to be a communicable disease within the plant. It's an insult to our intelligence to be asked to use it. In mid–February [1987] we received a shipment of 151 parts that had been cancelled in October; the system failed to issue a cancellation notice, and the plant was forced to pay for the mistake. Moreover, the engineering changes we made in the past six months may not have been accepted and changed, so we may be shipping unchanged parts. Our WCM efforts are being diluted because we're trying to make sure the system is doing what we want it to do, and in all cases, that's not true.

QUALITY PROCEDURES

Around 1975, there had been 24 inspectors in the quality department; currently, there were 4 inspectors in the plant of 135 hourly people. With the start-up of the PTC, a few more inspectors would be hired, but the quality control manager was planning to keep his staff small since machine operators and assemblers were required, for the most part, to inspect their own work.

In 1982, the plant began a weekly audit of at least two completed golf carts. The audit was conducted by a group of 8–10 people who changed weekly and represented such areas as production control, accounting, materials, and manufacturing engineering (hourly workers had not yet been involved). If there were problems with any of the machines, the manufacturing manager would call together the whole department to discuss them. Product quality was also evaluated by customers and dealers who filled out check sheets which accompanied each cart.

The quality group had also begun to implement shop floor controls similar to SPC in the mid-1970s. Manufacturing engineering was to conduct machine and operator capability studies while a quality control representative would determine what inspection equipment was needed, provide training for operators, and assume proper record keeping. Training programs in the saw and shear areas had been established. However, a budget crunch forced the quality control position to be terminated and resulted in the elimination of the shop floor control program. By February 1987, the quality control department had again decided to institute a program to improve quality at Council Bluffs. Rather than implementing SPC within the plant, however, this time they were focusing on purchased parts because they accounted for such a high percentage of the product cost.

IMPROVEMENT PROCESSES

Employees in the Council Bluffs plant could make suggestions for cost reductions by filling out a form describing their idea. Each month, the employee whose suggestion would save the most money won a reserved parking space for a month. Tickets to a local professional soccer team were also awarded, and in the past, savings bonds had been presented to quarterly winners. It was not unusual to have up to 30 ideas submitted per month. However, many lower-level employees viewed the cost improvement program negatively. One supervisor explained, "The hourly don't participate in the program because they don't know how easy it is to give a cost reduction suggestion. They don't like the paperwork, and some feel the engineers will take credit for their ideas." Another supervisor commented:

> The hourly got burned a long time ago. About 10 years back, when the program was new, people were excited about it; but people's ideas were not always accepted and there wasn't much follow-up or feedback about why ideas were rejected. Whether or not your ideas were accepted depended on what department you were working in. It's not like that anymore, but people remember it that way and hold a grudge. At least a lot of your best cost reductions get done anyway by the supervisors, so they're not totally lost.

Newly hired hourly workers, for the most part, were unaware such a program existed; it had not been mentioned during their orientation sessions.

The hourly workers at Council Bluffs also had the opportunity to voice their concerns to management at monthly "suggestion committee" meetings which were attended by four elected representatives from the hourly ranks, the employee relations manager, and the manufacturing manager. One longtime plant employee noted that the yearly elections, which began in 1968, used to be more competitive, but it had become increasingly difficult to get workers to run because they did not want to take on extra duties. Another employee remarked, "The committee no longer represents the hourly worker. It takes a lot of courage to represent the hourly sometimes, and you don't have that many people who are willing to do it. People are afraid to rock the boat."

Before each meeting, the representatives would spend an hour walking around their designated areas asking people for suggestions to bring to the committee. Examples of issues that had been discussed at previous meetings included safety, benefits, and minor maintenance concerns such as installing additional fans in the shop. In describing the suggestion committee process, one supervisor commented, "Supervisors get caught in the middle of it all quite a bit."

THE COUNCIL BLUFFS ORGANIZATION

The majority of Council Bluffs' personnel had been at the plant a long time. The current seven plant supervisors, who ranged in age from around 40 to 60 years old, had each worked at Council Bluffs for at least 17 years; three of them had been there for 25. They were working supervisors who dressed in jeans and helped out on the line; it was not unusual for a supervisor to drive the forklift or do actual assembly work. The supervisors had limited responsibility in areas such as discipline or scheduling vacations, however, since those types of activities were all done by the manufacturing manager. Lead personnel, who worked under the supervisors, did a similar job but on a smaller scale. They were responsible for scheduling, getting the right parts, and basically making sure things ran smoothly, notifying the supervisors when there were problems. The hourly workers, many of whom had been hired by Charles Manufacturing, averaged 15 to 30 years of seniority with the minimum of any worker being about 12 years (excluding new PTC hires).

Harrison had been brought to Council Bluffs because Transcorp upper management felt he was the type of leader needed to create a WCM facility. The fact that Harrison came from Cleveland posed a bit of a problem since anyone from headquarters was viewed as an outsider at Council Bluffs. Wary of Harrison when he first arrived, plant personnel naturally compared him to Longobardi. A manager noted that a major difference between the two was that Longobardi typically walked through the plant at least twice a day and would say hello to at least 90 percent of the people each week. Harrison indicated he had spent 30–40 percent of his time on the floor at Cleveland, but at Council Bluffs he was forced to spend a great deal of time in meetings, primarily due to the PTC. In addition, Harrison had no machining background which was seen as a limitation by some Council Bluffs employees. As one staff member explained, "To a large extent there's a problem trying to get Andy to understand the technical problems, for instance, how long it takes to tool." Still, plant personnel were impressed by Harrison's management style, claiming that he was "open and honest" and a "real people person."

Besides Harrison, another outsider at the Council Bluffs plant was employee relations manager Jim Roberts who arrived shortly before Harrison, in April 1986, to replace a staff member who had passed away. Roberts had previously been an industrial relations representative at Cleveland before moving into a labor relations position. His prior association with Harrison at Cleveland highlighted their outsider status. (The staff member with the next lowest level of seniority had been there for eight years while the others had been there over 10 years.)

THE PLANT'S INTRODUCTION TO
WORLD-CLASS MANUFACTURING

Shortly after Harrison joined the plant in September 1986, he organized a brainstorming session with his staff at which time each manager submitted a set of notes defining his or her views of WCM. In November, Manning conducted the first in a series of formal training programs. Each staff member, along with two exempt employees of his or her choice, was invited to attend (see Exhibit 5). The 18-member group took part in an 8-hour session followed by 10 hours of additional training on WCM concepts. The training was received with mixed reviews. For instance, one participant commented, "Some of this training is just hype. There's a point where you're overdoing it. The training was boring at first; at the end, it was like 'Let's get pumped up about the new buzz word.'"

In December, Harrison, the managers of manufacturing, quality control, purchasing, and materials, and the controller visited a company frequently cited as being world-class. They were disappointed to find it was just starting JIT and SPC and still had a lot of inventory; they concluded that the plant was not as world-class as had been suggested.

Also in December, all salaried manufacturing employees sat through a two-hour introduction to the WCM model in which JIT and TQC were emphasized. TQC was the subject of another six-hour training session held for salaried employees in January. Product engineering personnel participated in a six-hour course in December which focused on WCM concepts and the interactive role of the engineering department.

Following the initial training, Manning assisted the plant staff in developing a WCM implementation plan in early January 1987. Staff members also became participants in a World Class Model Plant Steering Committee along with Harrison, Manning, and the vice president in charge of the Innovation Group, Phil Clausen. Though Clausen was primarily responsible for overseeing the transfer of the PTC line to Council Bluffs, he became part of the steering committee since he was also responsible for new technology. (The WCM group had been meeting irregularly but planned to meet every two weeks starting in March 1987.)

The plant's hourly work force first heard about WCM in December when Transcorp CEO Benchimol held one of his regular quarterly meetings at Council Bluffs and mentioned that the facility had been chosen to become a model plant. In February 1987, the manufacturing manager presented a half hour talk to the hourly workers in which he introduced the

EXHIBIT 5 Council Bluffs Organization Chart

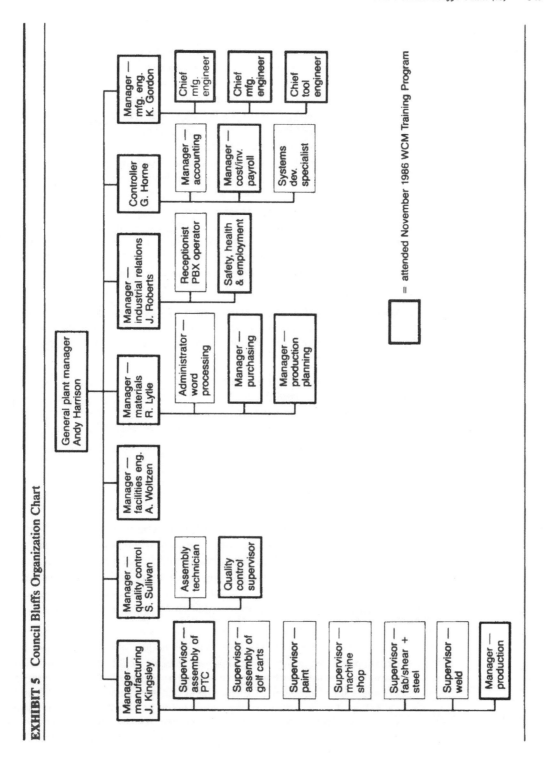

WCM model concept. Newly hired hourly employees would be exposed to the WCM philosophy during their orientation sessions, the first of which was to be held in March 1987. In addition, within their first month of employment, new workers were to receive 8 to 12 hours of classroom instruction to familiarize them with shop operations. There were plans for a two-hour abridged program on WCM to be held in March or April for current hourly employees which would present material similar to that covered in the new employee orientation.

For the most part, however, on-the-job training (OJT) would be the primary resource for training both old and new hourly employees in job performance. The supervisors, who in February had participated in an OJT workshop, would be responsible for making sure their workers learned about set-up, quality control, safety, preventive maintenance, and similar issues. A limited amount of classroom training would be used to supplement the OJT, but there was some concern about the ability of the long-service workers to handle that type of training. One manager explained why:

> Many of our workers at the plant today came from the old Charles company. Charles paid at the bottom of the pay scale so he got low scale, uneducated people. Many were illiterate, and one individual only had a third-grade education. Most of those people have since retired, but a few still remain.

Aware of the potential problem the older workers might face, the industrial relations department was being very careful about hiring new employees whom they felt would be able to deal with classroom training. There was also a concern that some managers would be reluctant to release employees for training.

INITIAL ACTIONS

In early 1987, the plant began its effort to become world-class by first focusing on its vendors, mainly because purchased parts accounted for 73 percent of the total product cost plus an additional 5 percent for processing costs. As a result, the purchasing department had begun working with vendors to try to encourage them to improve the quality of incoming parts. The purchasing manager explained the plant's underlying philosophy:

> The first thing to do is to clean up our own house. We must set examples for our suppliers. We're not being slow, we're being deliberate. We singled out a motor manufacturer with whom we do a lot of business and whose motors used to have a lot of problems, to make them our world-class partner. We invited them to visit us in January so they could get to know who we are and what we want to do. On March 2, we plan to bring them back and update them on what we've done here since January. Once we get their quality where we want it, the

cost of the product also has to go down. After we go after quality, we'll start going after delivery.

The purchasing manager had also devised a potential new delivery program that would better fit the demands of a JIT manufacturing system. JIT would require daily shipments of small loads of materials from numerous vendors, which would create a bottleneck at the plant's four inbound bays. Council Bluffs had been using vendors from all over the country, but the new plan called for choosing suppliers from an eight-state area around Iowa. Using Transcorp's own truck fleet or a common carrier such as Federal Express, the plant would develop a route and have a truck go from one vendor to another and then return to Council Bluffs with a full load of materials. Vendors would be notified when the trucks would be arriving, which would probably be once a week. The concept was still in the preliminary stages of investigation, and the purchasing manager did not yet know what the overall cost of the program would be or how it would affect material handlers at the plant.

Other initial actions taken by the plant included removing the fences around the receiving area but maintaining the fence around the warehouse to keep people and traffic out. Finally, an agreement was made with the local tire vendor to have daily deliveries on racks which could be moved directly to the line, in contrast to the truckload deliveries which had been stacked on skids.

VIEWS OF WORLD-CLASS MANUFACTURING

The personnel at Council Bluffs had mixed feelings about the entire world-class manufacturing concept and what it would mean to the plant. Their comments follow:

Andy Harrison, Plant Manager

WCM has awakened me to the fact that over 75 percent of our costs are related to purchased parts—receiving, inspecting, rejecting, replacing. We've come to the realization that in order to have JIT, we need good vendor quality. Rejects shut down the assembly line. In this particular case, it's given me a better understanding that we need to focus on purchasing and vendor quality and improving vendor delivery. Most suppliers don't care about pleasing the customer. In this plant, we were low on volume so the supplier didn't care if he lost this business or not. We didn't have enough leverage with him to create a bond between Transcorp-Council Bluffs and the manufacturer.

Manufacturing engineering is another key group we need to improve. Right now they are causing a lot of wheel spinning. For instance, we've had to run 16 tapes for just one part on the plasma machine. That's too many. We're doing something wrong. We should be able to do better than that. They're sending out mixed signals to the shop by saying, 'It's not quite the way we want it, but use it anyway.' That interrupts messages about quality. The manufacturing people or the manufacturing engineering people are forced to compromise which sends poor messages, especially to the new employees.

One problem, however, with all the changes in becoming a world-class manufacturer, such as JIT and so on, is that they tend to take the fun out of work. Managers and workers like to have a crisis to handle every 30 days or so. Supervisors won't change overnight. Some will have to retire because we won't be able to change their attitudes. Some of the staff probably feel the same way.

John Kingsley, Manufacturing Manager

WCM is not anything any different than what we've been doing for the past four or five years basically. All we have now is the visibility and a few more resources than before. We have the same goals as we did then. Without the stigma of WCM, we'd still be doing the same things with the PTC line. It's a little frustrating at times because we felt we were already doing a lot of things. JIT is a philosophy change. The problems don't come from the hourly people or the supervisors. The people higher up are set in their ways and philosophies change a lot slower. People at the top don't see the need to pass on information to the middle, while the middle must pass on information to the bottom. So, the middle ends up lacking information. It's frustrating.

One of the things I'm concerned about is lowering the decision making to the lowest levels. The main idea is that the guys on the floor know the best ways to fix a problem, but that's not always true. The majority can't make economic, technical, or broadscale decisions. They can only make minor suggestions that help themselves or the guy next to them on the line. I'm anti-quality circles. Someone still has to do the physical work; meetings take away from production time. You have to be careful people don't end up spending too much time meeting in small groups. I like doing things informally; if you have a problem, we'll work it out. That's the way we do things around here.

Quality Control Manager

The old inspector role from the 1950s is wrong. The Japanese attest to what I'm saying; they don't have any inspectors. My goal is to do away with the inspection department as we traditionally know it. I see it as a small department with auditors to audit the finished product, inspect tooling prior to production, check the hydraulics, monitor our vendors, and act as troubleshooters.

A Supervisor

I don't know what that means, world-class. No one in the shop really talks about WCM. It's the same old thing we've been doing with a few new directions here and there. I hear all the buzz words and I know what we're shooting for.

A Supervisor

JIT, I'm an old hand at that. A lot of times JIT parts were what we had even though it wasn't meant to be that way. But, you ought to have safety stock there. If one of the other plants goes on strike for a day or a week, it would kill me. We're already behind because the JIT parts are not here in time.

Also, with WCM everybody's supposed to be an inspector. WCM is more or less involving fab—quick changeover and no setup time in the shop, which will increase productivity. Quality is the biggest area we're working on. The tooling hasn't improved like some of the other areas, even though tooling helps with quality. World-class also means we have to pick up our pace in the shop; it means higher efficiency. We have to make the older folks work faster. We have to push people harder.

A Manager

Some of the things they've been talking about we've been doing for years because of our low volume. But, WCM does mean you'll be concerned about the people you employ. It's different from what's going on now. I think an outfit needs to be concerned about employees and their lives, and it hasn't been that way. I think that will change with Andy.

One year from now, you'll see a lot of changes. Number one because of management attitude. Industrial relations is hiring new employees and urging them to think about WCM. A couple of key management people will have to change their philosophy or they're going to be out the door. This plant will also have to become less interested in the productivity form of measurement everyone's comfortable with because it may not be the right one. Transcorp is going to have to change. We will probably run up against the Transcorp and Maxwell brick walls even if we manage to change the philosophy in the plant.

Turnaround Management

Companies in America's industrial base have frequently shut down unprofitable operations, assuming the situation has so deteriorated that it would be cheaper to move to a greenfield site. This approach, however, entails expenses in closing an operation (transferring equipment, relocation costs for key personnel, severance and pension liability for the laid-off work force, etc.) and in starting up the new site (purchasing real estate and constructing facilities, hiring and training a work force). After making this heavy investment, many managers have discovered that problems follow; if the site were in a less developed country, these often have been compounded by logistic and cultural difficulties. Hence, some companies have found (Mitchell, 1986; Greenhouse, 1987) that with tightening capital and the dollar's decline, revitalizing existing facilities is a more viable alternative. Managers confronted with such a situation generally face the challenge of breaking a "downward spiral," that is, where one problem leads to another and each is interwoven with the others. Exhibit 1 illustrates a classic spiral.

Many consider a "turnaround" the last step before an operation closes its doors because it lacks cash. This is but one type of turnaround situation, however. A turnaround situation exits whenever a failure or a series of failures has led to a climax that requires near-term action; recognizing that radical change is necessary for a facility to become more competitive can also constitute a need for a turnaround. Sloma (1985) identified four such stages:

1. Cash Crunch—The operation has run out of cash and also used up its line of credit.
2. Cash Shortfall—A cash crunch is forecast to occur within three to six months.
3. Quantity of Profit—An operation that has previously been profitable is experiencing an erosion of profit.

EXHIBIT 1   Downward Spiral

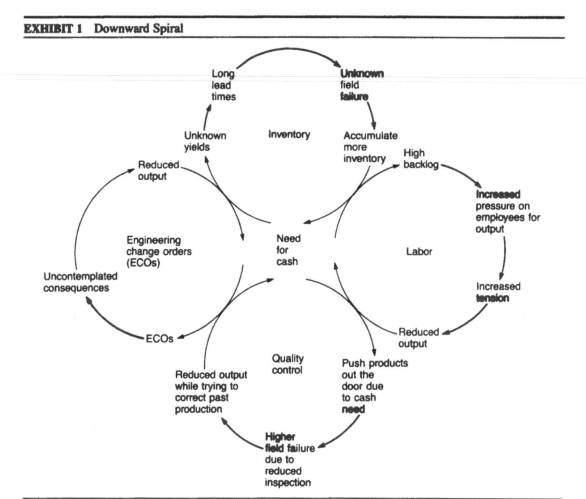

4. Quality of Profit—The time frame for concern may be a year or longer, but a firm has a decreasing value of the ratio of the operating pretax income divided by net sales.

The main difference in the stages is the sense of urgency to take action.

TYPICAL TURNAROUND ACTIONS

Confronted with a turnaround, operations managers have three potential areas to attack—product, process, and people (Sloma, 1985). This means choosing between the types of changes explored in the previous module on workplace change. The task facing the operations manager is which lever to use when an immediate crisis must be fixed, but the fix must lay the groundwork for future sustainability. Often these two conflict, which leaves the turnaround manager walking a tightrope.

Managers in turnaround situations, particularly at the plant level within large corporations, encounter a number of balancing issues:

1. Meeting short-term measures while laying the groundwork for the future—As has been shown in the module on workplace change, introducing change leads to a short-term productivity dip. Turnaround managers can ill afford any further dips during the crisis.

2. Focusing on change while maintaining ongoing operation—Although correcting a turnaround situation means systemic change, short-term output must continue. Energies and resources must be split between meeting short-term customer requirements and setting groundwork for the future.

3. Motivating the work force in face of a crisis—Turnarounds typically require cutting costs and one of the quickest ways to cut immediate expenses is through work force reductions. Turnaround managers must therefore ease the exit of those leaving while motivating the "survivors."

CHARACTERISTICS OF TURNAROUND MANAGERS[1]

Due to the urgency of the situation, turnaround situations demand managers who thrive on chaos. Many turnaround managers are also outsiders: It is common wisdom that an incumbent manager will have difficulty in taking necessary steps to revitalize the operation. Even if that manager is aware of the need for drastic change, many of the existing problems may have been created, or exacerbated, during that manager's tenure within the operation, and, thus, the incumbent may be identified as part of the problem. Establishing credibility in the eyes of the work force is critical for any new manager, but turnaround managers often do not have time to establish credibility within the organization. In many cases, then, a turnaround manager's credibility must be based on reputation: previous experience in related industries or on success in other turnaround situations. As outsiders, successful turnaround managers must, therefore, possess the following abilities:

1. Learning quickly in a new situation—The ability to ask questions that help relate new situations to previous experiences is vital; turnaround managers must quickly tap into the existing knowledge base. Successful turnaround managers must also be able to cross-check the validity of what is being said against their previous experiences and current observations.

---

[1]Much of this section is adapted from the note on "Turnaround Management" (Harvard Business School case #9-686-057), originally prepared by D. Daryl Wyckoff.

2. Conceptualizing—Successful turnaround managers must be able to assimilate and organize large volumes of data from many sources, test it, and determine its relevance to the situation at hand. This includes identifying relevant patterns and then formulating the information into clear, easily communicated concepts to guide oneself and others.

3. Making quick decisions—Many managers are able to identify most of the problems inherent in a turnaround situation, but the most successful turnaround managers are able to prioritize the problems and establish a logical sequence of activities in a project-like plan. They concentrate on a few critical problems first, holding less important problems for treatment later.

4. Communicating effectively—A high degree of skill in interpersonal relations is critical in turnaround situations. Turnaround managers must excel not only in their ability to elicit information but also in their ability to communicate the content, structure, sequence, and priority of tasks needed to achieve the turnaround. They must recognize that the success of the turnaround hinges on broad participation throughout the organization. During the implementation phase, successful turnaround managers tend to spend a large proportion of their time (60 to 80 percent) on communications: coaching, listening, evaluating, and encouraging others.

5. Managing stress—Turnaround managers must manage not only the troubled operation but also high levels of personal stress. In turnaround situations, goals are usually ambitious and insecurity at all levels of the organization is high. The two biggest challenges to mental stability are "overload"—inundation with information and tasks—and the psychological threat arising from the state of disorder. Turnaround managers must, therefore, be able to cope with uncertainty and disorder.

OVERVIEW OF MODULE

The cases within this module cover all the stages and issues raised above. The *Erie Redesign* was undertaken to save the Erie plant which was not meeting corporate expectations on profits. The case outlines the reasons for the operation's decline and broadly overviews the type of changes potentially required for a turnaround. The reading, "Reviving a Rust Belt Factory," provides another perspective on the turnaround process.

The Wabash facility, described in *Elastrocraft (A)*, is suffering a cash shortfall and soon shall be in a cash crunch (at least based upon internal transfer prices). Here, the strategies and actions of two seasoned turnaround managers (one at the division level, the other at the plant level) are explored. Their efforts, however, have not been sufficient to make the operation profitable. The managers must now determine whether the operation can be saved or whether it should be shut down with the work transferred to a lower labor cost greenfield site.

Simpson Pump and Valve, another operation not meeting corporate expectations for profits and losing market share, completes the turnaround cycle by considering what happens once a turnaround is complete. The strategy pursued during the turnaround is no longer appropriate, and the plant manager must find a way to convince the division manager that a change is necessary.

Finally, *New York Malleable and Ductile* is in the most critical stage of a turnaround, experiencing a major cash crunch, with creditors on the verge of closing the operation down. The case describes in detail the historical neglect of the foundry but leaves it totally up to the reader to design a plan to save the operation from bankruptcy.

REFERENCES

Greenhouse, Steven. "Revving Up the American Factory." *New York Times,* January 11, 1987.

Mitchell, Cynthia. "Some Firms Resume Manufacturing in U.S. after Foreign Fiascoes," *Wall Street Journal.* October 14, 1986, p. 1.

Sloma, Richard A. *The Turnaround Manager's Handbook.* New York: The Free Press, 1985.

The Erie Redesign

The corporation had been very successful in going south and in designing new facilities from scratch. They'd pick a management staff and a high potential product, design the plant for that, close down the rust-bucket plant and move it. Never in the past had they opted to stay with a facility.

Our approach was to get into their conscience and say that we, as the management of a large corporation, had the responsibility to make sure we had fully assessed the situation. Had we given the existing operation full chance to be competitive? We supported it with statistical data showing that our costs could change in a way that would minimize the difference between going south and staying and redesigning our facility.

Dan Sherry, plant manager at Riverstone Incorporated's Erie, Pennsylvania facility, thus summarized the factors that led to Riverstone's approving the Erie Redesign Program. The Erie factory manufactured calenders, machines used in the making of cloth, rubber, or paper. The material was pressed between a calender's three to four rollers to make it smooth and glossy (see Exhibit 1).

Prior to the redesign approval, 15 Riverstone plants had been shut down; for instance, in the early 1980s Riverstone had closed a foundry in

EXHIBIT 1   Example of Calender

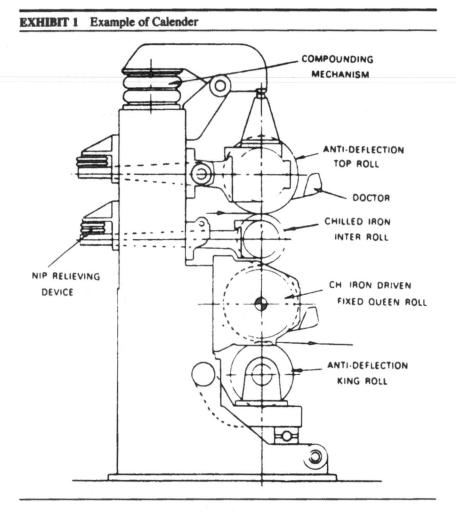

Detroit, Michigan, after investing $4 million in it. Hence, having convinced corporate management to give the Erie plant a second chance, Sherry knew he had to make the turnaround work. The Erie program would be measured against three goals: 1) a 50 percent improvement in quality; 2) a 30 percent reduction in cost over three years, including the consolidation of seven buildings into one; 3) attainment and maintenance of its annual operating plan (AOP). The redesign, which began in October 1984, was scheduled to be completed in October 1987. It was now September 1986, and though many successful changes had occurred at the plant, numerous challenges remained.

THE ERIE PLANT

In 1892 two Scandinavian immigrants founded the Erie Machine Works in Erie, Pennsylvania. The company, known as a "jack of all trades," rapidly expanded into a totally integrated manufacturer with nearly 12,000 employees.

In 1967 Riverstone Incorporated acquired the Erie Machine Works from the original owners and began investing money in equipment development, especially in electronics. Riverstone, which had sales of $10.2 billion in 1984, was one of the largest corporations in North America with 121 manufacturing plants and 347 other facilities around the world. At the time of the takeover, there were about 2,000 employees at Erie.

Erie's product lines were hit hard by the recession in the mid-1970s, causing plant employment to fall to the hundreds. Layoffs began to affect people with 25 years' seniority and over, and the bargaining unit[1] was reduced from 3,700 to approximately 750 employees. The plant was about ready to close its doors when a labor strike in Riverstone's Indianapolis facility led the corporation to transfer a major portion of that plant's business to Erie because of its good labor relations and excess capacity. The new product line, calenders, had been developed, manufactured, and marketed under Riverstone's Industrial Tools Division, which was headquartered near Chicago.

The new business led to a seven-day operation with laid-off workers recalled and new employees hired. In 1978–79, the plant shipped $130 million, and in the first quarter of fiscal year 1979–80, employment levels were high: 1,067 hourly and 261 salaried employees. But during this time, the value of the American dollar began to rise and there was increased competition from foreign manufacturers. In 1982 foreign sales, which constituted a major portion of the division's sales, dropped from 60 percent to 10 percent.

To counter fluctuations in foreign exchange rates, the division had begun to explore overseas production. In 1981, for instance, the Industrial Tools Division expanded to 50 percent its participation in a joint venture in Japan which manufactured Riverstone's products under license.[2]

Domestic competition was also intensifying. In 1981, Gordon International Machining, the largest American producer of calenders and one of the three largest in the world, began developing and operating what it

---

[1]Shortly after the Riverstone acquisition, the United Steel Workers of America (USW) replaced an independent union that had been in the plant.

[2]In 1984 Riverstone acquired the machine tools division of a French company and signed a licensing agreement in the People's Republic of China.

claimed was one of the most advanced, robotized machining lines in the country at its Atlanta, Georgia, plant. The computer-driven line produced precision rolls and cylinders at cost savings of up to 50 percent. Gordon also transferred some of its U.S. products to French, British, and Mexican plants, and created licensing agreements with China. (By 1985 it had decided to centralize most of its machining in its automated Georgia plant and was planning to invest $25 million in the facility over three years.)

Around the time these problems were being felt at Erie, Dan Sherry was appointed plant manager. Sherry, who had worked for Riverstone for 13 years, started as a corporate auditor and then became the Erie plant controller, a position he held for three years. As controller, Sherry became the plant manager's "right-hand man." Subsequently, when the manager was approaching retirement, Sherry expressed a desire to take over his position. After undergoing a 1-year training process with the plant manager, Sherry became the eighth plant manager (in 13 years) on January 1, 1982.

For Sherry, who had no previous manufacturing background, becoming plant manager was a big step. One staff member noted that Sherry's lack of manufacturing background was a weakness in the short run, but a strength in the long run. The primary measurement of business tended to be financial, which was Sherry's strength. As Sherry himself explained:

> It was scary to step in from being controller of the plant to being plant manager. It was an ideal move personally, but after I sat down and started asking questions I said, "What am I doing here?" Much of the volume had covered the inefficiencies in the plant, and fixed costs had been a very small percentage of the total. Then the bottom fell out and problems began to surface. What I brought was a balance to what strengths were here. We are great implementers, we are great mechanics and technicians, but our strengths weren't in the total business of bringing it all together. I think that's the background I picked up in finance.

THE DECISION TO REDESIGN

By October 1983, Erie management realized that customers were dissatisfied with the quality of delivery and lengthy installation time of the calenders. Costs were high and sales were dropping; down 36 percent in 1983 from 1981. Division management decided to move toward less vertical integration at the Erie facility and sent several representatives, referred to as the "undertakers" by many plant employees, into the plant to oversee a downsizing of the operation. The plating and heat treating operations were outsourced because a higher level of quality and a lower cost could be obtained from vendors. In addition, several buildings at the Erie

EXHIBIT 2 Employment Statistics

	Hourly (By Fiscal Year Quarter)	Salaried (By Fiscal Year Quarter)
Fiscal year 82/83	704	226
	632	223
	641	194
	624	179
Fiscal year 83/84	627	177
	598	178
	635	177
	647	158
Fiscal year 84/85	537	162
	525	163
	539	166
	446	153
Fiscal year 85/86	382	142
	346	141
	350	128

site were torn down. Layoffs at the plant were continual from that time onward. (See Exhibit 2 for employment statistics.)

Soon thereafter, a division-level study was initiated to determine whether the company would do better if the Erie operation moved to a greenfield site with lower wage rates and a less expensive yet more modern facility. Concurrent with the greenfield study, Sherry also convinced division management to investigate what it would take to make the Erie operation as profitable as a greenfield site.

In January 1984, Theodore Thomas Consulting (TTC) was hired to do a study. It conducted a six-week, in-depth process analysis of the facility which included interviewing a cross-section of managers at all levels, administering attitude and leadership surveys, and observing how people spent their time. The study revealed five major areas that needed to be improved in the plant.

1. Improving capacity utilization through better identification of upcoming workload and improved balancing of load to capacity:
   a. Machine operations needed rerouting.
   b. Machine downtime needed to be improved.
   c. The number of setups needed to be reduced.
2. Strengthening systems, procedures, and support functions:
   a. Inventory.
   b. Maintenance.

  c. Quality.

  d. Engineering change orders.

  e. Forecasting (marketing interface).

  f. Outsourcing.

3. Developing management and supervisory effectiveness:

  a. Pacing was occurring as the workload reduced.

  b. Supervisors spent only 6 percent of their time, on average, initiating contact with employees, giving instructions and guidance, or receiving performance and quality information.

  c. A survey revealed that the predominant leadership style was high task/low people.[3]

4. Planning and implementing consolidation of current manufacturing, assembly, and administration activities.

5. Identifying opportunities and laying the foundations for the application of available "state-of-the-art" technology.

The results of the studies revealed that the two options were basically equal: A 27 percent cost reduction was attainable in the greenfield site, while a 24 percent reduction could be achieved if the Erie plant were redesigned. Due to similar cost estimates, experience, and knowledge within the Erie work force, and a belief that it was time for the corporation to begin experimenting in the turnaround process, corporate management opted to keep the Erie facility open. (During the same time period, corporate management also initiated two other turnaround projects.)

The approval for the turnaround, referred to as the "redesign," was given on May 1, 1984. Division management established the goals, including a 30 percent reduction in costs within three years. In return, the corporation would provide funds to continue the introduction of new automated equipment which had been approved during the downsizing process and to contract outside consultants to help facilitate the redesign efforts.

THE REDESIGN PROCESS

"We recognized up front we couldn't pull this off without someone who had more experience than we had. We needed someone to help us with the systems and technology as well as the human resources side of it," noted Sherry. As a result, a joint team of TTC consultants and managers from the plant was set up. The internal redesign team members, compris-

---

[3]High task/low people managers assume that people and production concerns are mutually exclusive and that people do not like to work and must be coerced; such managers tend to give only negative feedback to employees and rely predominantly on systems to obtain results.

ing four Erie employees and led by Sherry, managed the redesign activities, with the consultants providing focus and facilitation, especially in human resource issues. Sherry chose Paul Nash, quality assurance manager, to be on the team because he was a newcomer to the plant, having transferred to Erie in 1982 from one of Riverstone's electronics divisions. The other members of the group were: a manufacturing engineering representative, who offered an historical perspective of the operation having come up through the bargaining unit and became the facilities specialist for the redesign; an IE supervisor who was selected for his analytical skills and became the team's materials specialist; and the former quality manager (prior to Nash) who had acted as a quality circles facilitator[4] and had extensive knowledge of the plant.

Because of budget appropriation delays, the redesign program did not officially begin until October 1984. Sherry began the process by holding a series of off-site staff meetings (two days a week for three weeks) to review the plant's products and processes and to identify which were critical and which could be eliminated. While these meetings were taking place, the remaining members of the redesign team prepared for an October 25, 1984 kickoff meeting, called the "Commitment Meeting." A diagonal slice of the organization with a total of 40 representatives from each functional area and shift was chosen to meet and prioritize the list of issues identified by the staff. Hats, coasters, pencils, and other paraphernalia adorned with the Erie Redesign Logo (shown at the beginning of the case) were given to the participants. The logo, created by the redesign team, was an Erie Railroad locomotive, a symbol that was both positive and cautionary: positive, because the Erie Railroad had local flavor and appeal; cautionary, because the railroad had gone out of business by ignoring the need to change. The curved tracks in the logo symbolized movement, first ahead and then into the future.

The first redesign activities got underway shortly thereafter. (See Exhibit 3 for a schedule of activities.) The plant staff reviewed the issue prioritization done at the commitment meeting and assigned the redesign team to work on the top 10 issues. One of the first actions the team took was to set up natural work groups consisting of 6 to 12 individuals throughout the plant to work on specific problems. These employees, a heterogeneous mix from different levels of the plant such as engineers, operators, and maintenance workers, were believed to have the best chance of finding a successful resolution to their assigned problem. The group was given a specific time frame in which to complete its assignment

---

[4]The division had set up quality circles at the Erie plant two years before the redesign program began. Originally, there were 15 circles, led by a full-time quality circle facilitator. In 1986 seven quality circles were still functioning, and their work was being incorporated into the redesign effort.

EXHIBIT 3   Redesign Schedule of Projected Activities

	Sep	Redesign goals and objectives met Bldg 21 vacated and surplus equip sold Heavy machinery in bldg 6 operational
	Aug	
	Jul	
	Jun	Interior office renovations completed
	May	Contract negotiations completed
	Apr	Operator process control fully implemented
	Mar	
	Feb	
1987	Jan	Cultural programs self perpetuating
	Dec	
	Nov	Preventive maintenance program started
	Oct	Tool management system operational OIP and OES training completed
	Sep	Exterior office renovations completed
	Aug	Pay and progression system designed
	Jul	Cylinder cell operational
	Jun	Switch gear and compressor air operating
	May	
	Apr	Buildings 25, 38, 23, 26 vacated
	Mar	Frame cell operational
	Feb	Shipping/receiving/assembly relocated Bldg 6 dock and material system installed
1986	Jan	
	Dec	Daily and weekly shipment target met
	Nov	
	Oct	Vendor bar stock program complete
	Sep	
	Aug	
	Jul	Status update to corporate management
	Jun	Buildings sold with lease back agreement
	May	Detailed layout of building 6 complete
	Apr	
	Mar	RTP assembly relocated to building 6
	Feb	Eccentric cell released to production
1985	Jan	Status update to corporate management
	Dec	Building 6 block layout complete
	Nov	Planning and success monitoring approach
1984	Oct	Kickoff and commitment meeting

and once that was achieved, a formal presentation was made to the staff explaining how the problem would be measured on an ongoing basis.

The selection of the initial projects was also based on achieving quick results (savings achievable within 15 weeks) to show workers in the plant, as well as personnel at the division and corporate level, that the Erie facility could make changes and would be successful. In addition, many of the projects focused on dispelling the "we/they" syndromes that existed within the plant between union and management, manufacturing and engineering, the assembly floor and the stock room, etc. About every four months, the plant staff continued to select a series of problems that the plant was experiencing and groups were formed to deal with them.

From the outset, the redesign team instituted a communication strategy to keep people in the plant informed about the activities taking place. A monthly bulletin and a special bulletin board were used; newspapers, published periodically both at the plant and division level, frequently contained articles about the plant turnaround. Sherry explained: "We learned from experience quickly that you had to use a multitude of communication vehicles, that a bulletin board or a plant newspaper wasn't enough. So we started doing some employee meetings." First came quarterly employee meetings, but Sherry soon moved to weekly vertical meetings, attended by about 25 people including managers, supervisors, hourly employees, and union representatives. The meetings' purpose, Sherry said, was to "sit down and go over what's going on, with the idea that the participants take that back to their work place and tell their fellow employees what's happening."

In July 1985, an executive steering committee made up of division- and corporate-level personnel was organized to assist and guide the plant through the redesign and to procure internal resources. This committee met quarterly to review the plant's progress. In addition, beginning in September 1985, the staff, the redesign team, and the union committee met weekly to review the status of the redesign program and to provide direction for it.

ORGANIZATIONAL CHANGES

Prior to the plant redesign, Erie management consisted primarily of individuals who had been promoted from within the facility. But as the redesign proceeded, Sherry eliminated four staff positions and replaced most of the remainder. Originally, he explained, "I thought I'd have the right people up front and just roll with it and be a winning team." Yet, he soon discovered, with the help of Amy Howard, a TTC consultant assigned full time to the Erie plant, that some of his staff members did not have the technical and interpersonal skills needed to carry out the redesign program; others were unable to deal with the changes that were occurring.

In the past, if someone were not performing well the person would be moved laterally or downgraded, but with the redesign, a poor performer was terminated. To aid in the development of the new staff, Howard spent 80 percent of her time one-on-one with Sherry and each of his staff members, giving feedback on their behavior and interactions with the other staff; the remainder of her time was split evenly between running management development workshops (e.g., how to delegate, how to organize effective meetings) and project management.

Sherry's division manager, a former Erie plant manager, played a major role in helping select new staff members, most of whom came from other Riverstone plants or other companies. (See Exhibits 4 and 5 for the Erie Plant organization chart, before and after the redesign.) Alex Swain, a 25-year veteran with Riverstone, was brought on board as the human resources manager in 1983. Swain, who had helped to close two other Riverstone plants noted, "Some people view me as the hatchet man, which often helps in my relationship with the union and my relationships in the Erie plant in general." With the introduction of automated technology into the plant, Eric Walton, who had been a member of the division manufacturing engineering staff, was hired as the manufacturing engineering

EXHIBIT 4   Organization Chart—Pre-Redesign

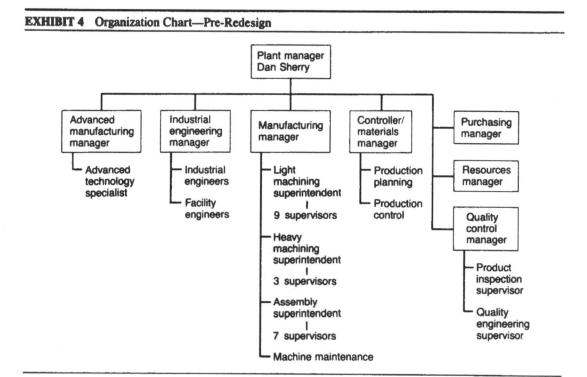

EXHIBIT 5 Organization Chart—September 1986

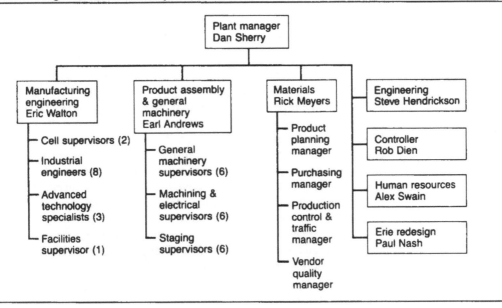

manager in July 1985. In his previous position he had been involved with the research and capital appropriation process for any project costing $2 million or more and thereby was part of the group purchasing the new automated equipment for the Erie plant.

Earl Andrews, the manufacturing manager, had been a long-time plant employee who moved up through the ranks into that position. In late 1985, however, a plant manager from a sister plant was brought in to replace Andrews and become Sherry's backup so that more of Sherry's time could be dedicated to the redesign program. Andrews was given responsibility for the assembly and machining areas, reporting to the new manufacturing manager. After six months, that manager was terminated and Andrews returned to Sherry's staff with the responsibility of managing the assembly and general machining areas, while Walton was assigned responsibility for the manufacturing cells (discussed below).

Another long-time Erie employee also switched roles during the redesign. Steve Hendrickson, who had spent 20 of his 30 years with Riverstone in Erie, had previously been manager of industrial engineering. When the revitalization program began, he was assigned what was initially an undefined role—plant engineer responsible for developing on-site engineering support at Erie and acting as a liaison between Erie and division engineering (located in a Chicago suburb). His role evolved into obtaining engineering approval for ideas and changes generated by the shop floor at Erie. This typically meant spending about three days a month at the division offices. Hendrickson remarked:

> My primary goal of the past two years has been to build the respect and confidence of the division engineering group. Communication used to be poor, and there wasn't much respect at all. People always felt engineering dictated what had to be done.

The remaining staff members were new to the company. The plant controller, Rob Dien, had worked in that capacity for another large company before joining the staff in November 1986. In January 1986, Sherry selected Rick Meyers, an MBA from Wharton, as the materials manager, choosing him over someone whom the division had wanted to hire. Because inventory and raw materials costs were growing relative to other types of direct costs, Sherry believed that materials carried more weight than manufacturing and that ultimately the materials manager should be responsible for the flow in the shop.

Another organizational change was the elimination of the management layer between the plant staff and the first-line supervisors. As of September 1986, there were two remaining midlevel managers (previously called superintendents) who were listed as general supervisors on the organization chart but who only had hourly employees reporting to them. Because the redesign had not yet been completed and more than one building was still in use, however, many of the first-line supervisors still took directions from the general supervisors rather than from the staff. As one union official noted, "The change [in reporting relationships] only happened on paper. There are still people there doing the same job as before."

In commenting on all the changes that had occurred, Sherry noted:

> I used to be just "one of the guys" and played softball with the people in the plant. With the redesign, I've had to lay off some very close friends. That's been very draining. I don't think, though, I would have been able to lead the redesign if I had come from the outside because I wouldn't have been sensitive to what motivates the people within the plant. I think my track record within the plant allowed people to trust me and to realize that I was personally committed.

THE MANUFACTURING OPERATION

Before the redesign, the Erie plant consisted of 800,000 square feet spread over seven buildings. The operation included the machining of major components (main frames, connecting cross bars, cylinders, rollers, and eccentrics[5]), the subassembly of frames, and the final assembly of calen-

---

[5]An eccentric was a piece of metal that held the rollers in place and worked like a cam.

ders. In addition, there was a small fabrication area which produced much of the guarding and other miscellaneous sheet metal parts.

Work areas were poorly utilized and transporting work-in-process was slow. The costs of heating all the buildings and normal maintenance and upkeep were high. To reduce overhead costs, which accounted for half of the product cost, (direct labor was 10 percent; material was 40 percent), the redesign included moving all of the manufacturing operations into a two-story, 365,000 square foot building, 257,000 square feet of which was prime manufacturing space. The first floor (or basement) would house the storeroom, tool room, receiving and inspection areas, and facilities and maintenance shop. The machining, fabrication, and assembly would all be located on the main floor. Both the interior and exterior of the building were refurbished, and a new roof, windows, and an automated material handling system were installed. The first department was relocated in March 1985. Other areas followed, and by September 1986, everything except heavy machining (frames, cylinders, etc.) and calender assembly had been moved. The vacated buildings had been put up for sale but, in May 1986, were taken off the market because of a potential for new business at Erie.

In the past, machining areas were organized as a typical job shop with lathes, boring mills, or drill presses each grouped in a separate area. In 1983 prior to the start of the redesign, division management decided to move toward cellular manufacturing in which a group of machines was dedicated to a particular component part. With the exception of three new highly automated cells, all cells would consist of existing equipment. A typical cell comprised two or three milling machines and a drill press. Although the drill press might be used for only one hole and, hence, stand idle most of the time, it was kept within the cell to enable the part to be machined from start to finish in only one handling. This was Erie's version of just-in-time manufacturing. The socket cell, the first to be installed, reduced material handlings from 14 to 1 and cycle time from two months to two days. Long term, the plan was to have approximately 15 cells, with three cell supervisors on the day shift and two on the second shift. In addition, there would be a general machining area made up of a small group of stand-alone machines; these would produce small volume parts that did not warrant a separate cell or would perform a variety of special tasks, such as repair, prototype machining, or engineering changes to stock parts in inventory.

Frames and calenders were assembled on a conventional line in one of the outer buildings. Once transferred into the consolidated building, all assembly would be done in a stall build where presses would be assembled in total at one work station by a team of operators. (Andrews preferred to call it "random assembly" rather than stall assembly, which he felt implied stagnation.)

The systems within the plant had also been significantly altered. Data collection systems were installed throughout the shop along with capacity and materials requirements planning (MRP) to help plan and schedule material, parts, and assembly. Operator process control (OPC), a form of statistical process control (SPC), had also been introduced into the manufacturing cells. A key element of the redesign, a preventive maintenance program, had not yet been established. Maintenance was still being done when employees were on vacation, even if it was not scheduled, so that the machine would be free when the operator was available. One supervisor noted:

> In my opinion, we don't have the staff to do preventive maintenance. To some extent the operators do it, but there are some things an operator shouldn't have to do. For instance, an operator should't have to change the hydraulic oil cleaning coolant because it's time-consuming and the machine is not used, and that's not utilizing the operator's time effectively. I think one crew going around doing the servicing would be more economical.

THE NEW PLANT LAYOUT

Developing a new layout for the whole plant was not only one of the central projects of the revitalization, it also became the key event in getting plantwide involvement in the redesign process. Early in the program, Riverstone's corporate management was pushing for a layout, so the redesign team put together a block layout in January 1985. Initially, the redesign team had hoped to get the line people involved in designing the layout, but they were not interested. In June, the team created a detailed layout and then made a videotape for all managers and operators to encourage them to consider their own departments' design. In September 1985, the team set up committees from within the plant, including union representatives, to approve different segments of the layout. (See Exhibit 6 for a facilities layout of the Erie plant.)

The facilities rearrangement had not been without its problems. At one point, to meet a commitment made to the executive steering committee to empty out a warehouse, a portion of the frame assembly area was temporarily relocated because the assembly's permanent location had yet to be freed up. The only space available required people to work in a recessed area. One manager commented, "We play checkers all the time, moving areas one, two, or three times. I wonder if it wouldn't have been easier to build a new facility right here in Erie instead of moving buildings around here."

NEW TECHNOLOGY

Erie's first step toward a fully integrated manufacturing system was the eccentric cell, which consisted of a computer numerical control (CNC) lathe, a Cincinnati T10 work center with a daisy wheel pallet exchange, a

EXHIBIT 6 Facilities Layout

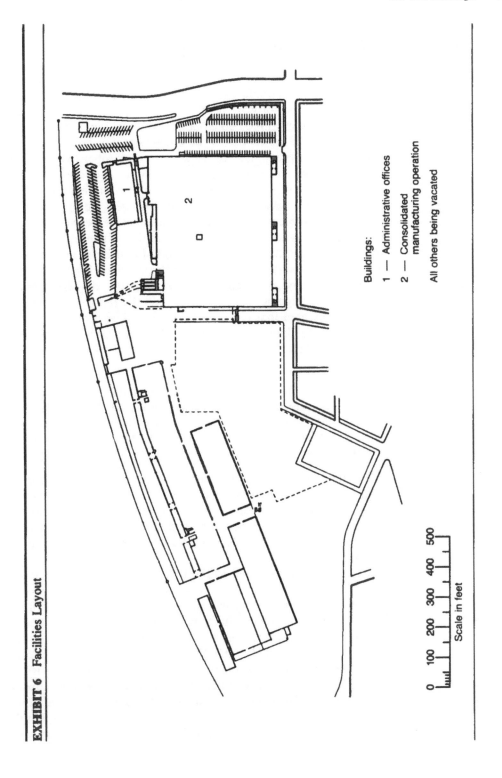

Buildings:

1 — Administrative offices

2 — Consolidated
   manufacturing operation

All others being vacated

Scale in feet

0   100   200   300   400   500

CNC universal grinder to do inside and outside diameter cuts, a coordinate measuring machine (CMM), and an overhead gantry robot with a vision system. The cell, purchased for $1.7 million in 1984, ran three shifts a day with one operator for each shift and produced 16 different designs.

The eccentric cell had caused problems since it first began operating. The robot and the software in the controller were purchased as a turnkey, with the vision machine being bought separately on the advice of the robot vendor. The remainder of the equipment was bought as stand-alone machines. The vision system worked with the robot to orient the eccentric and position it onto the grinder. Although the robot vendor had initially claimed it was possible to work with the different pieces of machinery, combining seven languages into one cell led to major technical difficulties. Problems developed at the interface between the robot and the machines as well as among the machines themselves. Also, vendor software support had been limited: On site at Erie, no technician from the vendor had stayed for more than three weeks. Finally, the robot was not accurate enough to orient the part properly. By September 1986, an operator was aligning the eccentrics, and the robot was being removed. It was discovered that the savings that could be achieved by having the robot do the loading and unloading of parts within the cell were offset by the operators' free time during the machine cycle. Reflecting on the learnings from the eccentric cell, Sherry explained:

> We've pushed the rest of our projects back so that we will do a better job up front now in specifying what our requirements are. The first time, we brought the cell into the plant before it ever worked. The cell we have on order now is being installed and tested at the vendor site so that we don't have the problem here in our factory.

The new cell Sherry referred to was the frame cell, scheduled to be installed in November 1986 at a cost of $8 million. This cell would consist of a CMM, two horizontal boring mills (with room for a third), and pallet and tool stands for loading and unloading parts and tools. There would be 120 tool pockets and 40 different tools which would be loaded or unloaded by robots.

The supervisor assigned to the frame cell noted that he accepted the job on one condition: that the mistakes from the first cell would not be repeated. He wanted to participate in the design of the cell, and as a result, made a number of trips to the vendor during the initial negotiations to review the blueprints and be assured that the system met the plant's needs. Although he had been influential in the cell design, he had had less say in the selection of its operators because of seniority provisions within the union contract. The cell operators would be joining the supervisor for a four-week training session at the manufacturer's site where they would receive a combination of hands-on and classroom teaching, but there was no guarantee that those trained would not be "bumped" out of the frame cell in upcoming work force reductions. The union contract permitted employees with 15 years' service to displace the least senior employee in any classification they believed they were qualified to perform, and they were given 10 days to meet the job requirements. As a result, operators had nothing to lose by trying jobs with which they were not familiar since they would be laid off anyway if they were not qualified to stay on the job. A few maintenance workers, who would be dedicated to the cell, would also receive maintenance electronics training, and there would be special training for programmers. The vendor training was estimated to cost about a quarter of a million dollars.

OTHER MAJOR CHANGES

Several other major changes were also taking place at the Erie plant. The make-buy philosophy switched from purchasing only 25 percent of the components, such as electrical controls, to outsourcing 75 percent. (The 75 percent of parts that were bought accounted for about 50 percent of overall material cost.) The new philosophy called for the creation of manufacturing cells to manufacture complete parts fitting into the following criteria: critical (proprietary) parts to Erie products (e.g., side frames, cylinders); parts with critical tolerances; parts that were capital intensive or required unique equipment; parts which could be manufactured in-plant with significantly lower cost; and parts that normally had to be made with insufficient lead time. An underlying premise was not to bring in low volume work just because the plant needed hours to absorb overhead.

Accounting procedures and standards also had to be revised to fit the changes. One area experimented with removing labor standards from the shop floor paperwork. Operators still had to record the quantity of pieces they produced each day, but since they no longer had access to the standards, the tendency to pace their output to the standards was eliminated. Many operators knew the past standards, but because of the broadening of classifications and the combining of jobs, many were unaware of the amount of time needed to complete the total job.[6] By removing the standards, productivity increased by 15 percent and run efficiencies jumped from 75 percent to over 100 percent.

A final outcome of the redesign was a change in attitude toward grievances. Between 1966 and 1979 there had been over 3,000 grievances, excluding disputes over incentive payments. The union won 14 out of the 15 cases that went to arbitration in 1978, at a time when the average success rate nationwide was 20 percent. From 1980 to 1983, the plant aver-

[6]In an attempt to increase employee flexibility, a new job classification system which reduced the number of classifications from 173 to 30 was negotiated in March 1984.

aged 80 grievances. In 1984 there were five grievances and in 1985 only two; as of September 1986, there had only been one grievance filed at the plant.

CURRENT STATUS

With the redesign scheduled to end in October 1987, strategic planning for the next three years began in July 1986. Planning was intended to be participatory, so lower-level managers were brought into the process. The union was invited to participate but declined. Sherry's objective was to gain a consensus from the entire Erie management team (all managers and supervisors within the plant) on the three-year strategic plan. The planning process included dividing the managers into small task groups; these would create task statements, intended to be the driving force for an action plan, and measurements to monitor their progress (see Exhibits 7 and 8 for the proposed task statements and measures). As of September 1986, the management team was still struggling with whether these were the appropriate tasks and measurements on which to focus.

Another challenge involved maintaining current measurements. The 30

EXHIBIT 7   Proposed Manufacturing Task Statement

We will utilize our assets to provide maximum flexibility to customer demand simultaneous to the achievement of manufacturing the highest quality product at the best value for our market.

It is acknowledged that the simultaneous occurrence of the components of our task will be difficult to achieve. To enhance our market position in every potential business opportunity, our assets, employees, and product flow must be in sync with customer requirements.

To this end, we must streamline our facilities to make us the most effective operation possible. We must keep open the lines of communication and keep information flowing from the newest employee straight through to our customers. We must poise our facilities to respond to customer requirements with predictable operational flow and a well-trained, dedicated work force utilizing adaptable manufacturing processes, dynamic materials systems, and effective supplier support programs. We must implement a sound capital investment strategy, opting only for an outlay which proves monetarily supportive of our manufacturing envelope.

Therefore, our trademarks must be product quality, predictability, and performance. In order to be the "Supplier of Choice," we must achieve the balance between cost effective asset utilization and the flexibility mandatory to customer satisfaction.

EXHIBIT 8 Plant Measurements Proposal

I. Return
  A. Primary Measurement
    1. Return on Investment: (Mark-up Factor)(COS)—Actual Plant Cost Divided by Net Assets
    2. Cost of Sales (COS)
  B. Supporting Measurements
    1. Inventory Investment: Days of Inventory vs Plan
    2. Asset Turns vs COS
    3. Total Variance vs Plan

II. Quality
  A. Primary Measurement
    1. Cost of Quality vs Cost of Sales
    2. Planned COQ/COS
  B. Supporting Measurements
    1. Percent Purchased Lot Acceptance
    2. Rework & Scrap Money
    3. Total Price of Non Conformance
    4. Customer Perception
        (a) Customer Index
        (b) Verbal Statement from market place

III. Predictability/Responsiveness
  A. Primary Measurement
    1. Plant On-Time Delivery (Percent of Delivery Money)
  B. Supporting Measurements
    1. Quoted Lead Time vs Actual Available
    2. Shop Floor Control—Deviation to Plan
    3. Inventory Accuracy vs Target
    4. Average Stock—Pull Success
    5. Percent Overtime vs Plan

IV. Facilities Focus
  A. Primary Measurement
    1. Compatibility Matrix (Products vs Criteria)
  B. Supporting Measurements
    1. Square Footage Utilized vs Hours/$ Produced

V. Productivity & Flexibility of Employees
  A. Primary Measurement
    1. Average Output per Employee (COS Divided by Employee Expense)
    2. Average Output per Employee (COS Divided by #Employees)
  B. Supporting Measurements
    1. Shop Productivity versus Plan
    2. Office Productivity versus Plan
    3. Hourly Flexibility (Percent Qualified per Classification/Skill)
    4. Salaried Flexibility (Percent Qualified per Function/Skill)

percent cost reduction had been left ambiguous as to whether it meant a 30 percent reduction in standard cost or overall costs. The plant chose to measure the former, adjusting for changes in volume and inflation (see Exhibit 9). As a result, it was able to meet that objective, but as one manager noted, "If they [division management] had defined it, we wouldn't have made it." Despite all the disruptions caused by the redesign program, quality costs as a percentage of cost of sales had been reduced by 75 percent. The facility had also been successful in achieving its AOP goals up until June 1986, when it missed its production schedule by one week, and July 1986, when it was delinquent on sales dollars promised on an interplant delivery to another Riverstone facility. The plant quickly recovered and was back on schedule in August. Reasons cited for the problems included the havoc created by the facility consolidation project, some of the make-buy decisions, and the lack of controls on many of the outsourced parts. Sherry laid out the weaknesses of each functional area as he saw them and told each manager to take specific actions to make sure that no more goals would be missed, noting that "it [missed measurements] would not happen again." Apparently, this would not be easy, especially in light of a forecast of continued layoffs, a critical component in achieving further cost reductions. One supervisor noted:

> It's very difficult to keep people motivated. People say, "Am I next?" What I try to pass on to my people is that, at this point in time, we as a whole, my area, are still here. There are no guarantees that we will be here in a month. But, I can assure them that if we give up and we don't keep scrap down, keep productivity up, and keep improving, we won't be here. If we don't all pitch in there's no guarantee that any of us will stay here.

A final challenge confronting the plant was what would happen once the TTC contract ended and Howard left in October 1986. The redesign

---

EXHIBIT 9   Cost Reduction Performance

	Total	*Percent of Total*
1984 standard cost (weighted average unit cost)	$ 53,600	100.0
Economics 8.0 percent	4,300	8.0
Volume reduction	7,700	14.4
Performance improvements through Sept. '86	− 8,700	− 16.2
Performance improvements identified for FY87	− 4,600	− 8.6
FY87 overhead reductions	− 700	− 1.3
FY87 material cost reduction	− 1,700	− 3.2
	$ 49,600	92.5
Total reduction from adjusted 1984 standard cost		29.7

team had already disbanded, and its members had returned to the line organization with the exception of Nash who had stayed on as the redesign manager. (His former role as quality manager had been eliminated with the responsibility for quality assurance being assigned throughout the organization.) Nash, who had acted as Howard's backup in project management, was to assume her role, but there was a concern that the staff would revert to their old habits after Howard was gone. One manager explained:

> When Amy leaves, that will be one of our toughest challenges. It will be hard to keep the participative management. You have to work at it. There's a heavy risk in the first year that people will change back to the way they were, moving back to the old directive style of management, because the dictatorial style is easier than the participative style in many ways. I think it will put a burden on Dan. If nobody says a word about it, my guess is it will slip. The next year is going to be even more challenging than the last.

Reviving a Rust Belt Factory

Dean M. Ruwe
Wickham Skinner

By the late 1970s, Copeland's Sidney, Ohio plant was a candidate for shutdown. Labor relations could hardly have been worse: in just six years, the number of plant workers fell from 3,000 to 1,000, a blow for a town of only 20,000. Exasperated managers removed Copeland's most profitable, high-volume products to dedicated facilities in other states. This only intensified the Sidney plant's difficulties: it manufactured the leftover products at excessive cost, with high inventories and poor delivery.

Copeland was not alone. American industry had built hundreds of large factories during the 1950s and 1960s. The principle was economy of scale; the architecture, an industrial engineer's dream. These giant installations often started out to make only a few products. As markets matured, managers endeavored to satisfy their customers' specialized needs by making product lines more varied and production more flexible. In consequence, large factories grew even larger; buildings sprouted annexes, warehouses, and workshops. Assembly lines for new products were often. piggybacked onto existing ones, not always with happy results.

We had originally made compressors for specialized refrigerators, such as supermarket display cases. Skilled employees made the compressors in teams under workshop-type conditions; orders would come in for two or

three at a time. In the early 1950s, Copeland built a new plant, about 600,000 feet square, mainly to produce compressors for standard air conditioners. Orders routinely came in for 2,000 or 3,000 units at a time. To service this market, Copeland introduced highly mechanized assembly lines and "flow" operations.

In fact, 1950s managers were so enamored of the newer and seemingly more efficient techniques of mass production that they decided to mesh much of the line that made refrigerator compressors with the mechanized lines making air-conditioner compressors. Warehouses and repair services were consolidated. It seemed a good idea at the time.

Clearly, the intention at headquarters was to share overhead: the plant expanded three times during the next 15 years. The markets it served, however, had very different requirements for quality and delivery. Even more important, two such different products required different process and manufacturing technologies. In the case of the air-conditioner compressors, management's mission was to prevent disruptions on the line. But with the refrigerator compressors, its mission was to manage inevitable disruptions.

Problems with the process led to problems in other areas. Even the smartest managers were overwhelmed.

In the early 1980s, the Sidney plant's volume production hit an all-time low. On top of that, the combined production processes created extra work and frustration for everybody at the plant:

Reprinted with permission from *Harvard Business Review*, May–June 1987, pp. 70–76.

- Employees filed many grievances. Absenteeism rose, and many workers refused overtime.
- Top management pressed the supervisors hard to keep machines and assembly lines running. This forced them to rush the supply of parts, take shortcuts in employee training, and attend only superficially to the handling of grievances and other employee complaints. Putting out fires became the supervisors' overriding occupation.
- Plant management was frustrated with headquarters' shifting priorities. No one ever seemed to have enough time to try to remedy the basic things: quality, training, vendor development, tool repair, maintenance, record keeping, and communication with workers.
- To increase production and reduce costs, factory management sought out the longest product runs and batched orders based on forecasts, some of which proved too hopeful. The result was excess inventory and confused vendors.

As other corporate departments grew impatient with the Sidney plant, some of us became resigned to closing it down. Workers could see that basic problems needed solving and could not understand why managers were so helpless.

What to do? Certainly the temptation grew to discontinue operations, but what about the future of Sidney's workers and the life of the small town? We made up our minds to keep things going but with a drastic rise in efficiency. We explored and debated many ideas. In the end, we decided to bet on the concept of the "focused factory."

Help on Its Way

At the heart of this idea is the simplification of the manufacturing task by limiting each factory to one or two specific markets. That way, management avoids the conflicting demands inherent in multiproduct facilities serving multiple markets. Rather, a few clear objectives define production management's mission.

When in 1981 we decided to focus the Sidney plant, we recognized that commercial refrigeration and residential air-conditioning would continue to be its two major markets, and we decided that reorganization could pivot on the differences between them.

The commercial refrigeration market demands high-quality, high-reliability, high-durability products. Refrigeration compressors must run 24 hours a day, year-round. If they fail, the refrigerated products will spoil. Moreover, commercial refrigeration applications are often custom designed; they are specialized products, made in small quantities and in configurations nearly impossible to forecast. Copeland's customers win their orders by being able to deliver finished refrigerators in the shortest possible time. The compressor manufacturer's lead time is comparably short and becomes part of the formula of competition.

To withstand the long, hard hours of operation, Copeland's refrigeration compressors have a heavy, cast-iron body that houses the compressor motor and serves as the frame for its mechanical parts. The compressors' motor capacity can be as high as 40 horsepower, and the compressors themselves can weigh as much as 400 pounds. Because of the importance of servicing them in the field, the machines have many access ports and covers, which are sealed with machined and gasketed surfaces. Workers who make refrigeration compressors, therefore, must be able to machine parts to exacting tolerances and pay close attention to detail when assembling components.

The residential air-conditioning market, in contrast, demands high-volume production. Customers can forecast their needs with considerable accuracy; annual blanket orders are commonplace, with monthly or quarterly releases against those orders. At the same time, variability in the product line is limited. The manufacturers of air-conditioning systems offer few models, and consumers almost never ask for custom design. All of this means longer lead times for the compressor manufacturer.

Compressors for air conditioners are rather small, weighing 50 to 75 pounds, and they operate infrequently, except in summer. Being light duty, the components can be made smaller and simpler than refrigeration components. The compressor is enclosed in two halves of a thin metal shell, seam welded to protect the contents from outside contaminants. There is no opportunity for field service. The manufacture of air-conditioning compressors does involve some highly skilled work—close-tolerance machining of the components, leak-tight welding of the shell, and a good alignment of the motor on the compressor frame. Yet the high volume means that many of these processes can be automated.

So the processes and the markets for the two main product groups were quite different. It made perfect sense to continue to think of air-conditioning compressors in terms of a flow operation—long runs, repetition, reduced setups—and no sense at all to think of refrigeration compressors this way. Management concluded that two factories would be necessary and that they should be as different as the products themselves.

But it is one thing to recognize that the existing plant is a raft of compromises and quite another to design two effective new structures.

Before we committed anything to paper, we had to ask ourselves two questions:

1. What would we have to be particularly good at in each focused factory?
2. What in the old system would be particularly difficult to change?

To ensure high quality and quick delivery, refrigeration compressor manufacture requires the most experienced, qualified, and committed workers—people who can be counted on for good judgment in assembling a multitude of parts. To produce residential air-conditioning compressors, Copeland could rely on equipment and process technology to manufacture high-quality parts with a lower-skilled work force. At the same time, workers needed to take training to learn the closely controlled procedures for the compressors' final assembly.

If refrigeration compressors require greater workmanship, air-conditioning compressors require employees with a more disciplined temperament.

As to difficulties, we were certainly daunted by the prospect of changing the old production and inventory control system in the Sidney facility. It had been set up to handle a wide variety of products—too wide for optimal operation in either projected factory. Yet we thought that splitting the materials system in two would be too big a project to tackle along with everything else. We also recognized how difficult it would be to rearrange some 500 machine tools while normal production was going on. In addition to a large number of single-purpose machine tools, there were also a number of "monuments," which seemed an awful task to relocate. These included a 25-year-old, 17-station transfer line, a heat-treating facility, and several old five-stage parts washers.

Yet particularly difficult, it turned out, was communicating to employees the benefits of dividing the plant into two focused factories— a point to which I'll soon return. None of our employees had any in-depth knowledge of the customers or the requirements of the marketplace. This was our fault, we concluded, but also an inherent problem in a big, overloaded, unfocused plant, where workers could not know if the piston they were making was headed for an office window or a supermarket aisle. Nor were middle managers completely sold on the concept of factory focusing. It was essential to gain everybody's enthusiastic support.

After drawing up plans for the relocation of production equipment and conveyors, the company decided to hold a three-day seminar for key managers at an off-site location.

We enlisted the help of several Harvard Business School professors to lead us in a number of case studies designed to illustrate the various problems and benefits of factory focusing. The managers in attendance were from manufacturing operations as well as from human resources, finance, engineering, and domestic and international marketing. Discussion of the cases brought out a range of problems we would not have otherwise anticipated.

Employees Consider the Plan

Any successful factory focusing project requires the support not only of management but also of hourly workers. Changes are impossible to bring off when managers and unions are adversaries.

By 1981, several management changes in the human resource and manufacturing departments had laid the groundwork for improvement. The heads of these departments had encouraged supervisors to do a better job of listening to workers and contributing to the administration of the organization. Supervisors began to act on grievances more promptly in cooperation with representatives of the hourly work force. After Copeland management made the decision to focus, it held meetings throughout the factory to discuss not only focus issues but also operating problems. Gradually the credibility of the management team grew, and an atmosphere of trust and mutual respect began to permeate the factory.

A signal event in this trust-building process occurred early in 1981. Because of seasonal capacity restraints at another factory, we found it necessary temporarily to assemble different, later model air-conditioning compressors at Sidney. Actually, our employees had assembled this compressor during its introductory phases, but minor changes to it now meant they needed retraining. The employees involved were asked (not told) to attend this session on a paid, overtime basis one Saturday morning. We had plenty of supervisors, engineers, and managers on hand to answer every question that the assemblers raised. We had a long morning break and served fresh coffee and doughnuts in the factory on a cloth-covered table.

I explained why we needed to do this extra work and how long we might expect product demand to continue. I stopped at the individual work stations and thanked each employee for the trouble he or she took to be present. One assembler told me, "I've never seen anything like this at Copeland before." I knew we were on our way!

Of course, this all sounds easier than it was. What certainly helped was Copeland's announcement at the start that it was willing to invest $4 million to revitalize the Sidney plant. The initial relocation of hundreds of machine tools was a manifest demonstration of company commitment.

By the end of 1982, we had completed the basic plans for two focused factories in the Sidney facility. We sought suggestions from employees concerning workplace locations and layouts, and we implemented many of their ideas. But a good many workers began to tell us their fears. Considering the difficult issues still unsolved, their anxiety was not unreasonable—nor was our own.

At first we had thought that having a separate work force, along with a separate union organization, for each factory was essential to the focused factory concept. We wanted our employees to take a broad view of their jobs: to think about how the products they made would ultimately be used, and by whom. We wanted them to know that, say, the piston they made was going into a Hussmann refrigerator case that would cool dairy products in a Safeway store. We wanted the union to consider itself a responsible agent of the production and marketing process.

Early on, however, we reluctantly recognized that our employees, members of the International Union of Electrical, Radio, and Machine Workers, would simply not accept this division. In the first place, they were loath to split their bargaining unit in two. But even more important was their acute fear that focusing would undermine the seniority system. This requires some explanation.

For our part, seniority presented a serious threat to the whole focusing effort. Under the old system, a layoff meant that a senior employee with low skills would stay on while a higher skilled worker with a shorter tenure would have to leave. Quite apart from our desire to have workers identify powerfully with the products they made, it was obviously essential to keep the most skilled workers for the refrigeration factory. We wanted people who might become familiar with acceptable variations and tolerances in the parts they were making and the components we were carrying. This would be impossible if we routinely had workers "bumping" each other or shuttling back and forth between two such different factories.

With the participation of the union leadership, we developed a concept of "protected classifications" in each of the factories. The most highly skilled workers would be protected from seniority privileges. We found that, in any case, many of the high-seniority people were already concentrated in high-skill jobs. And since union leadership had approved the protected classifications concept,

we thought this amounted to union approval and that the problem was licked. This was a hasty judgment.

At a board meeting, a director raised the question of whether the entire union membership should vote to ratify the contract changes. We knew we had the union leadership's support, and we hesitated to risk trying to get a majority vote from a larger group of employees. Nevertheless, we decided to run the risk.

And so we scheduled large group meetings in the local high school auditorium. On Wednesday we distributed detailed information; on Thursday we answered questions. Again, union leadership expressed their support for the idea. The vote was set for Friday.

At a meeting held just before the vote, the board of directors passed a resolution that, with heavy support from the work force, we would go ahead and divide the factory. Those of us most involved in the planning insisted that a small majority would not represent the level of support needed. What we needed, we said confidently, was 70 percent to 75 percent approval.

The union membership rejected the idea by a 15-vote margin.

This was extremely discouraging. Should management and union leadership pursue the concept any further?

The union leaders told us they still supported the idea and thought we should try for approval a second time. Discussions following the vote, they said, indicated that many people were still confused about the contract change. We were limiting normal seniority prerogatives, after all, and people were bound to find this particularly threatening.

We decided to plunge ahead with an extensive series of small group meetings where we could explain the provisions of the existing contract as well as the provisions of the proposed contract in depth.

Most of the questions union members raised

were predictable enough and easy to sympathize with. "What will happen to me if one of the factories closes altogether?" "What happens if I'm assigned to the air-conditioning plant and there's a big layoff? How will my seniority be enforced?"

In conducting all these meetings—they eventually amounted to 30—the plant manager and the director of employee relations made sure to leave ample time for questions. Some restrictions in the bumping system would exist, to be sure, but they would lead to higher quality and lower costs. I remember remonstrating with one distraught employee, "In the long run the protected classifications protect your job too, because they'll keep our plant in Sidney, and we'll be able to compete worldwide."

Just before Christmas in 1982, the union held a second vote on the concept of factory focus. This time the union membership approved the idea by a 3 to 1 margin. The next week we moved the first machine line.

The Reorganization

Several huge machines placed constraints on the layouts for the two new factories. We worked around these constraints, however, and constructed the two factories within one building. Nearly every weekend during 1983, we were moving big machine tools around the factory.

This was an unusual aspect. Companies normally handle moves faster. Since this situation involved changing entire systems, work assignments, and procedures, however, Copeland decided that moving equipment over many months would be less risky for customer commitments than a massive short move.

We were concerned about our ability to move 500 old machine tools and successfully get them back into production. Through years and years of use, these tools had gradually degraded, and in many instances they seemed incapable of holding the new machining tolerances.

In previous facility reorganizations at Sidney, outside companies had contracted to do much of the moving and reinstalling. Our maintenance people told us that this had frequently irritated them because, in the move, machines often went awry. The workers said they wanted to participate more in the Sidney revitalization. We agreed that their knowledge of the machine tools and processes would be very helpful, and so our own people did most of the work.

When the workers encountered problems with a particular machine installation, it was easy for them to talk to the maintenance employees and straighten the problem out. Over time, the maintenance people and other factory workers identified closely with the successes as well as the problems we were having with the machine tool relocations.

During the first several weeks of 1983, we moved some of the old machines as an experiment to see if we could get them back into production. We set them on new bases and leveled them and, to our amazement, this tune-up not only got them back into production but also got them running more precisely and more capably than they had before. It looked as though one of our major concerns had been unwarranted.

While we could move individual machine tools and lines in a weekend, or sometimes overnight, we couldn't handle the assembly area the same way. Major revisions in the assembly flow required us to reorganize the assembly area during our normal summer shutdown. Because the move was so complex, we consulted with the union and agreed that we would shut the plant down for an extra week, which gave us enough time to complete the assembly area reorganization and test the relocated equipment and revised line layout.

By the end of 1983, we had relocated and requalified all the machine tools; the factory's machining capability had never been better. We had totally separated the two factories' assembly areas and constructed new offices for the separate plant management staffs. These offices represented our commitment to operate as two separate factories. Further evidence of the separation was an eight-foot-high wall of concrete blocks that snaked through the factory and delineated the refrigeration from the air-conditioning compression facility. Each factory had its own paint and color scheme.

This separation is important both psychologically and operationally. Unless the plants are truly different—in layout, staff, management, and product—they probably won't develop their distinctive competences.

By the end of 1983, Sidney was operating at full production in two factories with separate management teams, separate work forces, and separate objectives.

About a year into the project, management at Sidney finally recognized that, to maximize efficiency, each business should have its own, custom-designed system for production scheduling, control, and inventory management. The umbrella system was not meeting the requirements of either business.

Before the focusing of the Sidney factories, the materials for the air-conditioning and refrigeration compressors were stored together. The storerooms were disorganized, and materials handling and identification were complicated. Focusing the factories allowed us to separate the inventory. In this new environment, storekeepers and materials handlers developed expertise in the identification, location, and organization of their own parts. Their jobs became easier, they became more productive, and the quality of their work improved.

Incidentally, the focusing showed us that our computerized master-scheduling techniques were inappropriate in the low-volume factory. We solved the problem by hiring one employee, familiar with the product and the factory, to do the master scheduling manually, on a part-time basis. This change resulted in smoother production throughout the factory and a much smoother demand on our vendors.

In the refrigeration factory, the employees started to view the orders as individual events, so each order took on a special meaning and identity of its own. Because the orders were so clearly delineated, the managers and schedulers could track each compressor through the plant. They knew the status of the order at all times. They could quickly identify and resolve any parts shortages or quality issues.

How it Turned Out

Our primary objective in the Sidney project was to resuscitate the facility through improved quality and delivery and reduced inventories and costs. Since the final machine was moved more than three years ago, the factories have run at—or beyond—the expected level. The Sidney factories have improved in four main areas:

Quality

As the focused factory concept jelled, we found people more and more willing to talk about high standards. They became involved in the product and in the customer's needs; the machinists and assemblers worked more confidently. Now, stopping either assembly line to correct a quality problem is both encouraged and expected. The quality outlook extends beyond the Sidney workers to the vendors.

Since the start of the focused factory, product quality, as measured by in-warranty failure rates, has improved dramatically; the failure rates of products manufactured in 1984 were less than half those of products manufactured in the early 1980s. Because the focused factories are easier to manage, the plant staff and supervisors have extra time to meet. One supervisor actually tore apart a troublesome compressor and placed it on a display table. He then got the workers together, discussed the problem with them, and they collectively determined how it could be prevented.

Delivery and Lead Time Improvement

In the early 1980s, delivery times for the Sidney products ranged from six to eight weeks; after the focusing, the lead times were cut in half. A survey in late 1982 showed that the Sidney plant was meeting less than half the promised delivery dates. In 1985, more than 98 percent of the deliveries were made within the promised week. In 1986, the refrigeration factory had no past due deliveries. We set up a special program for deliveries to the West Coast and increased our market share by 50 percent in that area. The shorter lead times and improved delivery performance have won Copeland numerous orders.

Labor-Management Cooperation

Worker involvement in the planning and implementation of the focus concepts at Sidney created a team atmosphere that has continued to thrive. First-line supervisors have gained skills in dealing with their people. Grievances per hundred employees have declined by two-thirds. Unplanned absenteeism has declined from about 4 percent to about 2 percent. When measured on the basis of time between injuries and time between lost-time accidents, plant safety has improved by a factor of two. The last two labor contract negotiations went through without a work stoppage.

This is something new. Until the change in the plant, four of the last six contract negotiations had led to a strike.

Clearly, one of the benefits of factory focus is that running the business becomes simpler; the company needs fewer people in the organization, particularly in the salaried and indirect hourly ranks. In 1983, Copeland needed 120 salaried people to operate the two Sidney factories. In 1985, it needed only 80—a reduction of 30 percent.

The number of indirect hourly workers at Sidney fell about 30 percent as well. Because we had always paid so much attention to direct hourly performance, our workers had al-ways been very productive. Between 1983 and 1985, therefore, their performance improved only by 10 percent. These reductions in salaried and hourly labor have breathed life into the new factories. Some workers did lose their jobs; the point is that nearly two-thirds did not.

Inventory Reduction

In 1981, the Sidney plant stored about $24 million of inventory. By 1985, we had cut that level practically in half to $13 million, evenly divided between the two factories. We have reduced inventory dollars and improved inventory balance and quality. The effects of this progress show up in delivery performance, lead times, and return on investment.

The story of the Sidney factory has a remarkably happy ending: an old, complex, dying factory came to life in a new form. And the concept of the focused factory was at the heart of the resuscitation. Less than 10 percent of industrial companies have actually developed a manufacturing task from their competitive strategy, as Copeland did. By dividing a complex factory into separate, manageable parts, managers and employees could simplify their tasks and zero in on the performance criteria most important for the company's success.

Copeland created two separate management teams for the Sidney factories. To improve the organization's overall cost structure, these teams reluctantly laid off a number of indirect salaried and hourly workers. The indirect work that does remain, however, is simpler and more rewarding. The union went along.

And so, in addition to proving the worth of the focused factory idea, the Sidney factories show that American industrial workers can contribute to and support cost reduction and quality improvement efforts. In so doing, they become involved in the business and interested in the customer. They also become proud of their work.

Elastocraft (A)

On March 31, 1983, at 9:30 P.M., Jim Plant, the plant manager of Elastocraft's Wabash, Indiana operation was faced with one of the toughest decisions of his 14-year career with the company. He was scheduled to meet Marvin Isles, his manager, the next morning at 9 o'clock and recommend that the unprofitable Wabash plant should either remain open or be closed and relocated to a greenfield site. His decision would be based on whether he believed the plant could be turned around despite the three previous years of financial losses.

A decision to close the plant would cause hundreds of workers to become unemployed. At the end of the first quarter of 1983 the Wabash operation employed 377 people. In 1981, there were 550 employees; there had been 1,320 in 1979, 10 fewer than the all-time high in 1976. The town would be affected as well: the Wabash plant was the town's major employer and source of taxes. Most employees were born, reared, and resided in Wabash County which had a population of 35,000; 15,000 lived in the town of Wabash. Because of layoffs at Elastocraft and other local businesses, there were few job opportunities in the community, whose unemployment rate was well over 9 percent.

Plant was also concerned that recommending a closing would be admitting failure. Eighteen months earlier he had been transferred from another Elastocraft operation to Wabash as the new plant manager and had subsequently attempted, jointly with Isles who had been brought on eight months later, to turn the plant around through strategic reduction of salaried and hourly employees (layoffs and terminations); wage freezes; im-

This case was prepared as the basis for class discussion rather than to illustrate either effective or ineffective handling of an administrative situation. Certain names have been disguised.

Copyright © 1986, Revised 1988, by the President and Fellows of Harvard College, Harvard Business School case 9-687-041.

plementation of statistical process control (SPC); and the exclusion of unprofitable product lines. Despite these efforts the company continued to lose money. It was projected that there would be a $1.3 million loss for the first quarter of 1983.

ELASTOCRAFT

Elastocraft, a producer of industrial rubber products, was one of International Products's many businesses. International Products was a highly diversified corporation with 1982 sales of $2.3 billion and profits of $18 million. The corporation's businesses ranged from providing goods and services for defense and aerospace projects to radio and television broadcasting, soft drink bottling, airline transportation, and movie production. Elastocraft had nine operating divisions: wallcovering, reinforced plastics, coated fabrics, plastic film, plastic extrusions, building systems, polymers, athletic products, and engineered elastomers.[1] With over 6,700 employees, Elastocraft had 20 manufacturing plants in the United States, Canada, Mexico, and the Republic of Ireland. Its 1982 sales totaled $498 million. (See Exhibit 1 for an organization chart.)

The Wabash plant belonged to Elastocraft's Engineered Elastomers division, which produced extruded and molded rubber products. Wabash was also the site of division headquarters where engineering, sales, and centralized purchasing functions were located. In addition to Wabash there were five other plants in the division: Batesville, Arkansas; Welland, Ontario; Marion, Logansport, and Peru, Indiana. Each plant was a profit center that had its financial statements analyzed several different ways each month: net income to net sales; operating income to net sales; and return on average assets employed. (See Exhibit 2 for an historical summary of the key measurements.)

THE WABASH PLANT

The Wabash plant was one of the largest industrial rubber production facilities in the world. It had been purchased in 1936 and by 1981 covered 600,000 square feet of land. The 200 different rubber compounds the plant produced were used in over 1,000 products (see Exhibit 3), making Wabash's products list greater than the other five plants' products lists combined. In addition to rubber compound mix, which was sold to the Marion

---

[1]Elastomers were materials which at room temperature could be stretched to at least twice their original length and, upon immediate release of the stress, return quickly to approximately their original length.

EXHIBIT 1   The International Products Organization

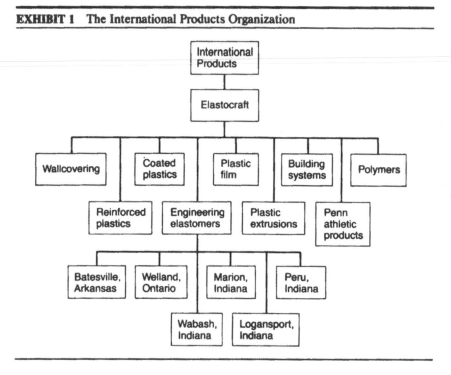

plant, the products were separated into four areas according to their manufacturing process: extruded products were made via the extrusion process, and molded products were made via injection, transfer, or compression molding. Exhibit 4 lists many of the key operations.

THE DECLINE OF THE WABASH PLANT

The plant had been profitable and maintained a good reputation from 1969
through 1978, with profits peaking in early 1977. But by mid-1977 profits
began rapidly decreasing; by the end of 1979 the Wabash plant was at
breakeven. The following year ended with a loss of over $2 million. (See
Exhibits 5 and 6 for income statements and balance sheets.) The deterioration had actually begun in 1976 with the oil embargo and energy crisis
and subsequent downturn in the auto industry. At that time the product
line had been heavily concentrated on that industry. The situation was
worsened by the granting of large wage increases following a 10-day
strike.

Between 1976 and 1981 Elastocraft had five different presidents, while
the Wabash plant had an equal number of plant managers. During that
time, Wabash redirected its product focus from the auto industry to the

EXHIBIT 2   Summary of Wabash Plant's Performance Against Key Measurements

	1982													1983	
	Dec.	Jan.	Feb.	Mar.	Apr.	May	Jun.	Jul.	Aug.	Sep.	Oct.	Nov.	Dec.	Jan.	Feb.
Net income/net sales (percent)															
Monthly	(25.9)	(9.4)	(6.7)	(10.3)	(9.4)	—	(6.3)	(10.0)	3.4	(15.3)	(20.8)	36.0	(35.9)	(12.5)	(5.7)
Year-to-date	(25.9)	(16.9)	(13.5)	(11.9)	(11.3)	(9.3)	(8.9)	(9.0)	(7.7)	(8.7)	(9.6)	(6.3)	(35.9)	(22.7)	(16.2)
Operating income/net sales (percent)															
Monthly	(29.6)	(9.4)	(6.7)	(10.3)	(6.3)	(6.3)	(6.3)	(10.0)	3.4	(19.2)	(20.8)	32.0	(33.4)	(12.3)	(9.3)
Year-to-date	(29.6)	(18.6)	(14.6)	(13.6)	(12.0)	(11.0)	(10.3)	(10.2)	(8.8)	(9.7)	(10.5)	(7.2)	(33.4)	(21.5)	(16.8)
ROAE* (percent)															
Monthly	(27.5)	(11.8)	(7.8)	(11.8)	(11.8)	—	(7.8)	(11.9)	4.0	(15.9)	(20.0)	35.8	(25.9)	(13.1)	(4.4)
Year-to-date	(27.5)	(19.6)	(15.7)	(13.8)	(13.3)	(11.1)	(10.6)	(10.9)	(9.2)	(10.3)	(11.3)	(7.3)	(25.9)	(19.7)	(14.6)

*ROAE = $\dfrac{\text{Annual income (before interest, after tax)}}{\text{Average net assets employed}}$

Net assets employed = Total assets − Current liabilities

EXHIBIT 3   The Wabash Plant's Product Line

	1981 Annual Sales (in millions)
Mixed compounds*	$ 1.2
Molded	
• Vibration isolation—automotive	
Silent bloc	8.3
Mold bonded—motor mounts	2.1 10.4
• Other automotive	
Free rubber bushings	5.1
Wiper blades	1.9
Brake diaphragms	6.8
Piston cups	1.1 14.9
• Nonautomotive, free rubber	1.4
• Construction products	3.3
• Agricultural—hayrollers	2.9
—haytines	.4 3.3
• Missile liners	.5
Subtotal molded	33.8
Extruded	
• Automotive	
Dense rubber	1.3
Sponge rubber	.4
Dual-durometer	18.9 20.6
Total sales	55.6

*Mixed compound sales to the Marion plant increased to $2.2 million in 1982 with the transfer of the dual-durometer extrusions.

specialty-products rubber industry, with the intent of becoming a producer of a full rubber product line. This product shift lessened the impact of the automobile industry but created a drain on engineering and led to increased overhead expenses. Much of the engineering talent, which had been improving processes for rubber products supplied to the auto industry, was reassigned to developing low-volume specialty products such as rubber webbing for roadway construction. Also, many key people left the firm through better job offers, layoffs, and retirements, but much of their knowledge was not transferred to those remaining in the plant.

By September 1981 the morale and productivity of the employees were extremely low while incomes were very high. The average cost of production per man hour for 1981 was $24; in 1982 it was $31; and by the end of the first quarter of 1983 it had reached $42. Wabash's capacity utilization in relation to annual sales was 55 percent in 1981 and 50 percent the

EXHIBIT 4 Key Operations in Production Process

Product	Manufacturing Process	Equipment Type	Quantity	Average Age (years)	Capacity (percent used)
Mixed compound*	Mixing	Banbury mixer	4	22	64
Silent bloc	Transfer mold	Small hydraulic press	135	35	85
Free rubber	Compression mold	Small hydraulic press	8	20	50
Wiper blades	Compression mold	Large hydraulic press	4	24	90
Construction	Compression mold	Large hydraulic press	8	14	85
Hayrolls	Compression mold	4-post hydraulic press	5	7	35
Missile liners	Compression mold	Injection press, rotary table	1	9	50
Wiper blades	Injection mold	Injection press, rotary table	6	7	85
Diaphragms	Injection mold	Injection press, shuttle table	3	4	70
Motor mounts	Injection mold	Small Wabash press	35	20	75
Haytines	Transfer mold	Extrusion line	3	23	25
Piston cup	Extrusion	Extrusion line	1	14	25
Dense rubber	Extrusion	Extrusion line (2 extra per line)	3	5	67
Sponge rubber	Extrusion				
Dual durometer	Extrusion				

*Mixing of rubber compounds was primarily for internal manufacturing requirements until 1982 when dual-durometer extrusions were transferred to Marion. In 1982 Marion compound requirements used 10 percent of Wabash's mixing capacity.

EXHIBIT 5   Income Statements for the Wabash Plant
(Periods ending November 30, 1978–82)*

	1978	1979	1980	1981	1982
Net sales (in $ millions)	68.8	65.7	46.8	55.6	34.8
Material cost	31.4	31.5	20.9	24.1	14.5
Direct labor cost	6.9	6.2	5.5	7.2	4.0
Overhead cost	22.1	23.4	18.5	19.9	16.2
Manufacturing cost	60.4	61.1	44.9	51.3	34.7
Actual manufacturing gross profit as percent	8.4	4.6	1.9	4.3	.1
of sales	12.2%	7.1%	4.0%	7.8%	.3%
Distribution cost	.4	.5	.3	.4	.2
Operating cost (sales, administration, etc.)	2.8	3.6	4.2	3.4	2.4
Operating income (loss) as percent of sales	5.2	.5	(2.6)	.5	(2.5)
	7.6%	.8%	(5.6%)	.9%	(.7%)
Other (income) expense	(.2)	(.3)	(.6)	(.7)	(.8)
Interest expense	.6	.7	.7	.9	.5
Net income before tax as percent of sales	4.8	.1	(2.7)	.3	(2.2)
	7.0%	.1%	(5.8%)	4.6%	(6.3%)

*Sales for 1982 have been adjusted to include compound sales to Marion ($2.2 million in 1982). Interplant compound sales were eliminated from income statements during this period.

following year. The average hourly earnings of Wabash workers was over $2.50 greater than the industry average. The company's cost for each worker's pension was also over $2 per hour per employee greater than the industry average. As a result, Wabash's products were no longer cost competitive. In addition, the products repeatedly suffered in quality. Waste was a continual problem: by the end of 1981 waste had exceeded 21 percent of production costs. The following year it was the same. By the end of the first quarter of 1983 it was down to 14 percent but still significantly above the goal of 9 percent. Overall, the plant was in a state of disarray.

THE MANUFACTURING PROCESS

Making rubber compound and other rubber products was dirty work. It was common to see some employees covered from head to toe with black powder as a result of working in certain areas. The walls and ceilings of the plant were also covered with the powder, which in conjunction with dim overhead lights, made the interior of the plant dark and cloudy.

The plant's hundreds of pieces of equipment, most of which were very old, made continuous, rhythmic noises when in operation. The equipment included banbury mixers, presses, extruding machines, tumblers, and drop mills, and ranged in age from brand new (very few pieces) to 47 years

EXHIBIT 6 Balance Sheets for the Wabash Plant (1980–82)
(in millions)

	1980	1981	1982
Assets			
Current assets			
Cash	$ —	$ —	$ —
Inventories	3.4	3.2	2.7
Receivables	6.5	5.7	5.8
Other	—	—	—
Total current assets	9.9	8.9	8.5
Fixed assets			
Original cost	23.7	26.1	27.8
Accumulated depreciation	(12.9)	(14.0)	(15.0)
Total fixed assets	10.8	12.1	12.8
Prepaid and other	—	.1	—
Total assets	20.7	21.1	21.3
Liabilities			
Current liabilities			
Trade payable	2.1	.9	2.1
Other payables and accruals	2.5	3.0	2.5
Total current liabilities	4.6	3.9	4.6
Long-term debt	—	—	—
Intercompany accounts and other	18.8	16.9	20.7
Retained earnings	(2.7)	.3	(4.0)
Total liabilities and earnings	20.7	21.1	21.3

old. Preventive maintenance was performed on the equipment during the five-day Christmas shutdown period; otherwise, maintenance was done after breakdowns or improper operations, which occurred frequently.

The Mixing Operation

The rubber compounds used in the extrusion and molding processes, and sold to the Marion plant, were made from combining synthetic and natural rubber,[2] carbon black,[3] oil, clay, and wax in one of four banbury mixers.

---

[2]Natural rubber was obtained from the Hevea Brasiliensis tree. These trees, grown primarily on rubber plantations in the East Indies and Africa, were tapped to obtain the sap, called latex. The latex, which contained 65 percent water and 35 percent rubber, was coagulated with acetic acid. After the water was removed by squeezing, the coagulate was milled into sheets and dried. The sheets were pressed into bales and shipped.

[3]Carbon black was a pigment which acted as a stiffener, toughener, and antioxidant.

First, a master batch was prepared in one of the two banbury mixers designed for this purpose. From an upper level a worker loaded rubber and as many as 14 chemicals into a conveyor leading to a mixer that operated like huge taffy-pullers to combine the ingredients. Each mixer had a capacity of 450 pounds and could mix a batch in approximately six minutes. A master batch had a shelf life of three to six months. As scheduling dictated, a master batch was recycled through the mixers to produce the finish batch with a shelf life of one to four days.

After mixing, the compound was sent to drop mills where rubber sheets were formed. Antitack, a white powder, was applied to each sheet to prevent sticking. Then the sheets were sent to staging areas or the finished compound bank, which supplied the plant's production flow system. The bank also supplied rubber compounds to satellite plants.

Extrusion

The extrusion process was similar to squeezing toothpaste out of a tube. Strips of rubber compound were fed into a screw chamber then heated and compressed through a heated die. The extruded rubber was then cured in a saltwater bath, where it was sufficiently hardened to preserve the shape the die had given it. Finally, it was cut to specifications for the particular part.

The extrusion process was both a quick and inexpensive way of shaping rubber. The dies were low in cost and the process involved the least amount of natural material waste. In addition, it was the only process available for making rubber tubing and shapes that had numerous reentrant angles, for example channeling, which was placed around car windows.

Injection Molding

Sixteen parts, such as brake diaphragms, motor mounts, and rake tines, were made by the injection molding process. Rubber strips were fed into a pressure chamber ahead of a plunger. As the plunger moved foward, the rubber strips were forced into a heating chamber, where they were preheated. From here they were forced through the screw section where they were melted and the flow regulated. After leaving the screw section they passed rapidly through a nozzle and into a hot mold, where the rubber cured at the instant the mold filled. To complete one cycle of this process required only a few seconds. While the finished part was being ejected from the mold, the material for the next part was being heated.

Once the part was removed from the mold, an air gun was used to clean away the scrap.

The positive aspects of injection molding included quick curing cycles and dimensions that could be held better than with any other process; quality of the end product was excellent and there was less natural waste than with transfer or compression molding. But injection molding could be used only for a small number of parts due to geometric limitations, and the machines were very expensive—costing nearly $250,000 apiece.

Compression Molding

Compression molding was used to produce approximately 140 parts in the Wabash plant including wiper blades, bushings, and hay rollers. The presses used to make parts by compression had hot molds (cavities) in the bottom half. The molds were in the shape of the desired end-product. Large chunks of rubber were put into the molds and the top half of the press was pulled down, thereby compressing the rubber into the shape of the molds. Compression cycle time varied with the size of the part; larger parts took longer. Once the cycle was completed the part had to be manually unloaded. If a worker did not unload the part quickly enough, it would be damaged through overheating.

Compression molding had the second highest rate of natural waste; in addition, it required continual attention from the worker, who had to fill each mold manually, hold the top of the press down, and manually unload each mold. Therefore, this process was used primarily for low-volume parts.

Transfer Molding

The transfer molding operation began with sheets of rubber being placed on the bottom half of the press. The top of the press, which had all of the desired shapes and designs on its face, automatically came down like the top of a large waffle iron, compressing the rubber to its hot surface. Once the compression cycle was completed the end-products were automatically unloaded.

Transfer molding, which had the lowest tooling costs of the three molding processes, could accommodate the largest mold cavities. Although this had the highest rate of natural waste, more than half of the company's molded parts were made by it. This included cured vibration isolation inserts which were sold to the Logansport plant. The inserts, which accounted for 15 percent of Wabash's sales, were molded rubber products that had specific spring and shock absorbent qualities that affected the quiet ride of an automobile.

Waste

The material content was high (approximately 40–45 percent of the product cost), largely due to high levels of waste common in the extrusion and molding processes. There were three kinds of waste: natural scrap, compound scrap, and finish scrap. Finish scrap was cured rubber that had defects in the end-product because of manufacturing problems. Compound scrap was rubber unable to cure because it had overaged as a result of schedule changes or mold problems. Compound scrap also resulted from the plant's temperature being too hot. But the most common and therefore costly scrap was natural. Accounting for over 60 percent of the total waste, natural scrap was an unavoidable by-product of the production process. It was the rubber which flowed into the mold through the runner systems and trim or flash pad.

LABOR RELATIONS

The Wabash plant was represented by Local #626 of the United Rubber Workers (URW), which also represented two other plants in the area (Elastocraft's Peru plant plus another small company), each having separate contracts. From the initial recognition of the union in 1941 up through the early 1970s, there was a strong adversarial relationship between the company and the local. By 1970 the local had become quite militant and strongly opposed any policies the company put forth. At that time, there were approximately 900 grievances filed per year by a work force totalling around 1,200 people.

The 1970s were particularly violent. The year 1970 included a 10-day wildcat strike and a 7-week contract strike, the longest in the company's history. During the latter strike, three salaried employees filed a $1.1 million lawsuit against Ernie Pack, the union president, for involuntary imprisonment. They accused Pack of leading a striking group of workers who forced them to remain inside the plant for three consecutive days in fear of their lives. The suit was dropped at the end of the strike. In July 1975, another 10-day wildcat strike, with much picket line agitation and recriminations on both sides, led to the termination of 22 people because of violence. In 1976, an orderly 10-day end-of-contract strike was resolved by granting the same wages and fringes as had been granted by the Akron tire contract.[4] Although the contract was negotiated locally and the local union was not bound by the national tire agreement, this was a pattern that had been followed since 1967. The contract was renewed in

---

[4]The major tire manufacturers in Akron, Ohio, such as Goodyear, Firestone, and General Tire, tended to grant similar wage increases much like the auto industry.

1979 without a strike—again with wages and fringes similar to the Akron agreement.

By 1980 the cost-of-living allowance costs had reached approximately $3.74 per hour (on a base of $7.25 per hour). This had raised Wabash's hourly wage rate above that of its competitors in the industrial rubber market. Most of the competition, particularly in the extruded rubber products market, was nonunion or smaller unionized firms which did not follow the Akron wage pattern. (Higher wages were less a factor for the molded rubber products because the primary competition also pegged their wage rates to the tire pattern.)

In 1980 the division management approached the local union about freezing wages through the end of the contract. Pack had heard similar threats before, and with his encouragement the union overwhelmingly rejected the proposal. Six months later the company moved one of the major extrusion operations, the production of dual durometers, to the Marion plant. Unlike the Wabash local, Marion's had agreed to concessions of $1.55 per hour. This move eliminated 244 jobs at Wabash. By the end of 1980 only 565 people were employed at Wabash.

ERNIE PACK

Before joining Elastocraft in 1963 as a banbury mixer operator, Ernie Pack worked as a coal miner in his home state of Kentucky. In 1969, Pack, at the age of 39, was elected as the president of URW Local #626. As the president, he was still an employee of Wabash, accruing seniority and sharing in all of the same benefits as other hourly employees. Pack became a very popular president among the union workers, serving longer than any of his predecessors. In the 1978 and 1981 elections, Pack ran unopposed; his constituency looked upon him as a tough negotiator who was not afraid of management.

Many changes occurred under Pack's leadership. The membership grew to over 1,400 (with as many as 99 from the Peru plant and 50 from the other company). He became the first president ever to come to an agreement with management on a new contract without a strike. But Pack was not able to change as many things as he would have liked. At every contract negotiation since he became president, Pack asked Wabash's management if a union member could sit in on regular top management staff meetings, as a nonvoting member. This request was always rejected. In 1972 Pack proposed a nine-year union agreement, which guaranteed no strikes and a wage review every 18 months. If wages or any other issues could not be solved, an arbitrator would be brought in, with the union paying half of the costs. He made this proposal in an attempt to eliminate strikes every three years, which he felt hurt his constituents as well as the company. Strikes hurt the company because before the end of a contract, the company focused primarily on building a stock of inventory, which was expensive to carry. Workers were hurt because after the strike many of them were temporarily laid off until the strike bank was depleted and regular production schedules started to back up. Despite the reasoning, the nine-year proposal was also rejected.

JIM PLANT

In September 1981, Jim Plant was appointed plant manager for the Wabash plant and given the assignment to turn the operation around. Aware of Wabash's past five-year history with five different plant managers, Plant was not exuberant about the assignment.

Plant was a seasoned manager with experience in five previous plant turnarounds. He believed that his ability to work successfully on these projects stemmed from his having learned the rubber business from the bottom up. In 1947 Plant began his career as a banbury mixer operator in an Akron, Ohio tire plant. While working full time, he received his bachelor's degree and MBA through night school. In 1969 he joined the Wabash operation as the production superintendent in the plant's extrusion area. From 1972 to 1975 he served as plant manager of the Logansport operation. In 1975 he became the director of manufacturing, responsible for seven Elastocraft plants. Four years later, following a staff realignment by the new Elastocraft president, Plant was assigned to the position of plant manager of the Peru facility; a year later the Logansport plant was again added to his responsibilities.

Plant's first action at Wabash was a 10 percent across-the-board reduction in employment; he also replaced half the employees on the staff that he inherited. (See Exhibit 7 for the organization chart of the plant.) Then he set about restructuring the cost and measurement systems, discovering he had to rebuild the profit and loss statement from the bottom up. He also established daily measurements of dollars shipped, scrap, sales volume, products produced, and labor variances. The short-term effect of these changes was that by the end of 1981 the plant was at break-even. But the next two years were not forecasted to be as promising.

MARVIN ISLES

In May 1982 Marvin Isles was hired as Plant's immediate supervisor, with the title of president of the Engineered Elastomers division. Isles, age 35, had an MBA and majored in industrial engineering at General Motors Institute. He had also spent an additional seven years as an employee of General Motors. After leaving GM, Isles worked as a management consultant for a year. From there he went to work for an automotive supply

EXHIBIT 7 Organization Chart for the Wabash Plant, January 1983

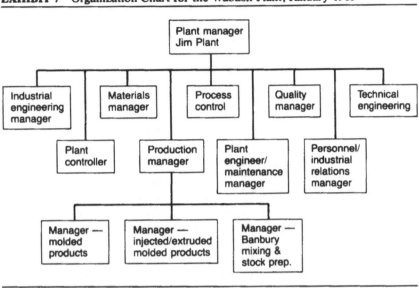

company where he was vice president of manufacturing and responsible for five plants. During his seven years there sales grew from $7 to $30 million, although for two of those years, the company had been close to bankruptcy. Following this stint, he began his career with International Products in November 1979 as the president of the Plastics Extrusions division. When he took over the division, it had sales of $20 million and was breaking even. Throughout his first year as the president there were constant rumors about the inevitable sale of the division because of its lack of profits. By the time he left, all the rumors had been squashed, as the division was realizing profits close to 25 percent of sales.

One of Isles' assignments was to make the ailing Elastomers division profitable. With this in mind, he saw the main issues for survival to be focused product lines, full use of the division's human resources, a strong technical base, improved quality, and an improved wage package. Isles began to take action immediately. Within six months of his hiring he replaced half of the 80 top division managers and ordered his staff to cut 20 percent of the division's employees.

He also made several changes to the division's strategic direction. He decided to refocus Wabash's product line on the auto industry, believing that it could supply the volume needed to turn a profit and no other single industry could offer that. Next, he embarked upon a multiplant supply strategy whereby there would be duplicate capacities across plants so that he would not become captive to any one facility and its union. He planned to balance this approach with his desire to focus each facility on specific products or processes.

But Isles realized further changes were needed. He felt that his toughest job would be teaching his operating staff how to manage in a competitive world. One approach he was considering was to implement an employee involvement program in one of his plants. From discussions with Kevin Benoit, Elastocraft's director of organization planning and development, he knew that top management would soon be encouraging divisional presidents to implement programs that got the workers more involved. Benoit had been hired in 1981 to develop and implement a training program for managers, teaching them how to manage human resources as assets. In Isles' mind this was employee involvement (EI), an approach he thought might be effective in the Wabash plant.

Isles' interest in EI dated to one of the first jobs he had with GM. He had managed a group of employees who were considered unemployable due to their poor attitudes, inconsistent work habits, and apathy. He found that once these workers were more involved in decisions that affected them, they became more productive. Thus, he suggested Plant do the same. Isles also believed Plant had a personable style that could be further developed by remembering employees' birthdays, congratulating them on company anniversaries, and making his walks through the plant more routine, so workers could depend on seeing him daily.

PLANT AND ISLES

Because divisional headquarters was located in Wabash, directly across the street from the manufacturing plant, Isles was easily accessible to Plant. Their operating styles tended to be complementary and they became a close-knit operating team. Plant felt extremely good about the support he received from Isles.

In conjunction with Isles' decision to refocus the product line, and with his approval, Plant decided to eliminate the unprofitable piston cup and seal product line that had accounted for $2.5 million in 1981 sales. They were extremely difficult products to make, requiring very high levels of maintenance. (In addition, they carried a high product liability because they were part of the brake system.) A savings of $500,000 was realized.

Plant and Isles also worked together in the 1982 labor contract negotiations. They sought and won a three-year freeze on wages, thereby terminating the cost-of-living adjustment (COLA). The wage freeze saved the company an estimated $500,000 (over the next three years), but was not sufficient to alleviate short-term losses. In July 1982, a month after the contract was signed, Plant had to reduce another 12 hourly and 17 salaried employees, saving another $500,000.

As a means of alleviating the waste problems Isles and Plant agreed that statistical process control (SPC) should be implemented. Thus, a training program on SPC, its uses and benefits, was developed for all supervisors and hourly employees. Despite all of these changes the Wabash plant continued losing money. In 1982 the loss was $2.2 million.

ANNOUNCEMENT OF WABASH CLOSING

During the first quarter of 1983 losses continued and in mid-March 1983, an additional 32 hourly (maintenance, shipping, and receiving) and 28 salaried employees were laid off. Although the layoffs contributed to a savings of $1 million, Isles announced to the union that due to continual financial losses the plant might have to be closed. He told them that he would prefer to remain in Wabash, but he had a viable option to relocate everything except the mixing operation to a new facility where wage rates would be lower. The mixing operation, which employed 44 people, would remain in order to supply compound to the Marion operation.

The union did not take this as a mere threat. With only 377 employees remaining (of which 280 were hourly) and the knowledge that International Products had recently relocated plants from Ohio and Pennsylvania to the South, resulting in 2,800 lost jobs, the union appeared willing to open negotiations for any kind of settlement that would keep the plant open.

In response, Isles asked Plant whether, in his opinion, additional wage concessions would be sufficient to turn the plant around. If Plant's answer was yes, he would have to supply a detailed plan of action on additional items needed to make the operation profitable.

The Simpson Pump and Valve Company

Mr. Bill Roth, general manager of Simpson's Warren, Indiana plant, returned to his office late in the afternoon of September 23, 1982, and wearily glanced through the pile of call memos, production reports, and notes from his secretary that covered his desk. It had been a long day, highlighted by the presentation he and his staff had made to a group of visiting corporate executives. He had at least another hour's worth of work ahead of him before he left his office and joined the group for dinner. And tomorrow, he noted as he examined his calendar, was going to be another busy one.

Some difficult decisions were going to have to be made in the next few days, he knew. Just a few minutes earlier he had confided to his personnel manager, Jake Thompson, as they walked back to his office after their presentation, "Two years ago I thought that if we survived the changes we were proposing then we could begin to relax a bit. But I'm beginning to realize that there's no end to the process we've set in motion. Looking back, the decisions we've made look like the easy ones. They're getting tougher as we go along!"

This case was prepared by Professor Robert Hayes as the basis for class discussion rather than to illustrate either effective or ineffective handling of an administrative situation.

COMPANY BACKGROUND

Simpson was the United States' largest producer of small water and fuel pumps for the auto, appliance, and industrial equipment industries. In mid-1982 it had somewhat over 40 percent of the U.S. market, but it was facing increasingly stiff competition from its two biggest U.S. competitors, as well as from Korita, a Japanese company. Worldwide it produced 12 million pumps (equivalent to about $200 million) per year, out of six plants. Three of them were located in the United States, one in Canada, one in Brazil, and one in Holland. It also had licensing arrangements with distributors throughout the world. One of these accounted for about 6 percent of the Japanese market.

Simpson's competitive vitality was much stronger than it had been three years previously, when it had been purchased by Gordon Enterprises, a large diversified firm whose total sales now exceeded $4 billion. At that time Simpson's situation had been deteriorating steadily for several years largely, it was felt, because of inbred management and overly conservative approaches to investment in new products and equipment. Simpson's top management group was therefore completely replaced, and Jim Marker was brought in to be its new vice president of manufacturing.

Marker was a tough, hard-driving man in his late 30s who had spent over 10 years in manufacturing with Chrysler. His assessment of Simpson's manufacturing capabilities when he arrived was gloomy: "The company was using basically the same type of equipment that Simpson had used for the previous 30 years. A substantial amount of that equipment, in fact, had come from its original plant in Simpson, Ohio—which had been closed down 10 years ago! In 1979 the three U.S. plants were dingy and dirty; housekeeping was poor, quality was suspect, and the level of efficiency was extremely low. They were also top-heavy with staff personnel. In short, they were typical of many industries in the U.S. today.

"Three years ago Simpson U.S. employed approximately 1,965 hourly workers to produce 30,000 pumps per day. In contrast, our major Japanese competitor was turning out about 30 pumps per day per worker, and they had half as many staff people supporting each worker. Moreover, Simpson had over 4 million pumps in finished goods inventory, and that inventory was growing daily. We were building more pumps than our salespeople could sell because our product was just too expensive. To offset some of these high manufacturing costs, we were importing 500,000 pumps from our subsidiaries in Brazil and Holland—because they could make these products, ship them from their plants, and land them in the U.S. cheaper than our U.S. plants could make them. To further add to Simpson's problems, Japanese competitors were shipping almost a million units per year into the U.S. market and that figure was growing steadily. All of our U.S. competitors were operating near breakeven, and as a result the marketplace was extremely price competitive. Our share of the U.S. market at that time was only 33 percent. Our return on capital was just 6 percent.

"Obviously, if Simpson was to survive it had to make some very sweeping changes; they had to be radically different from the approaches used in the past."

MARKER'S COMPARISON OF U.S. VERSUS JAPANESE IDEOLOGY

Marker's manufacturing philosophy had been shaped by two very contrasting influences. One was his former boss at Chrysler. "Technically, he was superb, and he was a born leader," Marker remembered, "but he sometimes pushed people beyond their limits. He just wasn't a very good manager of the 'people side' of manufacturing. Maybe that's why I have worked hard to emphasize that side of my managerial skills." The other influence was his exposure to a number of Japanese companies during the course of six trips to Japan made during his years with Chrysler.

"Japanese manufacturing operations maintain a policy of continual improvement. Tasks in the vicinity of 15–20 percent improvement are imposed, accepted and achieved each year. Industrial salvation is seen to depend more on the thousands of daily decisions made in countless factories and offices than any one step the government can take. In these efforts, it is important that senior managers not be perceived as being too distant from workers. They must be seen, and their policies understood. They must also be continually studying areas of the plant for potential improvements, and personally involve themselves in projects. This creates a good example, breeds respect, sets the policy of improvement, and breaks down barriers.

"But, above all, the key is the production worker. In the United States, from the time they are hired, workers are viewed basically as commodities. They are hired when sales levels increase, and promptly laid off when sales levels decrease. Even worse, when Simpson hired a person as a production operator, his or her job was implicitly defined as simply to produce parts. Those parts would be inspected by an individual from another part of the organization, the material for them would be brought to the worker by another individual from a different part of the organization, the equipment used in production would be designed by someone else, and it would be maintained by still another group. Workers weren't even responsible for setting up their own equipment.

"What Simpson was really telling their operators was that they had no responsibility for quality, for material, or for the operation of their equipment. Even worse, the company had totally alienated its production supervisors by removing from then any semblance of accountability for their area's proper functioning. Opportunities for conflict are everywhere in such a system. In fact, I came to call it 'The Production Conflict Syndrome' (see Exhibit 1A).

"In Japan, on the other hand, when a worker is hired into a company it is typically looked upon, by both management and the worker, as the beginning of a lifetime association. From the very outset Japanese workers are expected to make a substantial contribution to the quality and productivity improvement programs in their company. They are expected to analyze the process, the equipment, the facility, the material, and the product—and to propose improvements. Japanese organizations have developed an environment in which the worker is involved directly in the company's success or failure.

"U.S. management's attitude toward its workers has truly forced the creation of a third party—the union—to play the role of intermediary. Direct contact between management and workers in many instances has ceased to exist.

"Similarly, Quality Control in the U.S. company is a complex bureau-

---

EXHIBIT 1A The Production "Conflict" Syndrome

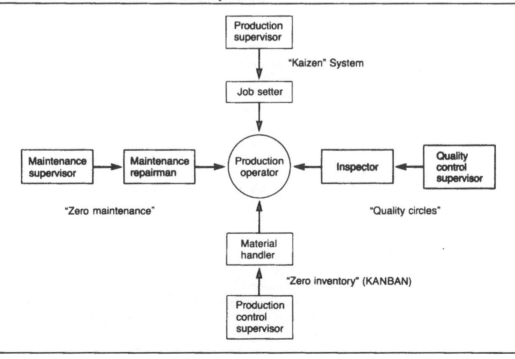

EXHIBIT 1B   The Participative Worker and Management

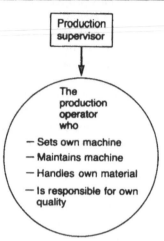

---

cratic function involving massive inspection routines. In Japan, it is the opposite: Workers perform their own quality checks. To help them, management provides workers with checking fixtures, gauges, and other quality verification media. More important, they are telling workers that they have full confidence in their ability to produce a quality product.

"A majority of U.S. industry also is not very good at analyzing the effect of, and implementing corrective action for, problems encountered in material flows. Seldom, if ever, does an hourly worker see a flow chart of his or her workplace. As a result, U.S. companies have much too large work-in-process inventories.

"In Japan, on the other hand, the flow of material within an operation is considered to be of primary importance. Each operation is analyzed, the location and amount of material are identified, and the distance the material is transported is charted—all of this by people in the plant. As a result of such analysis the Japanese have basically eliminated in-process inventory.

"Very seldom, if ever, do U.S. companies divulge the actual cost of a component part to the workers who make it. In general, employees are not aware, other than through the news media, if their company, let alone their plant, is profitable. Their ability to contribute to the profit and loss position of their operation is virtually ignored. In fact, in many instances when a company or a plant is in serious jeopardy and faces shutdown or bankruptcy, the hourly workers are the last to know. When a company does request assistance from its unions and labor force it is usually too late, and by that time there is often little left to save. So, during postmortems of the facility's (or company's) demise, labor can charge that management never intended to save it in the first place."

EFFECTING THE TURNAROUND

After studying the situation at each of the plants, Marker made a number of management changes. A new plant manager was hired from outside the company to take over the Ashton, Iowa plant and Bill Roth was promoted to the top job at the Warren plant. Roth had worked his way up through the plant organization, and had just turned 35 when given the job. "He didn't have a lot of formal education," commented Marker later, "but he was smart, worked like the devil and was a natural leader. Even more important, he was dying to make things better in the plant." A number of other changes in management personnel soon followed. Working with plant managers, Marker then began discussing Simpson's competitive situation with the workers. The message they tried to get across was that industry demand was stagnating, that new foreign competitors were entering the market, and that there just wasn't enough room for everybody. "If we were to be one of the survivors, we simply had to get our costs down." An MBO (Management by Objectives) system was established as a means for managing this cost reduction program.

In the first year of the MBO program, for example, over 40 staff personnel were eliminated from the Warren plant. Then attention turned to the hourly work force, in conjunction with an attack on the "Production Conflict Syndrome." During the second year, 15 quality control inspectors were released, and their place taken by four "Quality Assurance Auditors." The primary responsibility for quality was transferred back to the workers and their supervisors. A Quality Circles program was also set up, as a means for improving communication about quality problems.

"Each work station is now provided with the appropriate checking fixtures and gauges. Each operator is thoroughly trained in their use and in the reading of blueprints, and is given complete information about the level of quality expected for each operation," asserted Marker in mid-1982.

"The effect on our product quality has been encouraging. Our functional pump rejects have been reduced by over 40 percent and now represent less than 1.5 percent of the units produced. Our warranty cost per unit has been reduced 23 percent and one of our largest original equipment customers has informed us that our quality on their product is now so good that their pump warranty costs are almost immeasurable."

Concurrently, Marker was encouraging his plant organizations to work on work flow redesign, inventory reduction and maintenance. The key to these efforts was the "Kaizen" concept. "In Japan, the workplace layout is under constant review by a group of production workers called 'Kaizens,' who have full authority to alter layouts, processes, and methods. Literally translated, Kaizen means 'good change.' They are selected for their creativity, skill, and experience, and they typically serve in that capacity from 3 to 24 months, after which they return to their previous jobs.

"At Simpson we have made these permanent assignments and chosen people who have outstanding mechanical ability, motivation, and creativity. Their job is to identify areas requiring improvement, to cooperate with their fellow workers in determining what the improvement should be, and to help implement the change at Simpson.

"They were given the job of redesigning work flows and developing new processing equipment. They were also asked to design special lines for certain low-volume products. Two years ago the first 'Kaizen line' was started up at the Warren plant. The line was conceived, planned, and installed by the employees of the marine fuel pump area, in conjunction with their supervisors and two Kaizens. Today that line operates 32 percent more efficiently than did the previous work organization. There are now four Kaizen lines operating in our three U.S. plants (two of them at Warren), and all of them boast productivity gains in excess of 25 percent over previously established standards."

Managers, Kaizens, and workers also worked together to reduce the inventory in the system. "We divided the problem into four parts: raw materials inventory, parts purchased for resale, work-in-process, and finished goods. We were able to make reductions in all areas, but particularly the last two. The turnover of our finished good inventory, for example, has more than doubled and today our total inventory is 30 percent less, as a percentage of sales, than it was two years ago."

Obviously, the financial benefits associated with this inventory reduction were substantial, but the effects on work flow, working conditions and employee attitudes were even greater. "In the past, because we carried large amounts of inventory, the workers were surrounded by material, toiled in very poorly ventilated areas under apathetic lighting conditions, and in many instances had literally no perspective of what was happening in the rest of the plant. By reducing inventory levels the plants were able to remove most of the large racks and containers, which in many instances were stacked 16 feet high. In so doing we automatically improved the lighting and ventilation, and allowed the employees to observe their fellow workers in adjacent areas. More important, it created open areas which the plant could utilize for employee eating and recreation."

Efforts were also made to reduce material cost by producing internally parts that were formerly purchased from outside vendors. The designs of these parts were usually altered, and new production processes were used. In the presentation he made earlier that afternoon, Bill Roth had stated that "hardly a part made today is the same as it was three years ago, and over the past three years material savings of almost $7 million have been achieved through such efforts."

"The last segment of the 'Production Conflict Syndrome" is the maintenance problem," stated Marker. "We have devised a program which we call 'Zero Maintenance' for the purpose of eliminating the maintenance conflict. Historically, each of our plants carried a group of people categorized as maintenance personnel. Their sole purpose was to maintain and repair machinery. Now we believe that no one in the plant understands the machines better than the machine operator. Zero maintenance moves the maintenance function out of the maintenance department and into the production department, thereby tapping this capability. We are doing this by training each operator to perform the maintenance tasks on the equipment that he or she operates.

"This has benefit to the company, in that it allows machines to be repaired much faster, thereby reducing machine downtime. It also utilizes the machine operator in a productive way during the time that the machine is down. It has advantages to the employees also in that they are now much more highly trained and of significantly more value to themselves and to the company. Their pride in their own ability is significantly enhanced. We have recently instituted wage premiums to reflect the level of proficiency each worker has attained in this area.

"Admittedly this program is in its infancy. It will take us from three to five years to fully train all the operators in all our plants to perform all the maintenance of their machinery. However, substantial benefits are already being accrued in reduced machine downtime and lower maintenance support costs. For example, two years ago our three U.S. plants employed 120 maintenance personnel to produce 30,000 units per day. Today they employ 70 people to produce 33,000 units per day."

To reinforce these measures for improving cost, quality, and service to customers, Marker and his plant managers had been working to improve the morale and skill levels of their workers. The lighting in the plants was improved, walls and machines were repainted in colors selected by the workers, and straight broad aisles were opened throughout the plants. For the first the time families of the workers were invited into the plants as part of annual "Family Days" (planned and run by employees), which culminated in games, rides, and picnics. Supervisor and management training programs were also instituted to improve the professionalism of plant management.

In addition, each plant was developing sports facilities for its workers. The Warren plant, for example, had recently built a softball field in one corner of its land ("the best in the county," Roth had proclaimed it), and was contemplating building a complete "fitness center" in what had formerly been part of the finished goods storage area in the plant. This would include a basketball court, an exercise room, and eventually an indoor swimming pool.

THE SITUATION IN MID–1982

Jim Marker was justifiably proud of what he and his people had been able to accomplish over only three years. "While the sales in our industry have fallen by about 8 percent, our own sales have increased by over 7 percent. The major reason for our success is that we have been able to almost double our productivity—as measured by pumps per worker per day— and thereby keep the cost of our product almost constant during a three-year period when inflation has caused our costs to rise by over 30 percent. The wage rates we pay our workers have more than kept up with this inflation.

"But this increased demand has not been shared equally by all our plants. I have told them that they were in competition with each other, and the ones that were most successful in meeting their productivity goals would get the lion's share of the business. As a result, the Warren plant's production volume is up 50 percent over its level of three years ago. Similarly, the Ashton plant's production rate is up over 30 percent. On the other hand, our plant in Billings, Minnesota has experienced a drop in output and their work force is only a third of what it was in 1979."

During the past three years Simpson had seen three of its larger U.S. competitors withdraw from the business. One, a subsidiary of the General Appliance Company, had then signed a contract with Korita for over one million pumps a year. Another had closed down its production operations (its main plant was 150 miles to the north of Warren) and was buying pumps for Simpson for resale. "Both events had a beneficial effect," commented Marker. "First, they reinforced our message that the key to survival is continued cost reduction, and that the jobs of the many could only be preserved by eliminating the jobs of the rest. Second, losing that contract to Korita drove home that fact that we weren't safe yet. If the Japanese could beat us there, they could beat us in other places. We tried to get all our people to understand that we were serious, that anything was possible. Nothing was sacrosanct. The Warren plant is an outstanding example of what is possible.

THE WARREN PLANT

During the month of September 1982, the Warren plant had been operating close to its current capacity of 13,000 pumps per working day, equivalent to an annual sales volume of about $53 million. Although over 100 different models were produced, twenty accounted for 90 percent of the sales volume. About 500 people were employed at the plant, of whom 450 were production workers (see Exhibit 2).

The plant's average monthly sales, adjusted for inflation, were up 50 percent from their level in mid-1979 when almost 600 people had been employed. Moreover, roughly 10 percent of the work force was involved in activities (such as making parts in-house instead of purchasing them, and manufacturing products that had previously been produced in one of Simpson's other plants) that had not been carried on before. About four-fifths of the reduction in the work force had been obtained through natural attrition, and most of the remainder consisted of workers who had been fired for cause. Many of the workers who had been laid off at some point had eventually been rehired as the plant increased its rate of production.

One of Bill Roth's first activities upon taking over the plant had been to work on absenteeism. A three-stage warning process, culminating in dismissal, had been instituted and absenteeism in mid-1982 was running at only 2 percent, less than half of its level of two years earlier. Roth remembered his own attitude towards people who were habitually absent when he was working on the floor, and was not surprised that the tougher policy towards absenteeism seemed to be welcomed by most of the workers.

"We want to be able to provide lifetime employment to people who really want to work," he commented. "But first we have to get ourselves competitive. I think our workers understand this, and I haven't been able to detect any decline in morale because of the layoffs. They've seen our business expand, they've seen workers who had been laid off earlier hired back as openings became available, and they've seen their wages increasing steadily. We were one of the few plants in this region that was able to

---

EXHIBIT 2 Product Cost Structure (not adjusted for inflation)

	September 1979	*September 1982 (est.)*
Daily volume (units)	8,500	12,700
Number of employees	590	500
Fixed cost/unit	$ 1,212	$ 1,662
Variable cost/unit		
Labor and overhead	$ 3,685	$ 3,346
Material	4,940	4,328
Total manufacturing cost/unit	$ 9,837	$ 9,336
Other costs/unit	4,420	5,515
Total cost/unit	$14,257	$14,851

increase our wage rate more than the inflation rate this year." The average hourly wage at the plant was about $7, which was generally higher than the rate at other plants in the region.

In order to cushion the impact of layoffs, a strict plantwide seniority system had been followed: If 30 people were to be laid off, they would be those 30 who had the least seniority, no matter where they worked or what their skill level. This meant that each person laid off triggered a series of job changes, as the remaining workers bid for the open jobs. Roth estimated that four or five such changes occurred as a result of laying off a single person, and that it took at least two months for these changes to work their way through the plant.

A number of programs had been put in place to improve the quality of the products produced. Random samples were taken of products at various stages in the process, and demerits were assigned according to the severity of the defects observed. These demerits were translated into "Quality Ratings" for each department in the plant each month, and every six months the department with the highest quality rating was given a Merit Award and hosted to a steak dinner (cooked by Bill Roth and other plant managers). The plant's overall Merit Rating seemed to be improving over time, and its customer return rate was steadily falling. Roth estimated that the expenses associated with honoring Simpson's "Lifetime Guarantee" were over $2 million per year less than what they would have been given the warranty experience three years earlier. "And most of the products that are returned for warranty have absolutely nothing wrong with them," he stated. "The repairperson can't figure out what's causing the problems and so replaces the whole unit. The pump itself is usually not the problem."

DECISIONS FACING WARREN'S PLANT MANAGEMENT

Bill Roth's biggest concern was the scheduled layoff of an additional 30 people, including both staff and factory personnel, in early November. This had been agreed to by Roth and his management team as part of their MBO planning over eight months earlier. As is typical of most distant contemplated actions, it had not seemed as difficult then as it did up close. The Warren plant, in fact, had already absorbed successfully a series of mass layoffs, some of them larger than the one being planned. Moreover, in each of the past three years it had surpassed its MBO goals (usually with plenty of room to spare), even though at the time they were established they had seemed extremely difficult to attain.

As the number of people in the plant contracted, however, each layoff became more and more difficult to implement for several reasons. First, as the pool of low seniority people dried up, layoffs were increasingly affecting people who had been working at the plant for two or three years.

Many of these people were highly skilled and had become almost indispensable.

For example, the previous day the individual heading up his Manufacturing Engineering function had come in to talk to him about two people who were to be included among the 30 workers to be laid off. Both were in the plant's maintenance group, which had been able over the previous two years to transfer most of the plant's simple maintenance tasks to the work force and was therefore spending most of its time working on quite sophisticated projects: building new equipment or rebuilding existing machines so that they would work more reliably, produce higher quality products, or operate with less worker attention. These people were key to the plant's efforts to improve productivity and quality, and the two people who were slated to be laid off were among the most skillful in the whole group. Other workers with more seniority would bid for their jobs, of course, and would eventually achieve comparable skill levels, but in the interim the Maintenance Department's effectiveness would be crippled. The head of Manufacturing Engineering had argued that because of this situation in his area, as well as several other similar situations that he knew about in other areas, it was time to abandon the plantwide seniority system and move to some other system that would preserve management's flexibility to retain especially valuable workers.

The opposite problem would be encountered in the Shipping Department. These jobs were generally considered to be among the worst in the plant, because workers were subjected to continual pressure for delivery in the face of a widely fluctuating shipment schedule. It was almost impossible to automate the process, and little skill was involved. But it was essential to maintain morale and high standards in the department because a mistake there could not only neutralize all the good things that the rest of the plant had done, it could jeopardize the plant's customer relationships.

Most new workers, therefore, joined the plant in the shipping department and tried to work their way out of it as soon as possible. Because of this, a disproportionate number of the 30 people to be laid off would come from this group, and their place would be taken by low seniority workers who were being bumped out of their jobs in the chain reaction that would follow the layoff. Most of these new people would consider the move to the Shipping Department to be a clear demotion.

The case of Tom Rogers presented a similar dilemma. Roth noted that Rogers had requested an appointment to see him the following day, and knew what Rogers wanted to talk to him about. Rogers was one of his best, most dedicated workers. Since he had over five years of seniority at the plant he was not in danger of being laid off, but during the last layoff he had just missed being transferred from the day to the night shift (shift assignments were also governed by plantwide seniority). Rogers had come in to see him then and explained that such a change would be very difficult for him, because his wife had died tragically the preceding year and he was taking care of their three young children. This was possible as long as he was working the day shift, but a night shift assignment would impose great hardship on both him and his children. The fact that he wanted to see Roth again probably indicated that he had heard about the impending layoff and wanted to request that his special situation be taken into account when making the new shift assignments. Roth wondered how he should respond to such a request.

It was also becoming increasingly difficult to justify additional layoffs to the work force as a whole. Three years ago the plant's survival was clearly at stake. Demand was evaporating, workers were being laid off anyway, the the plant's profitability and return on capital were marginal. Sharing this financial and competitive information with the work force had been instrumental in enlisting their cooperation. Now, however, the situation was dramatically different. The plant was very profitable (its return on capital was almost 40 percent) and demand was increasing. In fact, his workers had been asked to work about 6 percent overtime during the current month in order to keep up with orders. Roth wondered what their reaction would be when the news about the new General Appliance contract was made public. He had just learned that afternoon, in the meeting with Marker and other corporate executives, that Simpson had won a contract for 1 million pumps during the next year from General Appliance, displacing Korita as its major supplier.

Yet Marker had insisted that continued productivity growth demanded continued layoffs, that the pressure imposed on the plant by taking workers away from it was the best way to motivate the search for more efficient ways of doing things. Reducing costs led, in turn, to increased sales which eventually allowed the plant to rehire most of those laid off. This kind of pressure had been effective in the past, but Roth was beginning to think that it might be time to relax that pressure a bit.

For one thing, it made unionization more attractive. Neither the Warren nor the Ashton plants were unionized (the Billings plant was), but the union had conducted unionization drives at Warren in each of the previous two years. Two years ago, 70 percent of the workers had voted against the union, but last year only 58 percent had voted against it. The plant was likely to encounter another effort in the coming year.

Moreover, his plan to build a comprehensive recreation facility was becoming increasingly difficult to justify. Relatively little money had been spent on building the softball field, but the gymnasium and fitness center would cost almost $200,000, and the proposed swimming pool would add at least another $100,000 to the total. He knew that it would be awkward to justify such expenditures to his workers after carrying out a massive layoff, particularly since they could point out that less than 30 percent of the workers currently made any use at all of the new softball diamond. Yet Marker felt that such facilities were essential to promoting the "family atmosphere" at the plant, and to making it a more rewarding place to work.

The rate of capital expenditure, in fact, was of increasing concern to Roth. Because of the pressure to reduce the work force, he had recently authorized funds to purchase (or build) equipment that would eliminate 10 jobs, primarily in the assembly area. Yet it had been impossible to justify this expenditure using the company's standard capital authorization procedures. He reviewed in his mind a recent conversation with his financial director, who had pointed out that in 1979 each dollar that had been authorized for capital expenditures had promised almost $3 in annual savings. By 1981 each dollar invested had promised only $1.50 in annual savings, and this year it had dropped below one-for-one. As he and his managers "skimmed the cream" of cost-reducing investments, the plant's investment base (which had been relatively low three years earlier because so much of its equipment had been almost fully depreciated) was rapidly expanding. Roth expected that this would soon have a detrimental effect on his return on capital, which was still one of the most important measures used to evaluate the plant's performance.

Roth also was beginning to note a distressing "stickiness" in his quality improvement and inventory reduction programs. Both had achieved significant improvement during their first two years, and both seemed to achieve only marginal improvements (and grudgingly at that) during the past year. Roth was beginning to think that both programs should be reviewed and altered, perhaps substantially. Major improvements in quality, productivity, and inventory reduction, in fact, might only be achievable through modifications in the approaches that had been used up to this point.

Modifying the approaches would take a considerable period of time, he knew. Yet the announcement of the proposed layoff had to be made within the next two weeks if the workers affected were to be given their usual three weeks notice. As he rose to leave the office for the night Roth noticed a picture of himself on the wall. It had been taken by one of his employees on the day his promotion to plant manager had been announced. He noticed, with a start, how much younger he looked in that picture than he had in the mirror that morning.

New York Malleable & Ductile

I learned more in the two years since I joined New York Malleable than I did in the 15 years since I graduated from college. It's been an extremely interesting but difficult situation. We are almost out of the woods on the financial problems so it's now time to fine tune the operation. When I took the job of operations manager three months ago, my assignment was to get the plant organized and to get manpower and scheduling under control. I've been so busy fighting fires that I haven't had time to sit down and analyze the operation and get to the root cause of some of our more pressing problems.

In October 1983, Ken Johnson joined New York Malleable & Ductile (NYMD) as the accounting manager. A year later, NYMD's operations manager walked out of the plant in the middle of a shift and quit in frustration over the president's management style. Because the company was short of cash, the operations manager was not replaced and Don Strasburg, president of NYMD, assumed the responsibilities. With sales declining and operational expenses on a rise, however, Strasburg found that he did not have enough time to devote to both sales and operations and decided to put Johnson in charge of both manufacturing and accounting, effective May 1, 1985. See Exhibit 1 for an organizational chart.

NYMD, a manufacturer of small malleable and ductile iron castings, was a wholly owned subsidiary of Prince Corporation, a small independently-owned company. NYMD was the second foundry Prince had purchased through a leveraged buyout. Prince, founded in 1971, owned three foundries in the United States and one in England. NYMD was Prince's least profitable foundry.

This case was prepared as a basis for class discussion rather than to illustrate either effective or ineffective handling of an administrative situation. Names and some data have been disguised.

EXHIBIT 1 Organization Chart, July 1985

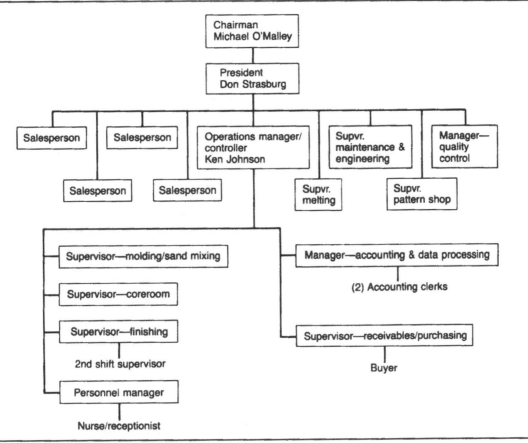

When Johnson took on the manufacturing job he knew that one of his major tasks was to get labor costs under control. The second quarter results (see Exhibit 2) heightened the urgency of the task. The hourly labor content (as a percentage of sales) had risen over six points. Johnson was unsure where to begin to attack the problem—whether the increase was due to the continual fluctuations in staffing, the supervisors, the work force, the production mix, or some other factor.

NEW YORK MALLEABLE & DUCTILE

NYMD began as a lock company, called Carbon Malleable, in Jamestown, New York in 1889. The plant, which slowly evolved into a jobbing foundry, was run by the initial family owners until 1945 when, close to

EXHIBIT 2   1985 Production Data

	1st Quarter	2nd Quarter
	Hourly Labor Content	
	(percent shipped to customer)	
Department		
Melting	1.24	1.57
Core room	3.03	3.23
Molding	7.81	8.75
Disamatic	1.65	2.02
Heat treat	.83	.99
Finishing	8.30	12.37
Total	22.86	28.93
	Dollars Produced	
Malleable	$1,592,926	$1,499,219
Ductile	21,886	232,886
Total	$1,614,812	$1,732,105
	Dollars Shipped to Customer	
Malleable	$1,449,081	$1,153,529
Ductile	202,636	345,072
Total	$1,651,717	$1,498,601

bankruptcy, it was purchased by Leo Spegiel, Sr. and Charles Lawrence. For four years, Spegiel and Lawrence commuted daily from Buffalo, New York (75 miles north of Jamestown) to manage the foundry, and because of this commuting time, the plant ran without managers for several hours each day. The union and workers became quite independent. In 1949, an operations manager was hired to "take the foundry back from the union." The foundry workers were represented by Local 389 of the International Molders' and Allied Workers' Union (AFL-CIO). The local also represented workers at five other foundries in the Jamestown area.

The early 1950s brought a name change to New York Malleable and the addition of a shipping room and docks. With the foundry working seven days a week for three consecutive years, the owners decided to build a new foundry in Blockville (10 miles west of Jamestown). Ignoring industry trends that indicated a downturn, they completed the new facility in 1956. During the next four years, the new foundry laid idle due to lack of volume as often as it was in operation.

On January 1, 1969 New York Malleable was purchased by Eastern

Manufacturing and Sales (EM&S). Spegiel and Lawrence were close to retirement, and the foundry required a large capital investment either to convert to electric melting or install air pollution controls to comply with newly passed environmental regulations. In 1970, EM&S invested $4 million to install an electric melter, a semiautomatic molding line, and an overhead sand conveyor system. The principal products were very small castings with an average weight of about one-quarter pound.

EM&S was organized under a "big brother system" whereby small plants, like New York Malleable, were managed by larger facilities. During the 1970s and early 1980s, overseer responsibility for the foundry shifted from one big brother to another. Beginning in the mid-1970s, the foundry also went through a series of plant managers. The first one resigned in frustration after a year and a half; the next remained until 1980. He was able to make the foundry profitable but at the expense of reinvestment back into the facility for maintenance and replacement of equipment. The next plant manager tried to rebuild the foundry (reinvest in equipment and improve the morale of the work force), but left after six months. At that point, New York Malleable was losing $200,000 each month on a monthly sales base of $680,000. EM&S began looking for a buyer to divest itself of the facility.

The metal casting industry was in a state of decline. Before the 1960s, the industry had been dominated by many small foundries with 30 to 35 foundries holding 40 percent of the market. Many smaller foundries did not have the money to invest in the pollution equipment required as a result of the environmental and safety regulations enacted during the 1960s and went out of business. As a result, in the 1980s, those same 30 to 35 foundries accounted for 75 percent of the tonnage produced.

New York Malleable specialized in castings weighing from one-half ounce to three pounds (see Exhibit 3), with the ability to produce 20-pound castings. Its primary competition was three small domestic foundries (one of which was also located in Jamestown, New York) and several Canadian and Japanese firms. Although General Motors operated one of the largest foundries in the United States, it was one of New York Malleable's larger customers because what GM considered to be small-volume, specialty castings, were high-volume items for New York Malleable.

PRINCE CORPORATION

Mike O'Malley, a son of an Irish immigrant who owned a small tavern in a Chicago suburb, graduated from Notre Dame with a degree in accounting. Following several years in the military, he joined Price Waterhouse as an auditor. Believing he could make more money in sales, he left Price Waterhouse after one year to become a sales representative for several

EXHIBIT 3   Advertisement Showing a Sample of NYMD's Product Line

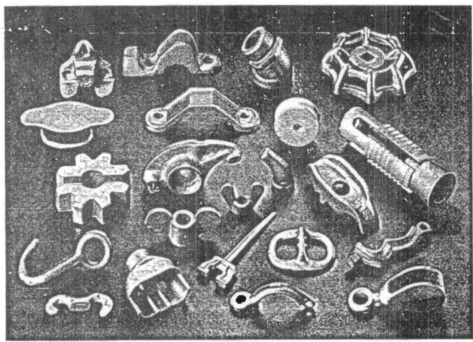

small is beautiful

small??? less than three pounds malleable specialists half ounce — two pounds

We've designed and modernized an entire foundry to manufacture your small castings in gray, malleable and ductile iron. To maintain our leadership in the small castings market, we have installed a new Disamatic molding line, complete with core setter.

castings weighing less than 20 pounds

Our variety of molding lines allows us to offer our customers the unique opportunity of purchasing virtually any size release quantities and our machining facilities will finish your castings complete . . . you don't even have to touch them. Save time and money, let us handle your service requirements.

We are specialists who know and understand small parts. Our unique sand properties insure surface finish previously unavailable in the foundry industry.

small castings are big business to us

casting manufacturers. After landing a couple of major accounts that provided steady income, he decided to buy a foundry. He read about one for sale in *The Wall Street Journal*. Located in northern Minnesota and built in 1968, the foundry had gone bankrupt and was owned by the federal government. O'Malley walked in and put a bid in on it. Since he was the only bidder, his offer was accepted and Prince Corporation was thereby created in the summer of 1971.

O'Malley hired Don Strasburg, a metallurgical engineer with 12 years' industry experience, to manage the foundry. Five years were needed to make the foundry break even, but thereafter it was extremely profitable. In May 1981, O'Malley decided there was enough excess cash to purchase another foundry. New York Malleable became his prime candidate because of its equipment and because it could be acquired through a leveraged buyout.

In October 1981, the purchase was close to being finalized and O'Malley sent Strasburg, who was to become president of the new subsidiary, on a house-hunting trip in the Jamestown area. The deal fell through, though, because they could not finalize it by year-end, one of O'Malley's initial stipulations for tax purposes. O'Malley then went back to EM&S and informed them that because of the foundry's high operating costs, he could not consider the purchase unless EM&S negotiated concessions from the union. These were achieved in March 1982, with the union agreeing to cut wages by $1.83 per hour.

The union agreed to concessions, in part, to prevent EM&S from selling the foundry to Prince. O'Malley had been counselled by his lawyers to "play hardball" with the union. At his initial introduction to the union membership, he walked into the union meeting "yelling and screaming and swearing." The union membership was so outraged that when EM&S promised if they agreed to the wage cut Prince would not be the buyers, they quickly ratified the concessions.

O'Malley also told EM&S that there would be no deal unless EM&S consolidated its two foundries into the Blockville facility and added another holding furnace in Blockville. This was done in the summer of 1982, and on September 1, 1982, Prince acquired New York Malleable as a wholly owned subsidiary for an aggregate purchase price of $7,240,042, including the assumption of pension withdrawal liabilities of $1,012,721.

NEW YORK MALLEABLE UNDER PRINCE'S OWNERSHIP

At the time that Prince acquired New York Malleable, the casting industry was experiencing a significant decline, due in part to foreign competition. Total industry sales declined from 20 million tons in 1979 to 9 million in 1983. The decline was also a result of a shift from iron to aluminum castings and the recession in the auto and truck industries, one of New

York Malleable's primary markets. Taiwanese foundries had a $10 per hour labor cost advantage over United States manufacturers. Freight, however, totaled $140 per ton, while electricity costs in Taiwan were double that in the United states, adding another $60 per ton. O'Malley believed he could successfully compete with foreign manufacturers by reducing the labor content of the foundry's output (which ranged from 45 to 85 labor hours per ton) to fewer than 20 labor hours per ton.

His immediate strategy was to convert New York Malleable from a low-volume, labor-intensive job shop into a high-volume, low-labor foundry. To do this, he purchased and installed a disamatic (automatic) molding machine. There were only 10 disamatics in use in the United States, and O'Malley believed the machine would provide him with the competitive edge he needed. In addition, he expanded the product line to include ductile iron,[1] which was experiencing a moderate market growth, and changed the name of the company to New York Malleable and Ductile (NYMD). (See Exhibit 4 for industry sales trends for malleable and ductile iron.)

Although Strasburg had been named president of NYMD and assigned responsibility of the entire operation, O'Malley spent 100 percent of his time during the first few months in the plant. He knew there were some difficult employment decisions to make, and he wanted the work force to hold him accountable for them rather than Strasburg. The first thing he did was to lay off 30 of the 70 salaried personnel. Unfortunately, an invitation to a dinner for those people who would be staying on the payroll was circulated to the invitees before the termination letters had been distributed. When O'Malley and Strasburg tried to find the people to give them their notice, many had already left. Two months later an additional 15 were laid off. EM&S had required New York Malleable to comply with its corporate bookkeeping procedures and to file extensive reports to their headquarter's staff. As a result, a large portion of the initial personnel reductions occurred in the accounting department. In addition, there was a reduction in operating personnel with the consolidation of the two foundries. The hourly work force was cut from 230 to 188 by year-end 1982. This was later reduced to 98 in June of 1984.

The second major action O'Malley immediately took was to raise prices, across the board, by 30–35 percent, and to refocus the sales effort on the higher-volume automobile and truck industries. Although NYMD lost some business, much of what was lost had been selling at a loss. Because EM&S had been unwilling to open its books to Prince prior to

---

[1]Malleable cast iron required prolonged annealing at temperatures around 1,600° F and very slow cooling to produce a uniform material of moderate ductility which could be machined at a low cost. By adding chemicals to the molten iron, ductile iron was less expensive to produce because it did not require the annealing cycle.

EXHIBIT 4 Industry Sales Trends

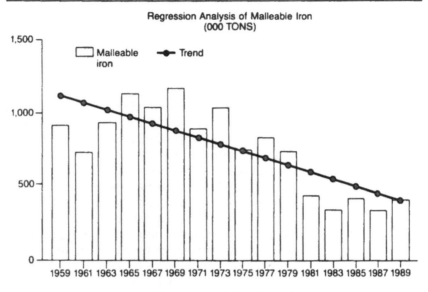

Regression Analysis of Malleable Iron
(000 TONS)

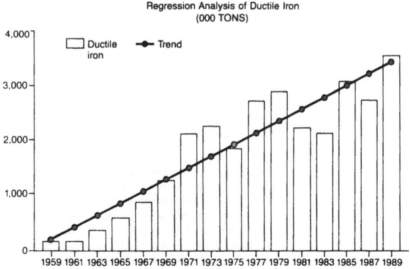

Regression Analysis of Ductile Iron
(000 TONS)

the final purchase agreement being signed, it had guaranteed Prince a specific level of sales from particular accounts. Unknown to Prince, however, part of that business was expected to be lost within a year, even without the price increase.

Due to operational losses and declining sales, NYMD was in continual cash-flow difficulties. By February 1984 the union went on strike, demanding reinstatement of their 1981 concessions and participation in an industry-wide pension plan. O'Malley was strongly opposed to the latter because he believed that the few industry survivors would be held liable for the rest of the foundries that went out of business. Although the wage concessions were ultimately given back, the union accepted a new pension plan. NYMD used the three-week strike to reduce inventories in the event the foundry was given back to EM&S.

NYMD's financial position quickly deteriorated further after the strike, to a point in August 1984 when the banks stopped payment on all withdrawals, including payroll checks. New York Power & Light threatened to disconnect electric service, and payments of all other obligations were past due. O'Malley threatened to sue EM&S for breach of contract on the guaranteed sales levels, which he believed had been misrepresented. EM&S agreed to restructure the acquisition agreement to include a 10-year interest-free loan and a cash contingency whereby payment was waived if there were a negative or zero cash flow. Meanwhile, O'Malley worked with an accounting firm to restructure the company and, in essence, went through bankruptcy procedures without formally filing for bankruptcy. Prince advanced $300,000 and EM&S advanced $350,000 to allow NYMD to continue operations. The plan was drawn up in August 1984, and O'Malley contacted all his creditors, to whom he owed a total of approximately $1.4 million, informing them that payment could not be made. The new financial plan proposed profitability by 1988, and by May 1985, initial progress had been made with a positive cash flow of $45,000 per month on a monthly sales base of $650,000. (See Exhibits 5 and 6 for a summary of financial results.)

Immediately after the restructuring in August of 1984, O'Malley spent 30 to 60 minutes with every hourly employee individually and asked, "What would you do differently if you ran the foundry?" He noted:

> It took a lot of time, but it was well worth it. Although a few of them were like talking to a casting, about 50 percent of them had good, solid ideas. I purposely interviewed the hourly people first because I didn't want to go into the interviews with any preconceived ideas. I had planned to talk with the salaried folks next, but I never got around to them. Don and I had a disagreement over the management of the foundry and what I perceived to be a lack of effort in gaining sales of ductile castings. In January (1985) I decided to let Don have a go at it in running the place, and I left to get out of his way.
>
> While I was there, though, we did a couple of things to try and get people excited about NYMD. There was a tradition of having an annual picnic, but the caterer we had the prior year never got paid and we had no money to hire someone else. I rounded up a couple thousand dollars just to cover the food costs and everybody pitched in to do the cooking. We also sent a bus load of hourly folks up to see one of our biggest customers. They saw a crate of castings coming into the shipping dock from Japan and got the message as to what we needed to do to make NYMD competitive.

EXHIBIT 5   NYMD Balance Sheets (December 31, 1983–84)

	1984	*1983*
Assets		
Current assets	$1,412,000	$1,793,000
Plant and equipment	3,492,000	4,476,000
Total assets	$4,904,000	$6,269,000
Liabilities		
Current liabilities	$1,796,000	$2,545,000
Long-term liabilities	3,427,000	6,814,000
Other	3,125,000	0
Equity	(3,444,000)	(3,090,000)
Total liabilities and equity	$4,904,000	$6,269,000

EXHIBIT 6   NYMD Income Statements (Years Ending December 31, 1983–84)

	1984	*1983*
Net sales	$5,947,000	$5,807,000
Cost of sales	5,891,000	5,778,000
Gross profit	56,000	29,000
General and administrative	1,853,000	2,045,000
Operating profit	(1,797,000)	(2,016,000)
Extraordinary gain	1,864,000	0
Net profit	67,000	(2,016,000)

My friends keep telling me to get out because NYMD will never make it. I know we have a disaster on our hands, but I believe in long-shot betting and plan to stay. I really believe American industry can make it and if I give up, it would be giving up hope. I'm concerned, though, about sales, which are down 5.6 percent from last year at this point.

THE MANUFACTURING OPERATION

The foundry operations were housed in the 148,000 square foot building in Blockville. (The original foundry in Jamestown, covering 110,000 square feet, had been converted to a warehouse.) As new operations manager, Ken Johnson assumed responsibility for the core room and the molding and finishing operations. Strasburg, who was also the foundry's technical expert, decided to retain control of quality control, the pattern shop, the melting operation, and maintenance and engineering. (See Exhibit 7 for a process flow diagram.)

Each casting required a pattern to create a mold cavity, which was a replica of the final product with a shrinkage allowance for the metal as it solidified and cooled. The pattern, made of wood or metal, also included a gate-and-runner system that permitted the molten metal to enter the mold and all air or gases to escape from the mold cavity prior to its being filled by the metal. The pattern shop, comprising four pattern makers and a supervisor, was in operation from 10 at night until 1 o'clock in the afternoon. (See Exhibit 8 for the distribution of employees and the typical work hours.)

Any casting containing hollow or reentrant sections also required cores. During the molding process, the cores were positioned within the pattern to block molten iron from areas that were to become holes in the casting. Cores were made of a sand mixture containing a binding material which was packed into a core box that contained a cavity of the desired shape. Since many of NYMD's castings required multiple cores, the core room, employing 17 people, provided 24-hour coverage for the foundry operations (melting, molding, and finishing).

The casting process began with the melting of steel scrap to produce the molten iron. For ductile iron, chemicals were added to the molten iron to produce the desired metal properties. There were two melting furnaces, which ran alternate weeks (when one was running, the other was being repaired), and two holding furnaces. The holding furnaces ran around the clock, every day of the year, except for an annual shutdown for maintenance and repair. The melting furnaces, however, were run primarily at night to take advantage of off-peak electricity rates. Each melter could produce 3.3 to 3.7 tons of molten iron per hour.

EXHIBIT 7   Manufacturing Process

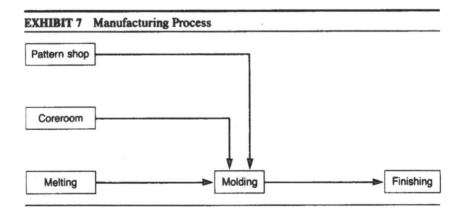

EXHIBIT 8   Hours of Work

Pattern Shop
  Supervisor   6:00 A.M.— 3:00 P.M.
  3 patternmakers   5:00 A.M.— 1:00 P.M.
  1 patternmaker   10:00 P.M.— 6:00 A.M.

Coreroom
  Supervisor   6:00 A.M.— 3:00 P.M.
  5 coremakers   10:30 P.M.— 6:30 A.M.
  7 coremakers   6:30 A.M.— 2:30 P.M.
  5 coremakers   2:30 P.M.—10:30 P.M.

Melting
  Supervisor   5:00 P.M.— 1:00 A.M.
  1 melter and 1 helper   4:15 A.M.—12:15 P.M.
  1 melter and 1 helper   12:15 P.M.— 8:15 P.M.
  1 melter and 1 helper   8:15 P.M.— 4:15 A.M.

Molding
  Supervisor   4:00 A.M.— 1:00 P.M.
  Group leader—disamatic   9:00 P.M.— 5:00 A.M.
  30 molders or pourers   4:15 A.M.—12:15 P.M.
  13 molders or pourers   12:15 P.M.— 8:15 P.M.
  12 molders or pourers   8:15 P.M.— 4:15 A.M.

Finishing
  Supervisor—first shift   6:00 A.M.— 3:00 P.M.
  Supervisor—second shift   3:00 P.M.— 1:00 A.M.
  12 soft iron sorters   6:00 A.M.— 3:00 P.M.
    (grind, finish, & inspect)
  10 plug and shell sorters   6:00 A.M.— 3:00 P.M.
  2 weight masters   6:00 A.M.— 3:00 P.M.
  2 quality checkers   6:00 A.M.— 3:00 P.M.
  1 truck driver   6:00 A.M.— 3:00 P.M.
  1 press operator   6:00 A.M.— 3:00 P.M.
  9 soft iron sorters   3:00 P.M.— 1:00 A.M.
  10 plug and shell sorters   3:00 P.M.— 1:00 A.M.
  4 furnace operators   7-day rotating shift
  4 blasting operators   7-day rotating shift

Maintenance & Engineering
  Supervisor   6:00 A.M.— 3:00 P.M.
  1 electrician   4:15 A.M.—12:15 P.M.
  1 mechanic   4:15 A.M.—12:15 P.M.
  1 group leader   7:00 A.M.— 4:00 P.M.
  1 all-around maintenance operator   7:00 A.M.— 4:00 P.M.
  2 helpers   7:00 A.M.— 4:00 P.M.
  1 all-around maintenance operator   12:30 P.M.— 8:30 P.M.
  1 electrician   2:00 P.M.—10:00 P.M.
  1 all-around maintenance operator   9:00 P.M.— 5:00 A.M.
    (and part-time disamatic operator)

The molten iron was transferred to the molding area by ladles, which were suspended from an overhead track and had to be manually steered and pushed. Four methods were used to pour the castings:

1. When the foundry had first opened, all pouring was done by a hand ladle on the "side floor." The area had enough space for 10 pourers, each of which could pour 2,200 pounds of molten iron per shift. In terms of the final product (after removal of all excess iron, such as gates and runners), this was equivalent to 900 pounds net weight of usable castings per pourer per shift. Since the side floor operation was extremely labor-intensive, NYMD had discontinued its use.

2. The majority of the molds were poured on the "pallet line," in operation since 1970. A molder formed a mold by forcing a sand mixture (delivered to the pallet line via an overhead conveyor system) around the pattern which had been placed within a metal or wooden box, or "flask." When the flask was separated and the pattern removed, the shape of the casting remained in the sand. If the casting required holes, one or more cores were inserted into the mold. The molds were then lined up, and a pourer filled them with molten iron. The pallet line had the capacity for 11 pourers who each poured 4,400 pounds per shift (2,100 pounds net weight per pourer per shift), but the overall capacity of the line had been reduced to accommodate the hand-pouring of some smaller jobs which had previously been poured on the side floor. In addition, one of the work stations was inoperable.

3. Although the automated disamatic molding machine had been in operation for almost two years, due to start-up problems it had only been utilized 15 percent of the time during the first year; its current utilization rate was about 65 percent. The disamatic had the capacity to produce 200 molds per hour with an average mold weight of 30 pounds. (The net-weight yield on the disamatic was approximately 52 percent.) NYMD operated the disamatic on a two-shift basis when producing ductile iron but only one shift for malleable iron.

4. The foundry also had a Breadsly Piper automated molding machine for larger castings. NYMD had not used it for over two years, however, because it was extremely inefficient.

Once the iron cooled, the molds were moved to a shaking area to remove the sand, leaving only the casting. In the finishing area, excess metal was knocked, blasted, or ground off. Malleable castings then proceeded to the annealing and heat treating ovens. Every 30 minutes, a tray of castings was pushed into an oven, which automatically pushed other trays within the oven along a conveyor belt. After 20 hours, a tray exited the other end of the oven. Castings were then straightened, cleaned, inspected, and shipped.

EXPANSION SHELLS

The production of expansion shells for the mining industry accounted for over 20 percent of NYMD's 1984 sales. The expansion shell (see Exhibit 9), used to hold up mine shaft roofs, had been produced in the Jamestown foundry for many years. At about the same time that Prince purchased NYMD, Bethlehem Steel, formerly a prime supplier of the shells, went out of that business. O'Malley considered expansion shells an excellent candidate for production on the disamatic and purchased Bethlehem

---

EXHIBIT 9   Description of Expansion Shells

Expansion Shells

C Series Shells

The Bethlehem C series of expansion shells are four-leaf type requiring a separate support device to assist in the preliminary stages of expansion. Once the shell is initially anchored or "set," the support device is no longer needed. As bolt rotation pulls the forged-steel plug deeper into the shell, the carefully designed relationship between leaf and plug taper causes the shell to provide optimum anchorage. After the shell is anchored, continued rotation develops the desired tension in the bolt.

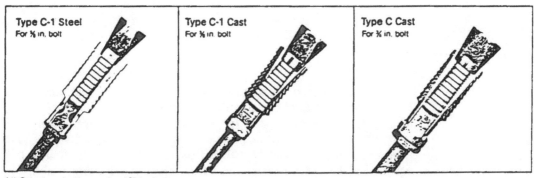

| Type C-1 Steel | Type C-1 Cast | Type C Cast |
| For ⅝ in. bolt | For ⅝ in. bolt | For ¾ in. bolt |

All C series shells require 1⅜ in. hole.

Steel's tooling. NYMD management believed, however, that they could make their own patterns in-house for the shells.

To increase sales and market share, NYMD priced its expansion shells 10–15 percent lower than the market price. They then attempted to produce the shells on the disamatic. O'Malley noted, "We basically ended up putting junk into the field because the shells were one of the first products we tried to make on the disamatic and we didn't know what we were doing." This resulted in a loss of business, and management finally decided in January 1985 to go back to the old method of manually pouring the shells on the pallet line.

Since the expansion shell project had been run by the operations manager and the metallurgist, both of whom had left in late 1984, no cause for the production problems was ever identified. (Like the operations manager, the metallurgist, also an MBA, was frustrated and left after several disagreements with the president. In late April, a process engineer from another defunct foundry was hired to replace the metallurgist but was assigned the task of introducing statistical process control into the foundry.) Supervisors and workers, who emphasized that they had never been asked for their input, gave the following opinions for the failure of the project:

1. The disamatic produced the shells at a much higher pace than the manual process, and the cooling line was not long enough for them to cool properly. As a result, when the castings hit the shakers, they cracked. To lengthen the cooling line, though, would require knocking down a wall and a major rearrangement of the equipment.
2. When the shells came out of the molding process, they were like glass and very susceptible to breakage. To avoid warping, the finishing operators had to position carefully each shell in a tray prior to annealing.
3. The same patterns had been used for 15 or 16 years and had begun to wear. In addition, it was felt that some major pattern changes were needed.

MANUFACTURING COSTS

At the time Johnson became operations manager, the material cost for each product was easily calculated because operators recorded both pour and net weight measurements for each casting. On average, materials accounted for 10 percent of the product cost. Actual labor costs by product, however, were not available. Based on total labor costs, hourly labor equalled 30 percent of sales, and when benefits and salaried labor were added, the number rose to 50 percent. Hourly staffing levels were determined by the production scheduler, who calculated the number of workers needed each day based on his estimation of the number of molds produced per worker.

Labor standards were determined by asking each supervisor how many pieces an operator in each department could produce per hour. In the molding department, the supervisor was able to track operator efficiency since the molders and pourers were on an individual incentive pay system, but the data were not used to determine standards. Similarly, a portion of the finishing department was on either individual or group incentives. There was no procedure, though, to track operator efficiency in the rest of the foundry. Johnson tried to have the supervisor in the finishing area make each operator log in how much time was spent on each piece, but the supervisor resisted, stating it would take too much time and would require an extra person to keep the log.

Because Johnson did not have detailed cost data, he asked his supervisors and some of the hourly employees (most of whom had worked in the foundry since the Spegiel and Lawrence days) to explain the sudden increase in labor costs. What he heard was a wide range of potential causes.

1. When the foundry was scheduled to shut down for the month of August for its annual furnace rebuild, Strasburg instructed the operation to clean up all work-in-process. (The foundry typically kept three weeks of work-in-process inventory for malleable products and one week for ductile, which totaled approximately $350,000.) Cleaning up meant putting labor into products that would sit in finished goods inventory and not be shipped until September. This had a significant impact in the finishing department, which was one of the most labor-intensive areas. In addition, the salespeople had begun to require the operation to hold a minimum amount of expansion shell inventory. Except for the shells, NYMD held very little finished goods inventory.
2. Since NYMD's survival was based on maintaining a positive cash flow, all measurements were based on sales volume. Strasburg was continually quoted as saying, "I don't care what it takes to get production up to so many dollars per day." Therefore, although supervisors were concerned about labor efficiencies, their primary concern was meeting daily shipment forecasts. Sales volume, however, was not viewed to be the proper measure for how efficiently the operation was running. It was too dependent on the price of the casting, which was directly related to its weight. A $1 casting required about the same labor as a $2 one. (The selling prices ranged from 24¢ to $10 per casting, with a minimum order price of $700.) Johnson felt that the tons produced per day had probably increased over the past year but because prices had fallen, the sales per day had declined.
3. Many attributed the rise in labor costs to a change in product mix. Some noted that lower-priced castings had been produced and, hence,

NYMD's total sales volume had dropped. Others cited the fact that some castings required multiple cores while others used none. Therefore, those castings that used several cores required more labor hours. The increase in the volume of shells also accounted for more labor usage. Finally, there was the mix between ductile and malleable iron. Due to the need for heat treating and additional finishing operations, malleable production was more labor-intensive. In addition, each time a changeover occurred, production time could be lost. Since approximately an hour and a half was needed to melt new iron in the furnace, the disamatic often had to stand idle. Ductile iron casting was also a new process for the foundry and was still in an experimental stage. As one molder noted:

> We've only been making ductile for a little over a year, and we still don't understand it. We lack a lot of the basic knowledge. When problems arise, there is no one to go to. There is no one with authority to make a decision when something goes wrong with the mixture of metal. We, the molders, don't know what to do because we really don't understand why you put certain chemicals into the mixture. I ask a lot of questions so I'm learning, but most don't bother to ask. There is very little communication between workers. If you want a different mix of sand, you have to shut your machine down and go back and tell the person who is mixing the sand. In addition, there is no way to go and get maintenance help, other than to stop everything and run after somebody.

4. Due to the lack of cash, NYMD had not been able to correct many of the equipment and building problems it had inherited from EM&S. There were no spare parts to fix machines when they broke down, which was frequently, and there was no time or people to do preventive maintenance. The maintenance personnel had to improvise ways to keep the equipment running, and between the time a machine initially went down because of a broken part and the replacement part arrived, several other problems usually arose. The continual equipment downtime aggravated the labor efficiency problem because operators would remain idle while maintenance tried to get the line up and running. To make matters worse, Johnson felt that the maintenance supervisor did not properly schedule the few people that he did have.

5. Related to the low level of maintenance was the lack of tools and equipment. At one point in the fall of 1984, there had been no money at all for supplies. Operators were using pipes, rather than hammers, to crack molds. O'Malley had set up a special account so that each supervisor was given $1,000 per month to use as he deemed necessary. In this way, the supervisor would share the decision-making responsibility of how the scarce dollars should be distributed. The special account was not sufficient, however, to meet all the needs. The molding area often had to run three shifts rather than two due to the limited number of patterns. (Each pattern cost approximately $1,000.)

6. The emphasis on volume and costs took its toll on quality. Hourly employees noted that the foundry had lost its concern for quality. They had to use the same sand mixture for the disamatic and the pallet line, but the two in fact required different mixtures; as a result, the sand was not strong enough for either. Another concern was feedback to the molders when problems arose. About a year after Prince purchased NYMD, in-process inspection of the castings as they came out of molding was eliminated. In January 1985, the inspection had to be reinstated because scrap was not being discovered until after the heat treat cycle. The molders also noted that in the past, they would both build the mold and cast it. Over the years the work was divided so that one worker made the mold while another one cast it. "As a result," commented one molder, "no one cares about the total process."

7. Johnson was concerned that the business had grown too fast and that the increase in sales may not have been worth the rapid increase in people. In June 1984, 98 people had shipped $450,000 per month; currently 150 people were needed to ship $600,000 per month. To win customers NYMD had to produce castings in an extraordinarily short turnaround. They did it, but at the expense of efficiency. To retain those customers, they had to maintain a three-to four-week turnaround which worsened efficiency. Another concern was the continual shifts in staffing levels to adjust for volume and machine breakdowns or for malleable and ductile iron changeovers. For example, when running ductile, 13 people were needed on the disamatic, while only nine were needed for malleable. When running malleable, eight of the nine also worked on the pallet line. Finishing, however, required fewer people during the ductile runs because heat treating was not required. The ductile lines ran 8 to 10 days each month for two shifts; malleable for the remainder of the month on one shift. Work hours also changed daily. Workers had to check each day to learn if they would be working the next day and what the hours would be.

8. Johnson was concerned about supervisory inaction. He was not sure whether it was a result of incompetence, indifference, or fear. He was inclined to believe it was the last, because he thought that the supervisors were a bit intimidated by Strasburg who had a tendency to "rant and rave." Every Monday and Thursday morning the supervisors met with Strasburg to set the daily shipment requirements, and as the supervisors walked into the conference room they often joked about whose turn it was to be "chewed out." One of Johnson's supervisors noted that the group had many ideas on cutting costs but were afraid to try them. In the past, when they had tried something new that had not worked, they were "chewed out" by Strasburg. The supervisors also voiced concern over changing priorities. They were unsure how to balance the need for shipping a particular sales volume per day

while reducing labor costs. The pressure to reduce labor had led them to cut personnel which, in turn, left supervisors with more "go-fering" to do.

JOHNSON'S FIRST THREE MONTHS

Johnson spent his first three months as operations manager trying to learn the process and getting to know the people in the manufacturing operation. His initial plan was to implement a data base management system to discern the number of cores needed per casting and the amount of labor required for each casting by operation. His objective was eventually to establish a formal production scheduling system throughout the foundry. He pointed out, "If you get organized, everything else falls into place."

As of July 1, 1985, he had yet to find time to start the data gathering process. During the last week of April, the personnel manager, who had anticipated that he would be given the operations job, unexpectedly resigned. As a result, Johnson had to assume his duties, which included daily meetings with the union representatives and chasing after benefit payments. He also found that he could not ignore the accounting department; he still became involved personally in preparing the monthly closings. He soon concluded he would not have the time to set up the information system and received permission to hire an assistant. His new assistant, also an accountant, was scheduled to start on September 3, 1985. Johnson noted that although his first assignment would be to develop cost standards and set incentives on those jobs that lacked them, he hoped to have his assistant fill in as personnel manager.

Author Index

Topic Index

T - #0065 - 071024 - C0 - 229/203/35 [37] - CB - 9780256068092 - Gloss Lamination